Perspectives in Ecological Theory and Integrated Pest Management

Since the early days of integrated pest management a sound ecological foundation has been considered essential for the development of effective systems. From time to time, there have been attempts to evaluate the ways in which ecological theory is exploited in pest control, and to review the lessons that ecologists learn from pest management. In the last 20 years there have been many developments within the contribution of ecological theory to integrated pest management, and the objective of this book is to capture some of the new themes in both pest management and ecology that have emerged and to provide an updated assessment of the role that basic ecology plays in the development of rational and sustainable pest management practices. The major themes are examined, assessing the significance and potential impact of recent technological and conceptual developments for the future of integrated pest management.

MARCOS KOGAN is Professor and Director Emeritus of the Integrated Plant Protection Center at Oregon State University.

PAUL JEPSON has been Director of the Integrated Plant Protection Center at Oregon State University since 2002.

Perspectives in Ecological Theory and Integrated Pest Management

EDITED BY
MARCOS KOGAN
AND PAUL JEPSON
Oregon State University

CAMBRIDGE
UNIVERSITY PRESS

CAMBRIDGE
UNIVERSITY PRESS

University Printing House, Cambridge CB2 8BS, United Kingdom

Cambridge University Press is part of the University of Cambridge.

It furthers the University's mission by disseminating knowledge in the pursuit of education, learning and research at the highest international levels of excellence.

www.cambridge.org
Information on this title: www.cambridge.org/9780521822138

First published 2007

A catalogue record for this publication is available from the British Library

Library of Congress Cataloguing in Publication data

Perspectives in ecological theory and integrated pest management / edited by Marcos Kogan, Paul Jepson. – 1st ed.
p. cm.
Includes bibliographical references and index.
ISBN-13: 978-0-521-82213-8
ISBN-10: 0-521-82213-0
1. Pests–Integrated control. I. Kogan, M. (Macros) II. Jepson, Paul C. III. Title.

SB950.P37 2007
632'.9–dc22

2007003606

ISBN 978-0-521-82213-8 Hardback

On May 14, 2004 we were shocked and saddened by the news that our good friend and colleague Ron Prokopy had died. He will not see published the excellent chapter he wrote for this volume in collaboration with his former student, Bernie Roitberg, but his legacy will live on. We dedicate this volume to Ron's memory for all that he contributed to advances in our knowledge of insect behavior and to progress in IPM.

Contents

Contributors

Miguel A. Altieri
Division of Insect Biology
University of California
Berkeley, CA
USA

P. J. De Barro
CSIRO Entomology
120 Meiers Rd
Indooroopilly
QLD 4068
Australia

Johann Baumgärtner
Center for the Analysis of Sustainable Agroecosystems (CASA)
Kensington, CA
USA
and
Population Ecology and Ecosystem Science
International Centre of Insect Physiology and Ecology
Nairobi, Kenya

Ring T. Cardé
Department of Entomology
University of California
Riverside, California 92521
USA

O. R. Edwards
CSIRO Entomology
Private Bag No. 5

Wembley, WA 6913
Australia

Fred Gould
Department of Entomology
Box 7630
North Carolina State University
Raleigh, North Carolina 27695-7630
USA

Geoff M. Gurr
Faculty of Rural Management
The University of Sydney
PO Box 883
Orange, NSW 2800
Australia

Andrew Paul Gutierrez
Ecosystem Science
University of California
Berkeley, CA
USA
and
Center for the Analysis of Sustainable Agroecosystems (CASA)
Kensington, CA
USA

Dieter F. Hochuli
School of Biological Sciences
The University of Sydney
NSW 2006
Australia

H. M. T. Hokkanen
Laboratory of Applied Zoology
Box 27
University of Helsinki
FIN-00014 Helsinki
Finland

Paul Jepson
Integrated Plant Protection Center and Department of
Environmental and Molecular Toxicology
Oregon State University
Corvallis, OR 97331
USA

George G. Kennedy
Department of Entomology
Box 7630
North Carolina State University
Raleigh, North Carolina 27695-7630
USA

Marcos Kogan
Integrated Plant Protection Center and Department of Horticulture
Oregon State University
Corvallis, OR 97331
USA

Richard Levins
Department of Population and International Health
Harvard School of Public Health
Boston, MA
USA
and
Cuban Institute of Ecology and Systematics
Havana, Cuba

I. Menzler-Hokkanen
Laboratory of Applied Zoology
Box 27
University of Helsinki
FIN-00014 Helsinki
Finland

Clara Ines Nicholls
Division of Insect Biology
University of California
Berkeley, CA
USA

Robert F. Norris
Weed Science Program
Plant Science Department
University of California
Davis, CA
USA

Antônio R. Panizzi
EMBRAPA
Centro Nacional de Pesquisa de Soja

Rod. Carlos João Strass
Caixa Postal 231
Londrina PR 86001-970
Brazil

Ronald J. Prokopy†
Department of Entomology
University of Massachusetts
Amherst, MA 01003
USA

Bernard D. Roitberg
Department of Biological Sciences
Simon Frasier University
Burnaby, BC V5A 1S6
Canada

Samantha L. Scarratt
Soil, Plant and Ecological Sciences Division
PO Box 84
Lincoln University
Canterbury
New Zealand

T. D. Schowalter
Entomology Department
Louisiana State University
Baton Rouge, LA 70803
USA

P. Sunnucks
Department of Genetics
La Trobe University
Bundoora, VIC 3086
Australia

Jason Tylianakis
Soil, Plant and Ecological Sciences Division
PO Box 84
Lincoln University
Canterbury, New Zealand

†Deceased May 14, 2004

J. C. van Lenteren
Laboratory of Entomology
Wageningen University
The Netherlands

Linton Winder
School of Biological Sciences
Faculty of Science and Technology
University of the South Pacific
Suva, Fiji Islands

Gary Witmer
USDA/WS National Wildlife Research Center
4101 LaPorte Ave.
Fort Collins
CO 80521-2154
USA

Ian P. Woiwod
Plant and Invertebrate Ecology Division
Rothamsted Research
Harpenden
Hertfordshire, AL5 2JQ
UK

Steve D. Wratten
Soil, Plant and Ecological Sciences Division
PO Box 84
Lincoln University
Canterbury, New Zealand

Preface

The dependence of integrated pest management (IPM) on sound ecological theory has been frequently reaffirmed by both IPM practitioners and theoreticians. Insect pests, diseases, and weeds still present us with enormous challenges in all global cropping systems, and we continue to be engaged in a struggle to understand the underlying drivers of their epidemiology and the most effective management strategies. Sustainable IPM systems in the future are going to depend on significant further advances in all the sciences and technologies that contribute to insect pest, disease, and weed suppression. Although IPM systems are deeply ecological in nature, no-one can argue that we have yet defined or formalized the ways in which ecological theory can be developed and exploited to maximize their effectiveness. The application of ecological ideas in the intensely practical realm of agriculture is a slow and difficult process. In this regard the book by G. H. Walter (2003) may serve as a preamble to this volume; although focused on insects, Walter's comments are equally applicable to plant pathogens and diseases. In it Walter states that "Insect ecology research for IPM purposes is represented by a rather grey area; the linkage between theory and practice is still not explicit." We think that some of the chapters in the present book offer insights arguing that ecological theory has already provided the foundation for some level of IPM. Unquestionably, however, much more research must be done to fully integrate ecological theory into IPM practice.

At their most fundamental level, IPM programs are directly linked to the spatial scales for which they are targeted. These scales range from single fields of a given crop, clusters of fields of the same crop, clusters of fields of multiple crops, multiple crop fields and the surrounding non-crop vegetation, complex landscapes, to entire watersheds and ecological regions. Consequently, the ecological processes at those spatial scales must be understood if IPM is to achieve the desired level of integration and meet the goals of the IPM program. Up until recently most IPM programs have been targeted to single fields of a particular crop. Sampling procedures and decision support systems have been developed for application at the field level, targeting a key pest or pest complex (arthropod pests, plant pathogens or weeds) in what we call level 1 IPM. The theory of single species population dynamics or, at best, the dynamics of host/predator interactions has found direct application in Level 1 integration for IPM programs. The complexity of ecological functions and processes at the community and ecosystems levels, however, have defied attempts to successfully advance IPM to higher levels of integration, particularly, the integration of control tactics that take into account interactions of multiple pests in different pest categories. The need to translate ever more robust ecological theory into implementable IPM systems has been perceived as one of the most serious constraints to the global adoption of IPM as the paradigm of choice in the protection of crops and domestic animals. This volume provides a collection of papers by some of the leading authorities in the synthesis of ecology and IPM.

In 1984, at the annual meetings of the Entomological Society of America, a symposium was held to assess the status of the ecological basis of IPM. The expanded symposium papers were published in the book *Ecological Theory and Integrated Pest Management Practice*, M. Kogan, editor (1986). On the occasion of the XXI International Congress of Entomology, held at Foz do Iguassu, Brazil, August 20–26, 2000, it seemed appropriate to revisit the subject. Much had changed, both in IPM and in ecology. Issues of conservation biology, biodiversity, and biological invasions now occupy the thoughts of many ecologists and the field of large-scale ecology has experienced major advances. More robust models for the population dynamics of single and interactive species are being developed, the role of competition in community dynamics and assembly is better understood, and the concept of metapopulations is gaining increasing attention. Community, ecosystems, landscape, and, more recently, ecoregion ecology studies have advanced with the incorporation of more powerful data collecting and analytical tools. This volume provides a collection of 16 chapters that incorporate some of the latest developments in ecology and behavioral ecology as applied to IPM, both conceptually and in real-life situations.

Much of the focus in these 16 chapters remains entomological, but chapters on weed ecology and IPM and on IPM for vertebrate pests provide compelling arguments supporting the need for more interdisciplinary research and the importance of advancing IPM to higher levels of integration. Finally, as stressed in chapter 1, it is vital to consider the human factor, not only in the magnitude of its negative impact on the environment, but also in its role as an engine for progress towards a more sustainable approach to the exploitation of nature and its resources.

Especial thanks are due to Karen Skjei for her careful pre-submittal editing of the manuscripts and formatting all chapters to comply with the publishers' guidelines.

Marcos Kogan and Paul Jepson
Corvallis, OR.
November, 2005

Reference

Walter, G.H. (2003). *Insect Pest Management and Ecological Research*. Cambridge, UK: Cambridge University Press.

1

Ecology, sustainable development and IPM: the human factor

M. KOGAN AND P. JEPSON

1.1 Introduction

If nature had a conscience she would name *Homo sapiens* her number one pest. Which other species, of the now assumed 10–30 million that inhabit the Earth, has caused more destruction, changed the natural landscape more deeply and extensively, exterminated more of the other species, or killed more of their own, than humans? But, ironically, we humans are, as far as it is known, the only species with a conscience. That conscience gives us the ability to classify and name the other millions of species. We readily call a "pest" any other living organism whose life system conflicts with our own interests, economy, health, comfort, or simply prejudice. The concept of a "pest" is entirely anthropocentric. There are no pests in nature in the absence of humans. From a human perspective, an organism becomes a pest when it causes injury to cultivated plants in fields, gardens, and parks, or to the products of those plants (seeds, bulbs, tubers) in storage. Outside agricultural settings, organisms become pests if they affect structures built to serve human needs, are a nuisance, or transmit pathogenic diseases to humans and their domestic animals.

It is our intention in this chapter to consider the roles of humans, as an animal species, in the global ecology and, by narrowing the focus, to project those roles into integrated pest management (IPM). Within a global context, humans often are lethal pests and the rest of nature is their prey. In the IPM context roles are reversed and a few thousand species of animals, plants, and microbes receive that unenviable label of pests; humans become the victims or targets of those pests. They deem the consequences of the pest species' struggle for survival to be an intolerable burden.

From their primate origins some seven million years ago, humanoids, first, and then early humans, lived in relative harmony with nature. As predators, scavengers, or, later, as hunter-gatherers, they were limited to subsisting on resources within their immediate surroundings. Whether sedentary or somewhat nomadic, their populations were naturally regulated and seldom exceeded the carrying capacity of their territory. But, at a certain point in their evolution, humans developed the ability to harness the properties of fire and invented tools that magnified their limited innate muscular power and physical attributes (Diamond, 2005). From then on they rapidly ascended to the top of the food chain. Reaching that status has had disastrous ecological consequences. Humans could kill other animals beyond their immediate needs for food. With more abundant food from hunting and the beginnings of agriculture with the domestication of both plants and animals, some 11 000 years ago, the gradual march to supremacy and overpopulation of the planet accelerated vertiginously. Success of agriculture provided the means to sustain human populations quickly approaching the limits of the carrying capacity of the planet. The consequent human population explosion has been a major cause of the ecological disasters that have shaken the planet in the past couple of centuries.

At the dawn of the twenty-first century the magnitude of the negative impacts of humans on factors that will determine the future of their own species and that of all life on earth seems to be beyond restraint. Despite the inherent resilience of most ecological systems, anthropogenic disturbances have left many systems with no hope for recovery (Brown, 2003). The following sections explore the ways in which humans have impacted the ecology of the planet. The picture is bleak and the outlook pessimistic. Humans are endowed with powerful intellects that allow visionary artists and scientists to produce sobering scenarios for the future of the Earth's biosphere. In their daily affairs, however, humans often are driven by the impulse for immediate satisfaction of their needs or desires. Little heed is given to warnings from those concerned with the survival and wellbeing of future generations. Ironically, policy makers in most developed countries are the least open to heeding those warnings. In Washington, as the economist Robert Samuelson wrote, "... there's already a firmly established bipartisan policy concerning the future: forget about it" (Samuelson, 2004).

But responsible humans cannot forget about the future. Responsible institutions and concerned individuals have made honest efforts to respond to the most serious threats that human civilization has faced to date. These threats do not stem from terrorist organizations or warmongering rogue nations; they stem from the unbridled, irrational human impulse to satisfy one's own short-term wants regardless of the long-term costs to the environment and to society. Reaction to those threats has sprung up as initiatives which aim to protect the planet for the benefit of future generations. In a capitalist, free initiative society, progress often is measured in terms of economic growth and development. China, although still officially under a communist regime, exploiting opportunities offered by the world capitalist free market economy, has advanced economically at a dizzying rate of over 9% a year (Fan *et al.*, 2003). This economic growth is, by most measures, considered progress and a sign of economic development even if it has brought about enormous social problems, loss of cultural values, and deterioration of the environment (according to the World Health Organization, seven of the world's most polluted cites are now in China). Material development is inherently linked to the notion of human progress. While quality of life continues to deteriorate due to air pollution, surface and groundwater contamination, depletion of non-renewable natural resources, and increasing disparity in the distribution of wealth, economists still express progress in terms of thousands of kilometers of new paved highways, number of automobiles per capita, or barrels of oil consumed per day. In agriculture, progress is measured in terms of thousands of hectares opened to cultivation, often at the expense of the destruction of precious tropical forest resources. The Food and Agriculture Organization of the United Nations (FAO) estimated that 15.2 million hectares of tropical forests were lost per year between 1990 and 2000, mainly in South America (the Amazonian rainforest), Africa, and Indonesia (FAO, 2001*a*). Much of the deforested land was converted to agriculture. Such market-driven global economy, in which short-term decisions often overwhelm concerns over long-term effects, leads one to question whether "sustainable development" is not, in fact, an oxymoron. Will development under an unplanned free market economy ever be sustainable? Despite justifiable skepticism, we must admit that if honestly and widely adopted, the principles of sustainable development represent the only hope for the long-term stability and survival of the Earth's biosphere.

The following sections discuss human negative impacts on the global ecology, and, in the second part of this chapter, the sustainable development model, with a focus on agriculture, serves as backdrop for a discussion of the role of IPM in sustainable development. In its own way and rather specific scope, IPM provides a paradigm whereby potential anthropogenic disasters can be attenuated, if not averted, when adopting a holistic view in decision-making management of processes that impact the environment.

1.2 The human factor in the global ecology

Modern technology has advanced methods to soften the impact of many natural disasters, but, as a rule, natural disasters are inevitable. Dams and drainage systems are built to regulate river flow and prevent flooding. Avalanche control is practiced in alpine regions of the world with considerable success in reducing fatalities and damage to mountain villages. Construction materials and architectural advances make buildings more resistant to earthquakes. Naturally induced forest fires often are controlled with fire retardants dropped by airplanes and the effective deployment of expert firefighting techniques. Early warnings help people to prepare for incoming hurricanes, tornadoes, and even tsunamis, saving lives, if not material goods. The absence of such warnings was a major reason for the 170 000 plus fatalities in the December 2004 tsunami in Southeast Asia. Ironically, however, anthropogenic disasters, which most certainly are evitable, dramatically increase with expansion of technologies that often are considered to be the very indicators of "progress." The impact of humans on the Earth's ecology has been the subject of numerous scholarly and popular volumes (Diamond, 2005; Goudie, 2000; Harrison, 1992). The following are some of the effects generally attributed to human actions as they impinge on the global ecology (Goudie, 2000). We stress those impacts that are closely linked to agriculture and pest management.

1.2.1 *Altering natural landscapes*

Up to the advent of agriculture, humans had negligible impact on local landscapes. Except for clearings to build huts for shelter, the hunting-gathering lifestyle of most pre-farming societies exerted little pressure on the structure and function of ecosystems. With domestication of plants and animals, agriculture brought about a major shift in the interactions between humans and their surroundings (Diamond, 1997). With about one-tenth of the Earth's land mass converted into agricultural or pasture fields (FAO, 2002), the impact of agriculture on natural ecosystems has been enormous. In many regions of the world the resilience of local ecosystems has allowed the replacement of diverse plant covers with a few crop species to remain viable for many centuries. Early Neolithic agriculturists occupied sites located in what are now Iran, Iraq, Israel, Jordan, Syria, and Turkey – the Fertile Crescent of Southwest Asia; in southeastern Asia, presently Thailand; in Africa, along the Nile River in Egypt; and in Europe, along the Danube River and in Macedonia, Thrace, and Thessaly. Early centers of agriculture have also been identified in the Huang He (Yellow River) area of China; the Indus River valley of India and Pakistan; and the Tehuacan Valley of Mexico, northwest of the Isthmus of Tehuantepec (Diamond, 1997; Advanced BioTech, 2004). These regions have been under continuous cultivation for six to ten millennia, even if several of them have suffered from considerable desertification and the areas under cultivation have shrunk. Also, in some of these regions, agriculture persists only with costly subsidies of water, fertilizers, and anti-erosion techniques. In other regions, with more fragile ecosystems, agricultural development resulted in the depletion of key resources (topsoils, nutrients, water) and the final creation of deserts where once existed a diverse plant cover.

There is evidence of repeated losses of agricultural productivity throughout history as a result of adoption of excessively intensive practices (McNeely *et al.*, 1995; World Resources Institute, 2000; Wood *et al.*, 2000). The Sahel region is a narrow band of West Africa between 15–18° N, with the Sahara to the north, and savannah and equatorial forest to the south. It extends from Senegal at the coast at about 15° W, across Mali and Niger, covering a surface area of 5.4 million km^2, with a population of 50 million. Although some suggest that the Sahara, encroaching into the Sahel, is a major cause of desertification, it seems more plausible that agriculture and the destruction of the natural plant cover account for the significant rates of desertification that is affecting some of the Sahel countries. It has been reported that in Niger 2500 km^2 of arable land is lost each year to desertification (Eden Foundation, 2000).

Population growth that consistently followed expansion of agriculture (or was it the other way around? – see Cohen, 1995) resulted in the establishment of totally new, much less diverse ecosystems. Many species of the original biota were decimated but some native and a few exotic species of plants and animals adapted to the new human-made agroecosystems. Some of these plants and animals became pests, from the human perspective. Alteration of landscapes has been a most serious problem associated with agricultural expansion in the world. The problem is aggravated at present due to heightened population pressure and rising standards of living in many developing countries (Brown *et al.*, 1999; Harrison, 1992; UN–EP, 2000). Activities associated with agriculture have contributed also to another major problem of profound ecological consequences, i.e. biological invasions.

1.2.2 *Promoting biological invasions*

Biological invasions have increased with modern means of transcontinental transportation, globalization of trade, and expansion of tourism. Besides the profound ecological impact of invasive species on natural and agricultural ecosystems, biological invasions are considered a serious threat to food safety and public health (Evans *et al.*, 2002; see also Evans, 2003). The specter of bioterrorism led to the creation of networks of scientists and laboratories dedicated to detecting and curtailing potentially damaging, maliciously introduced species (Meyerson and Reaser, 2003; Shalala, 1999). Since time immemorial, however, humans, intentionally or not, have moved species from region to region, country to country, continent to continent. By far the most dramatic and widespread invasions resulted from crop species transplanted from across continents around the globe. Many of these were positive transplants; but the same impulse to move plants for commercial or just curiosity purposes often had disastrous results (Pimentel, 2002).

It has been difficult to make a comprehensive assessment of the number and frequency of invasions worldwide. There are, however, useful regional studies that give an approximation of the magnitude of the problem (Kiritani and Morimoto, 1999; Wilson and Graham, 1983; US–OTA, 1993; Sailer, 1983; Sakai *et al.*, 2001; IUCN, 2006). According to Sailer (1983), of the over 2000 non-indigenous insects introduced into the USA, intentionally or not, 235 species have become serious agricultural and forestry pests and have caused cumulative losses of about 92.6 billion dollars between 1906 and 1991 (US–OTA, 1993). Table 1.1 provides historical examples of plant species that were intentionally introduced but which escaped control and became destructive invaders. Examples of introductions of animals also are common and well documented. These range from the almost criminal release or neglect of exotic animals imported as pets, to the careful and intentional release of species for biological control purposes which then switch hosts and become predators of desirable species (see Chapter 8 of this book). For example, 39 piranhas were confiscated in Hawaii in 1992; two having already been found in Oahu waterways. The piranhas were apparently mail-ordered from a dealer of dangerous pets, on the US mainland, and shipped to Hawaii through uninspected first class mail (DLNR, 2005). The intentional introduction of the cactus moth, *Cactoblastis cactorum*, for control of the invasive prickly pear, *Opuntia* spp., in Australia, was a classic example of successful biological control. The same species introduced into the USA for the same purpose, however, became a damaging invasive species in areas with a rich flora of cacti (Martin, 2005; Stiling, 2002).

With massive transcontinental trade of goods and multidirectional movement of people throughout the ages, humans have become inadvertent vectors of microbial organisms. Many of these organisms are causal agents of human and other animal and plant diseases. Diamond (1997) provides a compelling case for the impact of human vectored germs in the history of civilization. The current pandemic of HIV/AIDS demonstrates how humans continue to act as vectors of deadly diseases. Transporting microbial pathogens of plants, domestic animals, and humans due to ignorance or blatant disregard for the laws, borders on criminal behavior. The consequences to society can be enormous. In recent years two serious soybean pests endemic to Eastern Asia have been detected in the USA. The soybean aphid, *Aphis glycines*, was first detected in Minnesota in the summer of 2000, and is now well established throughout northern soybean growing areas. By 2003 there were widespread infestations throughout the Midwestern states and three Canadian provinces. Approximately 3–3.5 million acres were sprayed with insecticides in Minnesota alone (Venette and Ragsdale, 2004; Ostlie, 2005). The aphid, in addition to its damage potential to soybean, is an efficient vector of viral diseases of other major crops, such as potato (Davis *et al.*, 2004). The second pest is soybean rust, a devastating fungal disease, now spread into the USA, Argentina, and Brazil with the potential to cause a severe crisis in the world supply of this commodity (Hagenbaugh, 2004; Yorinori *et al.*, 2003).

Biological invaders often outcompete the native flora and fauna and drastically change the landscape. They become biotic contaminants of the environment. Just as grave, however, has been the enormous increase of abiotic contaminants of the biosphere due to humans' agricultural and industrial activities, as well as a consequence of their personal lifestyles (Matsunaga and Themelis, 2002).

1.2.3 *Contaminating ecosystems*

By-products of human agricultural and industrial activities, as well as those resulting from human personal lifestyles, often are extraneous materials which natural recycling processes are incapable of handling. These by-products accumulate as rubbish and toxicants on land, permeate into surface- and groundwater, and into the seas. The results have been amply documented and are of catastrophic magnitude (United Nations, 2005). Harrison (1992) suggested that the impact of humans on the environment could be computed by the formula: environmental impact = population × consumption per person × impact per unit of consumption. Often consumption per person is a measure of affluence. A good visual proof of this equation is a visit to a city dump. But the impact of even the garbage collected in New York City fades against the major sources of pollution and environmental contaminants that spew from industrial liquid discharges and smoke stacks, from automobile exhaust pipes, and from intentional

Table 1.1. *Examples of intentionally introduced plant species that escaped control and became destructive invaders*

Common name	Scientific name	Historical	Current status, USA	Reference
Giant hogweed	*Heracleum mantegazzianum*	Native to the Caucasus; introduced into Great Britain, Canada, and the United States.	Noxious weed; known to occur in New York, Pennsylvania, and Washington with infestations in Maine, Michigan, and Washington, DC. Planted as ornamental in the United States and possibly introduced for its fruit, used as a spice (golmar) in Iranian cooking. Watery exudate causes skin blisters with sunlight exposure.	APHIS, 2002
Japanese knotweed	*Fallopia japonica* var. *japonica*	Introduced from Asia as an ornamental in the 1800s.	Thickets can clog small waterways and displace other riparian vegetation, increasing bank erosion and lowering habitat quality for fish and wildlife.	Shaw and Seiger, 2002
Johnsongrass	*Sorghum halepense*	Mediterranean origin, introduced in 1830 as forage crop.	Competitive, poisonous grass. Invades crop, range lands, river beds, disturbed lands.	Royer and Dickinson, 1998
Kochia	*Kochia scoparia*	Introduced as garden ornamental from Asia in the early 1900s.	Weedy annual; drought tolerant, can invade both irrigated and dryland crops. Impacts potato, alfalfa, and wheat production.	Royer and Dickinson, 1998

Kudzu	*Pueraria montana* var. *lobata*	Introduced into USA in 1876 as part of the Japanese exhibit for the Philadelphia, Pennsylvania, Centennial Exposition. Adopted by American gardeners for ornamental purposes.	All Southeastern states. Most severe infestations in piedmont areas of Alabama, Georgia, and Mississippi.	Britton *et al.*, 2002
Leafy spurge	*Euphorbia esula*	Of Eurasian origin; introduced by accident or on purpose in 1927.	Spread to all US states except the South and Southeast. Serious invader of range lands in the prairie and Northeastern states.	Stein and Flack, 1996; Nowierski and Pemberton, 2002
Melaleuca	*Melaleuca quinquenervia*	Introduced into Florida from Australia, New Guinea, and New Caledonia in early 1900s as ornamental tree, timber source, and for drainage of wetlands.	By 1950s invaded marshes and prairies of the Everglades.	Pratt *et al.*, 2004
Purple loosestrife	*Lythrum salicaria*	Introduced from Europe into New England as an ornamental in the 1980s. Still sold as a landscape plant.	Most contiguous US states. Most serious as invader of wetlands in Northeast and upper Midwest, displacing native plants and wildlife.	Stein and Flack, 1996; Blossey *et al.*, 2001
Yellow star thistle	*Centaurea solstitialis*	Native of Eurasia	Aggressive invader of range lands, displacing native vegetation. Causes neurological disorders in horses.	Sheley *et al.*, 1999
Saltcedar	*Tamarix spp.*	Native of Eurasia. Introduced in 1800s by settlers as source of wood, shade, and erosion control.	Drains water from riparian woodlands in fragile desert ecosystems of the Southwestern USA.	Stein and Flack, 1996
Scotch broom	*Cytisus scoparius*	Introduced from Europe as ornamental.	Displaces forage and native plants and it is rejected by most livestock. Interferes with reforestation.	Coombs *et al.*, 2004

forest fires, to mention only a few. From an agricultural and IPM perspective, however, the primary concern remains the contribution of pesticides and fertilizers to global environment degradation.

Since 1942, when DDT (dichlorodiphenyltrichloroethane) ushered in the era of organosynthetic pesticides in agriculture and public health, it is estimated that over 90 million metric tons of toxic agricultural chemicals have been spread onto the Earth's surface or incorporated into its soils (conservatively based on estimates by the United States Environmental Protection Agency (US EPA) and the United Nations Food and Agriculture Organization (UN FAO) of world pesticide usage since the introduction of DDT). Granted, most of the properly applied chemicals degrade at various rates upon dispersal in the environment. Their half-life varies, however, from a few days to many years and there is mounting concern about the fate of stored obsolete or banned pesticides. It was estimated that more than 500 000 tons of pesticide waste stocks threaten the health and the environment in most developing countries (FAO, 2001*b*). Before degrading, however, pesticides often impact many organisms besides those against which they were intended (non-target effects, Metcalf, 1986; see also Chapter 16, this volume). One of the consequences of the environmental impact of pesticides, together with numerous other anthropogenic environmental contaminants, is the acceleration of species extinction (IUCN, 2005).

1.2.4 Accelerating species extinction

A remarkable trait of *Homo sapiens* seems to be the propensity to kill other species in numbers that far exceed humans' own needs for subsistence or survival. There is paleontological evidence that many species of large vertebrates became extinct shortly after colonization of new areas by humans. In his account of human evolution and history, Jared Diamond (Diamond, 1997) documented instances of large animal extinctions following colonization of isolated areas by humans far before population pressure caused humans to over exploit natural food resources, such as game, and industrial resources, such as timber for pulp and construction. Although the human role in other species extinction is not new, what is new is the rate at which human actions have accelerated extinctions and the enormous number of species affected. Some of the most species rich ecosystems, the tropical rain forests, are being destroyed at unprecedented rates (see below). Destruction of these ecosystems results in the irreversible degradation of highly specific habitats and the consequent demise of species adapted to those habitats.

Land conversion is one of the most important mechanisms that underlies the accelerated extinction rates of the global flora and fauna (e.g. Barbault and Sastrapradja, 1995; Purvis and Hector, 2000; Novacek and Cleland, 2001; Pitman and Jorgensen, 2002; Pitman *et al.*, 2002; Thomas *et al.*, 2004). Land conversion to agricultural and other human uses may compound the effects of global climate change to further reduce biodiversity (e.g. Warren *et al.*, 2001).

Mounting human population pressure has caused accelerated reduction of populations of large mammals through hunting to placate chronic protein hunger in impoverished societies in Africa and Asia. Much less morally justifiable, though equally devastating, has been poaching to satisfy demands of an affluent market for luxury items of dubious value (black rhino horns as aphrodisiacs or elephant tusks for ivory, for instance).

Massive extinctions result in loss of biodiversity. Loss of biodiversity is associated with the gradual destruction of natural ecosystems, both marine and terrestrial, biological invasions, pollution, and over-hunting. The GEO (Global Environment Outlook) 2000 (UNEP, 1999) report states that

> At the broadest level, biodiversity loss is driven by economic systems and policies that fail to value properly the environment and its resources, legal and institutional systems that promote unsustainable exploitation, and inequity in ownership and access to natural resources, including the benefits from their use. While some species are under direct threat, for example from hunting, poaching, and illegal trade, the major threats come from changes in land use leading to the destruction, alteration or fragmentation of habitats.

For example, two-thirds of Asian wildlife habitats have been destroyed with the most acute losses in the Indian subcontinent, China, Vietnam, and Thailand. In Latin America, the average annual deforestation rate during 1990–95 was 2.1% in Central and South America (UN-EP, 2000). It has been suggested that we live "amid the greatest extinction of plant and animal life since the dinosaurs disappeared some 65 million years ago, with species losses at 100 to 1000 times the natural rate" (Brown, *et al.*, 1999).

The ecological consequences of biodiversity loss, whatever its origins, include impairment of basic ecological functions (Loreau *et al.*, 2001; Sekercioglu *et al.*, 2004). The degree to which functional redundancy (i.e. the capacity of an ecosystem to lose elements of biodiversity, but retain its basic ecological function) exists within communities and functional groups but is still not fully understood (Hunt and Hall, 2002) and is likely to be highly variable in space and in time, and dependent upon the sensitivities and responses of organisms and functional groups in the disrupted system (Symstad *et al.*, 2003).

1.2.5 *Interfering with nutrient cycling*

Some of the most complex, delicate, and vital ecosystem functions involve energy and nutrient cycling, particularly water, nitrogen, and carbon. These natural cycles are being impacted by humans at rates that far exceed the buffering capacity of ecosystems. Although water and nitrogen cycles have already been severely affected, it is the human impact on the carbon cycle that seems to present the most serious, immediate, and long-term threat to the integrity of the biosphere. R. Houghton, of the Carbon Research Group, Woods Hole Research Center, reports that between 1850 and 2000 about 155 billion metric tons of carbon have been released to the atmosphere, mainly from worldwide changes in land use. The concentration of CO_2, the principal greenhouse gas, released into the atmosphere has increased by 30% since the middle of the nineteenth century, the beginning of the industrial revolution: most of this increase came from use of fossil fuels. About 25% of the total increase, however, came from changes in land use – the clearing of forests and soil cultivation for food production (Houghton, 2005). The increase in atmospheric CO_2 and other greenhouse gases has been the main cause of the recorded global warming of recent years.

1.2.6 *Altering the global climate*

Climate change no longer is a matter of opinion or speculation (Houghton *et al.*, 2001). The Intergovernment Panel on Climate Change, working on an update of the 2001 report, to come out in 2007, expresses the view that while uncertainties still exist, there is no doubt that the earth is heating up at an alarming rate. Only by including human activities can climate change models explain observed warming rates. Of great concern is the assessment of the extent of the changes and their potential impacts. Expected consequences of global warming trends include a shifting of climatic zones, changes in species composition and productivity of ecosystems, increases in extreme weather events, and impacts on human health (UN–EP, 2000). Tropical rain forests play a critical role in recycling CO_2 and are particularly vulnerable to climate change. Malhi and Phillips (2004) presented a sobering view in a synthesis of papers published under the theme "Tropical Forests and Global Atmospheric Change" (special issue of *Philosophical Transactions of the Royal Society, London B*, May 2004). They based their remarks on "the large-scale and rapid change in the dynamics and biomass of old-growth forests, and evidence of how climate change and fragmentation can interact to increase the vulnerability of plants and animals to fires." Based on field studies, it was apparent that

> changes in tropical forest regions since the last glacial maximum show the sensitivity of species composition and ecology to atmospheric changes. Model studies of change in forest vegetation highlight the potential importance of temperature or drought thresholds that could lead to substantial forest decline in the near future. During the coming century, the Earth's remaining tropical forests face the combined pressures of direct human impacts and a climatic and atmospheric situation not experienced for at least 20 million years (see also Malhi and Wright, 2004).

Thus continued increase in fossil fuel burning with consequent further emission of greenhouse gases and deforestation are driven by independent forces but combine to aggravate the risk of climate change. Both trends, accelerated rates of fossil fuel burning and deforestation, are likely to persist under pressure of a growing human population and greater affluence in developing countries of Asia and South America.

The consequences of climate change on crops are still being debated. Reports presented in a conference sponsored by the Royal Society on "Food Crops in a Changing Climate," however, seem to suggest that benefits of increased CO_2 levels to crops are less than previously thought and differ markedly among C3 crops (such as rice, wheat, and soybean) and C4 crops (such as maize sorghum, and sugar cane). Furthermore, some of the potential benefits of CO_2 fertilization are counteracted by the detrimental effects of increases in surface ozone (Bazzaz, 1990; Royal Society, 2005; Tuba, 2005).

1.3 *Homo sapiens:* a species out of control

Homo sapiens is the only species that has the capability of regulating its own population. Yet, for cultural, religious, or economic reasons, in vast regions of the world, human population is growing far beyond normal replacement levels and often exceeding the carrying capacity of the region. Advances in health care and nutrition have increased longevity in most industrial countries. Paradoxically, while natural population growth has leveled off, or even declined, in many of the richest countries, it continues to grow, most disgracefully impaired only by internecine wars, famine, and the epidemics of HIV/AIDS, in the poorest regions of the world (Brown, 2000; UNICEF, 2005).

Human population took 4 million years to reach the landmark of 2.5 billion in 1950. It more than doubled in just 37 more years (Brown *et al.*, 1999). Thomas Malthus (1798), in the late eighteenth century, was one of the first to call attention to the disparity between population growth rates and rates of increase in food production. The population issue continued to be debated as to whether agricultural development was the source of population growth or if it resulted from that growth and the consequent need to provide food for more people. Technological developments of the nineteenth and twentieth centuries may have delayed Malthusian predictions from becoming reality. But in some regions of the world the specter of Malthus' scenarios is already the sad reality. Even if demographers predict a leveling off of population growth to about 9 billion, in some countries human population quickly is exceeding the land's carrying capacity and irreversibly exhausting key finite resources. The figures that follow illustrate current trends in human demographics and disparities in population growth among the nations of the world (based on Brown *et al.*, 1999; Brown, 2000; FAO, 1999).

- Between 1950 and 2000, world population grew by 3.6 billion to reach the 6.1 billion mark. In other words, human population grew by 144% in 50 years.
- Another 2.6 billion people will be crowded onto the planet by 2050, at a rate of 80 million per year to reach 8.7 billion.
- In 59 countries, mostly in Africa and Asia, populations will double or triple by 2050 (e.g. Ethiopia, from 58 million in 1998 to 169 million in 2050) unless held in check by ever more frequent famines and diseases. Some of those countries are among the poorest in the world.
- India officially reached the 1 billion mark on 11 May 2000. Despite effective birth control measures, India should exceed China's population of 1.6 billion by 2040. Despite India's remarkable economic development of recent years, 50% of this huge population is illiterate; 50% of its children are under-nourished; and 33% of its people live below the poverty line.

Consequences of this uncontrolled population growth have been amply documented (Ehrlich, 1968; Ehrlich and Ehrlich, 1990; Cohen, 1995; Evans, 1998; Hinrichson and Robey, 2000; PRB, 2005). It has profound social impacts which transcend the scope of this paper. The ecological impacts, however, are relevant to the context of this book. The questions raised by these demographic projections are: (a) what are the impacts of humans, as key biotic components of most ecosystems, on the global ecology; and (b) whether sustainable development is still possible while human population continues to grow before it will eventually plateau at between 9 and 10 billion sometime after 2050.

There is ample evidence of ecological disasters caused by humans as they acquired the means to overwhelm competitors, either other humans or other animals. Such disasters occurred much before population growth accelerated to current critical levels (see citations in Diamond, 1997) but they have been vastly magnified as populations grew (see Section 1.2, above).

1.3.1 *Demographics and vital global resources*

Demographics build the pressure to increase food production, which in turn leads to the often unsustainable use of vital global resources (land, water, and energy) necessary to support that increase. This pressure demands questioning in ecological terms about the limits of the planet's carrying capacity, or as demographer J. Cohen put it: "How many people can the Earth support?" The essential requirements for human survival are the same as those required by any other animal species: food, water, air, and shelter; land and energy are the resources necessary to meet those requirements.

Food

Production of the major staples (i.e. all grains, including rice, maize, wheat, sorghum, millet, and other cereals; beans and pulses; and tuber crops, including potatoes, yams, and cassava – FAO, 2002) has increased significantly from 1961 through 1999 (Figure 1.1). Between 1950 and 1984 the rate of increase of grain production exceeded the growth rate of human population (Brown *et al.*, 1999). Annual harvest per person (grain availability) actually increased from 247 kg in 1979 to a peak in 1984 of 342 kg. It has since declined to about 300 kg per person in 1998.

Much of this remarkable rate of increase in global food production has been achieved through the increase in land area under cultivation, expanded water use, and the adoption of input-intensive production practices needed for maximum expression of the yield potential of "green revolution" crop varieties (Singh, 2000; Manning, 2001; Tilman *et al.*, 2002; Wikipedia, 2005). Management of available land, water, and many of the highly energy-dependent production inputs is in essence what the sustainability debate involves.

Land

The land area under agricultural production (crops and animals) worldwide was about 1.37 billion ha in 1994. To meet demand for increased

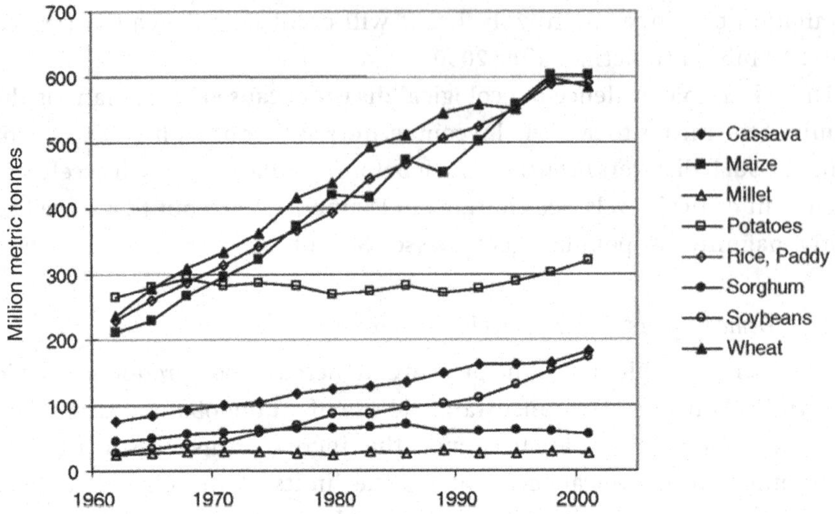

Total production–three year averages (FAOSTAT data, 2005)

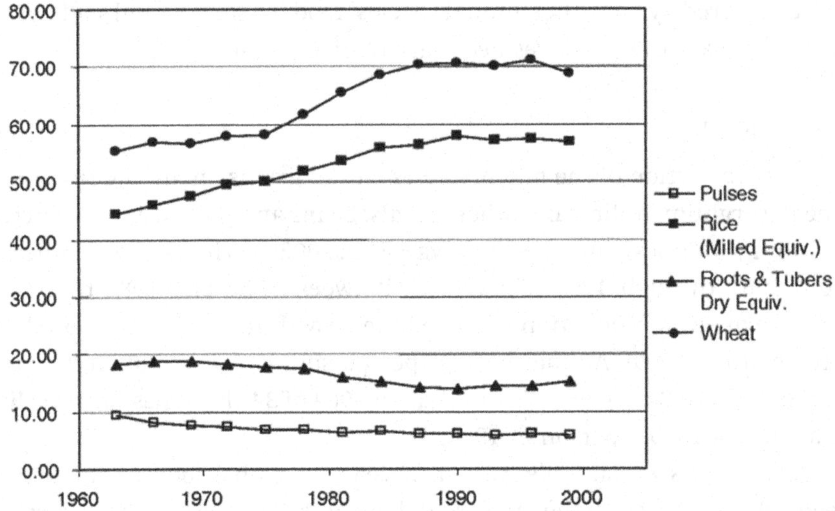

Production per capita per year in kilograms–three year averages (FAOSTAT data, 2005)

Figure 1.1. Production of major grain crops.

food production in developing countries, additional land is put into production through deforestation and use of marginal lands that need heavy subsidies of fertilizers and water and which are often steep hillsides, susceptible to erosion. Agricultural conversion alters the structure and functioning of natural ecosystems with a loss of native flora, loss of wildlife habitat; further potential biodiversity impacts from pesticide and nutrient exposure and run-off; and effects on the soil biota, soil quality, nutrient cycling and soil water relations which result from cultivation (Conway and Pretty, 1991; National Research Council, 1993; McLaughlin and Mineau, 1995; Mooney *et al.*, 1995*a*, 1995*b*; Matson *et al.*, 1997; Vitousek *et al.*, 1997; Tilman, 1999; Wood *et al.*, 2000; Tilman *et al.*, 2001).

The rate of deforestation in the recent past, in particular during the last 50 years, has been unprecedented in the history of the planet. In the beginning of the agricultural age, some 10K years ago, about 40% of the land area, or 6 billion ha, was covered with forests. In early 1990, there were 3.6 billion ha of forests left, and every year another 14 million ha are lost with 90% of the losses occurring in the tropics. In the Brazilian Amazonia, a fire set to clear land for agriculture consumed 5.2 million ha of forest, brush, and savannah, an area one-and-a-half times the size of Taiwan (Bright, 2000). Much of the Amazonian land that is cleared is put into ranching and some of the better land into farming. Expansion of soybean production in the Amazon region has been imputed as the main culprit in the destruction of forest and savannah. As soybean production expands cattle ranchers and producers of other staple crops are forced even deeper into the forest (Aide and Grau, 2004; USDA–FAS, 2004). A study by the World Bank has demonstrated that beef cattle production in the Amazon nets ranchers more than twice as much income per hectare than production in the more traditional areas of southern Brazil. With such an economic incentive it is difficult for the government to constrain squatters from continuing to encroach ever deeper into the forest (Margulis, 2004). Figure 1.2 shows the trends in population growth, area under production, and forested land. It is noticeable that forest area has declined with increase in food production due, significantly, to demographic pressure and increased demand of food-importing countries.

Despite efforts to increase the land base for agriculture, many parts of the world are reaching the upper limit for additional expansion. An index of land availability is the amount per person of world area from which grain is harvested; USDA data show a steady decline in the last 50 years (USDA, 1998; Brown *et al.*, 1999). Whereas crop land area has increased by about 19% over this half-century, human population increased by 132%. As a consequence the harvested grain area that was 0.24 ha per person in 1950 has dropped to 0.12 ha in 1998, about 1.5 times the area of a regular basketball court. Again the situation is worse in countries where populations continue to grow at about 3% per year and no more land is available for expansion. Extreme examples are Pakistan, Nigeria, and Ethiopia (Brown *et al.*, 1999).

In the developed countries, on the other hand, productive land is being lost at alarming rates due to urban sprawl and industrial development. Therefore, not only is new land increasingly scarce, but productive land is lost to

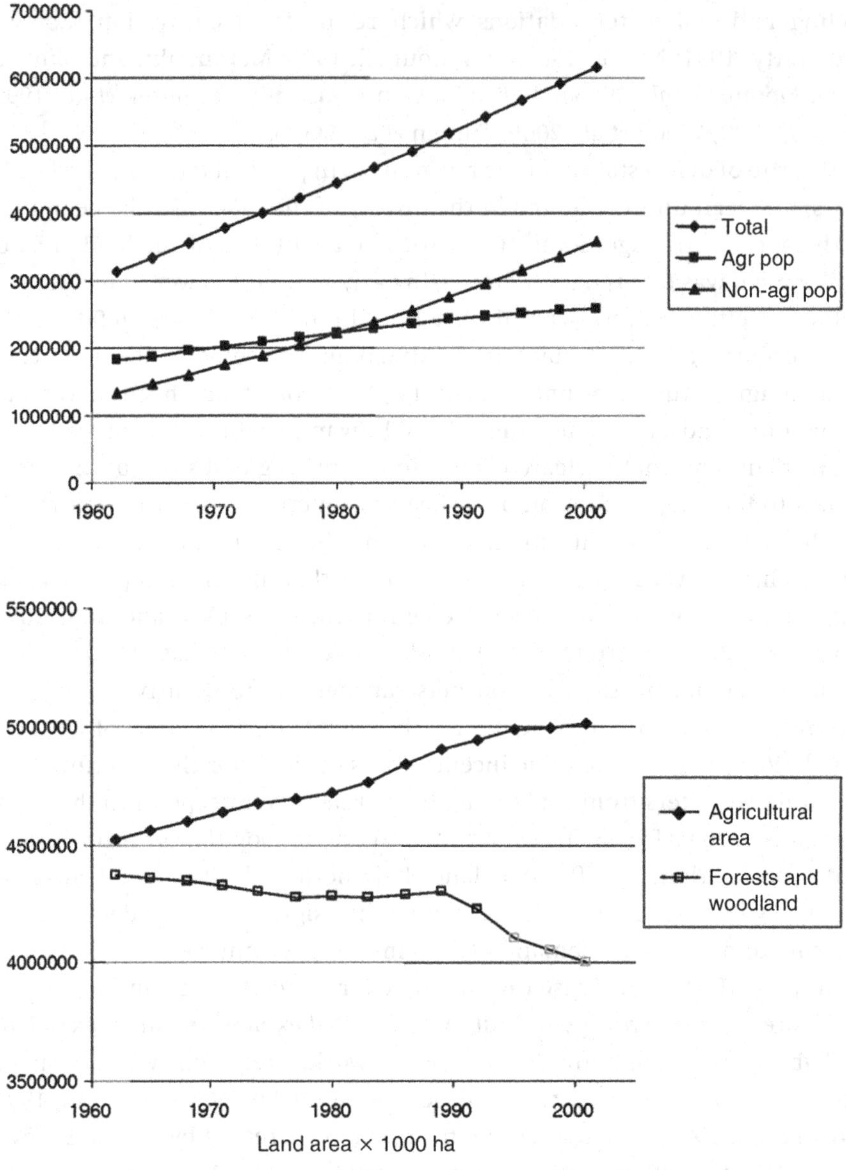

Figure 1.2. Trends in population growth, area under production, and forested land (FAOSTAT, 2002).

competing interests and to poor stewardship. To meet food production demands, it is not enough just to put more land into production. To maximize yields and to use marginal lands in arid zones there has been a large expansion of cropland area under irrigation around the world. Water, therefore, is the next bottleneck in the sustainability equation.

Water

According to GEO 2000 (UNEP 1999), population growth of the past 50 years, with consequent expansion of industrialization, urbanization, agricultural intensification, and water-intensive lifestyles is leading to a global water crisis. The area under irrigation worldwide has increased dramatically in the past 50 years. Agricultural uses continue to compete with industry and domestic use. It is estimated that about 70% of the water pumped out of surface and underground water sources is used for irrigation, and of the rest, 20% goes to industries and 10% for domestic use (Brown *et al.*, 1999; Seckler *et al.*, 1998). Aquifers in many parts of the world are being tapped at a rate faster than the rate of recharge. Although freshwater supplies vary enormously among the various regions of the world and even within the same country, it seems certain that more than a quarter of the world's population lives in regions that will experience severe water scarcity. Most severe water shortages are predicted for exactly those regions where population is expected to grow the most.

Energy

From an estimated peak production of about 25 billion barrels reached in 2005, world oil production will steadily decline and reserves should be finally depleted at current consumption levels by 2100. A recent report claims that the Exxon Mobil Corporation, the world's biggest company, by the end of 2004 held proved reserves of 22.2 billion barrels of oil, enough to continue production at current levels for about 14 years! Despite arduous protest of environmentally concerned citizens, the United States House of Representatives approved the opening of the Alaska Nature Preserve (ANWAR) to oil exploration. If oil is found, experts predict those reserves will be just enough to supply the US market for six months (Chance, 2005). Wise use of energy is critical for sustainable development. As dreadful as the prospect of impending depletion of oil reserves is, the oil industries of the developed countries seems to do little to promote deployment of alternative sources of energy and the widespread use of more efficient oil saving engines. As the developed world looks at these complex alternatives, the poorest people continue to extract energy from burning wood and charcoal which further accelerates the rate of destruction of natural woodlots and forests.

1.4 Sustainable development

It is worth reminding ourselves of the question asked by demographer Joel E. Cohen: "How many people can the earth support?" (Cohen, 1995).

There does not seem to be a simple answer. It all depends on the intrinsic cultural characteristics and level of affluence of the population. The ratio of food consumption by Americans and the people of India, for instance, is 5 to 1. If the world grain harvest reached 2 billion tons in the years ahead it would be enough to feed 2.2 billion Americans or 10 billion Indians, as suggested by Lester Brown of the Worldwatch Institute (Brown *et al.*, 1999).

Human population growth is at the heart of the sustainability issue because more people need more food, more houses, more energy and thus, more development. The population factor is pivotal to a discussion of sustainable development, agriculture, and IPM. These concepts arose from the growing awareness of the environment's fragility in the face of human interference. In its broadest sense, sustainable development is "development that meets the needs of the present without compromising the ability of future generations to meet their own needs" (WCED, 1987). The ability of future generations to meet their own needs already may have been irremeably compromised. Undoubtedly the days of plentiful supplies of fossil fuels are numbered; freshwater reserves in many countries are nearly depleted; fertile soils are gradually impoverished in many tropical countries; and the diversity of life on the planet is fast disappearing. Any hope for the future of humans on Earth depends on a drastic shift in the paradigm for development that places sustainability in the forefront. Development and deployment of IPM systems within the context of sustainable agriculture is a small but critical element of the overall sustainability equation.

The concept of sustainable development has evolved over a period of about 40 years. There have been several historical landmarks in the advancement of a global strategy to address the issue of human impact on the environment. In 1972 the United Nations Conference on the Human Environment was the first major meeting to look at how human activity was affecting the environment. The conference produced what became known as the "Stockholm Agreement" (UNEP, 1972), a declaration concerning problems of pollution, destruction of resources, damage to the environment, endangered species, and the need to enhance human social wellbeing. The conference stressed the need for countries to improve living standards of their populations and stated 26 principles which would determine that the development was sustainable.

In 1986, the United Nations established the World Commission on Environment and Development (WCED) to "study the dynamics of global environmental degradation and make recommendations to ensure the long-term viability of human society". The Commission was chaired by G.H. Brundtland, Prime Minister of Norway at the time. The Commission report (WCED, 1987) became the benchmark for analysis of development and the environment, and it popularized the expression "sustainable development."

The next major event in the world arena which focused on sustainable development occurred when representatives of more than 100 countries met in Rio de Janeiro, Brazil, in 1982 for the first international Earth Summit. The meeting addressed the urgent problems of environmental protection, social, and economic development. The following were some of the major agreements supported by the delegates to the summit: (a) the *Convention on Climate Change* – limiting emissions of the greenhouse gases carbon dioxide (CO_2) and methane (CH_4); (b) the *Convention on Biological Diversity* – giving countries responsibility for conserving biodiversity and using biological resources in a sustainable way; (c) the *Rio Declaration* and the *Forest Principles* – setting out principles of sustainable development and pledging to reduce deforestation; and (d) Agenda 21 – a plan for achieving sustainable development in the twenty-first century. Agenda 21 proposed that poverty can be reduced by giving people access to resources they need to support themselves. Developed nations agreed to assist others to develop in ways that minimize environmental impacts of their economic growth. Agenda 21 called on countries to reduce pollution, emissions, and the overuse of precious natural resources. Governments should lead these changes but the private sector and individuals also should be responsible for curtailing non-sustainable practices. It was emphasized that local actions could lead to the solution of global problems (United Nations, 1992).

The *Kyoto Climate Change Protocol* was created in 1997 when governments met in Kyoto, Japan to once more look at the problem of global warming. Previous agreements had tried to limit emissions of CO_2 to 1990 levels. As most countries failed to meet this target reduction, a new set of targets for the reduction of greenhouse gases was agreed upon. By 2012, emissions of six major greenhouse gases should be reduced to below 1990 levels (UN–FCCC, 1997). The US Department of Energy appointed a committee of energy experts to assess the consequences of the Kyoto Protocols on US economic development. The USA is not a signatory of the agreement (US–DE, 1998).

Ten years after the Rio Earth Summit, a conference was convened at Johannesburg, South Africa to review progress towards sustainable development (UN–DESA, 2002). The conference focused on a range of social and environmental issues, from poverty and access to safe drinking water and sanitation to the impact of globalization.

Sustainable development is, therefore, a multidimensional concept which reaches far beyond agriculture. It permeates all levels of human endeavor, economic, social, and cultural. Within the scope of this book, we have restricted the focus of this chapter to the sustainability of agricultural development.

1.4.1 *Sustainable development, agriculture, and IPM*

Sustainable agriculture and IPM are complementary concepts that emerged in the last third of the twentieth century. Sustainable development in connection with agroecosystems was defined as "the ability of an agroecosystem to withstand disturbing forces – particularly threats to its overall productivity" (Conway, 1993). Sustainable agriculture should create agroecosystems that are resilient. Other chapters in this volume deal with novel techniques and provide examples of practical implementation of sustainable agriculture and IPM systems (see Chapters 7 and 14 in this book). The following is limited to what we perceive as the formidable challenges to the achievement of success in both fields, in view of external pressures to maximize agricultural production, often at the cost of environmental integrity and long-term ecological consequences. The argument is offered that, if there is hope to successfully meet those challenges, IPM must advance to higher levels of integration. When IPM systems evolve to encompass the entire agroecosystem (level III IPM), IPM will share with sustainable agriculture its fundamental operational principles.

The long-term sustainability of agricultural production systems is affected by complex social, economic, and environmental constraints. Most of these constraints, if not directly caused, are aggravated by the explosive growth in human population and an uneven but general increase in standard of living worldwide. With the population issue remaining as the backdrop, how does the concept of sustainable development apply to agriculture? The moral and practical reasons to promote sustainable development were suggested by Robert Chambers, as cited by Barbier (1987):

> Poor people in their struggle to survive are driven to doing environmental damage with long-term losses. Their herds overgraze; their shortening fallows on steep slopes and fragile soils induce erosion; their need for off-season incomes drives them to cut and sell firewood and to make and sell charcoal; they are forced to cultivate and degrade marginal and unstable land. Putting people first, and enabling them to meet their needs, can be, then, to reduce these pressures, to reduce degradation, and to maintain potentials for sustainable agriculture and sustainable development at higher levels of productivity. And this in turn means that more people in future can have adequate, secure, and decent levels of living.

This statement, however, leaves out the role of "rich people" in the degradation of the environment. Their needs are met many times over and promotion of consumerism in developed countries is just as big, if not a bigger factor in the impact of humans on the environment, as the struggle of poor people to survive.

Sustainable agriculture as an extension of the more inclusive concept of sustainable development has been defined in almost as many different ways as IPM (Gold, 1999). A definition that seems to encompass the main components of what is commonly accepted as a sustainable agriculture system is provided by Rains *et al.* (2004) and slightly modified as follows: sustainable agriculture is a management strategy that seeks to meet current production and profitability needs while protecting the long-term ecological health and production capacity of a system through fostering and using inherent and renewable resources. The underlying sustainable development principle defined in the Brundtland report is immediately apparent in this definition. Five principles and goals should guide the sustainable management of agroecosystems:

a. prudent use of renewable and/or recyclable resources;
b. protection of the integrity of natural systems so that natural resources are continually regenerated;
c. improvement of the quality of life of individuals and communities;
d. sustained profitability to producers;
e. a land ethic that considers the long-term good of all members of the land community.

The agroecosystem should be the operational scale for a sustainable agriculture. These principles generally coincide with the basic tenets of IPM.

1.4.2 *Sustainable agriculture and IPM*

With a theoretical foundation in agroecology (Altieri, 1987; Gliessman, 1990; Vandermeer, and Perfecto, 1995; Gurr *et al.*, 2004), proponents of the sustainability concept for crop production and protection have found great affinity with the principles and approaches of IPM. Indeed, IPM provided both a conceptual and an implementation paradigm for sustainable agriculture. From an IPM perspective, the concept of sustainable agriculture offers a platform for propelling IPM to higher levels of integration (Kogan,1988; 1998; Prokopy and Croft, 1994; Prokopy and Kogan, 2003).

Since the late 1960s, when it was first adopted by entomologists, the concept of "integrated management" received widespread acceptance in a range of agricultural, industrial, and social activities; it is also fundamental to sustainable agriculture. Figure 1.3 shows the relationships of the components of a crop or livestock production system under integrated management.

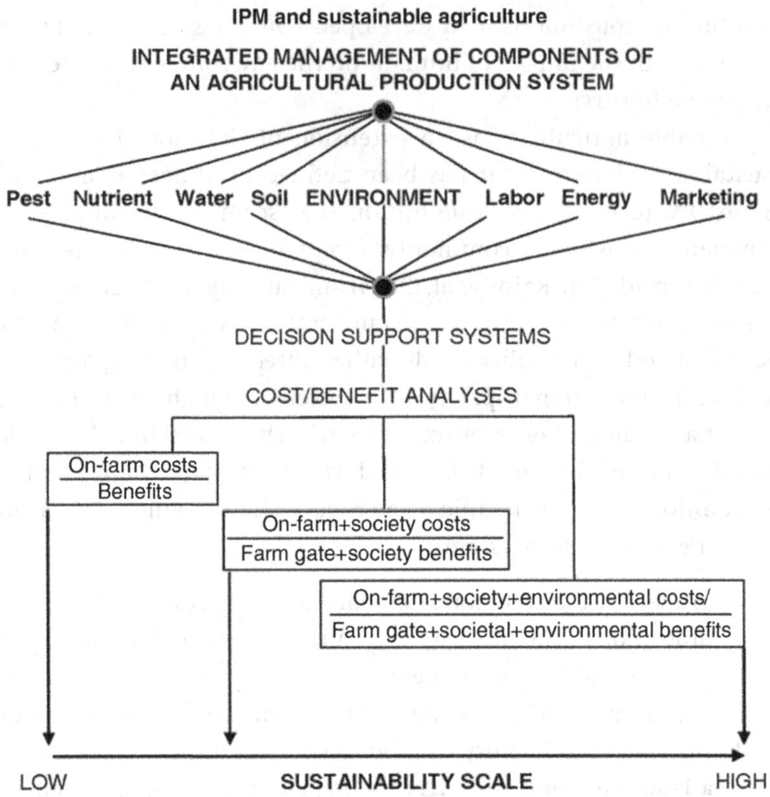

Figure 1.3. Relationships of the components of a crop or livestock production system.

The sustainability of an agricultural system requires all its components to be under levels of integrated management that take into account social and environmental impacts, in addition to the more traditional on-farm economic impacts. A key feature of the integrated management paradigm is an analysis of benefits and costs of management decisions. Sustainability increases along a continuum, depending on whether the costs and benefits are limited to the farm enterprise, or whether they also include costs and benefits to society and the environment.

1.5 IPM and food production

It should be a priority for human societies to produce enough food for their growing populations. With pest (arthropods and other animals, plant pathogens, and weeds) damage accounting for over 30% of losses in total agricultural production pre- and post-harvest, plant protection specialists and producers at the end of World War II were eager to adopt any technology with the potential to reduce those losses and, consequently, increase food output. That technology in the mid-1940s was the newly discovered organosynthetic pesticides. It is estimated that from 1947 until 1973, when their use was banned in the USA, 2.9 billion kg of chlorinated hydrocarbon insecticides were applied. Initial results were spectacular leading to predictions that "some insect pests will become extinct." At first there was little concern about possible side-effects, and, for a few years, none were anticipated. But soon it was discovered that over-use of pesticides resulted in one of the most pervasive intrusions of human-made chemicals into the environment (Metcalf, 1986).

The need to keep food production apace with population growth overshadowed concerns with environmental impacts. Creation of the network of CGIAR (Consultative Group on International Agricultural Research) international agricultural research centers and the subsequent introduction of high-yielding grain varieties that became the hallmark of the green revolution ensured that food production would remain ahead of food demands, at least in the foreseeable future. However, success of the green revolution varieties was, to a large extent, dependent on substantial subsidies of fertilizers and extensive use of pesticides, as many of the new varieties were susceptible to insect pests and pathogens. But the health and environmental risks of heavy reliance on pesticides were too real to ignore. Rachel Carson's 1962 book was a wake-up call that helped concerned scientists to pursue more aggressively the path which led to the adoption of IPM as the fundamental paradigm for pest control. Despite recent advances in pest control technologies and IPM program expansion worldwide, world crops in the 1990s still suffered about 30% losses to the aggregate impact of pre- and post-harvest pests, a level similar to those suffered at the beginning of the century (Schwartz and Klassen, 1981). These losses persist even while pesticide use continues to remain high worldwide (Figure 1.4) (Kogan and Bajwa, 1999).

1.5.1 *Anthropogenic disasters and the sustainability of IPM systems*

By definition, a pest control program under the IPM paradigm should be sustainable, although most current IPM systems still need to use inputs that are energy demanding, e.g., pesticides, motorized sprayers, and mechanical weed control implements, to mention only a few (Pimentel and Pimentel, 1996). After almost 30 years of IPM implementation there is an impressive record of achievements documented in many publications (Pimentel, 1991; Leslie and Cuperus, 1993; Metcalf and Luckmann, 1994; Dent and Elliot, 1995; Mengech *et al.*, 1995; Persley, 1996; Dhaliwal and

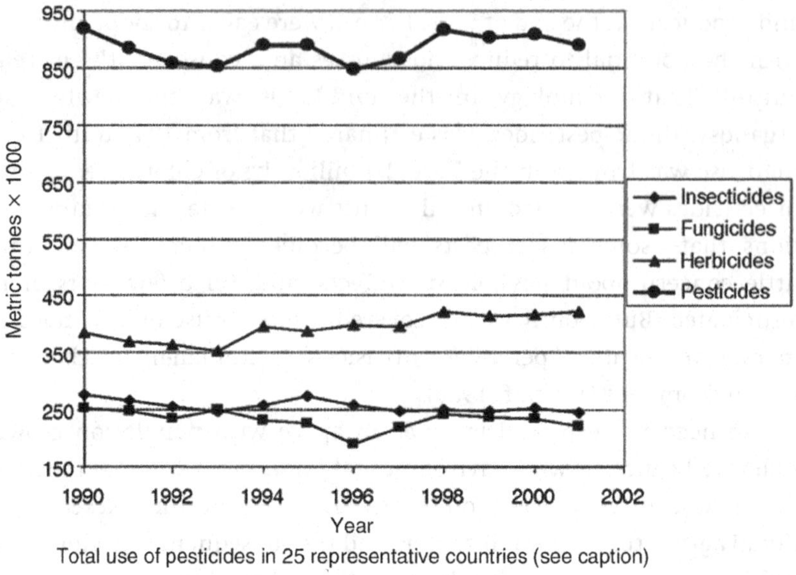

Figure 1.4. Pesticide use in 25 countries (Algeria, Austria, Bangladesh, Brazil, Columbia, Denmark, Ecuador, France, Germany, Greece, Honduras, Hungary, India, Italy, Republic of Korea, Netherlands, Norway, Pakistan, Poland, Portugal, Romania, Spain, Turkey, United Kingdom, USA), representing most continents and agricultural systems, over a period of 12 years 1990–2001. Overall quantities of pesticides used have remained steady, with minor declines in insecticide and fungicide/bactericide use and a slight increase in herbicides. (Kogan and Bajwa, 1999).

Heinrichs, 1998; Mariau, 1999; Meridia, *et al.*, 2003; Norris *et al.*, 2004; Norton *et al.*, 2005; Singh *et al.*, 2005; see also the series on crop IPM published over the past 30 years in *Annual Review of Entomology*, and the University of California IPM Handbook series). The permanence (= sustainability) of these systems, however, is constantly threatened by the same environmental, social, and economic pressures that also threaten the sustainability of agriculture as a whole. Those pressures come from resource limitations, from human-generated (anthropogenic) environmental disasters, and from societal pressures.

Impact of resource limitations

The pressure that rising populations exert on all components of agricultural production systems (air, land, water, and energy) has a direct impact on the sustainability of IPM. Plants growing on substandard soils are stressed and usually more susceptible to insect pests and diseases, and less competitive against invading weeds (Dale, 1988, Heinrichs, 1988). Scarce water supplies force growers to increase intervals between irrigation cycles or to reduce the amount of irrigation; certain arthropod pests, such as spider mites, thrive on drought-stressed plants (Holtzer *et al.*, 1988). Energy enters the IPM equation through control and monitoring equipment, and pesticides. The potential indirect effect of shortages would be felt through price increases for essential inputs. Cost is already a limiting factor for the crops of a few developing countries of Africa where pesticides are needed as part of the IPM system (Abate and Ampofo, 1996). Thus resource limitations which add to plant stress or restrict use of desired IPM tactics tend to aggravate the impact of insect pests and may eventually set back established IPM programs. However, the most serious impacts on the sustainability of IPM systems come from human activities that are independent of demography, albeit magnified by population pressures.

Impact of anthropogenic disasters

There is ample evidence to suggest that ecological systems are fragile and often respond to human interference in unexpected ways and with alarming intensity. Events in Honduras since the early 1970s illustrate the convergence of multiple anthropogenic factors such as shifting land use, climate change, over-use of pesticides, and deforestation, coupled with natural meteorological events (hurricanes) and malaria epidemics, to magnify environmental disasters and social upheaval. Bright (2000), based on a study by J. Almendares *et al.* (1993), described what may become known as "the Honduras syndrome". Over a period of about 20 years (1970–1990), Honduras suffered the consequences of a shift in agricultural production to increase its exports. Landowners in the south increased production of cattle, sugarcane, and cotton while those in the north actively cleared the tropical forest to plant banana, melons, and pineapple. In the south, intensive farming reduced the soil's water absorbency, so more rain ran off fields and less remained to evaporate back into the air. The drier air reduced cloud cover and rainfall and the region grew warmer. The median annual temperature increased by 7.5 °C between 1972 and 1990, to above 30 °C. In this drier environment mosquito populations dwindled and malaria cases declined. But soils also became impoverished and so people migrated north to the newly opened large plantations. The super-humid tropical environment in the north favored insect pests and diseases which were treated with massive pesticide applications (from 1989 to 1991, Honduran pesticide imports increased more than fivefold, to about 8000 tons.) The moist environment was also perfect for malaria mosquitoes which were suppressed temporarily by the insecticide drift around the plantations. The mosquitoes, however, eventually acquired resistance to most pesticides and populations quickly increased everywhere. Malaria spread within a human population that had become highly susceptible to *Plasmodium* due to reduced exposure in the south. From 1987 to 1993 the number of malaria cases in Honduras jumped from 20 000 to 90 000.

The final act in this drama was the result of a natural disaster, Hurricane Mitch. The hurricane in October 1998 stalled over the Gulf coast of Central America for four days, producing in mountain slopes, now deprived of a forest cover, massive mudslides which destroyed entire villages and 95% of Honduran agricultural production. This economic, social, and ecological disaster resulted from the confluence of diverse and seemingly unrelated factors such as shifting agricultural production practices caused by demands of the world market, internal migration of farm workers from malaria-free areas to areas of high incidence, climate change caused by intensification of agriculture and deforestation, and, finally, a violent hurricane with consequences greatly magnified by the clearing of steep slopes for plantations of export crops. The "Honduras syndrome" demonstrates how human actions can greatly magnify the impact of natural disasters. Although much of the ecological consequences of climate change and deforestation had been predicted, little was done to avoid the disaster. It seems that, from the way history repeats itself, little is learned from such tragic events.

Two categories of anthropogenic disasters potentially impact IPM systems: (a) those resulting from shifting environmental pressures, and (b) those resulting from shifting social pressures.

Shifting environmental pressures

Environmental factors most likely to affect IPM sustainability are climate change, biological invasions, and loss of biodiversity (discussed above in some detail, see Section 1.2). The relationship of greenhouse gases, particularly CO_2, to climate change has been emphasized. Agriculture, it seems, may have had both negative and positive effects on increased atmospheric concentrations of CO_2. Negative effects include (a) changing the amount of carbon stored in the vegetation of terrestrial ecosystems (deforestation and reforestation) and in soils; (b) burning fossil fuels in all phases of agricultural production and associated industries. Potentially positive effects include: (a) providing renewable energy resources to substitute for fossil fuels, e.g. sugar cane for ethanol production, and (b) producing energy from biomass which recycles carbon rather than allowing it to be released to the atmosphere (Pimentel and Pimentel, 1996).

Besides the pervasive impact that global warming will have on animal and plant ecology, the increase in atmospheric CO_2 concentrations may have direct effects on physiological processes (Royal Society, 2005).

Fruit IPM in the USA offers an example of potential impacts of climate change on the stability of a system. Fruit IPM has reached a high level of sophistication and adoption among producers (Prokopy and Croft, 1994). The management of the codling moth, *Cydia pomonella*, a key pest of apples, pears, and other fruit and nut-tree crops, is based on accurate monitoring following a biofix date, or the earliest date a set number of colonizing moths is detected in the orchard, and events predicted by phenology models (Coop, 2005). The pest usually has one and a half or two generations per year in the Pacific Northwestern states of the USA. A yearly increase of just 2 °C in average daily temperatures could cause a third generation to occur, forcing growers to spray exactly at the time when fruit is closer to harvest and spray restrictions most strict. Such a warming trend in the region could derail one of the most advanced IPM systems in the USA.

The potential economic impact of invasive species is not only when they become pests, but also results from restrictions imposed on imports by quarantine regulations. Invasive species can seriously impact established IPM systems. The invasion of Brazilian cotton fields in São Paulo and Paraná states by the boll weevil, *Anthonomus grandis grandis*, in the 1980s, set back some reasonably well-established IPM systems, as new spray schedules and new powerful insecticides were introduced to contain the insects.

Loss of biodiversity was discussed in Section 1.2.4. It is worth stressing, however, another kind of biodiversity which is being lost with potentially grave consequences to agriculture and IPM. In addition to the loss in species diversity, food crops around the world face an alarming narrowing of genetic diversity. With the advent of high-yielding, green revolution varieties, many local races were replaced by the new varieties. China reduced the number of planted wheat varieties from ca. 10 000 in 1949 to ca. 1000 in the 1970s (Brown *et al.*, 1999). Both loss of traditional land-races and loss of wild relatives make breeding of new varieties increasingly dependent on a limited genetic base. From an IPM perspective, the loss of genetic diversity increases crop plants' susceptibility to insects and diseases and reduces the chances for incorporating host plant resistance as a key component of IPM.

The relationship between biodiversity and stability of ecosystems is still being debated in ecological circles (Schowalter, 2000; see also Chapter 13, this volume). The debate has direct bearing on the argument about the importance of biodiversity for IPM. Altieri (1993) defended the argument that "biodiversity is a salient feature of traditional farming systems in developing countries and performs a variety of renewal processes and ecological services in agroecosystems." He argued that it is important to understand the role biodiversity plays in reducing pest problems, when vegetation management is used as a basic tactic in small-scale sustainable agriculture. The rising interest in use of cover crops, hedge management, and habitat enhancement for natural enemies, seems to support Altieri's argument, even if the techniques have limitations for large scale agriculture (see Chapters 7 and 14, and this volume).

Shifting societal pressures

Established IPM systems also are vulnerable to pressures that derive from real or perceived societal problems with pests and the techniques used for their management. Examples of such pressures are: (a) concerns about the safety of food supplies, particularly for infants; (b) the public debate about the introduction of genetically modified organisms (GMOs) in agriculture; and (c) the intermittent and increasingly less muffled rural/urban conflict.

Food safety in the USA, for instance, is monitored by several agencies. Analyses of over 8500 samples in 1998 showed no residues in 64.9% of domestic samples and 68.1% in import samples. None of the samples of grains, grain products, fruits, or vegetables had detectable residues above levels that violate current limits (US–FDA, 1999). The food supply appeared to be reasonably safe.

Whether real or perceived, public pressure in developed countries has led regulatory agencies to revise existing standards for food safety. Understandably, however, in most developing countries of the world the plight of chronic starvation often overwhelms concerns about food safety with regard to pesticides.

In the USA, the revision of food safety standards produced new guidelines under the *Food Quality Protection Act* of 1996 (US–EPA, 2003) regarding uses and tolerances of pesticides on food crops. The impact of these new regulatory laws on current IPM programs in the short run may be potentially disruptive. For example, removal of organophosphate insecticides (OPs) for use in fruit crops may increase incidence of codling moth and reduce effectiveness of established mating disruption programs. In the long run, however, FQPA may force producers and researchers to look harder for alternatives and propel IPM to higher levels of adoption and integration.

The technology to develop genetically modified crops has generated much interest within the academic, grower, and consumer communities, as well as within industry (Atherton, 2002; Khachatourians *et al.*, 2002; Shelton *et al.*, 2002; Liang and Skinner, 2004). Commercially developed transgenic crops were met with a mixture of excitement, controversy, and skepticism. The advent of cotton, corn, potato, soybean, canola, squash, papaya and many other field, fruit, and vegetable crop varieties genetically modified to incorporate δ-endotoxin-producing genes of the bacterium *Bacillus thuringiensis* was heralded by some as the next "magic bullet" in agricultural pest control. The level and tone of the criticism by those who fear a repeat of the DDT ecological disaster, however, has exceeded expectations. Is there a place for GMOs in IPM systems? It is questionable whether GMOs will be the magic bullet that some have anticipated. It is likely, also, that they are not the Pandora's box that some fear to open. The true role of GMOs will probably be yet another tactic in the IPM arsenal which, if used wisely, will provide new and potentially powerful strategic options. It is essential to remember that "integration" is the key term in the IPM equation. If GMOs are carefully integrated within IPM systems, taking into account resistance management prescriptions, interactions with other control tactics, and careful monitoring of undesirable side-effects, all within the context of ecosystem integrity, then the answer to the question is an emphatic "Yes," there is a place for GMOs in IPM systems.

The urban/rural dichotomy often is cause for friction and misunderstanding. Generations of urbanites, growing up in huge megalopolises distant from rural areas, are generally ignorant about how food is produced. Lack of understanding of farming practices and promotional tactics used by retailers lead grocery shoppers in big cities to demand fresh produce which is both unblemished and free of chemical residues. In many instances, cosmetic standards alone determine the level of pesticide use in a crop. Organic farming is helping to educate the public that a certain amount of bruising in an apple or a few aphids in a head of broccoli are acceptable. This lack of understanding is a key factor in the slow adoption of IPM practices, even if they fall short of guaranteeing the desired cosmetic standards for some crops. Food-exporting countries of the Third World must abide by the standards of their import markets. Even if farmers might adopt biointensive IPM practices for the local market, they would still be pressured to rely on chemicals for the export market.

Another area of potential rural/urban conflict that impacts IPM systems is the acceptance of the need to prevent biological invasions through eradication of incipient infestations of potentially serious pests. Large urban areas in the Pacific Northwest of the USA must be occasionally sprayed to eradicate nuclei of gypsy moth infestations. Eggs often hitchhike on trailers and campers of cross-country travelers. Should the gypsy moth become established in this highly forested region, the result would be disastrous and a real challenge to pest managers. Authorities in charge of spray programs are often the target of irate urbanite critics who often misunderstand the gravity of the problem and the generally safe control methods used in pest eradication programs in urban settings.

These aspects of the urban/rural dichotomy imply that IPM, within the broader context of sustainable agriculture, must consider the specific needs, demands, and biases of society as a whole. The question remains whether and how IPM may contribute to sustainable development in agriculture, despite formidable constraints imposed by human demands, cultural biases, and anthropogenic disasters that often set back well-established IPM systems. The following is an attempt to point towards directions that may be cause for cautious optimism.

1.6 IPM achievements and expectations

IPM can contribute significantly to reach the level of agricultural production needed to feed the 8 billion people expected to inhabit the planet in 2050, and still maintain harmony with the environment in a sustainable way. Effective IPM systems may prevent the 30% annual pest-induced production losses, thus adding 750 million tons of food grains, enough to feed 1.8 billion people at an adequate level of 400 kg per year. To reach this goal it will be necessary to intensify research into new and more effective and environmentally benign control tactics and put them to use in the development of advanced IPM strategies oriented to ecosystems' levels of integration. Progress has been made in all phases of IPM from diagnozing and identification, monitoring and surveying, to all basic control tactics, and a few innovative strategies that apply to spatial units larger than the single-crop field.

The recent literature provides examples of the effective integration of multiple tactics in IPM strategies. One such example illustrates how quickly changes can occur. The use of mating disruption for the control of the codling moth on apple and pear orchards in the Northwestern states of the USA was made possible by the identification of the moth's main pheromones in 1971 (Roelofs *et al.*,1971), and the availability of commercial formulations shortly thereafter. Slow release technology early on required dispersal early in the season of up to 1000 pheromone-releasing dispensers per ha and their replacement at least once by mid-season. The cost was prohibitive and growers were slow to adopt mating disruption technology. With the establishment of the area-wide IPM program for the codling moth in the western United States (Chandler and Faust, 1998; Kogan, 2002; Calkins and

Faust, 2003), technology was improved but still 1000 dispensers were recommended in 1996. By 1998 the rate had dropped to 500/ha, and in many areas only one application per season was needed. Companies began developing sprayable formulations of microencapsulated pheromones and mechanical dispensers (atomizers or puffers) that were effective at a rate of 2.5 per ha. The success of mating disruption in pome fruit orchards was made possible by its adoption over large contiguous areas with the cooperation of many neighboring growers. The new strategic approach became known as areawide IPM and it had since been adapted to other crop and pest complex situations.

Areawide IPM is a recent example of what can be accomplished with a modest strategic shift. IPM strategies must advance from the individual-field focus to models that consider entire ecological regions. Examples of such scale shifts are rare. The soybean IPM system for micro-river basins established in the state of Paraná, Brazil (Correa-Ferreira *et al.*, 1998), and the Hamabia project in several small discrete ecoregions in Israel (Ausher, 1999), are, however, excellent examples. Landis *et al.* (2000) stressed the importance of considering the structure of the landscape to enhance natural enemies and increase IPM sustainability. When IPM adopts the ecoregion (Bailey, 1998) as the fundamental operational unit for strategic planning, then it reaches level III integration; at this level IPM and sustainable agricultural development coalesce into a single powerful construct.

1.7 Overview and conclusions

Human population continues to grow at high, unsustainable rates, mainly in the poorest countries of the world. Poverty and underdevelopment often lead to environmental degradation. Poor people have few choices and in their struggle to survive they are driven to doing environmental damage with long-term effects (Chambers, 1983) – overgrazing cattle, farming steep slopes and marginal lands, cutting trees for firewood and for making charcoal. Population growth in most rich countries has stabilized, but mounting affluence leads to even greater exploitation of natural resources, due to excessive consumerism.

Although a global food crisis may have been temporarily averted, regional famines are occurring with increasing frequency, e.g. in Ethiopia (1991) and N. Korea (1997); one-sixth of the world population (1 billion) lives in a state of semi-starvation and 800 million are chronically malnourished. Every day 16 000 children die from hunger or hunger-related diseases (FAO, 2003).

The challenge to the agricultural community, i.e. producers, processors, distributors, the pesticide and the fertilizer industries, public and private researchers and educators, is how to expand food production without further depleting vital natural resources (land, water, forests, fossil fuels). Asia alone will require an increase in rice production of 40% by 2025, to feed the projected additional population. Is the need to feed 10 billion people compatible with a sustainable production system?

Planning for sustainable agricultural development requires a clear understanding of regional ecologies. Nature provides a blueprint for sustainable plant productivity in climax communities. Since the origins of agriculture, humans have replaced climax vegetation with a mix of crops and created artificial agroecosystems. Natural complex climax communities that once existed were replaced with much-simplified monocrop ecosystems. It seems that the greater the disparity between the original climax community and the crop community that replaced it (the ecological distance between the natural ecosystem and the introduced agroecosystem) the higher the vulnerability of the agroecosystem to adverse biotic and abiotic factors, including pest organisms. For example, the prairies occupy a vast area in the middle third of the North American continent. The great prairie of the Midwest was originally dominated by grasses, herbaceous legumes, and composites, with the relative dominance of certain species varying with local climatic conditions. Grasses and forbs are hardy and have evolved to tolerate extremes of temperature, periodic droughts, fire, and extensive grazing. A large portion of the original prairie ecosystem has been replaced by agroecosystems dominated by annual grasses (corn, sorghum, wheat, etc. and legumes) soybean, dry beans, and forage legumes. Although fire is not a factor in these agroecosystems, mechanical tillage (by humans) performs a comparable function. In both their basic botanical composition and functional dynamics, the small grain, corn, and soybean ecosystems of the American Midwest are within a short ecological distance from the prairie that preceded them. These systems are remarkably resilient and, if properly managed, can withstand considerable grazing by herbivorous arthropods before productivity is significantly affected.

By contrast, soybean in the state of Paraná and, more recently in the Amazon Basin, Brazil, was established in land previously covered with lush subtropical forests, in Paraná, and in savannah and rain forest, in the Amazon Basin. There is an enormous ecological distance between the forest climax and the new soybean ecosystems. Pest problems and problems with soil conservation in the crop often are severe. New pests flare up with certain frequency, requiring that producers and IPM specialists be on constant alert.

Although IPM systems have been successful in the region, these new pests are a threat to their stability.

To achieve higher levels of IPM integration within the context of sustainable agricultural development it may be necessary to redesign entire cropping systems. It may be necessary to test and implement new paradigms that take into account the ecological distance from the original climax plant cover in the region. This may become a source of conflict under the rules of the now dominant free enterprise system, because such large-scale planning will require greater involvement of the public sector and impose set parameters on the use of private property. In the absence of such planning, however, there is little hope that sustainable agricultural production can be achieved without large subsidies of inputs, both fertilizers and pesticides. A free enterprise system that is not tempered by societal longer-term needs and concerns pits humans' short-term interests against nature. In the words of John Ikerd (1999):

> Much of human history has been written in terms of an ongoing struggle of "man against nature." The forces of nature – wild beasts, floods, pestilence, and disease – have been cast in the role of the enemy of humankind. To survive and prosper, we must conquer nature – kill the wild beasts, build dams to stop flooding, find medicines to fight disease, and use chemicals to control the pests. Humans have been locked in a life and death struggle against "Mother Nature." We've been winning battle after battle. But, we've been losing the war.

The struggle to tame nature is not limited to fighting natural calamities and to extracting nature's products for nourishment and shelter. The struggle to conquer nature's obstacles to survival is part of the evolutionary process of any living species. Civilization, however, has created needs that go far beyond mere survival. To satisfy those needs humans have carried the fight to unprecedented dimensions. In some cases injury to the environment has been irreversible bringing many vital resources to a level beyond further sustainable use. Extensive human-made deserts around the world serve as constant reminders of the profound impact of humans on the environment. The lessons of those past and the many contemporary anthropogenic disasters must be quickly absorbed as sustainable use of remaining resources may still be possible. Quick action, however, is essential. Agricultural production in many parts of the world can return to more sustainable practices (Gurr *et al.*, 2004). Agricultural systems must be conceived that are more in balance with dominant ecoregional features. IPM also must be conceived at higher levels of ecosystem integration. At this level sustainable agriculture and IPM converge into a unified ecologically based agricultural production system potentially capable of meeting the demands to feed 10 billion people without further injury to the environment.

References

Abate, T. and Ampofo, J.K. (1996). Insect pests of beans in Africa: their ecology and management. *Annual Review of Entomology*, **41**, 45–73.

Advanced BioTech. (2004). *History of agriculture*. E-publication: www.adbio.com/science/agri-history.htm

Aide, M.T. and Grau, R. (2004). Globalization, migration, and Latin American ecosystems. *Science*, **305**, 1915–16.

Almendares, J., Sierra, M., Anderson, P.K. and Epstein, P.R. (1993). Critical regions, a profile of Honduras. *Lancet*, **342**, 1400–2.

Altieri, M.A. (1987). *Agroecology: The Scientific Basis of Alternative Agriculture*, Boulder, CO: Westview.

Altieri, M.A. (1993). Ethnoscience and biodiversity: key elements in the design of sustainable pest management systems for small farmers in developing countries. *Agriculture, Ecosystems and Environment*, **46**, 257–72.

APHIS. 2002. URL: www.aphis.usda.gov/lpa/pubs/poster_phhogweed.html

Atherton, K.T. (2002). *Genetically Modified Crops: Assessing Safety*. New York: Taylor & Francis.

Ausher, R. (1999). Promotion of areawide pest management (AMP) in Israel. *Phytoparasitica*, **27**, 83–4.

Bailey, R.G. (1998). *Ecoregions: The Ecosystem Geography of the Oceans and Continents*. New York: Springer.

Bajwa, W.I. and Kogan, M. (1996). *Compendium of IPM definitions*. Integrated Plant Protection Center, Oregon State University, Corvallis, Oregon, USA. Electronic database www.ippc.orst.edu/IPMdefinitions/home.html

Barbault, R. and Sastrapradja, S. (1995). Generation, maintenance and loss of biodiversity. In V. Heywood (ed.), *Global Biodiversity Assessment*. Cambridge, UK: Cambridge University Press. pp. 193–274.

Barbier, E. (1987). The concept of sustainable economic development. *Environmental Conservation*, **14**, 101–10.

Bazzaz, F.A. (1990). The response of natural ecosystems to the rising global CO_2 levels. *Annual Review of Ecology and Systematics*, **21**, 167–96.

Bright, C. (2000). Anticipating environmental surprise. In World Watch Institute, *State of the World: 2000*. New York: Norton. pp. 22–38.

Britton, K. O., Orr, D. and Sun, J. (2002). Kudzu. In R. Van Driesche, B. Blossey, M. Hoddle, and R. Reardon (eds.), *Biological Control of Invasive Plants in the Eastern United States*. Washington, DC: US Department of Agriculture, Forest Service Publication FHTET-2002-04. <http://www.invasive.org/eastern/biocontrol/25Kudzu.html>.

Brown, L. R. (2003). Plan B: Rescuing a Planet under Stress and a Civilization in Trouble. New York: Norton. Earth Policy Institute.

Brown, L. R. (2000). Challenges of the new century. In World Watch Institute, *State of the World: 2000*. New York: Norton. pp. 3–21, 203–7. PDF version: www.worldwatch.org/pubs/sow/2000/

Brown, L. R., Gardner G. and Halweil, B. (1999). *Beyond Malthus: Nineteen Dimensions of the Population Challenge*. New York: Norton. World Watch Institute.

Calkins, C. O. and Faust, R. J. (2003). IPM-areawide programs: overview of areawide programs and the program for suppression of codling moth in the western USA directed by the United States Department of Agriculture – Agricultural Research Service. *Pest Management Science*, **59**, 601–4.

Carson, R. (1962). *Silent Spring*. Greenwich, Connecticut: Fawcet Crest. 1994 edition, with introduction by A. Gore. Boston, MA: Houghton Mifflin.

Cavanaugh-Grant, D. and Dan, A. (1999). *The agroecology/sustainable agriculture program*. University of Illinois, Urbana-Champaign, Illinois, USA. Department of Natural Resources and Environmental Sciences.

Chambers, R. (1983). *Rural Development: Putting the Last First*. London: Longman.

Chance, N. (2005). The Arctic National Wildlife Refuge (ANWAR) - A Special Report. University of Connecticut, Storrs, Connecticut, USA. URL: http://arcticcircle.uconn.edu/ANWR/.

Chandler, L. D. and Faust, R. M. (1998). Overview of areawide management of insects. *Journal of Agricultural Entomology*, **15**, 319–25.

Cohen, J. E. (1995). *How Many People can the Earth Support?* New York: Norton.

Conway, G. R. (1993). Sustainable agriculture: the trade-offs with productivity, stability and equitability. In E. B. Barbier (ed.), *Economics and Ecology*. London: Chapman and Hall. pp. 46–63.

Conway, G. R. and Barbier, E. B. (1990). *After the Green Revolution*. London: Earthscan.

Conway, G. R. and Pretty, J. (1991). *Unwelcome Harvest: Agriculture and Pollution*. London: Earthscan.

Coombs, E. M., Markin G. P., and Forrest, T. G. (2004). Scotch broom. In E.M. Coombs, J. K. Clark, G. L. Piper, and A. F. Cofrancesco Jr., (eds.), *Biological Control of Invasive Plants in the United States*. Corvallis, Oregon: Oregon State University Press. pp. 160–4.

Coop, L. (2006). IPM Weather data and degree-days for agricultural and pest management decision making in the US. Integrated Plant Protection Center, Oregon State University, Corvallis, Oregon, USA. URL: http://pnwpest.org/wea/

Correa-Ferreira, B. S., Domit, L. A., Morales L., and Guimarães, R. C. (1998). Integrated soybean pest management in micro river basins in Brazil. *Integrated Pest Management Reviews*, **5**, 75–80.

Dale, D. (1988). Plant-mediated effects of soil mineral stresses on insects. In E. A. Heinrichs (ed.), *Plant Stress–Insect Interactions*. New York: Wiley. pp. 35–110.

Davis, J. A., Radcliffe, E. B., and Ragsdale, D. W. (2004). A new vector of PVY: Soybean aphid, *Aphis glycines* (Matsumura). *American Journal of Potato Research*, **81**, 53–4.

Dent, D. and Elliott, N. C. (1995). *Integrated Pest Management*. London: Chapman & Hall.

Dhaliwal, G. S. and Heinrichs, E. A. (1998). *Critical Issues in Insect Pest Management*. New Delhi, India: Commonwealth Publishers.

Diamond, J. M. (1997). *Guns, Germs, and Steel: the Fates of Human Societies*. New York: Norton.

Diamond, J. M. (2005). *Collapse: How Societies Choose to Fail or Succeed*. New York: Viking.

DLNR. (2005). *Hawaii Department of Land and Natural Resources*. URL: www.state.hi.us/dlnr/dofaw/consrvhi/silent/top10/

Eden Foundation. (2000). *Desertification – A threat to the Sahel*. Sweden: Eden Foundation. E-publication. URL: www.eden-foundation.org/project/desertif.html

Ehrlich, P. R. (1968). *The Population Bomb*. New York: Ballantine Books.

Ehrlich, P. R. and Ehrlich, A. H. (1990) *The Population Explosion*. Simon and Schuster, New York.

Evans, L. T. (1998). *Feeding the Ten Billion: Plants and Population Growth*. Cambridge, UK: Cambridge University Press.

Evans, E., Spreen, T., and Knapp, J. (2002). Economic issues of invasive pests and diseases and food safety, 2nd International Agricultural Trade and Policy Conference, Gainesville, FL.

Evans, E. 2003. *Economic dimensions of invasive species*. The Magazine of Food, Farm, and Resource Issues. E-publication: www.choicesmagazine.org/2003–2/2003-2-02.htm

Fan, S. and Zhang, X. (2002). *Growth, Inequity, and Poverty in Rural China: The Role of Public Investments*. Washington, DC: International Food Policy Research Institute.

Fan, S., Zhang, X., and Robinson, S. (2003). Structural change and economic growth in China. *Review of Development Economics*, **7**, 360–77.

FAO. (1999). *World Population 1998*. Rome, Italy: Food and Agriculture Organization of the United Nations.

FAO. (2001*a*). *State of the World's Forests – 2001*. Rome: Food and Agriculture Organization of the United Nations. URL: www.fao.org/forestry/index.jsp.

FAO. (2001*b*). *The ticking time bomb: Toxic pesticide waste dumps. News & Highlights*. Rome: Food and Agriculture Organization, United Nations. URL: www.fao.org/NEWS/2001/010502-e.htm

FAO. (2002). FAOSTAT – *Agricultural Data*. Rome, Italy: Food and Agriculture Organization, United Nations. URL: faostat.fao.org/faostat/collections?subset=agriculture

FAO. (2003). *State of Food Insecurity in the World: Monitoring progress towards the World Food Summit and Millennium Development Goals*. Rome, Italy: United Nations Food and Agriculture Organization. PDF version: ftp://ftp.fao.org/docrep/fao/006/j0083e/j0083e00.pdf

Gliessman, S. R. (1990). *Agroecology: Research in the Ecological Basis for Sustainable Agriculture*. New York: Springer.

Gold, M. (1999). *Sustainable Agriculture: Definitions and Terms*. Special Reference Briefs Series no. SRB 99-02, National Agricultural Library, Agricultural Research Service, US Department of Agriculture. E-publication: www.nal.usda.gov/afsic/AFSIC_pubs/srb9902.htm

Goudie, A. (2000). *The Human Impact on the Natural Environment*. Cambridge, MA: MIT Press.

Gurr, G., Wratten S. D., and Altieri, M. A. (2004). *Ecological Engineering for Pest Management: Advances in Habitat Manipulation for Arthropods*. Ithaca, NY: Cornell University Press.

Hagenbaugh, B. (2004). *Soybean disease finally hits*. USA Today. E-edition: www.usatoday.com/money/industries/food/2004-11-28-soybean-cover_x.htm

Harrison, P. (1992). *World agriculture: Towards 2015/2030 – summary report*. Rome, Italy: Food and Agriculture Organization of the United Nations.

Harrison, P. (1992). *The Third Revolution: Environment, Population, and a Sustainable World*. London, I. B. Tauris; New York, NY: Distributed by St. Martin's Press.

Heinrichs, E. A. (1988). Global food production and plant stress. In E. A. Heinrichs (ed.), *Plant Stress–insect Interaction*. New York: Wiley. pp. 1–33.

Hinrichson, D. and Robey, B. (2000). *Population and the environment: The global challenge*. American Institute of Biological Sciences, Washington, DC, USA. E-publication: www.actionbioscience.org/environment/hinrichsen_robey.html

Holtzer, T. O., Archer, T. L., and Norman, J. M. (1988). Host plant suitability in relation to water stress. In E. A. Heinrichs (ed.), *Plant Stress–insect Interaction*. New York: Wiley. pp. 111–37.

Houghton, J. T., Ding, Y., Griggs, D. J., Noguer, M., van der Linden, P. J., and Xiaosu, D. (eds.) (2001). *Climate Change 2001: The Scientific Basis. Contribution of Working Group I to the Third Assessment Report of the Intergovernmental Panel on Climate Change (IPCC)*. Cambridge, UK: Cambridge University Press.

Houghton, R. (2005). *Understanding the global carbon cycle*. Woods Hole Research Center. E-publication: www.whrc.org/carbon/

Hunt, H. W. and Hall, D. H. (2002). Modelling the effects of loss of soil biodiversity on ecosystem function. *Global Change Biology*, **8**, 33–50.

Ikerd, J. E. (1999). In harmony with Nature. Presented at AgriExpo '99, Columbia, MO. March 23, 1999. University of Missouri. Columbia, Missouri: URL www.ssu.missouri.edu/faculty/jikerd/papers/HARMONY.html

IUCN. (2005). *Species extinction: a natural and unnatural process*. World Conservation Union – International Union for the Conservation of Nature and Natural Resources (IUCN). PDF publication: www.iucn.org/news/mbspeciesext.pdf

IUCN. (2006). Global Invasive Species Database. Global Invasive Species Programme (GISP) in partnerships with the National Biological Information Infrastructure, Manaaki Whenua-Landcare Research, the Critical Ecosystem Partnership Fund and the University of Auckland. World Conservation Union–International Union for the Conservation of Nature and Natural Resources (IUCN). URL: http://www.issg.org/database/welcome

Khachatourians, G. G., McHughen, A., Scorza, R., Nip, W.-K., and Hui, Y. H. (2002). *Transgenic Plants and Crops*. New York: Marcel Dekker.

Kiritani, K. and Morimoto, N. (1999). Fauna of exotic insects in Japan with special reference to North America. In *Ecological Society of America Symposium*, Spokane, Washington, USA. Manuscript, 32 pp.

Kogan, M. (1988). Integrated pest management theory and practice. *Entomologia Experimentalis et Applicata*, **49**, 59–70.

Kogan, M. (1998). Integrated pest management: historical perspectives and contemporary developments. *Annual Review of Entomology*, **43**, 243–70.

Kogan, M. (2002). Areawide pest management. In D. Pimentel (ed.), *Encyclopedia of Pest Management*. New York: Marcel Dekker. pp. 28–32.

Kogan, M. and Bajwa, W. I. (1999). Integrated pest management: a global reality? *Anais da Sociedade Brasileira de Entomologia*, **28**, 1–25.

Landis, D.A., Wratten, S.D., and Gurr, G.M. (2000). Habitat management to conserve natural enemies of arthropod pests in agriculture. *Annual Review of Entomology*, **45**, 175–201.

Leslie, A.R. and Cuperus, G.W. (1993). *Successful Implementation of Integrated Pest Management for Agricultural Crops*. Boca Raton, FL: Lewis Publishers.

Liang, G.H. and Skinner, D.Z. (eds.) (2004). *Genetically Modified Crops: Their Development, Uses, and Risks*. New York: Haworth Press.

Loreau, M., Naeem, S., Inchausti, P. *et al.* (2001). Biodiversity and ecosystem functioning: Current knowledge and future challenges. *Science*, **294**, 804–8.

Malhi, Y. and Phillips, L.O. (2004). Introduction to theme: Tropical Forests and Global Atmospheric Change. *Philosophical Transactions of the Royal Society of London (B)*, **359**, 309–10.

Malhi, Y. and Wright, J. (2004). Spatial patterns and recent trends in the climate of tropical rainforest regions. *Transactions of the Royal Society of London (B)*, **359**, 311–29.

Malthus, T.R. (1798). *An Essay on the Principle of Population*. 1976 Norton Critical edition. P. Appleman (ed.). New York: W.W. Norton.

Manning, R. (2001). *Food's Frontier: The Next Green Revolution*. Berkeley, CA: University of California Press.

Maredia, K.M., Dakouo, D., and Mota-Sanchez, D. (2003). *Integrated Pest Management in the Global Arena*. Wallingford, Oxon, UK: CABI Publishers.

Mariau, D. (1999). *Integrated Pest Management of Tropical Perennial Crops*. Enfield, NH USA: Science Publishers.

Margulis, S. (2004). *Causes of Deforestation of the Brazilian Amazon*. World Bank paper 22. Washington, DC: World Bank.

Martin, T. (2005). *Cactus moth – cactoblastis cactorum Bergroth*. The Nature Conservancy: The Global Invasive Species Initiative. E-publication: http://tncweeds.ucdavis.edu/products/gallery/cacca1.html

Matson, P.A., Parton, W.J., Power, A.G., and Swift, M.J. (1997). Agricultural intensification and ecosystem properties. *Science*, **277**, 504–9.

Matsunaga, K. and Themelis, N.J. (2002). *Effects of Affluence and Population Density on Waste Generation and Disposal of Municipal Solid Wastes*. Earth Engineering Center, Columbia University, New York, NY, USA.

McLaughlin, A. and Mineau, P. (1995). The impact of agricultural practices on biodiversity. *Agriculture, Ecosystems and the Environment*, **55**, 201–12.

McNeely, J.A., Gadgil, M., Leveque, C., Padoch, C., and Redford, K. (1995). Human influences on biodiversity. In V. Heywood (ed.), *Global Biodiversity Assessment*. Cambridge, UK: Cambridge University Press. pp. 711–821.

Mengech, A.N., Kailash, K.N., and Gopalan, H.N.B. (1995). *Integrated Pest Management in the Tropics: Current Status and Future Prospects*. Chichester, UK: Published on behalf of United Nations Environment Programme (UNEP) by Wiley.

Metcalf, R.L. (1986). The ecology of insecticides and the chemical control of insects. In M. Kogan (ed.), *Ecological Theory and Integrated Pest Management Practice*. New York: Wiley. pp. 251–97.

Metcalf, R.L. and Luckmann, W.H. (1994). *Introduction to Insect Pest Management*. New York: Wiley.

Meyerson, L.A. and Reaser, J.K. (2003). Bioinvasions, bioterrorism, and biosecurity. *Frontiers in Ecology and the Environment*, **1**, 307–14.

Mooney, H.A., Lubchenco, J., Dirzo, R., and Sala, O.E. (1995*a*). Biodiversity and ecosystem functioning: ecosystem analyses. In V. Heywood (ed.), *Global Biodiversity Assessment*. Cambridge, UK: Cambridge University Press. pp. 326–452.

Mooney, H.A., Lubchenco, J., Dirzo, R., and Sala, O.E. (1995*b*). Biodiversity and ecosystem functioning: basic principles. In V. Heywood (ed.), *Global Biodiversity Assessment*. Cambridge, UK: Cambridge University Press. pp. 275–325.

National Research Council. (1993). *Sustainable Agriculture and the Environment in the Humid Tropics*. Washington DC: National Academy Press.

Norris, R.F., Caswell-Chu, E.P., and Kogan, M. (2004). *Concepts in Integrated Pest Management*. Upper Saddle River, New Jersey, USA: Prentice Hall.

Norton, G.W., Heinrichs, E.A., Luther, G.C., and Irwin, M.E. (eds.) (2005). *Globalizing Integrated Pest Management: A Participatory Research Process*. Ames, Iowa: Blackwell.

Novacek, M.J. and Cleland, E.E. (2001). The current biodiversity extinction event: scenarios for mitigation and recovery. *Proceedings of the National Academy of Sciences*, **98**, 5466–70.

Nowierski, R.M. and Pemberton, R.W. (2002). Leafy spurge. In R. Van Driesche, S. Lyon, B. Blossey, M. Hoddle and R. Reardon (eds.), *Biological Control of Invasive Plants in the Eastern United States*. Washington, DC: USDA Forest Service. pp. 181–94.

Ostlie, K. (2005). Status of soybean aphid as of June 10, 2005. *Insect & Insect Management*. University of Minnesota. E-publication: www.soybeans.umn.edu/crop/insects/aphid/aphid.htm

OTA. (1993). *Harmful Non-indigenous Species in the United States, U.S. Congress Office of Technology Assessment*, Washington, DC: Author.

Persley, G.J. (1996). *Biotechnology and Integrated Pest Management*. Wallingford, Oxon, UK: CAB International.

Pimentel, D. (1991). *CRC Handbook of Pest Management in Agriculture*. Volumes I–III. Boca Raton, FL: CRC Press.

Pimentel, D. (2002). *Biological Invasions: Economic and Environmental Costs of Alien Plant, Animal, and Microbe Species*. Boca Raton, FL: CRC Press.

Pimentel, D. and Pimentel, M. (1996). *Food, Energy, and Society*. Niwot, CO: University of Colorado Press.

Pitman, N.C.A. and Jorgensen, P.M. (2002). Estimating the size of the world's threatened flora. *Science*, **298**, 989.

Pitman, N.C.A. Jorgensen, P.M., Williams, R.S.R., Leon-Yanez, S., and Valencia, R. (2002). Extinction-rate estimates for a modern neotropical flora. *Conservation Biology*, **16**, 1427–31.

Pratt, P. D., Center, T. D., Rayamajhi, M. B., and Van, T. K. (2004). Melaleuca. In E. M. Coombs, J. K. Clark, G. L. Piper, and A. F. Cofrancesco, Jr., (eds.), *Biological Control of Invasive Plants in the United States*. pp. 268–70.

PRB (2005). *World population data sheet*. Washington, DC US Population Reference Bureau. 17 pp. E-publication: www.prb.org/pdf05/05WorldDataSheet_Eng.pdf

Prokopy, R. J. and Croft, B. A. (1994). Apple insect pest management. In R. L. Metcalf, and W. H. Luckmann (eds.), *Introduction to Insect Pest Management*. New York: Wiley. pp. 543–86.

Prokopy, R. J. and Kogan, M. (2003). Integrated pest management. In V. H. Resh and R. T. Carde (eds.), *Encyclopedia of Insects*. Burlington, MA: Academic Press. pp. 589–95.

Purvis, A. and Hector, A. (2000). Getting the measure of biodiversity. *Nature*, **405**, 212–19.

Rains, G. C., Lewis, W. J., and Olson, D. M. (2004). Sustainable agriculture: Definition and goals. In R. M. Goodman (ed.), *Encyclopedia of Plant and Crop Science*. New York: Marcel Dekker. pp. 1187–90.

Roelofs, W. L., Comeau, A., Hill, A. S., and Milicevic, G. (1971). Sex attractant of the codling moth: characterization with electorantennogram technique. *Science*, **174**, 297–9.

Royal Society. (2005). *Food crops in a changing climate: Report of a Royal Society Discussion Meeting held in April 2005*. Policy Document 10/05. London: The Royal Society. PDF publication: www.royalsoc.ac.uk/displaypagedoc.asp?id=13105

Royer, F. and Dickinson, R. (1998). *Weeds of Canada and the Northern United States*. Edmonton, Alberta: University of Alberta Press.

Sailer, R. I. (1983). History of insect introductions. In C. L. Wilson and C. L. Graham (eds.), *Exotic Pests and North American Agriculture*. New York: Academic Press. pp. 15–38.

Sakai, A. K., Allendotf, E. W., Holt, J. S. *et al.* (2001). The population biology of invasive species. *Annual Review of Ecology and Systematics*, **32**, 305–32.

Samuelson, R. J. (2004). Our kids will pay the bill. *Newsweek*, January 12, p. 41.

Schowalter, T. (2000). *Insect Ecology: An Ecosystem Approach*. San Diego, CA: Academic Press.

Schwartz, P. H. and Klassen, W. (1981). Estimate of losses caused by insects and mites to agricultural crops. In D. Pimentel (ed.), *Handbook of Pest Management in Agriculture*. Boca Raton, FL: CRC Press. pp. 15–77.

Sekercioglu, C. H., Dailey, G. C., and Ehrlich, P. R. (2004). Ecosystem consequences of bird declines. *Proceeedings of the National Academy of Sciences*, **101**, 18042–7.

Seckler, D. U., Amarasinghe, U., Molden, D., de Silva, R., and Barker, R. (1998). *World Water Demand and Supply, 1990 to 2025: Scenarios and Issues*. Research Report 19, Colombo, Sri Lanka: International Water Management Institute.

Shalala, D. E. (1999). Bioterrorism: How prepared are we? *Emerging Infectious Diseases*, **5**, 492–3.

Shaw, R. H. and Seiger, L. A. (2002). Japanese Knotweed. In R. Van Driesche, S. Lyon, B. Blossey, M. Hoddle, and R. Reardon (eds.), *Biological Control of Invasive Plants in the Eastern United States*. Washington, DC: USDA Forest Service. pp. 159–66.

Shelley, R. L., Larson, L. L., and Jacobs, J. S. (1999). Yellow star thistle. In R. L. Sheley and J. K. Petroff (eds.), *Biology and Management of Noxious Rangeland Weeds*. Corvallis, Oregon: Oregon State University Press.

Shelton, A. M., Zhao, J.-Z., and Roush, R. T. (2002). Economic, ecological, food safety, and social consequences of the deployment of Bt transgenic plants. *Annual Review of Entomology*, **47**, 845–81.

Singh A., Sharma, O. P., and Garg, D. K. (eds.) (2005). *Integrated Pest Management: Principles and Applications. Volume 1: Principles*. New Delhi, India: CBS.

Singh, R. B. (2000). Environmental consequences of agricultural development: a case study from the Green Revolution State of Haryana, India. *Agriculture, Ecosystems and Environment*, **82**, 97–103.

Stein, B. A. and Flack, S. R. (1996). America's least wanted: alien species invasions of US ecosystems. Arlington, VA: Nature Conservancy.

Stiling, P. (2002). Potential non-target effects of a biological control agent, prickly pear moth, *Cactoblastis cactorum* (Berg) (Lepidoptera: Pyralidae), in North America, and possible management actions. *Biological Invasions*, **4**, 273–81.

Symstad, A. J., Chapin, S., Wall, D. H. *et al.* (2003). Long-term and large-scale perspectives on the relationship between biodiversity and ecosystem functioning. *Bioscience*, **53**, 89–98.

Thomas, J. A., Telfer, M. G., Roy, D. B. *et al.* (2004). Comparative losses of British butterflies, birds and plants and the global extinction crisis. *Science*, **303**, 1879–81.

Tilman, D. (1999). Global environmental impacts of agricultural expansion: the need for sustainable and efficient practices. *Proceedings of the National Academy of Sciences*, **96**, 5995–6000.

Tilman, D., Cassman, K. G., Matson, P. A., Naylor, R. and Polasky, S. (2002). Agricultural sustainability and intensive production practices. *Nature*, **418**, 671–7.

Tilman, D., Fargione, J., Wolff, B. *et al.* (2001). Forecasting agriculturally driven global environmental change. *Science*, **292**, 281–4.

Tuba, Z. (ed.) (2005). *Ecological Responses and Adaptations of Crops to Atmospheric Carbon Dioxide*. Binghamton, NY: Food Products Press.

Turner, R. K. and Pearce, D. W. (1993). Sustainable economic development: economic and ethical principles. In E. B. Barbier (ed.), *Economics and Ecology*. London: Chapman and Hall. pp. 177–93.

UNDESA. (2002). *Johannesburg Declaration on Sustainable Development*. Rome, Italy: United Nations Department of Economic and Social Affairs – Division for Sustainable Development.

UNEP. (1972). *Declaration of the United Nations Conference on the Human Environment*. United Nations Environment Programme. E-publication: www.unep.org/Documents.Multilingual/Default.Print.asp?DocumentID=97&ArticleID=1503

UNEP. (1999). *Global Environmental Outlook – 2000*. (GEO 2000), Vol. 2000. Geneva: United Nations Environment Programme. E-publication: www.unep.org/geo2000/index.htm

UN–FCCC. (1997). *Kyoto Protocol., United Nations – Framework Convention on Climate Change*. Rome, Italy: Author.

UNICEF. (2005). *The state of the world's children 2005: children under threat*. Rome, Italy: United Nations. PDF publication: www.unicef.org/publications/files/EN_Summary.pdf

United Nations. (1992). *Rio Declaration on environment and development.* Rep. A/CONF.151/26 (Vol. I), United Nations – General Assembly. New York: Author. E-publication: www.un.org/documents/ga/conf151/aconf15126-1annex1.htm

United Nations. (2005). *UN atlas of the Oceans.* Rome, Italy: United Nations. E-publication: www.oceansatlas.org/

USDA. (1998). *Production, supply, and distribution.*, Washington, DC: U.S. Department of Agriculture. E-database: www.fas.usda.gov/psd/

USDA–FAS. (2004). *The Amazon: Brazil's final soybean frontier.* U.S. Department of Agriculture, Foreign Agricultural Center. E-publication, January 13, 2004. www.fas.usda.gov/pecad2/highlights/2004/01/Amazon/Amazon_soybeans.htm

US–DE. (1998). *Impacts of the Kyoto Protocol on U.S. Energy Markets and the U.S. Economy.* Washington, DC: U.S. Department of Energy.

US–EPA. (2003). *Food Quality Protection Act (FQPA).* E-publication: www.epa.gov/opppsps1/fqpa/

US–EPA. (2006). Pesticide Industry Sales and Usage. Washington, DC: U.S. Environmental Protection Agency. URL http://www.epa.gov/oppbead1/pestsales/.

US–FDA. (1999). *Residue Monitoring 1998. Food and Drug Administration - Pesticide Program,* Washington, DC: Author. Electronic database.

Vandermeer, J. and Perfecto, I. (2005). *Breakfast of Biodiversity: The Truth About Rain Forest Destruction.* Oakland, CA: Food First Publishers.

Venette, R. C. and Ragsdale, D. W. (2004). Assessing the invasion by soybean aphid (*Homoptera: Aphididae*): where will it end? *Annals of the Entomological Society of America,* **97**, 219–26.

Vitousek, P. M., Mooney, H. A., Lubchenco, J. and Melilo, J. M. (1997). Human domination of Earth's ecosystems. *Science,* **277**, 494–9.

Warren, M. S., Hill, J. K., Thomas, J. A. *et al.* (2001). Rapid responses of British butterflies to opposing forces of climate and habitat change. *Nature,* **414**, 65–9.

WCED. (1987). *Our Common Future. Report of the World Commission on Environment and Development,* (G. H. Brundtland, Chair). Oxford, UK: Oxford University Press.

Wikipedia. (2005). *Green revolution.* Wikipedia: the free encyclopedia. E-publication: http://en.wikipedia.org/wiki/Green_revolution

Wilson, C. L. and Graham, C. L. (1983). *Exotic Plant Pests and North American Agriculture,* New York: Academic Press.

Wood, S., Sebastian, K. and Scherr, S. J. (2000). *Pilot analysis of global ecosystems: agroecosystems.* Washington, DC: World Resources Institute. E-publication: http://sustag.wri.org/publications.cfm.

World Resources Institute. (2000). *World Resources 2000–2001: People and Ecosystems, the Fraying Web of Life.* Amsterdam: Elsevier.

Yorinori, J. T., Paiva, W. M., Frederick, R. D. *et al.* (2003). Epidemics of soybean rust (*Phakopsora pachyrhizi*) in Brazil and Paraguay from 2001 to 2003. *Phytopathology* **93** (supplement):103 (abstract).

2

From simple IPM to the management of agroecosystems

R. LEVINS

2.1 Introduction

In this brief essay I want to highlight some of the changes that have taken place in integrated pest management (IPM) over recent decades or that may become more important in our work:

1. from "magic bullet" to community level protection grounded in ecological theory
2. invasibility
3. from local to regional control of pests and vectors
4. redefining pest and disease resistance
5. preparing for uncertainty
6. relaxing the boundary conditions: from protecting a given crop or farm to asking what kind of agriculture is compatible with gentle technology and sustainability
7. the breeding of support species and guiding herbivore evolution
8. commercialization of IPM
9. a research perspective.

Early IPM was aimed at reducing the use of pesticides, especially the most toxic ones. Biological control was an obvious answer to the problem. Even before natural enemies could be used to replace the chemicals, ideas such as damage threshold or economic threshold allowed decisions to be made in each cropping season rather than by a fixed calendar schedule for spraying. This led to many ingenious monitoring methods to determine when intervention was required. Meanwhile, researchers sought out natural enemies of the pests and attempted to introduce them either to become established members of the community or through repeated release. Introduction of natural enemies is the stage of input substitution, in which a specific agent is used to control a particular pest. These biological agents were clearly gentler toward the environment and toward farm workers than the chemical pesticides but were conceptually similar and have been regarded as pesticides with wings, another kind of magic bullet. While their advantage was obvious, their disadvantages were the cost of introduction and their uneven effectiveness. The introduced control agent typically worked some of the time, in some places, and not in others. Sometimes the use of pesticides for other pest problems worked at cross purposes with biological control, eliminating the parasitoid that had just been introduced, or diminishing the natural controls that were already present. IPM practitioners are aware of this problem, and even though we still see papers entitled "IPM for the control of (pest x)," the practice almost always considers the effect of the recommended actions on other pests and their enemies. Thus interventions are increasingly directed toward the management of the ensemble of pests in the agroecosystem.

It makes sense that "if wasp X parasitizes pest Y, then its introduction will control pest Y." But common sense is not a sufficient guide through the labyrinth of interactions and feedbacks in the ecosystem of a cultivated field. Introduced control agents often are effective only some of the time (e.g. at the beginning of flowering or in dry weather) or in some places (on some crops but not others, in some localities but not others). But, perhaps more important, increased mortality of a pest does not necessarily reduce its population or the damage it causes. Therefore the individual-level observation that a predator or parasitoid consumes a prey is not equivalent to the population claim that predator or parasitoid can control the pest and the enemy that causes the greatest mortality does not necessarily exert that greatest control over the pest. The reason for this is that the initial reduction of a pest may reduce another, already present enemy more than enough to offset the benefit of the species which causes the most mortality. Further theoretical work is needed to identify the circumstances in which a biological agent is effective within a community of species. One example is the question of providing alternative hosts for pests to keep their natural enemies in sufficient numbers.

2.2 From magic bullet to community-level protection based on ecological theory

The old textbooks used to recommend as a universal rule the elimination of weeds, hedgerows, and border vegetation because they provided refugia for pests. However, it was later proposed that sometimes this vegetation may provide food for non-pest species which promote the build-up of natural enemies (Brown and Welker, 1992; Prokopy, 2003).

The provision of alternative resources and refugia for beneficial insects is one tactic toward this end. For example, the lion ant *Pheidole megacephala* is used on some 9000 hectares of sweet potatoes in Cuba as a predator of the sweet potato weevil, *Cylas formicarius elegantulus*. In monocultures of sweet potato, the lion ant has to be introduced each season since it does not tolerate well the exposure to direct sunlight before the vines themselves provide shade. But in polycultures of sweet potatoes with maize, the maize seems to facilitate the lion ant and control is maintained. The lion ant can maintain itself in banana plantings. Therefore when bananas and sweet potatoes are planted in alternating strips less than about 40 meters wide, the lion ants can forage among the sweet potato vines even before there is enough shade to nest there. Inoculative release of beneficial insects is an intermediate means of pest control between the more expensive inundative release and natural control where minimum conditions are provided to enhance the control agents or, at least, to not harm them. Since common sense plausible arguments lead to opposite conclusions, the question has to be posed differently with the help of a mathematical model.

I propose to analyze this question as the hedgerow problem. Figure 2.1 shows a model of a system with six variables: the crop plant; an alternative host; the pest on the crop; the pest on the alternative host; and the natural enemy attacking the pest in each of the habitats. I ask the question, what effect would a change in the abundance of the alternative host have on the yield of the crop? A change in the availability of the alternative host R_2 affects the crop R_1 by two pathways. The first pathway is from R_2 to H_2 as a result of herbivory, times the increase in predator due to its predation on the pest, times the movement of predator from the border to the crop, times the predation on the pest in the crop, times the herbivore damage the pest does to the crop. The upward and horizontal steps are positive, the downward one negative so that the net effect of this pathway is positive. The other pathway goes from R_2 to H_2 by herbivory, from H_2 to H_1 by movement, from H_1 to R_1 by herbivory. What is not obvious here, at the common sense level, is that this path must be multiplied by the resistance or feedback of the (P_1, P_2) subsystem, the self-damping of the predator. If this self-damping is strong then the short path is intensified, but if the predator has no self-damping then this path is nullified and the beneficial pathway by way of the

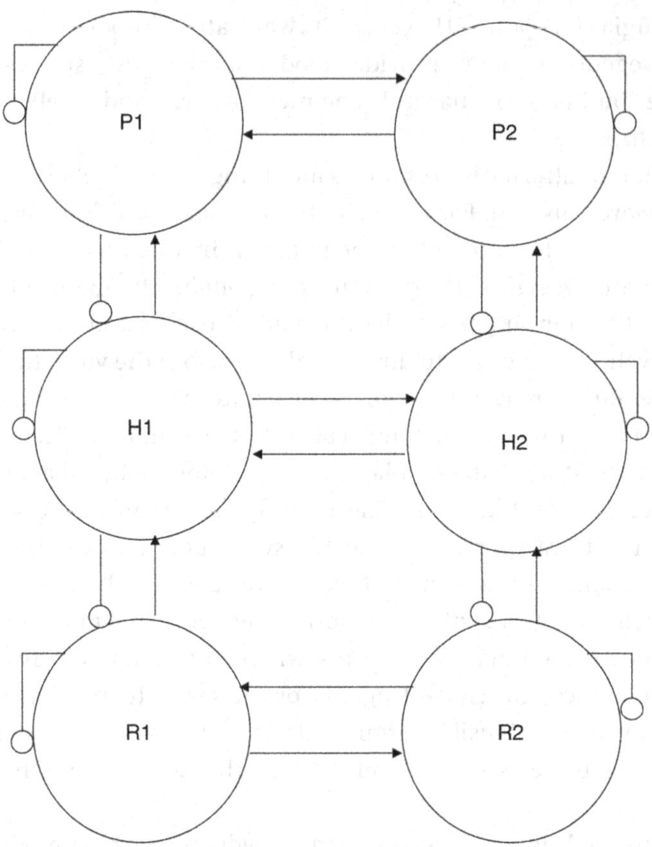

Figure 2.1. The hedgerow model. R1 and R2 are two resources for an herbivore. R1 is a crop and R2 is either another crop or a weed that serves as an alternative host for the herbivore. H1 and H2 are the populations of the same herbivore on the two hosts, and P1 and P2 are the populations of a natural enemy of the herbivore on the two plant types. A sharp-pointed arrow represents a positive effect of one variable on the other. The horizontal arrows refer to migration back and forth between the two hosts. Round arrowheads indicate negative effects. The vertical arrows indicate herbivory and predation or parasitism. The self-damping loops balance the migration exactly so that if there were no density-dependence or immigration into the system from outside the net feedback of the (P1, P2) subsystem would be zero.

predator predominates. The self-damping is increased if the predator migrates into the system, is subject to additional density dependence, or has other hosts beside the pest of interest.

The technical details of the calculations are explained in Levins (1998). In brief, the vertices of a graph are the variables of interest, usually populations of relevant species or biomass, or for perennial and tree crops the mobilizable photosynthate. The link from variable *i* to variable *j* is equal to $\partial(dx_i/dt)/\partial x_j$ evaluated at equilibrium, the a_{ij} of the interaction matrix. If variable *j* preys on variable *i* then a_{ij} is negative and a_{ji} is positive. A feedback loop is found by multiplying the values of the a_{ij} around the loop, and the value of a path P_{ij} is the product of the a_{ij} along a simple path (that does not intersect itself) from variable *j* to variable *i*. In qualitative analysis, the sign of the link is all that is needed. Thus a predator/prey loop is a negative feedback while competition or mutualistic loops give positive feedback. The loop from a variable to itself comes from density-dependence or immigration, but if the migration flows back and forth between variables within the system then the positive migration loop exactly balances the negative self-loops and the feedback of the whole two-variable system is zero unless some other process gives self-damping.

Therefore the self-damping of the variables in a community model of the agroecosystem is an important determinant of the outcome of interventions and is an object of investigation for theory and in the field.

Models of predator/prey interaction have to take into account the aggregation of pests, especially the more sessile kinds such as coccids. In work on Cuban citrus, Awerbuch *et al.* (unpublished manuscript) found that the Taylor exponent of aggregation is usually between 1 and 2 and decreases with the life stage of the scale *Lepidosaphes gloverii*. Aggregation increases the strength of density dependent regulation, since the rates of parasitism and fungal infection increase with the number of scales in a leaf, from 0.23 when there is a single scale to 0.64 when there are 15 scales on a leaf. It seems that the aggregation does not attract the parasitoids, but that once they find oviposition sites on a leaf they search it more thoroughly before leaving. Therefore interventions aimed at further reduction of the scale population by increasing mortality would be opposed by the feedback reducing natural mortality.

The models also indicated that temporal variation in the system is driven from below, from the condition of the tree, or weather acting directly on the fecundity of the scales, and secondarily on their natural enemies. This generates a positive correlation between host and parasitoid since an increase in scales supports more consumers. But spatial variation across trees and parts of the trees enters the system by way of the parasitoids, increasing mortality. Any change favoring the parasitoids reduces the population of their host, generating a negative correlation between them in space. If spatial and temporal data are combined, the correlation between parasitoid (or predator) and host (prey) will depend on the number of observations in each category. Therefore the analyses should be kept separate.

The correlation between pest and enemy will, of course, affect the effectiveness of the enemy. A negative temporal correlation indicates that the enemy comes after too long a delay. Roughly, the correlation is equal to the cosine of the delay (phase shift). Then we have to look for strategies to reduce that delay. It is here that the provision of alternative hosts for the enemy may be useful.

The spatial correlation both demonstrates that the enemy really does affect the pest population, and that it is not uniformly effective. We can cope with this unevenness in several ways. One is to determine the causes of uneven effectiveness and seek to remedy them. Another is to select an ensemble of natural enemies that separate by microniches so that the whole crop is monitored by some enemies everywhere. In the Cuban work cited above, the parasitoid was more effective on the undersurfaces of leaves, the fungi on the upper; the parasitoid in the upper strata of the tree, the fungi in the lower. There were also differences among the fungi in their distribution among cardinal points of the tree and proximal or distal position on the branches. The joint effect of the two or three fungi and one wasp gave fairly good coverage to the spatial microniches on the tree in space but less so in time.

2.3 Invasibility

A production system is a partly natural, partly anthropogenic community of plant, animal, and microbial species. The various species in that community interact through various mutualistic, predatory, and competitive pathways, creating the conditions for the maintenance of the community. Pests are species which either invade that community and are able to increase in sufficient numbers to cause damage, or are regular members of the community that sometimes escape from the mutual controls. Therefore one object of study in the design of agroecosystems is the invasibility of the community or its vulnerability to outbreaks of resident pests. A pest need not attack the crop directly. It may be a microorganism that suppresses the nitrogen-fixing cyanobacteria or a hyperparasitoid of beneficial parasitoids. Students of biodiversity are also concerned with invasions by exotic species as undermining conservation efforts, and programs for resisting desertification must also consider the invasion by desert species, so ecological research is now exploring the general problem.

In general, it seems as if a community that has greater species diversity is more resistant to invasion by plants (Brown *et al.*, 2001). In addition, fluctuations in resource availability create windows of opportunity for invasion (Davis and Pelsor, 2001). It is not always obvious whether the increased resistance is due to diversity per se or to the greater probability that in a more diverse community at least one of the species excludes the invader by itself.

A mathematical model of the problem of invasibility could represent a community by an interaction matrix of species that maintains either an equilibrium or a steady state with persistent average abundances. An environmental change appears in the model as a change in one or more parameters in the equations for species abundance. The derivative of the abundance of a species with respect to that parameter is then a measure of mutual control or invasibility, depending on whether the species of concern is represented as a member of the community starting from its steady state value or an invader beginning near zero abundance. We can solve for this derivative to get $\partial x_i/\partial C = \Sigma \partial(dx_j/dt)/\partial C\ P_{ij} F\{\text{complement of } P_{ij}\}/F_n$ where: $\partial(dx_j/dt)/\partial C$ is the direct change in the equation for species j due to changes in parameter C, P_{ij} is the product of the interaction terms $\partial(dx_h/dt)/\partial x_k$ along a path from x_j to x_i, the complement of the path is the subsystem left after removing the variables along the path, F is its "feedback;" the determinant of that subsystem multiplied by $(-1)^{m+1}$ if it has m disjunct cycles (loops) in the graph, and F_n is the feedback of the whole *n*-variable system. Summation is over all paths, from the equation where the parameter enters, to the target species. See Levins (1998) for technical details. Clearly the result depends on where the environmental change enters, the strength of the paths to the target variable, and the feedback of the remaining subsystem. If the feedback of the remaining subsystem (often a single variable) is zero or weak, that subsystem is a sink that absorbs impact. The denominator refers to the whole system. It can therefore be thought of as a measure of resistance. Thus a species may escape from control if the resistance of the whole community is weak, and our task is to design systems with strong resistance so that changes in conditions do not result in explosions of any one species. When we have specific invaders in mind we can work to strengthen the subsystems without that species. The resistance (variously labeled the "gain" or "feedback") of a community is therefore an object of research and intervention.

2.4 Regional control

The unit of IPM is almost always the individual farm, the locus of decision-making. The research reflects this reality and accepts the notorious difficulty of organizing and enforcing coordinated practices over a larger area. But control within the confines of a single farm has its limitations.

Relatively immobile pests can be controlled in small areas. Prokopy's apple orchard was only some 200 meters away from trees that were infected with a variety of insects, yet he was successful in minimizing pest damage (Prokopy, 2003). But for highly mobile pests the situation is different. If the pest can be destroyed or reduced after it has done its damage, and if it is a highly mobile pest, then there is no incentive for the individual farmer to carry out that practice. Next year's invasion of the farm will come from a regional pool of insects rather than from the ones that emerged from that same farm. If farmers agree to coordinate their planting dates, the better prices to be obtained outside of the agreed season serve as an incentive to break ranks.

There are a few cases of legally enforced prohibitions of planting during part of the year. In the Dominican Republic restrictions on planting periods for potential hosts of whiteflies (*Bemisia tabaci*) created a period of 75 days free of host plants. Enforcement was severe: host crops planted out of season were destroyed. The program was successful first regionally and then for the country as a whole (Morales *et al.*, 2000). Larger farms are more autonomous ecologically and can practice zones of quarantine or rotational schemes.

The new cooperatives in Cuba (the UBPC, Unidades Basicas de Producción Cooperativa), formed from the division of previous state farms, are still large enough to be the source of most of the pests, vectors, and beneficial insects that affect them, and coordination among them is relatively easy through the enterprise that remains from the older state farm system and serves for marketing, technical support, and advice. Here planning is possible on a scale compatible with the demography of most pests.

For pests and vectors such as the whitefly, the unit of epidemiology has to be a whole region over the complete cropping cycle. But herbivore damage and transmission of disease have different dynamics. Anderson (1991) has stressed this difference even when the same insect plays both roles. For instance, insects that harass a herbivore can reduce herbivore damage and drive it from a field, but harassment of a vector may cause it to visit more host plants and cause more damage. In collaboration with F. Morales and others, P. K. Anderson has been developing an international network of whitefly researchers. From the perspective of the regional control of whiteflies, the landscape consists of at least six types of patches (Table 2.1).

Thus frijole plants are a feeding host but *Bemisia* hardly lays any eggs on bean plants. Cotton and tomato are reproductive hosts, and grasses are neither. The cropping cycle allows whiteflies to move from one area to another, for instance from rainy season cotton to dry season beans, with some

Table 2.1. *Level of use of crop protection methods applied in Cuba*[a]

Crop	Chemical pesticides	Biopesticides	Release of natural enemies	Conservation and enhancement of natural enemies	Resistant varieties	Agronomic practices
Coffee	N	L	N	H	M	H
Citrus	L	M	N	H	M	L
Sugarcane	N	L	H	H	H	M
Tobacco	L	M	N	L–M	M	M
Pasture	N	H	H	L–M	M	L
Rice	M	M	N	L–M	H	M
Corn	M	L	L	L	N	L
Beans	M	L	N	L	H	L
Cabbage	L–N	H	N–L	M–H	N	L
Tomato	M	M	N	L	H	M
Sweet potato	N	H	H	L	M	H
Cassava	N	M	H	H	M	H
Potato	M	H	N	M	H	M
Plantain and banana	L	M	M	M	H	H

Level of use: N = none, L = low, M = medium, H = high.

[a]from Perez and Vazquez, 2002, p. 134.

intervening time on roadside *Malvaceae*. The category "not utilized and competent for the virus" is empty at present, but it is useful to know whether, if the insect began to use that host, it could spread the virus. In the Americas, the gemini viruses are limited to dicots, but in Africa one type invades sorghum. Therefore some examination of the host-range plasticity of the whitefly and of the viruses is needed.

Any strategy for regional control should start with modeling the landscape as a mosaic of different land uses that change with the seasons. On a reproductive host the whitefly population increases, but the proportion infected by the virus will decrease if that crop is not competent for the virus since there is no transovarial transmission. If the crop is also competent for the virus, the balance of mortality and recruitment can lead toward an equilibrium level of infection, but how far along this trajectory the population advances will depend on the length of the growing season as well as the parameters of birth, death, and infection. On a feeding host the population may decrease since there is mortality but no reproduction, but the proportion of infected flies may increase if the crop is competent, as flies acquire the infection from infected plants during feeding. On non-feeding hosts, the whiteflies rest for a while and then move on or die. After harvest the whiteflies seek new hosts, most likely with high mortality in transit. Here the layout of the land becomes important.

We do not know yet where density dependence operates in this system but it must act somewhere since the fluctuations in numbers of whiteflies are enormous and yet remain bounded. Thus mathematical modeling must include birth, death, and infection processes in a spatially and temporally structured system with a specific dynamics in each crop and the whole linked by sequential migrations of the vector.

2.5 Redefining disease resistance

Regional control of pests and vectors opens up a broader approach to the concept of resistance. An individual plant is resistant if it cannot be infected. A feeding host may be resistant if the incubation period for the virus is long enough, so that by the time new whiteflies are infected by infected plants, the development of the plant has made the infection innocuous. Yet it may be a source of infection for other crops. On a reproductive host, resistance might also be achieved if the plant slows down the development of the insect so that tertiary infection is too late to cause appreciable damage. Thus resistance is shifted from the individual plant to a population phenomenon. Finally, a region is resistant to the gemini viruses if the temporal sequence of crops is such that the rate of reproduction of infection falls below 1 for the agricultural year as a whole, including cycles on each host and mortality during migration. Since the incubation period, development rate, and infection rate depend on parameters of the plants as well as the flies, breeders can select genotypes which enhance population and regional as well as individual resistance. The length of the growing season for each crop variety becomes a demographic parameter for the pest. The spatial arrangement of crops creates a measure of the "viscosity" of the environment for that pest, the ease of movement from one host to another. A multilevel notion of resistance allows us more options for damage reduction.

2.6 Uncertainty

Agriculture is subject to several kinds of uncertainty. There has always been unreliability, greater in some regions than others, but always a problem. In recent years, not only has the average temperature and rainfall been changing but also the variance has increased. The uncertainty of climate change presents us with major uncertainty for pest management since the pests and their natural enemies may respond differently to new conditions.

There is also the uncertainty of pest outbreaks, sometimes associated with the weather but also due to long-distance transport of pests through commerce or by wind. The prices of agricultural commodities and inputs actually vary more than yields in response to natural conditions. Economic uncertainty through market conditions or changes in land tenure is often accompanied by changes in preferred crops and investments in agrotechnology, which in turn affect the pest array. Technical changes and replacement of varieties alter the vulnerability of the crop to pests and diseases. Sometimes measures undertaken to confront one problem create vulnerability to others. For instance, many commercial orchards grow Valencia oranges on sour orange rootstocks in order to resist *Phytophora*. But such trees are more vulnerable to tristeza disease (*Xanthomonas campestris*), so the trees have to be regrafted. Sometimes the opening up of new employment opportunities draws labor away from farming, changing the options available.

The overall result is that crop protection must confront uncertainty as a general problem. We have examined the problem of how other species cope with environmental change. There are four main modes, not necessarily mutually exclusive.

1. Detection of a problem, with rapid and effective response. Scouting and monitoring technologies have been improving, while economic threshold criteria have calibrated the responses to the state of the system. This is likely to improve in the future, perhaps reinforced by better reporting systems so that farmers over a whole region can be aware of emerging problems (regional alert systems are available already in certain regions – see Correa-Ferreira *et al.*, 2000).
2. Prediction. There is short-range prediction, as when a dry season indicates a greater whitefly problem. Longer-range prediction makes use of ecological and evolutionary principles. For instance, a decline in habitat for insectivorous birds or bats may permit more long-distance migration of crepuscular species, or the very great increase in the area sown to a given crop virtually guarantees that more herbivorous species will utilize it. Guy Bush's (1993) study of niche extension in the apple maggot fly and current monitoring of the speciation of the gemini viruses are other examples. One kind of prediction involves an increasingly sophisticated biometeorology as is practiced in Kentucky for the blue mold of tobacco and simulation programs to turn readings into forecasts. Another kind is based on evolutionary ecology and depends on more traditional methods of field work, niche delineation, population genetics of potential pests (assisted by DNA

sequencing) to identify evolutionary potential for overcoming our control methods or for invading new hosts.
3. Broad tolerance (horizontal resistance) so that it is not necessary to identify the pest problem in order to reduce vulnerability. The preservation and enhancement of biodiversity increases the number of potential bioregulators and reduces invasibility.
4. Prevention. Here we reach out into the environment to change the conditions that would lead to pest problems. Crop diversity, polyculture, and mosaic design of a farm or region are elements of prevention.
5. A mixed strategy, combining many methods. This is the only way of preparing for the responses of the pests to our interventions.

2.7 Expanded criteria and relaxed boundary conditions

The initial problem for IPM is, given a particular crop on a farm and the local land use pattern, how can we minimize pest damage or more precisely income loss due to pests? The crop has already been chosen by farmers under the influence of climatic and economic conditions. This is top-down IPM, in which management of an existing system is gradually improved. Bottom-up IPM starts with land that has not been used for some time for agriculture or at least for the crops that are intended. Then inputs are introduced only as needed, with reluctance, and an effort is made to minimize their impact on the ecosystem. In developing countries there is still a process of bringing new land under cultivation and adjusting the land use, demography, water resources and cropping systems to national goals. These may include production for export, local food security, employment, protection of watersheds, and other goals. There are also constraints: the production system must be sustainable, preserve local employment throughout the year and economic viability, preserve rural life, be compatible with the health of the farmers and consumers, provide buffers against the uncertainties both of climate and markets, and contribute to the conservation of biodiversity, water resources, and clean air. This is a daunting agenda that often runs into the contrary pressures of industrialized agriculture and national economic policies.

Sustainability can be thought of on different levels. The sustainability of a particular production system may require increasing intervention in the face of changing conditions. For instance, new pests appear or old methods lose their effectiveness and have to be replaced, or economic fluctuations make technically effective methods no longer viable. The goals of water conservation can be met in part by building the moisture-holding capacity of the soil. But as we define sustainability more broadly and consider not a particular set of crops, but rather food production and economic sustainability, a shifting set of practices and crops can be incorporated that makes use of longer-term succession. Thus annual crops can be planted between the trees in new fruit plantations, livestock introduced when the trees are tall enough so that the harvest is not damaged by the animals, and areas left free of a crop for several years to break the cycle of pest outbreaks. A mix of crops in a region may be desirable as a buffer against uncertainty even when a single crop may be most profitable in the short run. Traditional (under-shade) coffee plantings preserve biodiversity and can act as buffers for national forests.

The most self-conscious large-scale conversion to ecological agriculture in the world is now taking place in Cuba. The process has been detailed elsewhere (Funes *et al.*, 2002; Levins, 1993). This program includes major elements:

1. The integration of rural, suburban, and urban agriculture, with the more perishable crops grown closer to the consumers. The increase of vegetable production toward 3 million tons per year in a population of 11 million promotes a major improvement in the Cuban diet and reduces imports.
2. Geographic diversification, so that each province and even municipality provides much of its own food. Although this means growing crops where they do not give maximum yields, it provides a buffer against all kinds of uncertainty.
3. De-specialization, so that each farm (typically a cooperative of several hundred to several thousand hectares) is a mosaic of land uses that stabilize employment and grow a wide variety of food for consumption on the cooperative farm and for the region. Orchards, woodland, and pasture produce their own harvests but also provide refugia for pollinators and natural enemies of pests. Polycultures maintain soil fauna and flora. Nitrogen-fixing bacteria on the legumes and in the soil, mycorrhizae, earthworms, composting, and rotations provide much of the mineral needs and maintain organic matter. Pests are controlled by a diversity of predators, parasites, and insect diseases. Livestock is increasingly integrated into these systems, for instance for weed control in orchards. There is even a project to reduce the area in monocultures of sugar to allow for rotations, mosaics, and an emphasis on food production. The diversity of products also diversifies

the necessary tasks, making work more varied. The Cuban IPM program includes the use of *Bacillus thuringiensis* on 14 pests, 4 kinds of fungi for 17 pests, and 5 pathogens, insect control agents for 10 crops, neem preparations on 14 crops to control 25 pests, plus rotation, polyculture, and cultural practices. A summary is shown in Table 2.1 (taken from Perez and Vazquez, 2002).

4. Decentralization of management, to locate most of the decision-making within the assemblies of the new cooperatives. Within the cooperatives, individual members or small groups take on responsibility for particular pieces of land. The well-educated rural population has increased opportunity for combining physical labor with intellectual labor in making technical and economic decisions. The farms contract part of their production to the state and anything above that can be sold in farmers' markets. This is one factor that helps retain rural population and deter the urbanization that is giving us the megacities of much of the Third World. A major land law is now being debated throughout the country to define better the relations between the cooperatives and the state.
5. Reduction of pesticide and chemical use. More than half the food in Cuba is now grown organically.

2.8 Support species and herbivore evolution

Almost all of agricultural genetics is directed naturally enough towards crop improvement. The non-crop species which contribute to the harvest as pollinators, enhancers of soil fertility, or biocontrol agents are taken as we find them, and, if any of them are promising, they are introduced. However, it is also possible to apply the skills of plant and animal improvement to the support species. For instance, there are two ways of getting a parasitoid to control a pest. Either we find one that already controls the pest elsewhere and then select it to survive and reproduce in our conditions, or we choose a species which is already in our habitat and adapt it to parasitize the pest. This latter requires first of all an understanding of the behavior of the parasitoid in host selection, then the successful infection of the host and finally the synchronization of its development with that of the host. As starting material we should choose local species of genera which exhibit a broad host niche that includes species as similar as we can find to the pest. Eggs may then be inserted into hosts, which had been weakened beforehand through chemical or temperature treatment, in order to overcome resistance. From there we would design selection experiments to improve the rate of parasitism in unweakened hosts. Such a program would be a long one and would require public support. It would rest on a base of fundamental research in the physiology and evolution of parasitoid/host relations. Another direction would be the selection of self-destructing cover crops for weed suppression in the tropics. (We already have a start in the temperate zone with crops such as winter rye.) A third pathway aims at guiding the evolution of herbivores from pests to commensals. This might involve the exertion of selection to favor herbivores which utilize parts of the plant that do not affect the harvest. For instance, since the grain yield in maize and rice is produced by upper leaves, a technique that increases mortality of pests mostly on the upper leaves while leaving the lower leaves relatively safer would select for less harmful herbivores. Similarly the self-shaded leaves of tomatoes that respire but do not photosynthesize much could be left to herbivores with even some benefit to the yield.

2.9 The manager as part of the system

Starting with quantum theory in the 1920s, the idea has spread throughout the sciences that the experimenter is part of the experiment. In medicine this has led to the double-blind experiment to exclude the influence of the researchers' expectations on the outcome. In IPM, since the rules for intervention depend on the state of the system and also influence the system, a model for intervention should include the decisions of the manager.

This has not often been done in agroecosystems research. Vandermeer (1997) linked cultivation intensity, yield, and price in a dynamic system. Because of the non-linear feedbacks among these variables there may emerge more than one outcome or "production syndrome" that is a combination of intensity of cultivation, yield, price and ecological trajectory. Other models might result in permanent oscillation.

Suppose, for example, that a pest species exhibits a dynamic of the qualitative form shown in Figure 2.2 that plots pest abundance in one time period as a function of its abundance in the previous period. At low densities, the population grows exponentially, but above some threshold it becomes economical to intervene to reduce the pest population by a procedure that increases mortality. The curve then decreases sharply. This is a typical web map with many applications in ecology and economics. Its qualitative behavior is well known. There is an equilibrium population of pest where the 45° bisector intersects the descending limb of the curve. If the slope here is flatter than -1 the equilibrium is stable, but if it is steeper than -1 the equilibrium is unstable and pest abundance will oscillate. The oscillations

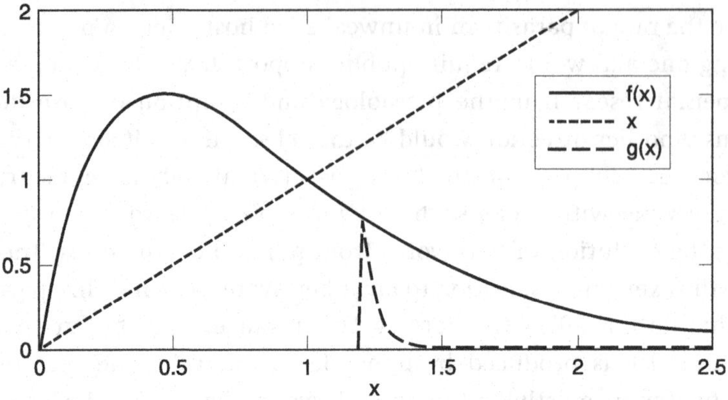

Figure 2.2. How economic thresholds can lead to chaotic dynamics. The solid line is a hypothetical population curve with self damping, $x_{n+1} = x_n \exp(a(1-x_n))$. It has an equilibrium at $x=1$ which is locally stable for $a<2$. But suppose that an intervention is initiated at $x=b>1$. Then survival decreases exponentially at the rate $\exp(-cx_n)$. For the appropriate values of the parameters, the process is now chaotic. This is determined as follows: the peak of the curve is at $1/a$. Its height is M. If $x_n = M$ then $f(M) = m$. M and m are the upper and lower boundaries of the region of permanence within which x will remain after at most three oscillations. The pre-images of the equilibrium y1 and y2 from which equilibrium is reached in one or two steps are calculated. If they are below m then there cannot be chaos but at most a cycle of length 2, while if they lie above m then the dynamic is chaotic (see Levins, 2001). When an intervention takes place starting at the threshold b, where b is between equilibrium and M, the new curve has the same equilibrium, M and pre-images but m becomes m2. If this is small enough (if the curve decreases fast enough) then $m2<y2$ and chaos results.

may be periodic or chaotic. They will be chaotic if the descending limb of the curve is steep enough (Levins, 2001). Thus it is possible for an IPM program determined by economic thresholds to provoke erratic oscillations in pest abundance. On the other hand, a strategy that lowers the rate of increase of the pest no matter what its abundance, so that the threshold is not reached during the growing season, is more likely to lead to a stable equilibrium. The design of a more resistant ecosystem would serve our purposes by lowering the whole curve of pest demography.

2.10 Commercialization

The very successes of IPM have opened up the potential markets for IPM products. The *IPM Practitioner* Journal of November/December 1995 lists more than a thousand pest control items in its *Directory of least-toxic pest control products* with some eight pages of suppliers. Many of these products are living materials (insects or microorganisms), which are raised industrially for release in the farm but do not become part of the community of organisms around the crops and have to be reapplied regularly. It is this need that often makes IPM more labor-intensive than conventional agriculture. On the one hand this is a very laudable development, as more research is directed toward products that are compatible with human health, maintenance of biodiversity, and sustainability. But it also carries the danger that IPM will evolve toward increasing numbers of inputs for farmers to buy, so that the value added in agriculture might still come increasingly from off-farm sources. Such risk is especially apparent in the applications of genetic engineering to agriculture and its uncritical acceptance by development planners as the new answer to world hunger. Genetic engineering promises to help production at the level of individual plants by introducing specific genes "for" particular traits. It is very attractive to investors since these products have to be purchased repeatedly by farmers. The criticisms of this strategy are both technical and social (see also Chapter 10) as listed below.

a. The more alien the introduced DNA, the less predictable its behavior. This behavior depends on where the new DNA is located in the genome and the more mobile the insertion within the genome.
b. The introduced pesticide resistance is a typical vertical resistance in that it can be the target of natural selection in the pests and lose efficacy as quickly as other elements of pest resistance, such as the rust-resistant genotypes of wheat that are useful only for a few years.
c. As genetic elements they can be spread through pollen and contaminate varieties and nearby wild species, thus increasing the pesticide exposure of the pests and the intensity of natural selection for resistance.
d. There is an indecent haste to patent marketable inputs to agriculture that downplays possible failures, or even harm, caused by the introduction, and devotes research money to protecting the investment.
e. Farmers are made more dependent on external inputs and developing countries are made more dependent on purchases abroad that increase foreign debt.
f. Research effort is diverted from ecosystem management to genetic magic bullets. The capture of the designation "modern biology" by molecular biology dismisses modern population and community

ecology, modern biogeography, ecological physiology, and the complexities of development.

g. Other goals of agricultural development such as protecting biodiversity are subordinated to a narrow focus on maximizing yield.
h. Such research no longer asks the question "what is the best way to protect the harvest?" but rather "what is the best product we can design to sell to the farmers?".

In order to redress this potential imbalance it is important to encourage publicly supported research into the design of agroecosystems that, as far as possible, are self-operating. Such research has to be carried out at the multiple levels of organization from the physiology of plants to social strategies for sustainability. Crop protection should at least be compatible with these goals (see also Chapters 8 and 15).

2.11 Research perspectives for IPM

The ultimate goals of IPM are not the pest, or even the crop, but the viability of rural life as a whole and a safe, sustainable food supply in the face of many kinds and increasing intensity of uncertainty. The central intellectual problem is the understanding of the behavior of complex, multilevel systems sufficient to design systems that are productive, economical, resistant to uncertainties, safe for the producers and consumers, and protective of the local and global environment.

Therefore it is important that publicly supported research maintains a balance among levels so that our field is not swamped by the glitter of "high-tech" molecular technologies that promise rapid returns. The analysis of population dynamics in heterogeneous communities, the local and regional geography of production and land use systems, microevolutionary responses to changing conditions, and the sociology of knowledge of pest problems are all elements of a balanced strategy in areas that may require theoretical detours which are unlikely to receive private support. Interaction with non-agricultural ecologists, population geneticists, and biogeographers should be strengthened. Collaboration with medical epidemiologists is surprisingly useful, since, despite the differences in the hosts and pathogens, the basic processes of exposure, colonization and disease, intervention, and microevolution are similar.

We even have to look at the organization of research: the gentler the technologies we work with, the more site-specific the processes we impact. Active participation by farmers in a broad, democratically organized research program is needed to combine the detailed, intimate knowledge that farmers have of their own situations with the comparative and generalized knowledge that requires some distance from the particular.

References

Anderson, P.K. (1991). *Epidemiology of Insect-Transmitted Plant Pathogens*. Dissertation, Department of Population Sciences, Harvard School of Public Health. p. 5.

Awerbuch, T., Gonzalez, C., Hernandez, D. *et al.* (2004). The natural control of the scale insect *Lepidosaphes gloverii* on Cuban citrus. *Newsletter of the Inter-American Citrus Network*, **21/22**, 40–1.

Brown, C., Fargione, J. and Tilman, D. (2001). *Species Diversity, Resource Competition and Community Invasibility: a Minnesota Grassland Experiment*. Ecological Society of America 86th Annual Meeting, Abstracts August 2001. See http://abstracts.co.allenpress.com/pweb/esa2001/document/?ID=26165.

Brown, M.W. and Welker, W.V. (1992). Development of the phytophagous arthropod community on apple as affected by orchard management. *Environmental Entomology*, **21**, 485–92.

Bush, G.L. (1993). Host race formation and speciation in *Rhagoletis* fruit flies (Diptera: Tephritidae), *Psyche*, **99**, 335–57.

Correa-Ferreira, B.S., Domit, L.A., Morales, L. and Guimarães, R.C. (2000). Integrated soybean pest management in micro river basins in Brazil. *Integrated Pest Management Reviews*, **5**, 75–80.

Davis, M.A. and Pelsor, M. (2001). *Fluctuating resources lead to predictable changes in competition and invasibility*. ESA abstracts 86th annual meeting, August 2001. See http://abstracts.co.allenpress.com/pweb/esa2001/document/?ID=24138.

Funes, F., Garcia, L., Borque, M., Perez, N., and Rosset, P. (eds.). (2002). *Sustainable Agriculture and Resistance: Transforming Food Production in Cuba*. Havana, Cuba: Food First Books, ACTAF (The Cuban Association of Agricultural and Forestry Technicians) and CEAS (Center for the Study of Sustainable Agriculture, Agrarian University of Havana).

IPM Practitioner. Nov. (1995).

Levins, R. (1993). The ecological transformation of Cuba. *Agriculture and Human Values*, **10**, 52–60.

Levins, R. (1998). Qualitative methods for understanding, prediction and intervention in complex systems. In D. Rapport, R. Costanza, P. Epstein, C. Gaudet, and R. Levins (eds.), *Ecosystem Health: Principles and Practice*. Blackwell: Oxford, UK, pp. 178–204.

Levins, R. (2001). The butterfly ex machina. In R.S. Singh, C. Krimbas, D. Paul, and J. Beatty (eds.), *Thinking About Evolution*. Cambridge University Press: Cambridge, UK, pp. 529–43.

Morales, Garzon, and Francisco, J. (2000). *El Mosaico Dorado Y Otras Enfermedades Del Frijol Comun Causadas Por Geminivirus Y Transmitidos Por La Mosca Blanca En America Latina*. Palmira (Valle) Columbia: Centro Internacional de Agricultura Tropical (CIAT).

Perez, N. and Vazquez, L. (2002). Ecological pest management. In F. Funes, L. Garcia, M. Borque, N. Perez and P. Rosset (eds.), *Sustainable Agriculture and Resistance: Transforming Food Production in Cuba*. Havana, Cuba: Food First Books, ACTAF (The Cuban Association of Agricultural and Forestry Technicians) and CEAS (Center for the Study of Sustainable Agriculture, Agrarian University of Havana).

Prokopy, R. (2003). Two decades of bottom-up ecologically-based apple pest management in a small commercial orchard in Massachusetts. *Agriculture, Ecosystems and Environment*, **94**, 299–309.

Vandermeer, J. (1997). Syndromes of production: an emergent property of simple agroecosystems dynamics. *Journal of Environmental Management*, **51**, 59–72

3

Populations, metapopulations: elementary units of IPM systems

L. WINDER AND I. P. WOIWOD

3.1 Introduction

Integrated pest mnagement (IPM) is directly concerned with manipulating potentially damaging pest populations and exploits the synergy between control strategies. The drive towards sustainable methods of crop production has demanded that we should minimize chemical use, energy inputs, the effects on non-target organisms, and harm to the wider environment. The development of a framework to conceptualize the processes that drive pest population dynamics will be crucial to this endeavor and allow us to devise increasingly sophisticated, successful, and sustainable management strategies.

Over the last 20 to 30 years there has been a revolution in ecological thinking on how populations function, with the growing appreciation of the spatial dimension and the importance of movement in population dynamics (Woiwod *et al.*, 2001). This revolution found its expression in the development of metapopulation theory, based on the concept of local populations linked together by movement (Hanski, 1999; Hanski and Gilpin, 1997). The metapopulation approach, and indeed the word itself, originated from models developed by Levins (Levins, 1969; Levins, 1970) but this conceptual framework only started to be applied widely in ecology from about 1990 onwards (Hanski and Simberloff, 1997).

The development of this theory has been driven largely by theoretical ecologists and its application by those working within conservation biology. This is perhaps unsurprising as early model formulations emphasized discrete subpopulations with high likelihood of turnover (i.e. extinction and re-colonization) and low levels of movement between patches, which is clearly directly applicable to many species of conservation concern. However, it is salutary to note that the metapopulation approach was originally conceived to investigate problems relating to biological control (Levins, 1969) and it is likely that there is much potential for its application to questions in IPM. Indeed, Wissinger (1997) notes that the development of optimal agricultural landscapes for pest control in annual crop systems will require a metapopulation approach.

The concept of metapopulation dynamics has been applied to agricultural and horticultural crop pests and we review some of these applications in this chapter. However, there have been virtually no attempts to extend its use to the development and testing of IPM strategies. In this chapter we attempt to identify how the theory of metapopulation dynamics could be applied to IPM problems in relation to agricultural systems subjected to a wide variety of agronomic practices and inhabited by pests and natural enemies with a wide variety of life history traits. We will largely be considering insects (and mites) in our examples, partly because we are entomologists, but also because most applied examples of metapopulation analysis are to be found amongst these pests.

3.2 The metapopulation concept

Ideas on what constitutes a metapopulation and which sort of organisms it should be applied to are still evolving. At its simplest, it is just a "population of populations" linked by dispersal (Hanski, 1991). The idea has been widely used as a framework for field studies but it was originally conceptualized as a model. The original Levins (1969; 1970) model was general; there was no spatially explicit arrangement of patches and their subpopulations and few assumptions about patch size or quality. Its simplicity provided a starting point for theoretical developments but limited its application to "real" situations that might be of relevance to IPM. As the concept became more widely adopted, models were developed which overcame many of the restrictive early assumptions. The "classical" metapopulation still maintains some core characteristics, particularly related to asynchronous subpopulation dynamics and linkage by regular but usually limited movement. The often quoted requirement of a substantial probability of extinction and re-colonization (turnover) of subpopulations within patches has not been seen as an issue in some recent models, where the effects of species interactions have been the primary concern (Hanski, 1991). Extinction and colonization (turnover) processes are, however, still built into the assumptions of most models.

Table 3.1. *Examples of predator-prey studies and evidence of metapopulation process*[a]

System	Movement important?	Metapopulation structure?	Coupled predator–prey?
Mosquitoes and various predators	Yes	No	No
Zooplankton and predator	Probably	Yes	No
Fireweed aphids	Yes	Yes, prey only	Probably not
Oak gall wasp and parasitoids	Yes	Yes, prey only	Partly
Barnacle and predatory snail	Yes	No	No
Intertidal snail	Yes	Yes, prey only	No
Island spiders and lizards	Yes	Yes	Partly
Olive scale and parasitoids	Probably	Probably, prey only	Yes
Cottony-cushion scale and vedalia beetle	Probably	Probably, prey only	Yes
Larch sawfly and parasitoids	Possibly	Probably	Yes
Greenhouse mites	Yes	Partly	Yes
Goldenrod aphid and lady beetle	Yes	Yes, prey only	Yes
Mites on apple saplings	Possibly	Probably	Partly

[a]References to individual studies may be found in Taylor (1991).

There is considerable disagreement about the general applicability of the metapopulation concept. Attitudes range from the belief that all species probably persist as metapopulations at an appropriate scale (Harvey *et al.*, 1992), to, perhaps the more general, view that true metapopulations only occur in a relatively small number of cases in the real world. Nachman (2000) stated that "almost all organisms are part of a metapopulation consisting of a finite number of subpopulations connected by gene flow via dispersal". Taylor (1991), however, reviewing 13 examples from the literature, concluded that examples supporting the presence of such processes do occur but that they are by no means universal (Table 3.1).

Harrison (1991), reviewing examples from conservation biology, concluded that the empirical literature presented few examples that fitted well the conventional view of metapopulation dynamics. Instead, she identified three alternative situations:

- persistence due to "extinction-resistant" mainland populations
- patchy populations in which dispersal between patches or subpopulations is so high that the system is effectively a single extinction-resistant population

- non-equilibrium populations where local extinction occurs to a species' overall decline

Intuitively, we might suspect that most highly mobile pest species fit well into the second of these classes. Consequently, it is possible that the classical metapopulation approach may not be appropriate for many pests which are controlled using IPM, as a rigid definition would usually exclude its use. This leads us to the question "whether a metapopulation approach is useful or not" posed by Hanski (1999) and suggests that it may be useful if it is valid to assume that space is discrete, that ecological processes take place at local and metapopulation scales, and that habitat patches are large and permanent enough to permit persistent local breeding over several generations. Before we can answer this question and assess the relevance of a metapopulation approach to IPM we firstly need to consider aspects of the agricultural system itself.

3.3 The agricultural system

Metapopulation studies in conservation biology are often conducted on "rare" species that have specific ecological requirements and whose habitat is fragmented into relatively small but highly distinct spatial units. Landscapes of interest to those developing IPM strategies are clearly considerably different. Indeed, crops often form the inhospitable "spaces" between semi-natural landscape fragments which support population patches of these rare or declining species.

In agricultural systems, pest species are characteristically abundant and attack crop areas of relatively large extent. Additionally, while in conservation biology habitat fragments are relatively stable temporally and spatially, the opposite is the case in agricultural systems where a variety of management practices cause regular perturbation in a largely artificial environment. Three aspects of the agricultural system demonstrate its complexity: (a) the spatial structure of farmland with its characteristic "mosaic" arrangement of landscape elements, (b) agronomic practices undertaken, and (c) the variety of ecological traits of individual species (of both pest and beneficial status).

3.3.1 *Spatial structure*

The agricultural landscape is comprised of elements that can be classified broadly as unmanaged natural or semi-natural (e.g. woodland), managed semi-natural (e.g. unimproved grassland and field boundaries), and farmed (e.g. improved grassland and arable) as well as other non-agricultural features, such as roads and gardens. The landscape is

Table 3.2. *The extent of some farmed and managed semi-natural landscape elements in regions of the United Kingdom*[a]

	England and Wales		Scotland		Northern Ireland	
Element	Ha (000s)	%	Ha (000s)	%	Ha (000s)	%
Improved grassland	4431	29.1	1051	13.1	568	41.0
Arable and horticultural	4609	30.3	639	8.0	59	4.3
Neutral grassland	444	2.9	168	2.1	254	18.3
Acid grassland	547	3.6	748	9.3	28	2.0
Calcareous grassland	38	0.2	27	0.3	1	0.1

[a]Source: Countryside Survey, 2000.

characterized not only by the proportional presence of these elements but also by their spatial arrangement. Field size and use, degree of aggregation between fields, connectivity due to field boundary type, and the complexity of adjacent landscape elements also characterize the landscape mosaic. These landscapes also differ structurally locally, regionally, and nationally, as shown, for example, for the UK in Table 3.2.

3.3.2 *Agronomy*

The degree of intervention within each landscape element will depend on management intensity. Hence, natural or semi-natural elements may be relatively intervention-free, while farmed elements will be subjected to cultural practices, pesticide applications, crop variety selection, and biological control strategies. All these are likely to influence population processes of both pests and their natural enemies, which we summarize in Table 3.3. Agronomic practices can be broadly categorized into those that cause spatial instability (e.g. rotation), temporal instability (e.g. cultivation), enhancement of population processes (e.g. maintenance of natural enemy populations or provision of monoculture for pests) and suppression of population development (e.g. by pesticide application, use of resistant varieties, or tillage). The consequences of these agronomic practices, however, will be species-specific, particularly with regard to pest and beneficial organisms.

3.3.3 *Ecological traits of pests and natural enemies*

The wide variety of ecological requirements and life history traits of pest and beneficial species found within the agricultural landscape further

Table 3.3. *Examples of agronomic practices that are likely to influence population and metapopulation processes within farmed landscape elements*

Practice	Population/metapopulation effect
Cultural	
Rotation	Temporally unstable habitat. Spatially unstable distribution of crop types. Disruption of poor disperser life cycles.
Planting and harvesting date	Synchrony in host plant development. Effect on establishment of pest subpopulations. Driver for synchronous pest subpopulation and metapopulation development. Discontinuity of host plant distribution.
Tillage	Disturbance of habitat. Disruption of life cycle for pest and beneficial species by physical perturbation of, for example, larval overwintering sites.
Monoculture	Creation of contiguous habitat for population processes.
Intercropping	Increased landscape complexity. Disruption of pest colonization. Increase in natural enemy activity. Overall suppression of pest subpopulations.
Agrochemicals	
Broad-spectrum pesticides	Suppression of subpopulations of both pest and natural enemies. Pest resurgence. Driver for recolonization processes within landscape elements.
Specific pesticides	Reduction in subpopulation of pests. Driver for dispersal of host-specific natural enemies. Disruption of predator–prey population processes.
Crop variety	
Resistance	Suppression of pest population growth rates. Possible increase in dispersal due to reduction in host plant suitability.
Biocontrol	
Structural (beetle banks etc.)	Increased structural complexity leading to establishment and maintenance of natural enemy subpopulations. Enhancement of natural enemies and suppression of pest populations.
Release of agents	Inundative release of natural enemies – "artificial dispersal", leading to suppression of pest subpopulations.

complicates the system. In Table 3.4 we illustrate this complexity by providing four examples of species found in agroecosystems and consider how the range of life histories might affect the applicability of the metapopulation concept. The examples include both natural enemy and

Table 3.4. *Examples of dispersal characteristics of two natural enemy and two pest species and their suitability for study within a metapopulation framework*

Dispersal	Organism	Characteristics	Suitability
	Natural enemies		
Good	Sheetweb and dwarf spiders *Linyphiidae*	Large scale aerial dispersal by ballooning. Movement by thermal and entrainment in weather systems.	Long- and short-range dispersal with largely undirected movement. Probably unsuitable for "classical" metapopulation approach.
Poor	Carabid beetle *Pterostichus melanarius*	Dispersal and movement essentially by walking. Mean displacement distance 2.6 to 5.3 m day^{-1} (Thomas *et al.*, 1998). Field boundaries act as barrier to dispersal.	At the landscape scale, subpopulations defined as separate fields. Dispersal between fields mediated by permeability of field boundaries and rate of displacement. Good candidate for metapopulation approach.
	Pests		
Good	Russian wheat aphid *Diuraphis noxia*	Winged forms able to disperse over 100s of miles using low-level jet streams. Small-scale dispersal also evident when seeking alternative host plants.	Extreme dispersal ability implies that subpopulations extend over very wide geographical areas. Difficult to use as model species for metapopulation studies
Poor	California red scale *Aonidiella aurantii*	Dispersal mainly in the form of new born "crawlers" typically moving about 1m although a very small proportion become airborne (Murdoch *et al.*, 1996). Most settle on tree in which they were born (DeBach, 1958).	Limited dispersal and behavioral characteristics suggest that metapopulation approach could be applied at the orchard scale.

pests which we have categorized as "good" or "poor" dispersers. This dichotomy is crucial in considering firstly the relevance of the metapopulation approach and secondly the scale at which these concepts may be applied. It should be noted, however, that the dispersal ability of species may be affected by non-habitat ("matrix") vegetation. For example, in a mark-recapture study, two species of flea beetles (*Apthona*) were shown to have differing dispersal abilities according to the vegetation matrix (Jonsen *et al.*, 2001). One species, *A. nigriscutis*, had a much higher ability to disperse into host plant patches when moving through a grass-dominated matrix than a shrub-dominated matrix, whereas *A. lacertosa* dispersal was similar in both matrices but at a lower rate than *nigriscutis*. Such results suggest that metapopulation dynamics may be strongly affected by the types of matrix habitat present within the landscape. Additionally, insect populations are influenced by the degree of habitat fragmentation. Tscharntke *et al.* (2002) identified key fragmentation characteristics as pure habitat loss, size, isolation, edge effects, quality, productivity, and landscape mosaic.

The concept of extinction, at anything but the most local of scales, is very difficult to conceive for ubiquitous, highly dispersive and highly reproductive species such as aphids. Conversely, some species, particularly natural enemies (e.g. predatory ground beetles) could conceivably become extinct at the scale of a single field due to isolation mediated by field boundaries and/or competitive exclusion.

Additionally, there is a temporal component to the suitability of landscape elements for pest and prey species. For example, field boundaries may be utilized by overwintering predatory beetles as refuges, spending the summer months within the fields themselves. Indeed, the importance of each landscape element will depend on the ecological requirements of each pest or beneficial species. Wissinger (1997) states that annual crop systems are "predictably ephemeral" habitats. Further, he characterizes types of insect life histories, namely "cyclic colonization", where insects that exploit predictably ephemeral habitats respond to disturbance by dispersing to refugia which are permanent, where they overwinter and then re-colonize the following year. Cyclic colonization requires spatial heterogeneity. His focus was on annual cycles of colonization between permanent refugia and a mosaic of crop fields. Invertebrates in agroecosystems that migrate cyclically include linyphiid spiders, aphids, Heteroptera, Homoptera, Neuroptera, braconid parasitoids, Coccinellidae, Carabidae, and Staphylinidae (Wissinger, 1997) which may display either local or long-distance dispersal, or a combination of both.

The role of landscape structure, fragmentation of habitats, and the interaction between pest and beneficial arthropods is clearly a major issue for those involved in developing integrated pest management strategies (Hunter, 2002).

3.4 The question of scale

In conservation biology, studies are usually conducted at the landscape scale (see Hanski, 1999 for examples). However, in pest control there are examples (given below) of studies ranging from consideration of dynamics with subpopulations on individual leaves to those that investigate the long-range migration of insects, all of which use the term "metapopulation" to describe observed processes. Hanski and Gilpin (1997) point out that there is a risk that ecologists will refer to aggregated populations at any spatial scale as a metapopulation and conclude that "it seems useful to us not to extend the metapopulation concept to patchy distributions of individuals at the local scale". It may be that the use of the metapopulation approach is not helpful: de Roos *et al.* (1998) demonstrated that local population dynamic processes do occur even when no distinct patch or subpopulation structure is evident (see Murdoch, 1994). Similarly, Amarasekare (2000) investigated (in a non-agricultural environment) interactions between the harlequin bug, *Murgantia histrionica*, and two parasitoids: *Trissolcus murgantiae* and *Ooencyrtus johnsonii*, and concluded that local interactions rather than spatial processes mediated parasitoid coexistence. It is also evident that, in some systems, there is a dynamic and largely ephemeral spatial structure (Figure 3.1). These systems are perhaps those that Hanski and Gilpin (1997) consider to be unsuitable for a metapopulation approach and alternative conceptual frameworks should be sought.

The metapopulation concept, when applied to systems in which subpopulations have been defined at the single plant level, is at a very different scale from the "classical" studies. However, such studies demonstrate processes of occupancy and extinction that are at least analogous to those found in classical metapopulation theory. The question of what is an appropriate scale is essentially the same as "what are islands?" posed by Simberloff (1986) in relation to the applicability of island biogeography theory. He went on to identify individual plants, small stands, fields, all fields in a region and a plant species as possible "islands". The scale at which studies are undertaken is of great importance because, if it is inappropriate, failure to identify processes may occur. For example, Ives *et al.* (1993) studied lady beetle (ladybird) and aphid populations on firewood at a range of scales (individual beetles on individual stems, populations of beetles on individual stems, and populations of beetles in populations of firewood stems). At the finest scale, beetle behavioral response was only weakly linked to aphid density with 4% to 11% variance explained. However, at the largest scale within 25 m field plots

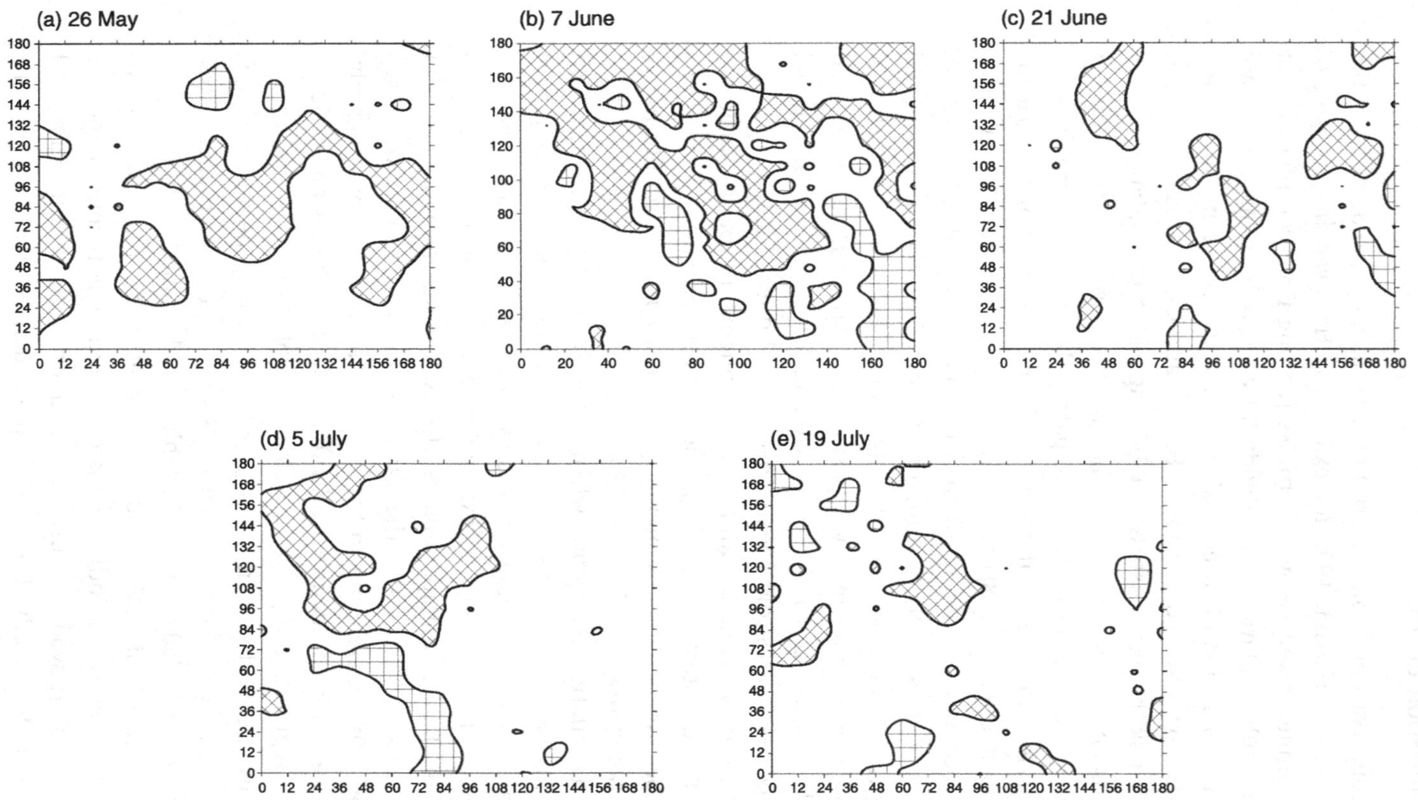

Figure 3.1. SADIE "red-blue plots" of field-scale distribution of grain aphid, *Sitobion avenae*, infestation in winter wheat recorded at two-weekly intervals during 1999 (Winder *et al.*, 2001). Plots are based on a 16 × 16 grid (with a 12m lag) of locations where aphid counts were done. Cross-hatched (blue) areas denote clusters of relatively low counts that form "gaps" while square-hatched (red) areas are the converse, namely clusters of high counts forming "patches". White areas denote space where the aphid distribution is unstructured. Gap and patch clusters are spatially and temporally unstable.

between 50% and 90% of variation in residence of beetles within plots could be explained by aphid density.

3.5 Metapopulation processes in pest systems from field studies

We now turn to some of the empirical studies which imply that metapopulation processes do occur in systems where IPM strategies are applied. Although there are few, if any, studies that explicitly link metapopulation theory and IPM, there are a number of examples relating to pest management in general that illustrate its applicability.

Nachman (1991) studied mite predator–prey interactions in a *Phytoseiulus persimilis–Tetranychus urticae* system in greenhouse cucumbers. Subpopulations were defined as those on individual plants within a greenhouse "metapopulation". He concluded that metapopulation processes were evident, with a characteristic "shifting mosaic" spatial structure. Similarly, Walde (1994) conducted a study where a field experiment in apple orchards was conducted to determine if a predator (*Typhlodromus pyri*)–prey (*Panonychus ulmi*) system was best described by single- or meta population dynamics. In this experiment, two factors were manipulated, namely the initial density of the prey (a subpopulation effect) and secondly the number of potentially interacting subpopulations (a metapopulation effect). He concluded that the most probable scenario for the observed predator–prey interaction was that *T. pyri* can drive *P. ulmi* locally extinct and that *P. ulmi* only persists at the orchard scale due to re-colonization events within or between orchards.

Strong, Slone and Croft (1999) used the metapopulation concept to investigate the control of the two-spotted spider mite *Tetranychus urticae* by the predaceous phytoseiid mite *Neoseiulus fallacies*. In that study individual leaves constituted subpopulations (a "patch") and a plant the metapopulation. Detailed observation of the mites' occupancy of individual leaves led to the conclusion that evidence of a metapopulation process was provided by: (a) the presence of empty patches (attributed to predatory activity); (b) re-colonization of an empty patch; (c) dispersal of predator and prey between patches; (d) asynchrony in predator–prey development (occupancy classes with older leaves empty, with pest only, with predator only or with pest and predators present); (e) direct evidence of local extinction; and (f) stability of the metapopulation. It was concluded that predator–pest interactions were stabilized at the scale of the whole plant due to asynchronous subpopulation dynamics. The authors also concluded that a strategy to promote predatory activity should retain the lower leaves on plants, conventionally removed, in order to enhance interplant dispersal thus creating a "metapopulation landscape".

However, not all studies demonstrate processes akin to those within metapopulations. Murdoch *et al.* (1996) studied a citrus pest, Californian red scale, *Aonidiella aurantii*, and its interactions with the parasitoid, *Aphytis melinus*, which is considered to control its populations. This was done by manipulating refuges within a "subpopulation" (i.e., an individual tree) and also by isolating individual trees to test for metapopulation effects. They recorded a decline in red scale densities of 60% when refuges were removed. However, neither removal of the refuge population nor isolation of individual trees increased the temporal variability of the scale population or led to drift in population density. They postulated that the reason for lack of metapopulation effect was synchrony of population fluctuations within the citrus grove and concluded that stability in the control population was not maintained by metapopulation processes (or refuges) but rather by small-scale processes. In a review of several studies on this system, Murdoch (1994) concluded that population regulation could not be accounted for by metapopulation dynamics.

3.6 Metapopulation processes in pest systems from models

Metapopulation studies are rooted in model development and this approach has been used in studies that are of relevance to the development of IPM strategies. In this section we illustrate how models which adopt a metapopulation approach have been used in this context.

3.6.1 *Pesticide application*

At the landscape scale Sherratt and Jepson (1993) considered localized "catastrophes" in the form of within-field pesticide applications when testing the hypothesis that a farm- or landscape-scale population might persist in the longer term as a result of metapopulation processes. Their simulation model incorporated the possibility of "local extinction" by catastrophic events and was parameterized with phenology, dispersal rate, and susceptibility to pesticide. They applied this model to a predatory ground beetle to investigate the impact of local and episodic applications of pesticides, firstly on predator's population alone and secondly in a system incorporating prey. They found that the persistence of the beneficial insect increases in sprayed fields if few other fields were sprayed, the rate of application was low or the pesticide has low toxicity. They also demonstrated that an optimal dispersal rate may maximize the beneficial's ability to persist and that predator/prey dispersal rates are important influences on the eventual population size of prey.

3.6.2 *Cultural control*

Ives and Settle (1997) provided an extension of the original Levins type model in order to investigate control strategies for mobile insect pests. The model incorporated an internal structure of subpopulations, dynamics of predators and prey, and more spatial realism by considering a one-dimensional array of fields. The metapopulation model showed that while synchronous planting led to lower pest densities in the absence of predation, asynchronous control strategies could be most effective in the presence of predators, dependent upon the level of dispersal and population dynamics of predator and prey.

3.6.3 *Resistance*

Caprio and Hoy (1994) used a stochastic model to demonstrate that metapopulation dynamics affected the development of resistance in the predatory mite *Metaseiulus occidentalis*. Models incorporating or excluding metapopulation dynamics and local extinctions indicated that the establishment of resistant strains was affected by processes increasing local homozygosity within patches. They went further and concluded that the development of resistance was accelerated in a metapopulation system when the mechanism was recessive.

3.6.4 *Biological control*

Kean and Barlow (2001) used a coupled map lattice to incorporate dispersal explicitly into a metapopulation model in order to investigate the population dynamic relationships between the pest weevil *Sitona discoideus* and the parasitoid *Microtonus aethiopoides*. From data collected in New Zealand, model output matched well the empirically observed findings of a successful biological control program. Additionally, the model enabled good simulation of unsuccessful biological control in Australia by advancing weevil oviposition in autumn and reducing parasitism rates among estivating weevils, reflecting local conditions. They concluded that the success of biological control, as well as local pest density may be a function of habitat and this might imply that the breaking up of extensive monoculture would be beneficial.

Skirvin *et al.* (2002) used a spatially explicit simulation model, based on a square grid containing 100 plant units, to investigate control strategies for ornamental nursery crops in a spider mite/predatory mite (*Tetranychus urticae*/*Phytoseiulus persimilis*) system. Control was assumed to occur when the pest was driven to extinction. Their model showed that "one-off" introductions of the predator rarely caused pest extinctions, even at high introduction rates, and that increased aggregation of *T. urticae* led to a greater extinction at lower predator introduction rates, but time to extinction was increased due to the time taken for the predator to cope with the increased pest density.

On a much larger scale, Holt and Colvin (1997) developed a model of the interaction between the Senegalese grasshopper (*Oedaleus senegalensis*) migration and its predator within a two-zone seasonal habitat. The model was based on Lotka-Volterra equations incorporating cyclical carrying capacity. Further, two subpopulations (north and south) were included, with asynchronous seasonality, predation, and *O. senegalensis* dispersal between the two zones proportional to its density. The model was used to demonstrate that migration did not necessarily lead to increased overall grasshopper abundance and this was attributed to the promotion of predator survival. The model also supported the hypothesis that oviposition in the northern zone may generate pest outbreaks.

3.6.5 *Crop rotation*

Halley *et al.* (1996) developed a simulation model for linyphiid spider dynamics for a one-dimensional agricultural landscape with different field types. The model demonstrated that landscape heterogeneity was important for survival and abundance of the spiders. The model incorporated rotation of crops, so that fields could act as both "patches" and "spaces". While spider populations were able to survive when entire landscapes were sprayed with pesticides, the introduction of a second "source" land type caused a dramatic increase in the likelihood of local population rescue. Halley *et al.* concluded that because of the dispersal power of linyphiid spiders, local extinctions are likely to be rare and short-lived in most arable landscapes whereas in large, synchronously sprayed regions a reduction in the degree of natural pest control may be evident.

3.7 Future prospects

Many of the studies using a metapopulation approach within the literature test for the presence of stability, either in the context of the continued existence of single species or in the stabilizing effect of prey and (usually) parasitoid interactions (Mills and Getz, 1996). Such studies have provided an insight into the fundamental ecological processes at work in such systems. It is a challenge to those working to develop IPM to use the same approach to develop an understanding of the systems within which we work.

The key question to address is whether pest suppression is increased when a spatially explicit approach to the development of IPM strategies is adopted. Kean and Barlow (2000), using a host–parasitoid model, demonstrated that a spatially explicit metapopulation system could produce higher host suppression than non-spatial models. Attributes of IPM systems could be tested using a spatially explicit approach in order to determine if similar effects are likely when a range of agronomic practices (e.g. crop rotation) is incorporated into spatially explicit models. Landscape-level host–parasitoid and prey–predator interactions can be stabilized by habitat patchiness (Reeve, 1988, 1990; Pacala *et al.*, 1990; Sabelis *et al.*, 1991; Wissinger, 1997) and management strategies influence this feature of the agricultural landscape. Walde (1994) summarizes the debate on how stability might occur in biological control systems as (a) that the predator–prey interaction is stable at a relatively small (although undefined) spatial scale due to mechanisms such as foraging behavior, or fine-scale physical or biotic heterogeneity or (b) that the interaction is unstable at a small scale, but stable at a larger spatial scale due to migration among subpopulations (i.e. metapopulation dynamics) and it is these processes that could be studied to investigate the effects of a range of IPM practices.

The answer to the question "whether a metapopulation approach is useful or not", posed by Hanski (1999), is often not a resounding "no" but "it depends on the species and scale". For some species, the "classical" metapopulation approach may be useful, because it is valid to assume that space is discrete, that ecological processes take place at local and metapopulation scales and that habitat patches are large and permanent enough to permit persistent local breeding over several generations (Hanski, 1999, p. 3). This definition applies over the range of spatial scales which might be of interest to IPM development, namely from the single plant upwards.

In fact, it is the complexity of the IPM system which is unique in the context of metapopulation studies. The creation of an agricultural landscape with its characteristic mosaic structure, combined with the agronomic practices that are applied has resulted in a system that is far removed from one that could be considered "natural". The discontinuities in time and space that are generated are attributes that metapopulation studies have not yet addressed. Conservation-orientated studies tend to concentrate on spatially stable landscape elements with temporal processes controlled by seasonal and climatic variation; this is certainly not the case for within-field subpopulations which are subject to a range of agronomic activities that cause dynamic changes in landscape elements (due to the adoption of rotation, cultivation, and pest control strategies). Complexity is increased further by each landscape element having a unique spatial context, and different intensity and range of agronomic intervention as well as a different complement of pests and natural enemies. However, the effect of these practices is difficult to predict and it is likely that a metapopulation modeling approach could be used to develop the understanding of how such factors interact. It is the perturbation of population processes for both pest and natural enemy that is the striking difference between "agricultural" and "conservation" biology.

We believe that in order to use a metapopulation approach for IPM the following issues, as set out in the subsections below, need to be considered.

3.7.1 *Developing the framework*

The metapopulation

The classical definition of metapopulation processes may be applied to a number of pest and pest/natural enemy systems. However, the strict adherence to the definition should not preclude the use of a metapopulation framework being used to address problems. This is probably of particular relevance to the process of extinction. Abundant and dispersive species are unlikely to go extinct at anything other than the smallest spatial and temporal scales. It may therefore be appropriate to relax the definition of metapopulation process, although studies need to clearly define the framework which they adopt.

Spatial terminology

Dungan *et al.* (2002) criticized the use of the word "space" in ecological studies and suggest that the use of concepts such as extent (the total length, area, or volume that is analyzed), grain (which could be average field size in relation to IPM, although definitions are varied), and resolution (the size of the smallest sampling unit) should be adopted. These terms may be used directly within metapopulation studies; extent defines the area within which the population exists, grain may relate a subpopulation, and resolution could be applied to the smallest definable subpopulation.

Spatial studies

Techniques such as SADIE and other allied statistics available to describe spatial pattern need to be used to understand the spatially explicit population structure of species of (see Perry *et al.*, 2002 for a useful review). Kean and Barlow (2001) note that "despite the abundance of theoretical evidence there have been few attempts to investigate the importance of spatial dynamics in real biological control systems". Without an understanding of the spatial pattern observed in the field it is difficult to see how models may be developed that are realistic enough to have an applied use, particularly when attempting to define what might constitute a subpopulation. Field-based studies that characterize distributions over a range of extents and grain sizes will allow a more considered view of spatial structure, echoing Levins' (1969) advice that "theoretical investigation is no substitute for entomological knowledge". The defined units that form subpopulations in many studies are often chosen because of their expediency rather than a credible conceptualization of subpopulation structure.

Modeling

The use of models that explicitly describe population dynamics within patch or subpopulations with movement of individuals between patches where local extinction may be absent or infrequent is probably the most suitable for IPM studies. Additionally, spatial and temporal instability may make it difficult to construct models and so an approach considering population dynamics in continuous space such as coupled-map lattice model may be most appropriate (Hanski, 1999). In the context of predator–prey, or parasitoid–prey processes, there may also be a need to increase the complexity of such models to account for the stochastic emergence (and disappearance) of patches mediated by external system factors (e.g. predation and parasitism by other organisms within the community, or environmental perturbation). It may also be possible to test the validity of models by combining the description of pattern identified (see section 3 above) with these spatially explicit models. It may be that pests that are habitat generalist, mobile, and highly reproductive might be difficult to study using the metapopulation approach. However, their populations might be controlled or suppressed by natural enemy species that do exhibit metapopulation structures, so they should not be excluded routinely from study. Systems in which dispersal characteristics between predator and prey vary considerably are probably common. de Roos *et al.* (1998) studied the interactions of predators and prey that are characterized by a scale difference in their use of space. Prey were assumed to occupy patches forming a metapopulation with low dispersal, whereas predators were assumed to be homogeneously distributed due to large-scale foraging or juvenile dispersal. In such systems there may be "partial adherence" to a metapopulation structure.

3.7.2 *Investigating the system*

Testing IPM systems

Ultimately we wish to understand how IPM may be refined to improve its effectiveness. Levins (1969) and Ives and Settle (1997) were interested in investigating the importance of synchronous or asynchronous pest control while Sherratt and Jepson (1993) suggested that their model could be used to test a range of pesticide application scenarios. Helenius (1997) argued that the regional rotation of crops could be used as an approach to control, although it is perhaps difficult to see how this could be practically adopted or tested rigorously before implementation. Shea and Possingham (2000) used a metapopulation approach allied with decision theory to test release strategies of biological control agents under a range of strategies. This investigative approach could be used to test management strategies.

Causation of synchrony/asynchrony

This should be studied across a wide range of agronomic practices to determine the cause of synchrony/asynchrony in population development of both pest and prey. Asynchrony in subpopulations is one of the key facets of classical metapopulation studies and it is interesting to hypothesize as to the effect of management practices on the dynamics of pest and natural enemy populations. Stabilization of populations is often achieved through immigration and emigration and spatially out-of-phase fluctuations. Agronomic practices (e.g. planting date) may cause synchrony in population processes whilst others (e.g. rotation) may result in asynchrony. What is the effect of this on the stability of pest and natural enemy populations?

Testing complexity

The interaction between IPM practices could be tested to determine their synergistic impact. It would be possible to test characteristics of systems that adopt IPM strategies (e.g. high crop diversity through mixtures in time and space, discontinuity of monoculture or the inclusion of features that support beneficial invertebrates) to determine their importance in stabilizing or perturbing pest population dynamics.

Multispecies models to investigate biodiversity in IPM systems

In densely populated countries, such as the UK and elsewhere in northern Europe, relatively little environment remains that has not been created or extensively modified by human occupation. In such cases biodiversity is often maintained through the maintenance of long-established farming practice. It would be worthwhile to investigate how assemblages of beneficial invertebrates are able to persist in modern agricultural landscapes and test scenarios that might cause enhancement of, or decline of, such populations.

In the 1986 version of this volume Daniel Simberloff contributed a chapter on "Island biogeography theory and integrated pest management". In his conclusion, he stated that "it appears that we have learned what lessons there are to be gleaned, and it is now important to move on to other approaches". It is interesting to note that in this volume, that chapter on Island biogeography theory has been replaced by one on "Populations and Metapopulations". This is an understandable progression as the metapopulation paradigm has almost completely replaced the former theory, at least in a conservation context. It often seems to be that applied pest control theory is running to catch up with ecological theory. However, pest species population dynamics, with all their spatial fluidity and extreme temporal fluctuations, often test such theories to their limits. Perhaps it is time for pest management scientists to take more of a leading role in future conceptual developments; it then might be intriguing to predict what the title might be that replaces this one in a future volume.

References

Amarasekare, P. (2000). Coexistence of competing parasitoids on a patchily distributed host: local vs. spatial mechanisms. *Ecology*, **81**, 1286–96.

Caprio, M.A. and Hoy, M.A. (1994). Metapopulation dynamics affect resistance development in the predatory mite, *Metaseiulus occidentalis* (Acari, Phytoseiidae). *Journal of Economic Entomology*, **87**, 525–34.

de Roos, A.M., McCauley, E. and Wilson, W.G. (1998). Pattern formation and the spatial scale of interaction between predators and their prey. *Theoretical Population Biology*, **53**, 108–30.

DeBach, P. (1958). The role of weather and entomophagous species in the natural control of insect populations. *Journal of Economic Entomology*, **51**, 474–84.

Dungan, J.L., Perry, J.N., Dale, M.R.T. *et al.* (2002). A balanced view of scale in spatial statistical analysis. *Ecography*, **25**, 626–40.

Halley, J.M., Thomas, C.F.G. and Jepson, P.C. (1996). A model for the spatial dynamics of linyphiid spiders in farmland. *Journal of Applied Ecology*, **33**, 471–92.

Hanski, I. (1991). Metapopulation dynamics: brief history and conceptual domain. *Biological Journal of the Linnean Society*, **42**, 3–16.

Hanski, I. (1999). *Metapopulation Ecology*. Oxford Series in Ecology and Evolution. Oxford: Oxford University Press.

Hanski, I. and Simberloff, D. (1997). The metapopulation approach, its history, conceptual domain, and application to conservation. In I.A.G. Hanski and M.E. Gilpin (eds.), *Metapopulation Biology: Ecology, Genetics and Evolution*. San Diego: Academic Press. pp. 5–26.

Hanski, I.A. and Gilpin, M.E. (eds.) (1997). *Metapopulation Biology: Ecology, Genetics and Evolution*. San Diego: Academic Press.

Harrison, S. (1991). Local extinction in a metapopulation context: an empirical evaluation. *Biological Journal of the Linnean Society*, **42**, 73–88.

Harvey, P.H., Nee, S., Mooers, A.O. and Mooers, L.P. (1992). The hierarchical views of life: phylogenies and metapopulations. In R.J. Berry, T.J. Crawford and G.M. Hewitt (eds.), *Genes in Ecology*. Oxford: Blackwell Scientific Publications. pp. 123–37.

Helenius, J. (1997). Spatial scales in ecological pest management (EPM): Importance of regional crop rotations. *Biological Agriculture and Horticulture*, **15**, 163–70.

Holt, J. and Colvin, J. (1997). A differential equation model of the interaction between the migration of the Senegalese grasshopper, *Oedaleus senegalensis*, its predators, and a seasonal habitat. *Ecological Modelling*, **101**, 185–93.

Hunter, M.D. (2002). Landscape structure, habitat fragmentation, and the ecology of insects. *Agricultural and Forest Entomology*, **4**, 159–66.

Ives, A.R., Kareiva, P. and Perry, R. (1993). Response of a predator to variation in prey density at three hierarchical scales: lady beetles feeding on aphids. *Ecology*, **74**, 1929–38.

Ives, A.R. and Settle, W.H. (1997). Metapopulation dynamics and pest control in agricultural systems. *American Naturalist*, **149**, 220–46.

Jonsen, I.D., Bourchier, R.S. and Roland, J. (2001). The influence of matrix habitat on Apthona flea beetle immigration to leafy spurge patches. *Oecologia*, **127**, 287–94.

Kean, J.M. and Barlow, N.D. (2000). Can host-parasitoid metapopulations explain successful biological control? *Ecology*, **81**, 2188–97.

Kean, J.M. and Barlow, N.D. (2001). A spatial model for the successful biological control of *Sitona discoideus* by *Microctonus aethiopoides*. *Journal of Applied Ecology*, **38**, 162–9.

Levins, R. (1969). Some demographic and genetic consequences of environmental heterogeneity for biological control. *Bulletin of the Entomological Society of America*, **15**, 237–40.

Levins, R. (1970). Extinction. *Lectures on Mathematics in the Life Sciences*, **2**, 77–107.

Mills, N.J. and Getz, W.M. (1996). Modelling the biological control of insect pests: a review of host-parasitoid models. *Ecological Modelling*, **92**, 121–43.

Murdoch, W.W. (1994). Population regulation in theory and practice – the Robert H. Macarthur Award Lecture presented August 1991 in San Antonio, Texas, USA. *Ecology*, **75**, 271–87.

Murdoch, W.W., Swarbrick, S.L., Luck, R.F., Walde, S. and Yu, D.S. (1996). Refuge dynamics and metapopulation dynamics: an experimental test. *American Naturalist*, **147**, 424–44.

Nachman, G. (1991). An acarine predator–prey metapopulation system inhabiting greenhouse cucumbers. *Biological Journal of the Linnean Society*, **42**, 285–303.

Nachman, G. (2000). Effects of demographic parameters on metapopulation size and persistence: an analytical stochastic model. *Oikos*, **91**, 51–65.

Pacala, S. W., Hassell, M. P. and May, R. M. (1990). Host–parasitoid associations in patchy environments. *Nature*, **344**, 150–3.

Perry, J. N., Liebhold, A. M., Rosenberg, M. S. *et al.* (2002). Illustrations and guidelines for selecting statistical methods for quantifying spatial pattern in ecological data. *Ecography*, **25**, 578–600.

Reeve, J. D. (1988). Environmental variability, migration, and persistence in host–parasitoid systems. *American Naturalist*, **132**, 810–36.

Reeve, J. D. (1990). Stability, variability, and persistence in host–parasitoid systems. *Ecology*, **71**, 422–6.

Sabelis, M. W., Diekmann, O. and Jansen, V. A. A. (1991). Metapopulation persistence despite local extinction: predator–prey patch models of the Lotka-Volterra type. *Biological Journal of the Linnean Society*, **42**, 267–83.

Shea, K. and Possingham, H. P. (2000). Optimal release strategies for biological control agents: an application of stochastic dynamic programming to population management. *Journal of Applied Ecology*, **37**, 77–86.

Sherratt, T. N. and Jepson, P. C. (1993). A metapopulation approach to modelling the long-term impact of pesticides on invertebrates. *Journal of Applied Ecology*, **30**, 696–705.

Simberloff, D. (1986). Island biogeographic theory and integrated pest management. In M. Kogan (ed.), *Ecological Theory and Integrated Pest Management Practice*. New York: Wiley-Interscience. pp. 19–35.

Skirvin, D. J., de Courcy Williams, M. E., Fenlon, J. S. and Sunderland, K. D. (2002). Modelling the effects of plant species on biocontrol effectiveness in ornamental nursery crops. *Journal of Applied Ecology*, **39**, 469–80.

Strong, W. B., Slone, D. H. and Croft, B. A. (1999). Hops as a metapopulation landscape for tetranychid–phytoseiid interactions: perspectives of intra- and interplant dispersal. *Experimental and Applied Acarology*, **23**, 581–97.

Taylor, A. (1991). Studying metapopulation effects in predator–prey systems. *Biological Journal of the Linnean Society*, **42**, 305–23.

Thomas, C. F. G., Parkinson, L. and Marshall, E. J. P. (1998). Isolating the components of activity-density for the carabid beetle *Pterostichus melanarius* in farmland. *Oecologia*, **116**, 103–12.

Tscharntke, T., Steffan-Dewenter, I., Kruess, A. and Thies, C. (2002). Characteristics of insect populations on habitat fragments: a mini-review. *Ecological Research*, **17**, 229–329.

Walde, S. J. (1994). Immigration and the dynamics of a predator–prey interaction in biological-control. *Journal of Animal Ecology*, **63**, 337–46.

Winder, L., Alexander, C., Holland, J. M., Woolley, C. and Perry, J. N. (2001). Modelling the dynamic spatio-temporal response of predators to transient prey patches in the field. *Ecology Letters*, **4**, 568–76.

Wissinger, S.A. (1997). Cyclic colonization in predictably ephemeral habitats: a template for biological control in annual crop systems. *Biological Control*, **10**, 4–15.

Woiwod, I. P., Reynolds, D.R. and Thomas, C. D. (2001). Introduction and overview. In I. P. Woiwod, D. R. Reynolds, and C. D. Thomas (eds.), *Insect Movement: Mechanisms and Consequences*, Wallingford, UK: CAB International. pp. 1–18.

4

Arthropod pest behavior and IPM

R.J. PROKOPY AND B.D. ROITBERG

4.1 Introduction

An ecological theory can be considered as a general explanation for an ecological phenomenon or pattern of observed ecological tendencies. A well-established ecological theory is usually one constructed upon a foundation of underlying ecological principles and verified hypotheses and may eventually take on the mantle of an ecological law (Lawton, 1999). Because ecology is broad in scope, numerous ecological theories have been generated (Lawton, 1999).

One suite of ecological theory of particular relevance to this chapter on arthropod pest behavior and integrated pest management (IPM) is the suite associated with population dynamics, or the rise and fall of populations over time and space. As interpreted by Lawton (1992), each species or even each population of a species may have its own peculiar dynamics, but essentially this plethora of dynamics can be reduced to variations upon a few common themes. Two key components represented in all such themes are the intrinsic rate of natural increase of the organism and its environment. With exceptions, most arthropod pests, especially pests of agricultural crops, tend to have above-average rates of natural increase (Southwood, 1977; Kennedy and Storer, 2000). Arthropod pests having high intrinsic rates of natural increase from one generation to the next and having several generations per year may become too numerous towards the end of a season for effective management by any single approach, including behavioral manipulation. Even so, for behavioral management to succeed, it is the need to address the structure of the present environment as well as its changing character in time and space that poses the greatest ecological challenge to pest managers.

Our intent here is first to describe some basic aspects of IPM and follow this with discussion of some basic elements of a behavioral approach to managing arthropod pests. We then go on to consider ways in which the state of the environment can influence the success of a behavioral management approach. Subsequently, we attempt to illustrate principal themes using selected examples in which different kinds of arthropod pests have been managed behaviorally. Finally, we draw upon our own experience with a particular pest insect (the apple maggot fly) and present a conceptual model of how we might approach analyzing the influence of environmental state variables on establishing protocols for behavioral management. Throughout, our emphasis is on manipulation of the foraging behavior of arthropod pests for food and ovipositional sites, not foraging for mates. The latter is treated by Carde in an accompanying chapter (see Chapter 5).

4.2 Integrated management of arthropod pests

In essence, integrated pest management is a decision-based process involving coordinated use of multiple tactics for optimizing the control of all classes of pests in an ecologically and economically sound manner (Prokopy and Kogan, 2003). Tactics available for IPM use include cultural management, host resistance, and other forms of genetic-based management, biological control, behavioral control, and pesticidal control. Whichever the tactic or blend of tactics chosen for use, it is important to recognize that arthropods are but one among several classes of pests, including microbial pathogens, vertebrates, and weeds, which ought to be considered when formulating an IPM approach.

Making ecologically and economically sound pest management decisions within an IPM framework is often a complex process that involves weighing effects of many kinds of uncontrolled variables on possible outcomes. Relevant variables include biological characteristics of targeted pests; levels at which pests can be tolerated; degrees of protection afforded by candidate management tactics; costs of management tactics; current and prospective future biotic and abiotic states of environments within which pest management tactics are applied; and an array of social, cultural, legal, and political elements which may override other variables.

As stated in the section 4.1, this chapter emphasizes the state of the environment and its relevance to the success of one particular approach – a behavioral approach – to arthropod pest management. Among the other variables affecting the process of IPM, this is the variable with which we have had the most combined experience with regard to influence on behavioral approaches to pest management.

4.3 Behavioral approaches to arthropod pest management

4.3.1 Point of departure

In any scientific endeavor, it is always helpful to start at the beginning. With respect to developing behavioral approaches to arthropod pest management, the most rewarding beginning point would be to study how the behavior of the arthropod in question is organized in time and space, preferably in natural habitats. This might be accomplished by choosing at least two different habitats populated by the arthropod and first conducting a census at periodic intervals of the precise location of individuals within the habitat. A second essential component of such a study would be to choose either individual pest arthropods, or individual subcomponents of the habitat, for focused observation over extended periods when the arthropod is found present in the habitat. Combining information from periodic censusing and focal observations should give rise to at least the outlines of a portrait of the activities of individuals in those habitats. If performed across several generations of the arthropod and several habitats, a revealing picture of the arthropod's foraging behavior for essential resources can be obtained. Such a picture should yield valuable insights into the range and ranking of resources, particularly food and ovipositional sites, sought and used by members of different demes or populations of the arthropod under study. Armed with insights from observational studies in the field, one may then carry out semi-field and laboratory studies to identify and quantify the precise nature of stimuli from higher-ranking, lower-ranking, and non-resource items that positively or negatively influence the arthropod and the nature of movement patterns used by the arthropod when acquiring a resource, or rejecting a non-resource item. Such studies ought to include, or be accompanied by, experimental manipulation to identify causal factors that determine the expression of behavior. Without experimental manipulation, one is reduced to relying upon correlation or degree of association among variables, an exercise that can yield unsatisfactory or even misleading results when attempting to modify or explain the behavior of a pest for management purposes, especially a pest which exhibits complex behavior.

Possession of solid observational information from field studies, in conjunction with follow-up studies involving experimental manipulation to identify causal factors, can serve as a foundation upon which to build a variety of potential behavioral approaches to managing arthropod pests. Lack of such a foundation may deny or compromise opportunity for successful behavioral management.

4.3.2 *Nature and deployment of stimuli used in behavioral management*

As described by Foster and Harris (1997), manipulation of a pest's behavior for purposes of pest management may be conceived primarily as use of stimuli that either stimulate or inhibit a behavior and thereby change its expression. In some cases, the stimuli used may be natural. Manipulation includes such practices as arranging particularly attractive cultivars or species of plants as barriers to intercept or concentrate immigrating pests (trap cropping) (Hokanen, 1991); arranging unattractive plants in association with principal crop plants to disrupt host finding or host acceptance (intercropping, undersowing, companion planting) (Finch and Collier, 2000); and breeding for biotypes of desired plants or animals that are unattractive or unacceptable to pests (antixenosis) (Kogan, 1994). In the great majority of cases, however, stimuli used in behavioral manipulation are artificial equivalents or variants of natural stimuli presented in subnatural, natural, or supranatural amounts. As pointed out by Foster and Harris (1997), deployment of defined artificial stimuli permits greater flexibility and control in behavioral manipulation than does use of undefined or natural stimuli. This chapter focuses exclusively on use of artificial stimuli in behavioral manipulation.

Artificial stimuli for behavioral management of pests are either chemical or physical. Chemical stimuli include odor stimuli operative at some distance from the source and contact-chemical stimuli operative upon arrival. Physical stimuli include visual and acoustical stimuli (rarely mechanical stimuli) that usually operate at some distance from the source.

Deployment of an artificial stimulus may attract or repel a pest engaged in resource-finding behavior at some distance, positively stimulate or deter a pest engaged in resource-examining or resource-acceptance behavior upon arrival, or mask arthropod perception of stimuli emanating from a resource. Management of pests responding to a deployed stimulus can be achieved by various means. Individuals attracted by an odor, visual, or acoustical stimulus can be trapped on arrival or otherwise killed on arrival by touching a contact-type insecticide incorporated with the stimulus or applied to the locale of the stimulus. Toxic or debilitating sublethal effects of insecticide can be enhanced by deploying a contact feeding stimulant in association with the attractive stimulus to encourage ingestion of insecticide, or at least increase the frequency of contact or prolong the period of contact with insecticide.

Concepts advanced by Bernays (2000) in regard to fundamental aspects of host selection by insects would suggest that, for management of pest arthropods which are resource specialists and exhibit great sensitivity to one or a few positive host-specific stimuli, synthetic attractant or positive contact stimuli might need to be applied in a particularly high amount in order to achieve contrast with levels characteristic of the resource, though not in such high amount as to compromise effectiveness of traps or insecticide. For management of resource generalists, which often exhibit lesser sensitivity than do specialists to positive stimuli of prospective hosts but greater sensitivity to negative stimuli of non-hosts, a lower amount of synthetic attractive or positive contact stimuli may suffice to achieve an effective level of contrast with natural host stimuli. Individuals repelled from arrival at a resource or deterred from accepting a resource through deployment of a chemical, visual, or acoustical stimulus should refrain from using the resource in the treated locale. Affected individuals may, however, move to and accept less protected resource items or, after undergoing sensory adaptation or central habituation to the artificial stimulus, no longer be repelled or deterred. Thus, repellents and deterrents are often most successful when deployed in association with attractants or positive contact stimulants in what have come to be known as deterrent-stimulo or push-pull behavioral management systems (Prokopy, 1972; Miller and Cowles, 1990; Borden, 1997). Masking of resource stimuli can be achieved by application of natural or synthetic equivalents, or analogs of resource stimuli, which are applied in sufficiently high concentration over a sufficiently broad area to effectively prevent a pest individual from finding targeted resources.

Examples highlighting all of these approaches to behavioral management of a variety of arthropod pests are given in Bjostad *et al.* (1993); Foster and Harris (1997); Carde and Minks (1997); and Hardie and Minks (1999). The examples used in this chapter will be a selected subset relevant to the major themes we wish to illustrate.

4.3.3 *Some practical considerations*

Perhaps the foremost practical consideration in behavioral pest management is that the targeted pest should be a key pest or one with populations that repeatedly exceed tolerable levels if unmanaged. Because behavioral management is usually more informationally and operationally challenging and more expensive than alternative management approaches, especially during development preceding refinement, it is not employed routinely against secondary or occasional pests.

Behavioral pest management almost invariably results in use of less pesticide against a key pest and should have a less negative effect on beneficial arthropods and non-target organisms than does a purely pesticide-based approach to management. This inherent appeal must be weighed against possible build-up of secondary and occasional pests if key pests are managed behaviorally and no effective biological, cultural, or other control measures are taken against secondary or occasional pests that may rise to higher pest status after the decline of key pests.

In addition to these considerations, Foster and Harris (1997) put forward the following as desirable attributes of candidate stimuli for use in behavioral manipulation of arthropod pests:

a. accessibility – the stimulus ought to be presented in a form suitable for detection by the pest
b. definability and reproducibility – the more precisely the stimulus can be defined, the more precisely it can be reproduced artificially
c. controllability – the better the ability to control various parameters of a stimulus, particularly its intensity and duration, the greater the probability of achieving successful behavioral manipulation
d. specificity – the more specific a stimulus is in affecting a behavior of a pest, the more likely it is to be perceived above background level and succeed in manipulating that behavior
e. practicability – use of the stimulus ought to be as simple as possible, affordable, and not affect non-target organisms.

According to these attributes, it is not surprising that odor stimuli have predominated over contact chemical, visual, and acoustic stimuli in behavioral management of pest arthropods.

4.3.4 *Dynamic variables affecting pest foraging behavior*

Even when an apparently practical system has been devised for behaviorally managing an arthropod pest, the effectiveness of that system over time and space can not be ensured unless careful consideration is given to major variables that affect behavioral decisions of resource-foraging individuals. As developed by Mangel and Clark (1986), Houston *et al.* (1988), Mangel and Roitberg (1989), Minkenberg *et al.* (1992), and others, a dynamic variable approach to foraging behavior decisions encompasses joint consideration of (a) the genetic background of the forager (e.g. population origin); (b) the current physiological state of the forager (e.g. degree of hunger, degree of reproductive maturity, mating status, age); (c) the current informational state of the forager (e.g. memory of type, abundance, quality, or distribution of resources affecting behavior); and (d) the state of the current environment with respect to structure of habitat, essential resources, enemies, competitors, and abiotic factors.

Any or all of these variables can have a profound effect on the degree to which a forager can be prevented by behavioral management tactics from injuring a protected resource. For example, a forager from a population different from populations that served as the foundation for building a behavioral management system might, owing to a different genetic structure, be comparatively unresponsive to the artificial stimuli deployed for management but still highly responsive to stimuli from the would-be protected resource. A forager with a high egg load might be less inclined than a forager with a lower egg load to rank the relative power of encountered stimuli and thus fail to be won over to an artificial attractant deployed to protect a resource. A forager with feeding or ovipositional experience on a particular host may have developed what loosely could be considered a "search image" for that host and be comparatively unresponsive to artificial attractant stimuli that differed from search image stimuli. If stimuli from a would-be protected resource were a close match to search-image stimuli, effectiveness of the behavioral management system could be compromised. As a final example, the nature of the protected resource could change in time and space (e.g. through growth) and become increasingly competitive with the artificial stimuli deployed to protect it.

All four types of variables (genetic, physiological, informational, and environmental) are distinct entities. Yet from both an ultimate (evolutionary) and proximate (immediate time) perspective, the state of the environment can be considered as a core variable shaping impacts of the other variables. Hence, our focus in this chapter is on the state of the environment as a principal factor underlying the degree of success of behavioral approaches to pest management.

4.4 Influence of environmental state on success of behavioral management

The environment of an arthropod pest population is comprised of several biotic elements such as resource items, non-resource items, natural enemies, and competitors as well as several abiotic elements such as temperature, humidity, rainfall, wind, and light. The quality and quantity of each of these elements may vary considerably over space and time. Moreover, each species or biotype of pest insect exhibits a characteristic set of behavioral traits that influence the distance over which individuals travel to seek a resource item, the distance at which a resource item can be detected, and the degree of specialization in resource acceptability. Various aspects of interplay between these behavioral traits and environmental state in shaping the foraging behavior and population dynamics of insects are explored by Southwood (1977), Kareiva (1983), Bell (1991), Hunter and Price (1992), Margolies (1993), Kennedy and Storer (2000), and Landis *et al.* (2000).

Paramount to the development and practice of strategies and tactics for behavioral management of insect pests is consideration of spatial and temporal scales of environmental structure. Spatial scales that are relevant to effective behavioral pest management may be as great as entire ecoregions or even entire continents or as small as single fields or single buildings on a farm. Relevant temporal scales may embrace several generations of a pest and, if a farmscape, several successive plantings of one or more types of crops or involve only a single generation of a pest and a single type of crop having a narrow time frame (perhaps only a month or a week) of vulnerability to pest damage.

Our intention here is to explore effects of environmental state on behavioral pest management by dealing first with effects on practices intended primarily to suppress pests at distant origins of population development and then with effects on practices intended primarily to provide on-site protection of valued resources against damage by pests. Most of our examples involve pests and practices that have received an amount of attention sufficient for appreciation of one or more environmental variables which may impact the success of the practices used.

4.4.1 *Suppression of pests at distant origins of population development*

In some behavioral pest management programs, the proximate aim is to reduce or eliminate pest populations at distant sites of population origin, with the ultimate aim of ensuring that emigrants are too few to cause unacceptable injury to valued resources in other locales. In some cases, achievement of effective suppression also provides on-site economic or health benefits.

Mediterranean fruit flies

Perhaps the most extensive efforts ever undertaken to manage a pest using a behavioral approach have involved aerial and ground spraying of proteinaceous bait containing insecticide to reduce or eliminate populations of Mediterranean fruit flies, *Ceratitis capitata*. Adults lay eggs in developing fruit, where the larvae feed, eventually causing fruit decay. To attain fecundity, female medflies require frequent access to protein (or nitrogen), which can be acquired by feeding on protein-rich fruit, bird feces, honeydew or bacteria growing on decomposing fruit (Yuval and Hendrichs, 2000). Protein such as NuLure or Mazoferm used in bait sprays both attracts medflies and stimulates alighting flies to feed on bait droplets and thereby ingest insecticide. Such bait sprays have been used for more than three decades on several continents, either to suppress medflies in areas of extant populations prior to release of sterile males or establish zones (sometimes hundreds of kilometers in breadth and tens of kilometers in depth) beyond which invading medflies cannot spread (Steiner, 1969; Roessler, 1989; McQuate and Peck, 2000).

Given the widespread implementation of this behavioral management approach, it is surprising that essentially no quantitative ecoregion-level research has been conducted on effects of biotic and abiotic environmental factors on the success of bait sprays. Nonetheless, some relevant research at the level of habitat patch has been conducted. It shows that there can be a high to moderate degree of reduction in fly response to bait spray droplets in an environment containing high-quality natural sources of protein (Prokopy *et al.*, 1992; McQuate and Peck, 2000). Also, bait sprays are more effective if dry weather rather than damp or rainy weather follows application (McQuate and Peck, 2000), perhaps in part because the water component of bait spray droplets itself becomes attractive to thirsty flies (Cunningham *et al.*, 1978). Further, under aerial application in coffee plantations in Guatemala, foliage of overstory shade trees may intercept most bait spray droplets before they reach understory host coffee plants, thereby limiting access of medflies to the droplets (McQuate and Peck, 2000). Finally, variation in the structure of local habitat (e.g. variation in the density, distribution, or quality of natural food or larval host plants) may have a substantial (but poorly understood) differential effect on the performance of bait sprays applied in alternating sprayed and unsprayed swaths versus total-coverage sprays (Steiner, 1969; Howell *et al.*, 1975; McQuate and Peck, 2000). Local availability of natural food and/or host plant resources could act to reduce the tendency of flies to disperse and call for use of total-coverage bait sprays, whereas local lack of such resources could promote fly dispersal and allow use of alternating sprayed and unsprayed swaths. To effectively address the influence of local resource availability on devising strategies for optimal application of bait sprays against medflies, research is needed on distances at which medflies can detect resource stimuli and distances over which medflies move to acquire resources under varying conditions of local resource availability, terrain and weather.

Tsetse flies

In Africa, tsetse flies of the genus *Glossina* are widespread across several countries, where they bite native animals, domestic livestock, and humans, injecting often lethal trypanosomes into their victims and adversely affecting social and economic development (Leak, 1999). Over the past decade or so, three principal approaches to tsetse fly management have achieved prominence: aerial spraying of insecticides, application of the sterile insect technique, and behavior-based management (Leak, 1999). The latter has been aimed principally at *G. moristans* and *G. pallidipes* and involves an attracticidal approach that has proved to be particularly effective in Zimbabwe.

As described by Vale *et al.* (1988), vertical visual targets consisting of 100 cm × 80 cm sheets of black cloth bordered on either side by 100 cm × 70 cm sheets of black netting (in later years, sheets of blue cloth) were impregnated with the insecticide deltamethrin, baited with synthetic equivalents of attractive host odor (octenol plus either acetone or butanone) and deployed at the rate of 3–5/km^2 of infested habitat in Zimbabwe. Tsetse flies alighting on the insecticide-treated black cloth died quickly, resulting in a greater than 99% reduction in the tsetse population in the 600 km^2 area of target trap deployment over the 10-month period of trap use. Subsequently, attracticidal traps of this or modified types have been used to establish (a) zones in which tsetse flies are suppressed to a point where they pose little or no threat to the treated locale and (b) barriers that prevent dissemination of flies into fly-free locales (Muzari, 1999).

In the most thorough study of environmental effects on varying outcomes of a behavioral pest management approach of which we are aware, Vale (1998), building on earlier studies by Williams *et al.* (1992), Paynter and Brady (1992), Griffiths and Brady (1995), and others, provides an in-depth account of the influence of a range of biotic habitat and abiotic factors on the degree to which tsetse flies detect and respond to attracticidal traps similar to the above. Drawing upon experiments that involved assessment of differences in the construct of natural habitat as well as experiments involving artificial equivalents of natural habitat, Vale found the following.

In areas lacking natural hosts or artificial host odor sources, tsetse flies in open space preferentially move towards patches of vegetation that typically might harbor host animals. Finding no host, flies engage in local flights until they arrive at a gap in the vegetation, such as a path used by animals. They proceed to fly along the gap or path searching for a host, traveling a few to several hundreds of meters per day. Flies can detect the odor of a host at distances up to 90 m. An attracticidal trap placed in closed vegetation attracts very few flies. It becomes substantially more attractive when placed in a clearing amidst vegetation, especially if there is a gap in the vegetation downwind of the trap, and is maximally attractive (olfactorily detectable by flies at 50 meters or more) when in open terrain with scattered bushes in the vicinity (15 meters or so away) and little vegetation 25 cm or taller that might interfere with fly detection of attractive odor plumes or visual recognition of a trap (detectable at distances up to 12 m). In all cases, the species composition of vegetation seems to have far less effect than the architectural arrangement of vegetation. Maximum possible exposure of attracticidal targets to sunlight is crucial to heating of odor dispensers sufficient to ensure that a biologically effective amount of odor is disseminated at a time when tsetse flies are most active (during sunny periods). Under biotic and abiotic conditions maximally conducive to attracticidal trap detectability and performance, traps placed at 250 m intervals (about 4 traps/km^2) can provide highly effective suppression or create a highly effective barrier, as shown above by Vale *et al.* (1988) and later by Muzari (1999).

Mountain pine beetles

The most damaging pest of lodgepole pine in western North America is the mountain pine beetle, *Dendroctonus ponderosae*, which in combination with symbiotic fungi kills host trees by damaging the phloem. One means of preventing small infestations from expanding beyond existing borders and causing outbreaks in distant areas is removal of all beetle-infested trees within the confines of the infested locale. Several circumstances may hinder timely execution of this demanding practice (Borden *et al.*, 1983). As an alternative that is used routinely in British Columbia, Canada, lodgepole pine trees are baited in grid-like fashion (50 m apart) with a blend of three semiochemicals: the host tree volatile myrcene and two aggregation pheromones, the female-produced trans-verbenol and the male-produced exo-brevicomin (Borden *et al.*, 1983). Beetles within the infested area preferentially colonize the baited trees, which are then either logged and taken to a mill where the beetles are killed in the debarking process or else are treated in situ with insecticide to kill the beetles (Borden, 1989). The net effect is reduction or prevention of population spread into uninfested areas.

The structure of the habitat in which this attracticidal practice is carried out can have a significant influence on the selection of trees for baiting and on the performance of the bait. Mountain pine beetles are unable to overcome host tree defenses (such as secretion of resin) unless they attack a tree in very large numbers (Raffa *et al.*, 1993). Although there can be substantial inter-tree and inter-seasonal variation in host susceptibility to attack (Raffa, 2001), the trees most susceptible to attack are those stressed by old age, drought, lightning strike, wind damage or disease. Such stressed trees, especially ones of large diameter, are the best candidates for affixing attractive semiochemical blend to achieve maximum concentration of foraging beetles (Gray and Borden, 1989; Preisler and Mitchell, 1993). Even so, local habitat variables, such as tall growth of forest floor vegetation, can compromise the plume structure of blend odor to an extent that reduces baited tree attractiveness (Borden *et al.*, 1983). When successful, such an attracticidal tree approach for concentrating mountain pine beetle attack can effectively reduce numbers of beetles produced (and emigrating from behaviorally managed tree patches) to levels below thresholds required for overcoming defenses of distant trees (Borden, 1993), especially distant trees growing in relatively thin stands (Preisler and Mitchell, 1993). In addition to local habitat structure, temperature can affect the extent to which emergence of mountain pine beetle adults is synchronous and, if sufficiently high, can result in diffuse emergence. This reduces the likelihood that, at any one time, the beetle population will be sufficiently large to effectively colonize distant trees (Logan and Bentz, 1999).

4.4.2 *Protection against pest damage at sites of valued resources*

In most behavioral pest management programs, the principal aim is to provide on-site protection of valued resources against damage by pests. In some cases, the majority of pests originates at a distance from the valued resource and arrives as immigrants, whereas in other cases the majority originates within the valued resource as progeny of a previous generation. Approaches to behavioral management at valued resource sites include (a) using attractants or arrestants to encourage pests to alight upon, examine, or consume stimulants that are coupled with an item or practices causing pest mortality, (b) using inhibitors that deter pests from damaging a valued resource, and (c) using inhibitors in combination with attractants in a push-pull or deterrent-stimulo fashion.

Western corn rootworm

In the central region of the United States, the western corn rootworm, *Diabrotica virgifera*, is a major pest of maize. Eggs are laid in soil adjacent to maize plants, and larvae feed upon and often severely damage the roots. For decades, western corn rootworms were managed using broadcast application of insecticide against larvae, sometimes in combination with every other year rotation of maize with soybeans. However, development of rootworm resistance to many insecticides, coupled with development of extended diapause, which allows eggs to overwinter for more than one year, has compromised these practices (Chandler *et al.*, 2000). The net result has been extensive within-field emergence of adults that threatens any newly planted maize.

As a substitute management practice, the US Department of Agriculture initiated in 1996 an areawide behavioral management program for western corn rootworm encompassing four states (Chandler *et al.*, 2000). The program involves aerial or ground spraying of maize plants with a bait (SLAM®) consisting of an adult feeding stimulant (buffalo gourd powder) and an insecticide (carbaryl). Far less insecticide is needed when combined with the feeding stimulant than when applied alone.

The performance of bait sprays is markedly affected by the state of the maize crop. As shown by Weissling and Meinke (1991), adult foraging activity is greatest in upper parts of maize plants (where preferred food such as maize silk and maize pollen predominates) and decreases gradually toward the ground. Droplets or particles of bait that adhere to the uppermost foliage of maize plants are much more effective than those adhering to lower parts or falling to the ground. Even when in optimal position, however, bait particles do not compete effectively with natural sources of adult food during the flowering period of maize, thereby rendering ineffective the application of bait sprays during the flowering period (Weissling and Meinke, 1991). Given the lack of any attractant incorporated into SLAM®, it is not surprising that SLAM® applied in alternating sprayed and unsprayed swaths of 23 m is not nearly as effective as total coverage sprays (Chandler, 1998). However, even one of the most attractive semiochemicals known for western corn rootworm females (the plant kairomone 4-methoxycinnamaldehyde) did not prove effective, when combined with feeding stimulant, in luring females just a few meters away, possibly owing to competition from odor of natural food sources (Lance, 1993). The recent discovery of more powerful feeding attractants for western corn rootworm (Hammack, 2001), coupled with further study of adult movement in response to feeding stimuli, could open the way to application of bait sprays in swaths in future years.

Cotton boll weevil

One of the most damaging pests of cotton in Central and North America is the boll weevil, *Anthonomus grandis*. Females deposit eggs in developing bolls of cotton, where larvae feed and damage internal tissue.

Among a wide variety of approaches used to manage this pest, deployment of attracticidal bait sticks is receiving increased attention (Spurgeon *et al.*, 1999). Such devices consist of a vertically emplaced cardboard tube coated with visually attractive, yellow-pigmented clay that is impregnated with a mixture of feeding stimulant (cottonseed oil) and insecticide (malathion). Aggregation pheromone (Grandlure) is affixed to the tube (Spurgeon *et al.*, 1999). When placed outside the edges of cotton fields to intercept adults immigrating from overwintering sites in outlying habitats, bait sticks (or earlier versions of attracticidal traps) are reported to have provided very effective control of low-density boll weevil populations, although not medium- or high-density populations (Leggett *et al.*, 1988; Hardee and Mitchell, 1997).

Wind has recently been found to have a major effect on the performance of Grandlure-baited traps (and presumably therefore bait sticks). As indicated by Sappington and Spurgeon (2000), wind speeds exceeding 5 km/hour or so markedly inhibit upwind movement of boll weevil adults and strongly diminish the probability that adults can track Grandlure to its source. Presence of adjacent woods or brush lines significantly dampens the negative effects of wind on boll weevil response to Grandlure, especially on leeward sides, and can result in threefold greater levels of arrival at traps (Sappington and Spurgeon, 2000). Thus, habitat architecture can affect patterns of wind, which in turn can affect the performance of attracticidal bait sticks for direct control of boll weevils.

The density of boll weevils in cotton fields is in part a product of the degree of survival of adults in overwintering habitats outside of the fields. Recent studies by Pierce and Ellers-Kirk (1999) indicate that in locales such as Texas and New Mexico, where irrigated cotton is grown in desert valleys, the ground beneath the comparatively sparse vegetation serving as overwintering quarters is subject to extreme spring heating and desiccation, which greatly reduces spring survival of overwintered adults. Outlying overwintering habitats of this sort can reduce numbers of immigrating boll weevils to a level conducive to successful use of attracticidal bait sticks for control.

Ambrosia beetles

On the west coast of North America, two species of Ambrosia beetles, *Gnathotrichus sulcatus* and *Trypodendron lineatum*, are major pests of stored logs at sawmills. The adults invade the logs and in the process introduce symbiotic fungi, which serve as the exclusive food source of the beetles but stain the wood, greatly reducing its value in the marketplace (Lindgren, 1990). The adults are initially guided to susceptible logs by kairomones, especially ethanol, and subsequently by aggregation pheromones such as sulcatol or lineatin. In perhaps the most commercially successful program of mass trapping for control of any scolytid, ethanol and aggregation pheromene are deployed in traps positioned strategically at lumberyards to protect susceptible logs (McLean and Borden, 1977; Lindgren and Fraser, 1994). Most invading beetles are captured by the baited traps, but some are drawn to adjacent logs (or slabs), which serve as natural traps and can be processed before damage or population build-up can occur (McLean and Borden, 1977).

The age and condition of stored logs at a sawmill can have a strong influence on trapping protocol and prospects for effective behavioral management. The most preferred logs are those which have been exposed to rainfall for several months after being felled (Kelsey, 1994), largely because water absorbed by logs gives rise to ideal conditions for production of the attractant ethanol (Kelsey and Joseph, 1999). Recently cut logs and those protected from rainfall are far less likely to attract beetles (Kelsey and Joseph, 1999). Adjustment of trap density and trap distribution according to degree of susceptibility of stored logs to beetle attack can enhance the effectiveness of traps for beetle control at sawmills.

Aphids

Worldwide, aphids are important pests of vegetables, primarily due to their ability to transmit viruses that can cause severe reduction in yield. Following takeoff from plants on which nymphal life has been spent, winged aphids respond positively to ultraviolet wavelengths of skylight and may do so for many minutes or even hours (flying tens of kilometers or more) until undergoing a change in physiological state, which results in a reduced tendency to disperse and an increased tendency to alight on plants (Moericke, 1955). At this point, aphids become more prone to respond to longer wavelengths of light reflected by plants than to shorter wavelengths, as from skylight (Kennedy *et al.*, 1961). This basic knowledge of aphid behavior was put to practical use for aphid management by Kring (1964), who showed that incoming aphids could be deterred from alighting on host plants by surrounding plants with aluminum foil that reflected ultraviolet light and stimulated aphids to keep on flying. Because insecticidal control of aphids has often proven ineffective in preventing the spread of vectored plant viruses, the behavioral management approach of protecting plants against alighting aphids using underlain ultraviolet-reflective mulch has become widespread.

The efficacy of this inhibitory or disruptive type of approach for aphid management can be highly dependent on the type of crop grown and the state of its development. Peppers, for example, have received longer protection against virus than have tomatoes, apparently because pepper plants are smaller and permit a greater total amount of ultraviolet reflectance from mulch (Kring and Schuster, 1992). Similarly, squash and pumpkins have received effective protection against aphid transmission of virus for two to four weeks following plant emergence or transplantation but not thereafter, primarily because plant growth eventually obscures ultraviolet reflectance of mulch (Brown *et al.*, 1993; Brust, 2000). Nonetheless, the amount of protection gained was sufficient to markedly enhance yield and be cost effective.

Southern pine beetle

In the southern United States, the southern pine beetle, *Dendroctonus frontalis*, is the most destructive pest of pine forests, attacking and killing standing trees of several different pine species. During local outbreaks, spots of infestation progress in a predictable manner, wherein trees rapidly change status from uninfested to newly infested to brood-producing and finally to dead as the wave of infestation advances through a forest (Payne and Billings, 1989). A successful strategy for managing local infestations involves application of verbenone (an inhibitory, anti-aggregation pheromone) to uninfested trees to disrupt attack at the head of the infestation coupled with felling of nearby infested trees behind the head to disrupt visual orientation of adults emerging from brood-producing trees (Payne *et al.*, 1992; Borden, 1997; Strom *et al.*, 1999).

Although very effective in reducing local populations below levels necessary for mounting successful attacks on uninfested local trees, this strategy does not take into account the ultimate fate of beetles dispersing from suppressed infestations. As shown by Cronin *et al.* (1999), about 10% of beetles emerging from natural untreated infestations disperse 100–500 m in searching for new hosts, and this percentage increases to nearly 40% among beetles emerging from disruptant-treated locales. Experience has shown that such dispersers are too few to pose a threat to uninfested trees within 100–500 m of a treated locale (Payne *et al.*, 1992), but they may contribute significantly to the growth and destructiveness of other populations already established in the vicinity (Cronin *et al.*, 1999). Hence, the density of other populations of conspecifics in the forestscape, coupled with movement capability and propensity of beetles in a disruptant-treated locale, could be decisive in determining the areawide effectiveness of a disruption approach to management of southern pine beetles.

Cherry fruit flies

In parts of Europe, a key pest of cherries is the European cherry fruit fly, *Rhagoletis cerasi*. Adults oviposit into developing cherries, where larvae eat the fruit flesh. Following egglaying, a female deposits a trail of marking pheromone on the fruit surface that deters further oviposition (Katsoyannos, 1975). This marking pheromone has been identified, synthesized and evaluated as a spray application to prevent fly damage to cherry fruit (Aluja and Boller, 1992).

Although at present the marking pheromone is too expensive and not sufficiently rainfast to be recommended for widespread commercial use, there is yet another kind of weather effect that can influence the success of this inhibitory type of approach to cherry fly management. As reported initially by Prokopy (1972) and verified by Roitberg and Prokopy (1983) for *Rhagoletis pomonella* flies (which likewise deposit marking pheromone on host fruit after egglaying), two or three consecutive days of rainy or otherwise unfavorable weather is sufficient to deny any opportunity for oviposition. Meanwhile eggload is enhanced and female readiness to oviposit increased to such an extent that marking pheromone alone (diminished by effects of rainfall) is unable to discourage females and protect fruit against egglaying when the weather again becomes favorable for oviposition. Thus, certain abiotic conditions, via effects on insect physiological state and eggload, can significantly lessen the probability that an approach to pest management, based solely on disruption of behavior, will succeed. Even under continuously favorable weather, however, foraging cherry flies eventually oviposit in pheromone-treated fruit in some locales within orchards if no additional means of management are used (Boller, 1993). This is especially likely to occur in the case of cultivars having prolonged ripening periods, where for some weeks a proportion of the crop is highly susceptible to oviposition.

Onion maggot

In the northern hemisphere, the onion maggot, *Delia antiqua*, is a major pest of cultivated onions. Females oviposit in soil adjacent to onion seedlings. The larvae feed on developing onion tissue, causing decay. Traditionally, insecticides have been used for control, but increasing cases of resistance to insecticides have spurred a search for alternative approaches. The most promising behavior-based alternative, although not yet operational, involves a push-pull approach. In concept, it consists of disrupting oviposition by distributing formulated ovipositional deterrents such as cinnamaldehyde (Cowles *et al.*, 1990) or pine oil (Ntiamoah *et al.*, 1996) on soil next to onion seedlings, coupled with diverting or attracting disrupted females to worthless culled onions (abundant at most commercial operations) (Finch and Eckenrode, 1985; Miller and Cowles, 1990), possibly aided by addition of attractive host volatiles such as dipropyl disulfide (Dindonis and Miller, 1981; Vernon *et al.*, 1981). Culled onions can then either be plowed deeply under the soil or removed from the field and placed in deep piles to prevent adult emergence (Finch and Eckenrode, 1985).

Several factors, discussed by Miller and Cowles (1990) and Cowles and Miller (1992), have been found through experimentation or are hypothesized to affect the outcome of a push-pull approach to controlling onion flies. First, the ultimate level of crop damage cannot exceed 3% to 5% to be economically sustainable. Calculations suggest that for every tenfold increase in population density of females in an onion field, the ratio of difference between the attractive stimulus (culled onions) and the diverting stimulus (ovipositional deterrent) must increase by a similar tenfold level if competition from onion seedlings is to continue to be averted and if economic protection is to be attained. Ultimately, a population of 10 000 females in an onion field (typical for Michigan) would require an attractant:deterrent ratio of about 500:1. Environmental conditions favoring still higher populations of females, such as continuous availability of culled onions resulting from several consecutive years of growing onions on the same ground (Martinson *et al.*, 1988), might be prohibitive. Second, although onion flies are capable of responding to the odor of dipropyl dissulfide at distances of 100 m or more when outside of host patches, host odor may play a lesser role in discovery of individual host plants by flies foraging within a patch (Judd and Borden, 1988). Even so, the comparatively high mobility of most females, capable of moving 100 m or more per day via multiple short flights in an onion field (Martinson *et al.*, 1988), favors discovery of culled onions for oviposition. Third, to be economical to use, deterrents must withstand adverse environmental conditions (high wind and rainfall) throughout the approximately six weeks that onion seedlings must be protected against ovipositing adults. Finally, the spatial arrangement of onion culls in relation to onion seedlings is recognized as important, although not yet studied outside of a large enclosure setting. Perhaps more than any other behavioral pest management approach, a push-pull approach requires intimate understanding of effects of environmental state variables, as well as characteristic movement patterns of pest insects, if it is to succeed.

4.5 Building a conceptual model for approaching behavioral management of apple maggot flies

In the previous section, we discussed a number of environmental parameters that can impact efficacy of behavioral management of pest insects. From our evaluation of several case studies, one point is clear: habitat structure is important to behavioral control efficacy, but every case seems to be unique in how such effects are mediated. Are there any principles or rules that can guide our work or must we start from scratch for each and every pest organism? Added to that difficulty is the fact that habitat structure is a multivariate entity, and it is not obvious which subcomponents are essential to consider for successful behavior-based control.

We contend that the best way to study the influence of habitat structure is from the perspective of the focal organism. Furthermore, by employing concepts from evolutionary biology and life history theory, we can evaluate a seemingly disparate set of structural parameters using a common currency, fitness. In that way, we can ask how pest insect behavior evolves in response to habitat structure, i.e. what is the adaptive nature of pest behavior in habitats with a particular structure (e.g. spatial distribution of resources – Roitberg and Mangel, 1997). Mangel (1994) and Lima and Zollner (1996) provide reviews of this approach that is coined spatial behavioral ecology.

Below, we describe the spatial and temporal behavioral ecology and agricultural importance of the apple maggot, *Rhagoletis pomonella*. We then describe our approach to behavioral control of this pest in an ecological context.

The apple maggot is a native pest of hawthorn that currently attacks a limited number of hosts, including apple, sour cherry, and rose hips. It is univoltine, spends most of its life in pupal dormancy but about 30 days as an adult, has relatively low mobility and fecundity (compared to tropical tephritids), and tends to form relatively stable and permanent populations at local sites. Populations can be found on hawthorn and on feral or abandoned apple trees. It is these populations that cause the greatest concern to apple producers (see below).

Apple maggot females utilize two types of food sources in nature: carbohydrates and proteins that support somatic and gametic function, respectively. Several tactics have been employed by pest managers to exploit fly resource requirements. First, ammonia-producing bait sprays exploit protein-seeking behaviors of flies (Mohammed and Ali Niazee, 1989).

Second, sticky red spheres exploit fly visual search for oviposition sites (Prokopy, 1968). Third, synthetic host fruit odor (e.g. butyl hexanoate) can be added to spheres to increase their attractiveness (Reissig *et al.*, 1982).

There are two ways in which treatments could reduce apple maggot damage: (a) decrease the survival of females, and (b) decrease per capita output (oviposition rate). The approach we describe using odor-enhanced red sphere traps functions primarily through the former, i.e. we attempt to trap females before they cause economic damage to fruit (Prokopy *et al.*, 2000). Thus, our goal is to minimize the time it takes for a fly to find a lethal red sphere trap.

It is highly likely that habitat structure plays a role in the efficacy of odor-enhanced red spheres in controlling apple maggot flies. At the simplest level, the arrangement of apple foliage could reduce/increase efficacy if it makes traps more difficult or easier to locate, respectively. At a higher level, the arrangement of trees could reduce or exacerbate rates of fly movement among trees. A more subtle effect could come into play if high quality fruit are difficult to locate and flies become more willing to accept fruit that they normally reject based upon hardness or sugar content. These examples are just a few among many that illustrate the complexity of interactions between a fruit fly and its environment. To deal with such complexity, we use spatially explicit life history theory to organize experiments and empirical studies (see Roitberg, 2000).

Our approach is predicated on an assumption that apple maggot fly foraging for food and oviposition sites has evolved to maximize lifetime reproductive success, our index of fitness (see Roitberg *et al.*, 2001). Moreover, we assume such behavior has evolved in nature (i.e. in non-agricultural settings) where habitat structure, including the distribution of food and oviposition sites within and among trees, could differ considerably from that found at managed orchards. Thus, as noted in an earlier section, it is essential to initiate behavioral control research by carefully observing pest behavior in a natural environment. The next step in the process is to develop predictive models of pest behavior in nature. Once we have determined how a fly will respond to particular ecological circumstances in nature, we can then alter its environment (e.g. trap numbers and placement) to optimize control. Below, we characterize non-agricultural settings within which apple maggot fly behavior most likely has evolved.

4.5.1 *Life history model*

Our model (see equation 4.1 in section 4.5.1.3) considers how individual females accrue fitness as a function of "decisions", in particular life history and ecological settings. Developing this model requires that we characterize both the individual and its environment.

Characterizing the fly

Rhagoletis pomonella females are described by four states: energy (ζ), protein (π), egg (ε), and age (t). In our model, we define 10 discrete energy states that vary from maximum (9) to minimum (0). We assume that energy is used for maintaining soma, i.e. it does not directly impact gametic function. Thus, when energy state drops to zero or less (0), the individual "starves". Note that we allow energy state to vary continuously and we solve for the impact of non-integer energy state via linear interpolation (Mangel and Clark, 1988).

Protein is assumed to be the key external limiting source for egg production (see VanRanden and Roitberg, 1996). As above, we define 10 discrete protein states that vary from maximum (9) to minimum (0). Also, as above, we allow protein state to vary continuously. It is important to note that protein state can change as a function of two processes: (a) protein is converted to eggs at some rate ($\zeta \leq 0$) so long as egg state is below maximum, and (b) protein satisfaction increases during each feeding act by a value of *v*.

Similarly, egg state (ε) varies on a daily basis from minimum (0) to maximum (18). As above, egg state increases at same rate (μ) as eggs mature so long as protein is available and decreases by one during each oviposition event.

Finally, we characterize each individual fly by her age. We assume that females live up to a maximum of 20 days post-maturation (based on Dean and Chapman, 1973). The likelihood of surviving long enough to die of old age depends on the odds of surviving during particular events that we describe below.

As a final note, we break time into four-minute units: the total time required for a fly to assess, exploit, and pheromone-mark a fruit. Each day of a mature female's life is composed of four sections: morning (07:00–09:00), wherein only food foraging can occur; midday (09:00–17:00), wherein food and oviposition sites can be exploited; late day (17:00–19:00), wherein only food foraging can occur; and night (19:00–07:00), wherein flies rest and mature new eggs if they harbor sufficient protein (based on Prokopy *et al.*, 1972).

Characterizing the habitat

The environment can be characterized in terms of food and oviposition sites. We characterize these resource distributions in a hierarchical manner i.e. at the multi-tree patch (T_h), at the individual tree (T_i), and within-tree levels (T_w). To keep the problem simple, we divide these levels into three categories based upon availability and quality: poor (p), moderate (m) and good (g). For example, a patch of wild trees (T_h) may be mostly bereft (p) of protein-rich food but could still harbor individual trees or parts thereof that harbor substantial fruit resources (g). We assume that flies can assess their overall probability of locating resources based upon the concentration of odor emanating from these sites.

Key events

The final part of the theory employs a description of possible events that a fly might experience, the likelihood that such events might occur, and the outcome of particular fly behaviors as a function of fly state and circumstance. The theory is shown in equation 4.1. There are two major circumstances to consider, i.e. among and within trees. When a fly chooses to abandon a tree, we assume that it makes its decision to alight on another tree based upon tree size, distance, and cultivar. There are two costs to emigration: the energetic cost of flight and the increased risk of predation. Insects in flight are much more obvious and vulnerable than when foraging or resting on trees (see Roitberg and Mangel, 1997). In the dynamic model shown in equation 4.1, the first lines contained within the square bracket express all the events that can happen within a tree along with their probabilities, whereas the last line shows the fitness that an individual can expect from emigrating. The max term indicates that the "remain" vs. "emigrate" decision is made based upon expected lifetime fitness from these two major behavior categories. The decision with the higher pay-off should be chosen.

Within a tree, several mutually exclusive events may occur with particular repercussions. Lines 1 through 8 in equation 4.1 indicate these events, respectively. First, with probability p_{storm}, inclement weather forces the fly to curtail its foraging for the rest of the day. The probability that such an event occurs is 0.0014/time unit or once in 5 days. During that period, protein, energy and egg states change due to intrinsic (i.e. physiological) processes and the new states are indicated by primes. Time advances ϕ units. Second, with probability p_{breeze}, high-velocity wind prevents a fly from foraging for a single time unit. The probability that such an event occurs is 0.08/time unit or 12 times per day. During that period, protein, energy, and egg states change and are indicated by double primes, i.e. these changes over a small time interval differ from those that occur over the longer interval associated with storms. Should no storms nor breezes occur, several kinds of encounters are possible:

1. Nothing encountered – With probability p_{null} neither food nor fruit odors are encountered, time moves ahead 1 unit and internal states change as above.
2. Protein encountered – The fly perceives protein odor nearby with probability p_{pro}. Here we use methodology from stochastic dynamic programming to calculate the lifetime (i.e. future) pay-off for seeking the food site vs. rejecting the site. This is indicated by the second max term in equation 4.1. The first pay-off comes from rejecting the protein and thus is similar in structure to that when nothing is encountered. The second pay-off comes from accepting protein with the concomitant changes in protein state noted by a triple prime. The calculations consider six different kinds of feeding sites that a fly might encounter: high and low concentrations as well as high, medium, and low volumes, with concomitant time costs for exploiting the meal and changes in egg and energy states.
3. Fruit encountered – The fly perceives the presence of fruit at close range. Again, we use stochastic dynamic programming to calculate the lifetime (i.e. future) pay-off for alighting on fruit vs. continuing search. For natural conditions, the calculations consider the different kinds of fruit that a fly might encounter (immature, mature uninfested vs. mature infested) in a given habitat, with probability of encounter being p_{imm}, p_{hfru} and p_{ifru}, respectively. These probabilities will vary across sites and at different times of the season but can be described in a site-specific manner (see Roitberg and Mangel, 1997 for tephritid flies in wild rose habitats). When flies encounter fruit, they will invest time assessing fruit and possibly also ovipositing and pheromone-marking fruit should the fruit prove acceptable. To solve for the evolutionary implications, we must calculate the fitness pay-off from ovipositing. For example, flies that oviposit in an uninfested mature apple at a wild site would have a very high chance of their eggs and subsequent larvae surviving, giving a fitness increment f_h of 0.9, whereas oviposition in immature or infested mature fruit of the same variety might give lower fitness increments, f_{im} and f_i, of 0.2 and 0.3, respectively. These later two pay-off values are due to low larval survival caused by deleterious plant secondary compounds and/or low levels of essential nutrients and by larval competition within fruit,

respectively. These values will vary depending upon the variety and size of the host fruit as well as host species (e.g. hawthorn vs. apple). We assume that acceptance of a fruit for oviposition will engender a time cost of 1 unit and an egg cost of 1. As above, the decision to accept an infested fruit is predicted when acceptance of low quality fruit enhances lifetime reproductive success (van Alphen and Visser, 1990). This decision to accept or reject is indicated by the max term. This event can occur only during the middle part of the day.

4. Protein and fruit encountered – Here a fly encounters protein and fruit stimuli simultaneously with probability p_{both}. Again, we use stochastic dynamic programming to consider pay-offs from three mutually exclusive responses: reject both, seek food, and seek fruit. Calculations are made as above.

All of the state variables, events, and repercussions can be assembled into the dynamic optimization equation as shown in equation 4.1.

$$
\begin{aligned}
&F(\xi,\pi,\varepsilon,t,T) \\
&= \max \begin{bmatrix}
p_{storm}\rho^{\phi}F(\xi',\pi',\varepsilon',t+\phi,T)+ \\
p_{breeze}\rho F(\xi'',\pi'',\varepsilon'',t+1,T)+ \\
p_{null}\rho F(\xi'',\pi'',\varepsilon'',t+1,T)+ \\
p_{pro}\max\{\rho F(\xi'',\pi'',\varepsilon''t+1,T);\ \rho F(\xi'',\pi''',\varepsilon'',t+1,T)\}+ \\
p_{hfru}f_h+\rho F(\xi'',\pi'',\varepsilon''-1,t+1,T)+ \\
p_{ifru}\max\{f_i+\rho F(\xi'',\pi'',\varepsilon''-1,t+1,T);\ \rho F(\xi'',\pi'',\varepsilon,t+1,T)\}+ \\
p_{imm}\max\{fimm+\rho F(\xi'',\pi'',\varepsilon''-1,t+1,T);\ \rho F(\xi'',\pi'',\varepsilon'',t+1,T)\} \\
p_{both}\max\{ff+\rho F(\xi'',\pi'',\varepsilon''-1,t+1,T);\ \rho F(\xi'',\pi''',\varepsilon'',t+1,T); \\
\qquad \rho F(\xi'',\pi'',\varepsilon'',t+1,T)\}
\end{bmatrix}; \\
&\rho\delta^{\theta}F\left(\xi^{iv},\pi^{iv},\varepsilon^{iv},t+\delta,T\right)
\end{aligned}
\tag{4.1}
$$

The equation is solved via a technique referred to as backwards induction (Mangel and Clark, 1988). The procedure yields a truth table that provides optimal response for all possible combinations of state variable values. In other words, the table provides deterministic solutions for organisms that live in stochastic worlds. Another way to visualize this table is as a kind of policy based upon simple rules, i.e. under this set of conditions do this, but under another set do that. Recall, however, that this policy is for fly behavior in the wild. Thus, such rules will have evolved in respect of resource distributions found in natural settings. Again, this tells us that it is extremely important to elucidate habitat structure in nature. For example, host fruit may be highly patchily distributed within clusters of wild apple trees and lead to a very different emigration "policy" than if such fruit were to be uniformly distributed.

There is still one critical element that requires development before our approach can be employed for agricultural systems. In the next section, we utilize the aforementioned in-nature truth table in a series of forward iterations.

4.5.2 *Forward iterations*

Now that we have a set of conditional behaviors that optimize fly performance in nature, we can ask how flies might behave in orchards and how managers could exploit their behavior. We do this by simulating individual flies in orchards from the moment they arrive until death ensues. Individuals may or may not encounter resource cues during each foraging period, as determined by random draw from an encounter distribution. At each point, each fly is assumed to follow the policy from the aforementioned truth table. There are two important aspects of within-population variability to keep in mind. Individual flies may arrive at orchards with different energy and different protein states. Thus, even if they were to perceive the same cues, they might respond differently as a function of difference in position in the truth table. Second, as a result of stochastic events, two flies of identical condition that enter the orchard could soon deviate from one another on the basis of chance and response. For example, imagine that two identical energy-deprived flies forage within a tree wherein one fly locates high-energy food while the other fails to find food. Soon thereafter, both flies encounter fruit odor. Under such conditions, the former fly might attempt to maximize reproductive success by seeking fruit whereas the latter should reject fruit to avoid risking starvation.

4.5.3 *The orchard scenario*

Here we deal specifically with immigrant flies. We do so for two reasons. First, in managed orchards, immigrant flies pose a much higher threat numerically than do residents (Prokopy *et al.*, 1996). Second, immigrant flies are likely to differ from residents in terms of physiological state and spatial distribution. As a result, behavioral exploitation of these two fly classes will depend upon different suites of tactics as described below. The same procedure can be used to determine management policy for resident flies, but the parameter values would obviously differ.

Immigrant flies arrive at the periphery of a commercial orchard having abandoned their wild sites. Fly arrival rate will be a function of two key parameters: distance from wild sites and structure of surrounding habitat. All else being equal, fly arrival rates can be ranked from low to high for open fields, hedgerows, and open forest, respectively (Hoffman *et al.*, 2002).

These habitat-specific arrival patterns permit us to refine our goal of minimizing "arrival time to trap" in a spatially explicit manner. In particular, the focus now is on perimeter-row trees of an orchard. If ringing perimeter rows of orchards with odor-baited red spheres prevents females from penetrating into interior rows, then managers can focus their efforts on a subset of the agroecosystem. As such, savings from limited environmental disturbance and direct dollar investment can be substantial.

For perimeter-row trapping to be effective, perimeter trees must cause flies first to alight on perimeter trees and second to remain long enough to ensure capture on traps. Here, we return to our life history theory to evaluate management options. Recall that females trade off search for food and oviposition sites and that fly response to particular resource cues should depend upon fly physiological state. These responses will ultimately impact residence times (e.g. Prokopy *et al.*, 1994) and thus likelihood of capture by traps. Rull and Prokopy (2000) found that the vast majority of immigrant females arriving at perimeter-row apple trees has mature eggs. This is not surprising. If protein were limiting at wild sites, we would not expect that flies would be drawn to orchards by host fruit odor or its synthetic equivalent.

Our truth table predicts that fruit-deprived but not protein-deprived flies should respond most actively to fruit odor. Indeed, Rull and Prokopy (2000) found that food odors did not enhance trap capture rates in orchards. Thus, we choose to line-treat perimeter-row apple trees with fruit-odor-baited traps to attract/arrest immigrants.

The next step is to determine the density of traps to use on perimeter-row trees. Recall that our goal is to trap flies before they exit perimeter-row trees and move to the interior of the orchard. For each additional trap added to the perimeter row, apple crop value should increase by some amount as a function of reduced infestation rate of interior trees. The optimal trap density is that wherein the marginal value equals the marginal cost (Hindelang and Dascher, 1985) (Figure 4.1). The marginal cost line has a slope of zero, i.e. it has the same incremental cost for each trap added. The marginal value curve will be determined by degree of reduction in oviposition rate as trap density increases and depends upon degree of fly response to trap odor. The best way to calculate these curves is from our forward iterations for different parameter combinations (Roitberg, 2004). Key parameters include tree size (e.g. Rull and Prokopy, 2001), tree spacing (Green *et al.*, 1994; Rull and Prokopy, 2001) and tree cultivar (Hoffmann *et al.*, 2002).

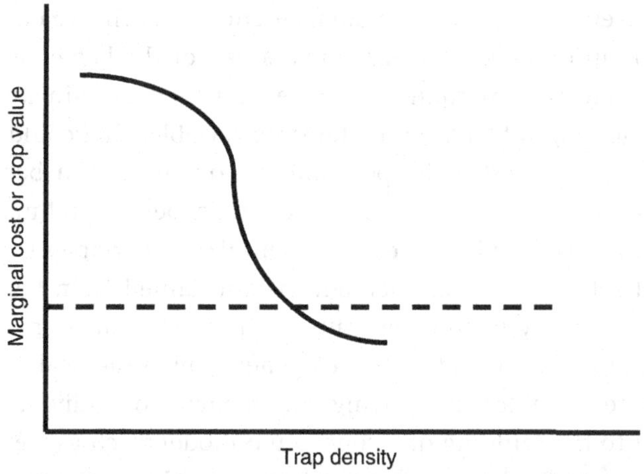

Figure 4.1. Hypothetical marginal cost (dashed) and marginal crop value (solid) curves for trap deployment tactics in managing apple maggot flies. The marginal cost line is linear and parallel to the x-axis because each additional trap increases cost by an equal increment. By contrast, the marginal crop value line is curvilinear. The initial portion of this curve is nearly flat, indicating that each additional trap reduces damage by decreasing but very similar amounts. At the midway point, there is a great drop off in further increments to crop value when more traps are added. This indicates that there is little additional benefit to adding more traps because, for example, fly preference for artificial versus natural fruit odor changes as a function of artificial odor source density. The optimal density of traps occurs where the two curves intersect.

Unfortunately, response curves for insects are notoriously non-linear and frequently multidimensional (Roitberg, 2000). This makes the preceding marginal value curve extremely difficult to determine empirically. As we move from orchard to orchard, with concomitant change in parameters, it always becomes a fresh challenge to assign a trap deployment protocol that provides effective if not optimal control. Even so, an ecologically based life history theory provides the necessary flexibility for establishing relevant trap deployment protocols. Thus, we return to our initial conjecture that any behavior-based management program depends upon knowledge of the structure of the pest's environment and how that structure impacts pest fitness.

4.6 Conclusion

In this chapter we have attempted to relate ecological theory to behavioral management of pest arthropods by focusing on ways in which the state of the environment (a key component of population dynamic theory) can affect the success of a behavioral management approach. We emphasize that to engage in an organized observational study of the behavior of pest individuals in a natural environment is the best way to gain an initial appreciation of the range of environmental state variables that could impact upon the behavior of a pest arthropod and the outcome of a behavioral management program. Follow-up studies involving experimental manipulation of essential resources and other environmental state variables in natural and seminatural settings can complement insight gained from behavioral observations in nature by uncovering cause and effect relationships. Such relationships, along with consideration of genetic, physiological-state and informational-state variables influencing the behavior of individuals, can then be woven into the building of a conceptual model (such as we present here for apple maggot flies) for approaching behavioral management of the pest in question.

To illustrate the variety of ways in which the state of the environment might influence the outcome of a behavioral management program, we have drawn upon studies of ten different insects (apart from apple maggot flies) that have been the target of behavioral management. For some of the insects (Mediterranean fruit flies, tsetse flies, and mountain pine beetles), the proximate aim has been to suppress populations at distant sites of population origin, with the ultimate aim of ensuring that emigrants are too few to cause injury elsewhere. For other insects (western corn rootworms, cotton boll weevils, ambrosia beetles, aphids, southern pine beetles, fruit flies, and onion maggot flies), the overarching aim has been to provide on-site protection of valued resources.

Studies on the insects highlighted here have revealed a broad range of environmental state factors that have been found to affect the success of behavioral management. These include the biotic factors of quality or structure of the resource items targeted for protection, quality, or structure of competing non-targeted resource or non-resource items in the habitat, and the density of conspecifics. They also include the abiotic factors of rainfall, wind, direct sunlight, and temperature.

Perhaps the most useful theme to emerge from our endeavor in constructing this chapter is that there is in fact no one theme which characterizes the nature of the impact that environmental state variation can have on the success of behavioral management of any given pest arthropod. To invoke the sentiments of Lawton (1992) noted in the Introduction, at a very basic level each species of animal or plant has its own peculiar kind of population dynamics that characterizes the rises and falls of its populations. In our judgement, so also does each species of pest arthropod exhibit its own peculiar set of behavioral and biological traits and each is impacted by environmental variables in its own peculiar way. Coping with these peculiarities, especially in conjunction with a broad range of dynamic economic, social, cultural, and legal variables that can positively or negatively impact any given approach to managing a pest arthropod, is a challenge to those who engage in behavioral pest management.

References

Aluja, M. and Boller, E. (1992). Host marking pheromone of *Rhagoletis cerasi*: field deployment of synthetic pheromone as a novel cherry fruit fly management strategy. *Entomologia Experimentalis Applicata*, **65**, 141–7.

Bell, W. J. (1991). *Searching Behaviour: The Behavioural Ecology of Finding Resources*. London: Chapman and Hall.

Bernays, E. (2000). *Plant–Insect Interactions: A Synthesis*. Abstract Book I: XXI International Congress of Entomology, Brasilia, Brazil; Bjostad, published at Washington, DC, USA. pp. 8–13.

Bjostad, L. B., Hibbard, B. and Cranshaw, W. (1993). Application of semiochemicals in integrated pest management programs. In S. D. Duke (ed.), *Pest Control with Enhanced Environmental Safety*. American Chemical Society. pp. 199–218.

Boller, E. (1993). Current research on fruit fly host marking pheromones. In M. Aluja and P. Liedo (ed.), *Fruit Flies: Biology and Management*. New York: Springer-Verlag. pp. 195–8.

Borden, J. (1989). Semiochemicals and bark beetle populations: exploitation of natural phenomena by pest management strategists. *Holarctic Ecology*, **12**, 501–10.

Borden, J. (1993). Strategies and tactics for the use of semiochemicals against forest insect pests in North America. In B. A. Leonhardt and J. L. Vaughn (eds.), *Pest Management: Biologically Based Technologies*. American Chemical Society, Washington, DC, USA.

Borden, J. H. (1997). Disruption of semiochemical-mediated aggregation in bark beetles. In R. T. Carde and A. K. Minks (eds.), *Insect Pheromone Research: New Directions*. New York: Chapman & Hall. pp. 421–38.

Borden, J. H., Chong, L. J. and Fuchs, M. C. (1983). Application of semiochemicals in post-logging manipulation of the mountain pine beetle, *Dendroctonus ponderosae*. *Journal of Economic Entomology*, **76**, 1428–32.

Brown, J. F., Dangler, J. M., Woods, F. M. *et al.* (1993). Delay in mosaic virus onset and aphid vector reduction in summer squash grown on reflective mulches. *Hortscience*, **28**, 895–6.

Brust, G. E. (2000). Reflective and black mulch increase yield in pumpkins under virus disease pressure. *Journal of Economic Entomology*, **93**, 828–33.

Carde, R. T. and Minks, A. K. (1997). *Insect Pheromone Research: New Directions*. New York: Chapman & Hall.

Chandler, L. D. (1998). Comparison of insecticide bait aerial application methods for management of corn rootworm. *Southwestern Entomologist*, **23**, 147–59.

Chandler, L.D., Coppedge, J.R., Edwards, C.R., Tollefson, J.J. and Wilde, G.E. (2000). Corn rootworm area-wide management across the United States. In K.H. Tan (ed.), *Area-Wide Control of Fruit Flies and Other Insect Pests*. Malaysia: Penang, Penerbit Universiti Sains. pp. 159–67.

Cowles, R.S. and Miller, J.R. (1992). Diverting *Delia antiqua* oviposition with cull onions: field studies on planting depth and a greenhouse test of the stimulo-deterrent concept. *Environmental Entomology*, **21**, 453–60.

Cowles, R.S., Miller, J.R., Hollingworth, R.M. *et al.* (1990). Cinnamyl derivatives and monoterpenoids as nonspecific ovipositional deterrents of the onion fly. *Journal of Chemical Ecology*, **16**, 2401–28.

Cronin, J.T., Turchin, P., Hayes, J.L. and Steiner, C.A. (1999). Area-wide efficacy of a localized forest pest management practice. *Environmental Entomology*, **28**, 496–504.

Cunningham, R.T., Nakagawa, S., Suda, D.Y. and Urago, T. (1978). Tephritid fruit fly trapping: liquid food baits in high and low rainfall climates. *Journal of Economic Entomology*, **71**, 762–3.

Dean, R.W. and Chapman, P.J. (1973). Bionomics of the apple maggot in eastern New York. *Search Agriculture* (Geneva, N.Y.), **3**, 1–64.

Dindonis, L. and Miller, J.R. (1981). Host finding responses of onion and seedcorn flies to healthy and decomposing onions and several synthetic constituents of onions. *Environmental Entomology*, **9**, 467–72.

Finch, S. and Collier, R.H. (2000). Host-plant selection by insects – a theory based on 'appropriate/inappropriate landings' by pest insects of cruciferous plants. *Entomologia Experimentalis Applicata*, **96**, 91–102.

Finch, S. and Eckenrode, C.J. (1985). Influence of unharvested, cull-pile and volunteer onions on populations of onion maggot. *Journal of Economic Entomology*, **78**, 542–6.

Foster, S.P. and Harris, M.O. (1997). Behavioral manipulation methods for insect pest management. *Annual Review of Entomology*, **42**, 123–46.

Gray, D.R. and Borden, J.H. (1989). Containment and concentration of mountain pine beetle infestations with semiochemicals: validation by sampling of baited and surrounding zones. *Journal of Economic Entomology*, **82**, 1399–405.

Green, T.A., Prokopy, R.J. and Hosmer, D.W. (1994). Distance of response to host tree models by female apple maggot flies, *Rhagoletis pomonella* (Walsh) (Diptera, Tephritidae) – interaction of visual and olfactory stimuli. *Journal of Chemical Ecology*, **20**, 2393–413.

Griffiths, N. and Brady, J. (1995). Wind structure in relation to odor plumes in tsetse fly habitats. *Physiological Entomology*, **20**, 286–92.

Hammack, L. (2001). Single and blended maize volatiles as attractants for diabroticite corn rootworm beetles. *Journal of Chemical Ecology*, **27**, 1373–90.

Hardee, D.D. and Mitchell, E.R. (1997). Boll weevil: a summary of research on behavior as affected by chemical communication. *Southwestern Entomologist*, **22**, 466–91.

Hardie, J. and Minks, A.K. (1999). *Pheromones of Non-Lepidopteran Insects Associated with Agricultural Plants*. Wallingford, UK: CAB International.

Hindelang, T.J. and Dascher, P. (1985). *The IBM/PC Guide to Marginal Analysis*. Wayne, PA: Banbury Books.

Hoffmann, S., Mittenthal, R., Chandler, B. *et al.* (2002). Influence of odor bait, cultivar type and adjacent habitat composition of performance of perimeter traps for controlling apple maggot flies. *Fruit Notes*, **67**, 20–4.

Hokanen, H.M.T. (1991). Trap cropping in pest management. *Annual Review of Entomology*, **36**, 119–38.

Houston, A., Clark, C., McNamara, J. and Mangel, M. (1988). Dynamic models in behavioral and evolutionary ecology. *Nature*, **332**, 29–34.

Howell, J.F., Cheikh, M., Harris, E.J. *et al.* (1975). Mediterranean fruit fly: control in Tunisia by strip treatment with a bait spray of technical malathion and protein hydrolysate. *Journal of Economic Entomology*, **68**, 247–9.

Hunter, M.D. and Price, P.W. (1992). Natural variability in plants and animals. In M.D. Hunter, T. Ohgushi and P.W. Price (eds.), *Effects of Resource Distribution on Animal–Plant Interactions*. New York: Academic Press. pp. 1–12.

Judd, G.J.R. and Borden, J.H. (1988). Long-range host-finding behavior of the onion fly, *Delia antiqua* (Diptera, Anthomyiidae) – ecological and physiological constraints. *Journal of Applied Ecology*, **25**, 829–45.

Kareiva, P. (1983). Influence of vegetation texture on herbivore populations: resource concentration and herbivore movement. In R.F. Denno and M.S. McClure (eds.), *Variable Plants and Herbivores in Natural and Managed Systems*. London: Academic Press. pp. 259–89.

Katsoyannos, B.I. (1975). Oviposition-deterring, male-arresting, fruit marking pheromone in *Rhagoletis cerasi*. *Environmental Entomology*, **4**, 801–7.

Kelsey, R.G. (1994). Ethanol and ambrosia beetles in Douglas fir logs with and without branches. *Journal of Chemical Ecology*, **20**, 3307–19.

Kelsey, R.G. and Joseph, G. (1999). Ethanol and ambrosia beetles in Douglas fir logs exposed or protected from rain. *Journal of Chemical Ecology*, **25**, 2793–809.

Kennedy, G.G. and Storer, N.P. (2000). Life systems of polyphagous arthropod pests in temporally unstable cropping systems. *Annual Review of Entomology*, **45**, 467–93.

Kennedy, J.S., Booth, C.O. and Kershaw, W.J.S. (1961). Host finding by aphids in the field. III. Visual attraction. *Annals of Applied Biology*, **49**, 1–21.

Kogan, M. (1994). Plant resistance in pest management. In R.L. Metcalf and W.H. Luckmann (eds.), *Introduction to Insect Pest Management*. New York: Wiley & Sons. pp. 73–128.

Kring, J.B. (1964). New ways to repel aphids. Connecticut Agricultural Experiment Station. *Frontiers of Plant Science*, **17**, 6–7.

Kring, J.B. and Schuster, D.J. (1992). Management of insects on pepper and tomato with UV-reflective mulches. *Florida Entomologist*, **75**, 119–29.

Lance, D.R. (1993). Effects of a nonpheromonal attractant on movement and distribution of adult *Diabrotica virgifera*. *Environmental Entomology*, **22**, 654–62.

Landis, D.A., Wratten, S.D. and Gurr, G.M. (2000). Habitat management to conserve natural enemies of arthropod pests in agriculture. *Annual Review of Entomology*, **45**, 175–201.

Lawton, J.H. (1992). There are not 10 million kinds of population dynamics. *Oikos*, **63**, 337–8.

Lawton, J.H. (1999). Are there general laws in ecology? *Oikos*, **84**, 177–92.

Leak, G.A. (1999). *Tsetse Biology and Ecology*. Wallingford, UK: CAB International.

Leggett, J.E., Dickerson, W.A. and Lloyd, E.P. (1988). Suppressing low level boll weevil populations with traps: influence of trap placement, grandlure concentration and population level. *Southwestern Entomologist*, **13**, 205–16.

Lima, S. and Zollner, P.A. (1996). Towards a behavioral ecology of ecological landscapes. *Trends in Ecology and Evolution*, **11**, 131–5.

Lindgren, B.S. (1990). Ambrosia beetles. *Journal of Forestry*, **88**, 8–11.

Lindgren, B.S. and Fraser, R.G. (1994). Control of ambrosia beetle damage by mass trapping at a dryland log sorting area in British Columbia. *Forestry Chronicle*, **70**, 159–63.

Logan, J.A. and Bentz, B.J. (1999). Model analysis of mountain pine beetle personality. *Environmental Entomology*, **28**, 924–34.

Mangel, M. (1994). Spatial patterning in resource exploitation and conservation. *Philosophical Transactions of the Royal Society of London B*, **343**, 93–8.

Mangel, M. and Clark, C.W. (1986). Towards a unified foraging theory. *Ecology*, **67**, 1127–34.

Mangel, M. and Clark, C.W. (1988). *Dynamic Modeling in Behavioral Ecology*. Princeton, NJ: Princeton University Press.

Mangel, M. and Roitberg, B.D. (1989). Dynamic information and host acceptance by a tephritid fruit fly. *Ecological Entomology*, **14**, 181–9.

Margolies, D.C. (1993). Adaptation to spatial variation in habitat: spatial effects inagroecosystems. In K.C. Kim and B.A. McPheron (ed.), *Evolution of Insect Pests*. New York: Wiley & Sons. pp. 129–44.

Martinson, T.E., Nyrop, J.P. and Eckenrode, C.J. (1988). Dispersal of the onion fly and larval damage in rotated onion fields. *Journal of Economic Entomology*, **81**, 508–14.

McClean, J.A. and Borden, J.H. (1977). Suppression of *Gnathotrichus sulcatus* with sulcatol-baited traps in a commercial sawmill and notes on the occurrence of *G. retusus* and *Trypodendron lineatum*. *Canadian Journal of Forestry*, **7**, 348–56.

McQuate, G.T. and Peck, S.L. (2000). Suppression of Mediterranean fruit fly populations over mountainous areas through aerial phloxine B-protein bait sprays: regional medfly program in Guatemala. In K.H. Tan (ed.), *Area-Wide Control of Fruit Flies and Other Insect Pests*. Penang, Malaysia: Penerbit Universitati Sains. pp. 409–18.

Miller, J.R. and Cowles, R.S. (1990). Stimulo-deterrent diversion: a concept and its possible application to onion maggot control. *Journal of Chemical Ecology*, **11**, 3197–212.

Minkenberg, O.P.J.M., Tatar, M. and Rosenheim, J.A. (1992). Egg load as a major source of variability in insect foraging and oviposition behavior. *Oikos*, **65**, 134–42.

Moericke, V. (1955). Uber das Lebensgewohnheiten der geflugelten Blattlause. *Journal of Applied Entomology*, **37**, 29–91.

Mohammed, A. B. and Ali Niazee, M. T. (1989). Malathion bait sprays for control of apple maggot. *Canadian Entomologist*, **92**, 464–7.

Muzari, M. O. (1999). Odour-baited targets as invasion barriers for tsetse flies: a field trial in Zimbabwe. *Bulletin of Entomological Research*, **89**, 73–7.

Ntiamoah, Y. A., Borden, J. H. and Pierce, H. D. (1996). Identity and bioactivity of oviposition deterrents in pine oil for the onion maggot, *Delia antiqua*. *Entomologia Experimentalis Applicata*, **79**, 219–26.

Payne, T. L. and Billings, R. F. (1989). Evaluation of s-verbenone applications for suppressing southern pine beetle infestations. *Journal of Economic Entomology*, **82**, 1702–8.

Payne, T. L., Billings, R. F., Berisford, C. W. *et al.* (1992). Disruption of *Dendroctonus frontalis* infestation with an inhibitor pheromone. *Journal of Applied Entomology*, **114**, 341–7.

Paynter, Q. and Brady, J. (1992). Flight behavior of tsetse flies in thick bush (*Glossina pallipides*). *Bulletin of Entomological Research*, **82**, 513–16.

Pierce, J. B. and Ellers-Kirk, C. (1999). Overwintering habitat influence on boll weevil trap captures in New Mexico. *Southwestern Entomologist*, **24**, 183–92.

Preissler, H. K. and Mitchell, R. G. (1993). Colonization patterns of the mountain pine beetle in thinned and unthinned lodgepole pine stands. *Forest Science*, **39**, 528–45.

Prokopy, R. J. (1968). Visual responses of apple maggot flies; *Rhagoletis pomonella*; orchard studies. *Entomologia Experimentalis Applicata*, **11**, 403–22.

Prokopy, R. J. (1972). Evidence for a marking pheromone deterring repeated oviposition in apple maggot flies. *Environmental Entomology*, **1**, 326–32.

Prokopy, R. J. and Kogan, M. (2003). Integrated pest management. In V. H. Resh and R. T. Carde (eds.), *Encyclopedia of Insects*, New York: Academic Press. pp. 4–9.

Prokopy, R. J., Bennett, E. W. and Bush, G. L. (1972). Mating behavior in *Rhagoletis pomonella*. II. Temporal organization. *Canadian Entomologist*, **104**, 97–104.

Prokopy, R. J., Papaj, D. R., Hendrichs, J. and Wong, T. T. (1992). Behavioral responses of *Ceratitis capitata* flies to bait spray droplets and natural food. *Entomologia Experimentalis Applicata*, **64**, 247–57.

Prokopy, R. J., Cooley, S. S., Prokopy, J. J., Quan, Q. and Buonaccorsi, J. P. (1994). Interactive effects of resource abundance and state of adults on residence of apple maggot flies in host tree patches. *Environmental Entomology*, **23**, 304–15.

Prokopy, R. J., Mason, J. L., Christie, M. and Wright, S. E. (1996). Arthropod pest and natural enemy abundance under second-level versus first-level integrated pest management practices in apple orchards: a 4-year study. *Agriculture, Ecosystems and Environment*, **57**, 35–47.

Prokopy, R. J., Wright, S. E., Black, J. L., Hu, X. P. and McGuire, M. R. (2000). Attracticidal spheres for controlling apple maggot flies: commercial orchard trials. *Entomologia Experimentalis Applicata*, **97**, 293–9.

Raffa, K. F. (2001). Mixed messages across multiple trophic levels: the ecology of bark beetle chemical communication systems. *Chemoecology*, **11**, 49–65.

Raffa, K.F., Phillips, T.W. and Salom, S.M. (1993). Strategies and mechanisms of host colonization by bark beetles. In T. Schowalter and G. Filip (eds.), *Beetle-Pathogen Interactions in Conifer Forests*. San Diego, CA: Academic Press. pp. 102–28.

Reissig, W.H., Fein, B.L. and Roelofs, W.L. (1982). Field tests of synthetic apple volatiles as apple maggot attractants. *Environmental Entomology*, **11**, 1294–8.

Roessler, Y. (1989). Insecticidal bait and cover sprays. In A.S. Robinson and G. Hooper (eds.), *Fruit Flies: Their Biology, Natural Enemies and Control*. Amsterdam: Elsevier. pp. 329–35.

Roitberg, B. (2004). From parasitoid behaviour to biological control: Applied behavioural ecology. *Canadian Entomologist*, **136**, 289–97.

Roitberg, B. (2000). Threats, flies and protocol gaps. Can behavioral ecology save biological control? In A. Ives and M. Hochberg (eds.), *Parasite Population Biology*. Princeton, NJ: Princeton University Press. pp. 254–65.

Roitberg, B., Boivin, G. and Vet, L. (2001). Fitness, parasitoids and biological control. *Canadian Entomologist*, **133**, 429–38.

Roitberg, B. and Mangel, M. (1997). Individuals on the landscape: behavior can mitigate differences among habitats. *Oikos*, **80**, 234–40.

Roitberg, B.D. and Prokopy, R.J. (1983). Host deprivation influence on response of *Rhagoletis pomonella* to its oviposition-deterring pheromone. *Physiological Entomology*, **8**, 69–72.

Rull, J. and Prokopy, R.J. (2000). Attraction of apple maggot flies, *Rhagoletis pomonella* (Diptera: Tephritidae), of different physiological states to odour-baited traps in the presence and absence of food. *Bulletin of Entomological Research*, **90**, 77–88.

Rull, J. and Prokopy, R.J. (2001). Effect of apple orchard structure on interception of *Rhagoletis pomonella* flies by odor-baited traps. *Canadian Entomologist*, **133**, 355–63.

Sappington, T.W. and Spurgeon, D.W. (2000). Variation in boll weevil captures in pheromone traps arising from wind speed moderation by brush lines. *Environmental Entomology*, **29**, 807–14.

Southwood, T.R.E. (1977). Habitat, the template for ecological strategies? *Journal of Animal Ecology*, **46**, 337–65.

Spurgeon, D.W., Coppedge, J.B., Raulston, J.R. and Marshall, H. (1999). Mechanisms of boll weevil bait stick activity relative to pheromone traps. *Journal of Economic Entomology*, **92**, 960–6.

Steiner, L.F. (1969). Control and eradication of fruit flies on citrus. *Proceedings of First International Citrus Symposium*, **2**, 881–8.

Strom, B.L., Raton, L.M., Goyer, R.A. and Meeker, J.R. (1999). Visual and semiochemical disruption of host finding in the southern pine beetle. *Ecological Applications*, **9**, 1028–38.

Vale, G.A. (1998). Responses of tsetse flies to vegetation in Zimbabwe: implications for population distribution and bait siting. *Bulletin of Entomological Research*, **88** (Supplement 1), 1–59.

Vale, G.A., Lovemore, D.F., Flint, S. and Cockbill, G.F. (1988). Odour-baited targets to control tsetse flies, *Glossina* spp., in Zimbabwe. *Bulletin of Entomological Research*, **78**, 31–49.

van Alphen, J. J. M and Visser, M. E. (1990). Superparasitism as an adaptive strategy for insect parasitoids. *Annual Review of Entomology*, **35**, 59–79.

VanRanden, E. and Roitberg, B. (1996). Eggload affects superparasitism in the snowberry fly. *Entomologia Experimentalis Applicata*, **79**, 241–5.

Vernon, R. S., Judd, G. J. R., Borden, J. H., Pierce, H. D. and Oehlschalager, A. C. (1981). Attraction of *Hylemya antiqua* in the field to host-produced oviposition stimulants and their non-host analogues. *Canadian Journal of Zoology*, **59**, 872–81.

Weissling, T. J. and Meinke, L. J. (1991). Semiochemical-insecticide bait placement and vertical distribution of corn rootworm adults: implications for management. *Environmental Entomology*, **20**, 945–52.

Williams, B., Dransfield, R. and Brightwell, R. (1992). The control of tsetse flies in relation to fly movement and trapping efficiency. *Journal of Applied Ecology*, **29**, 163–79.

Yuval, B. and Hendrichs, J. (2000). Behavior of flies in the genus *Ceratitis*. In M. Aluja and A. L. Norrbom (eds.), *Fruit Flies: Phylogeny and Evolution*. Boca Raton, FL: CRC Press. pp. 429–58.

5

Using pheromones to disrupt mating of moth pests

R.T. CARDÉ

5.1 Introduction

Many kinds of insects rely on sex pheromones to locate mates, with attraction often occurring over distances of ten meters or more. The prospect of achieving direct population control of insect pests by applying synthetic copies of these attractants to a crop has long intrigued entomologists and chemists. Because odor communication is mediated by miniscule quantities of pheromone, it was imagined that application of relatively small amounts of synthetic pheromone, perhaps a fraction of a gram of pheromone per ha per day, would readily interfere with mate location by "confusing" the responders. But testing the feasibility of mating disruption requires synthetic copies of these chemical messages, and it was not until the late 1960s and early 1970s that advances in techniques for characterizing pheromone structures facilitated the identification of the pheromones of many of the world's most damaging pest insects. The availability of synthetic pheromones in turn enabled field testing of this method of pest control, beginning with Gaston *et al.* (1967).

Female-emitted pheromones that induce mate-location behaviors in males are now known for hundreds of moth species (Arn *et al.*, 2000), and for insects in many other groups (Mayer and MacLaughlin, 1991). Although some moth species have pheromones consisting of a single compound, most moth pheromones are blends of two to four components. The majority of compounds utilized are unbranched alkyl chains, 10 to 18 carbons in length, most always with an even number of carbons, one or two double bonds, and with a terminal acetate, alcohol, or aldehyde moiety. Compounds of this type are abbreviated here with the following notation system: 14:Ac = tetradecyl acetate; Z11–14:Ald = (*Z*)-11-tetradecenal; and E8, E10–12:OH = (*E,E*)-8, 10-dodecadien-1-ol. References to the identification of moth pheromones and pheromone antagonists considered in this review can be found in the web-based *Pherolist* of Arn *et al.* (2000).

Pheromones need to be formulated to prevent degradation in the field and to release them slowly into the atmosphere. A variety of formulations has enabled entomologists to undertake field evaluations of their efficacy in mating location and ultimately crop protection. Nearly all of the successful demonstrations of mating disruption have been with moth species and it is on this group that this review will focus. Reviews of the development of this method and the underlying principles of behavioral manipulation include those of Roelofs (1979); Bartell (1982); Minks and Cardé (1988); Cardé (1990); Cardé and Minks (1995); Sanders (1996); and Ogawa (2000). Comprehensive reviews of mating disruption also are provided in volumes by Mitchell (1981); Kydonieus and Beroza (1982); Justum and Gordon (1989); Ridway *et al.* (1990); and Howse *et al.* (1998).

Generally, mating disruption has many advantages: (a) it is highly specific to the target pest; (b) it is safe for workers; (c) it does not interfere with biological control agents; and (d) it is non-damaging to the environment. Its great specificity to the target species, however, also means that cost of application must be justified by the economic damage caused by the particular pest against which the disruptant treatment is aimed.

Application of pheromone for mating disruption is an important component of many current integrated pest management (IPM) programs, and one purpose of this review is to summarize selected examples of successful behavioral manipulation. The development of efficacious formulations and successful application protocols for some pests has been a long and sometimes vexing process. A second intent of this review is to explore the behavioral and ecological foundations for such variability in efficacy. For example, mating disruption has been spectacularly successful for some species of moths, even in initially dense populations, and consequently this technique has been quickly incorporated into IPM programs used for these pests. In other species, however, mating disruption has proved effective in preventing crop damage only if populations were low when treatment was initiated. Use of this method on such species requires careful population monitoring and guidelines for application. For a few important moth pests, such as some of the heliothine moths, mating disruption has not yet proved efficacious, despite considerable research efforts (not all of which have been published) over the past 30 years. Such intractable pest moths are often highly dispersive and mating disruption may succeed only when applied to very large areas (> 100 km^2), as in areawide management (AWM) programs.

The explanations for such wide differences in efficacy seem, in part, linked to how mating disruption works in different species. In turn, which mechanisms of disruption are operative are greatly influenced by the type of formulation that is used (Weatherston, 1990). The topic of how to formulate disruptants will be considered first, followed by a consideration of mechanisms of disruption. Then, selected examples will be used to illustrate successful cases of mating disruption, as well as efforts with several important pest species for which this method has so far been impracticable.

Two salutatory cautions: first, having an efficacious control technique does not ensure that it will be incorporated into an IPM program. Adoption of any control program is influenced by a variety of socioeconomic factors and by the need to manage all pests on a crop, rather than whether or not mating disruption protects the crop against a particular target species. Second, mating disruption usually is not a "stand-alone" technology. For many species, it is necessary to apply some conventional pesticides to reduce the population to a level appropriate for mating disruption. Despite these two general limitations, mating disruption now plays an important role in protecting many crops from some of their most damaging moth pests.

Although this review considers only mating disruption of moths, there have been many investigations of direct control by male annihilation with a high density of pheromone-baited traps. In the first successful example, Trammel *et al.* (1974) achieved commercially acceptable levels of control of the redbanded leafroller (*Argyrotaenia velutinana*) in a large-scale demonstration by use of one trap per standard-sized apple tree (≈75–100 traps ha^{-1}). Other examples demonstrating the potential of this technique include Mafra-Neto and Habib (1996) and Zhang *et al.* (2002). One example of successful implementation of mass trapping is for control of the brinjal borer (*Leucinodes orbonalis*). Having 100 traps per ha along with hand removal of infested shoots produced approximately 50% more marketable fruit than by insectidicide treatment alone (Cork *et al.*, 2003). (Other work failed to find an effect of invested shoot removal in insecticide-treated plots.) Generally, however, the mass-trapping method would seem impractical for most moths, even if it is efficacious, owing to the high cost of traps, lures and their deployment. In contrast, mass trapping of palm weevils (Alpizar *et al.*, 2002; Oehlschlager *et al.*, 2002), provides population control and has been commercially successful.

The principles described for moth pests will undoubtedly apply to management of some other insect groups. To date, however, there have been few examples of successful tests for disruption of pheromone-source location that are directly comparable to moths (but for a beetle example, see Polavarapu *et al.*, 2002; Sciarappa *et al.*, 2005). Management of forest bark beetles by use of semiochemicals mediating aggregation mainly seeks to interfere with host colonization (see review by Borden, 1996). Prokopy and Roitberg in an accompanying chapter (see Chapter 4) consider additional non-moth examples of successful population manipulation by pheromones and other semiochemicals, particularly cases connected with foraging for food and ovipositional resources, as well as sex attraction and aggregation mediated by pheromones.

5.2 Formulation types

To disrupt orientation, a synthetic pheromone usually is applied to the crop as a formulation, often contained in some sort of plastic matrix that is designed to protect these generally labile compounds from degradation (via isomerization, oxidation, polymerization, etc.) while gradually releasing pheromone into the atmosphere (Weatherston, 1990). Example types and their characteristics are provided in Table 5.1. Most formulations increase their emission of pheromone as temperature increases. An ideal formulation would maximize pheromone release when moths are mating (typically night-time when temperatures are coolest), and compensate for seasonal temperature effects. The only delivery systems with a temperature-insensitive characteristic are devices that emit pheromone as an atomized spray; these also can be programmed to tailor emission to set times of the day or night.

An ideal formulation would be expected to release pheromone at a constant (that is zero-order) rate, until its reservoir of pheromone was completely exhausted, with the longevity of the formulation being appropriate for the crop-pest system. These latter release characteristics are difficult to achieve and most formulations release the disruptant at a rate that decreases gradually, while retaining some disruptant at the end of their effective period. The ideal formulation also would be easily and quickly applied at low cost.

Formulation types also differ markedly in how evenly they disperse pheromone into the crop canopy. Formulations that should result in relatively homogeneous patterns of dispersion generally are those such as microcapsules (Cardé *et al.*, 1975; Campion *et al.*, 1989) that are applied to crop

Table 5.1. *Formulation types, with typical density and method of application, typical longevity, and probable mechanisms of disruption*[a]

Formulation	Density ha^{-1}	Application	Longevity	Mechanisms
Atomizer	<1–several	Hand placed on stake or hung on tree	Season-long	Sensory interference; camouflage
Sealed plastic tube	Hundreds	Hand	Season-long	Sensory interference; camouflage
Open-ended hollow fiber, laminated plastic flakes	≈10,000	Specialized equipment, aerial	Weeks; season-long	Sensory interference; camouflage; competition
Microcapsules	Many millions	Conventional spray	Days to several weeks	Sensory interference; camouflage
Attracticide	≈1000	Specialized equipment	Weeks	Direct toxicity; impairment of orientation; competition

[a]All of the mechanisms assume that the formulation contains a synthetic copy of the natural pheromone. Sensory interference combines the mechanisms of adaptation of peripheral receptors and central nervous system habituation into a single category. The amounts released per hectare for most formulations are generally similar, in the range of 100 mg per day per hectare, except for attracticides, some formulations of which release as less than a mg per day per hectare.

canopy with the same equipment used to apply conventional pesticides. For example, with microcapsules applied to apple, a density of hundreds of capsules per cm^2 leaf or bark can be achieved (Waldstein and Gut, 2003). Other formulation types are comprised of point sources. Hollow plastic fibers that are open at one end (Scentry® fibers, see Doane and Brooks, 1981) and plastic flakes (Hercon® flakes, Miller *et al.*, 1990) can emit sufficient pheromone to be attractive point sources. A typical rate of application is about 1 fiber per m^2 or 10 000 per hectare. Fiber and flake types of formulation require specialized application equipment, but have the advantage that they can be applied aerially, and they can be combined with an insecticide in the sticker to create an "attracticide" effect or used in "dual applications" (pheromone with an insecticide overspray, Staten *et al.*, 1996). Another point-source formulation is sealed plastic tubes (Isomate® "ropes",

Flint *et al.*, 1990), which are applied at densities of about 250–500 per hectare. These point sources also can be attractive to males, if the pest species is attracted to release rates of pheromone that are much higher than emitted by a pheromone-emitting – or "calling" – female moth. A major difficulty with the use of sealed tubes is that such dispensers generally are hand applied to the crop, and this can be a logistical challenge (finding cost-effective labor) for crops that are grown over large areas, or with difficult placement requirements, such as a tall walnut tree. Another formulation type applied at low density is an aqueous paraffin emulsion applied to a tree trunk (Atterholt *et al.*, 1999).

Other types of release devices store pheromone along with an inert carrier under pressure in a canister and release an atomized mist of pheromone in timed bursts (for three versions of these systems, see Shorey and Gerber, 1996; Fadamiro *et al.*, 1999; Issacs *et al.*, 1999). Such aerosol devices have the advantage of protecting pheromone from degradation and being able to mete out pheromone precisely; they also can be set to emit disruptant only at times of the day or night when the insects are seeking mates. These devices have the disadvantage of being expensive and labor intensive to deploy, and therefore they are used at low density, on the order of one per hectare. At such sparse dispenser densities, disruptant is released into the crop in an uneven pattern and gaps in coverage can result. To date some of these devices also have a high failure rate of mechanical parts.

Thus differing formulation types produce very different patterns of pheromone dispersion in the field. Microcapsules sprayed throughout the canopy are expected to generate a relatively even distribution of airborne disruptant within the foliage without notable gaps. However, microcapsules need to be applied with a sticker to help keep them on the plant surface; even so, they can be susceptible to being washed off by rain or irrigation. Hollow fibers also should produce a relatively even coverage throughout the canopy (unless the canopy is dense, as in cotton late in the growing season, and the application is to the top of the canopy), but these point sources also generate plumes of pheromone that can be attractive to males. Sealed-tube dispensers may emit pheromone at rates that are hundreds or thousands of times as high as the rate from individual females. Males encountering plumes near such dispensers can experience much higher local concentrations of pheromone than males in a crop treated with microcapsules. The subsequent ability of males to orient along plumes from females may be very different in males that have been exposed to constant but low concentrations of pheromone from sprayable formulations, compared to males that have had only brief encounters with plumes of high concentration. These differences in distribution and concentration are dependent on the type of formulation and its distribution in the canopy, and not per se on the rate of application in terms of amount of active ingredient (AI) per hectare. Effective disruption with a given formulation, however, certainly can be contingent on the amount of AI per hectare.

5.3 Effects of foliage on dispersion of disruptant

Although the type of formulation used sets the initial pattern of dispersion, the characteristics of the foliage canopy into which disruptant is released greatly modulates its eventual distribution in a crop. For example, when pink bollworm pheromone is hand applied in an Isomate® rope to the base of a cotton plant at the "pin-square" stage (when cotton has its first flower bud), the plant is less than 30 cm high, and most of the field is bare ground. By the end of the growing season, the cotton is over 1 m in height and its canopy is closed. The Isomate® rope remains at the base of the plant and by season's end there is little windflow within the dense canopy. Released pheromone must diffuse upward to reach the top of the canopy, where most of the mating of pink bollworm takes place (Kaae and Shorey, 1973). Other formulation types are applied aerially and presumably result in a quite different pattern of odor dispersion. Scentry fibers, for example, are applied aerially at 10–14 day intervals; by the middle of the growing season, active fibers (those recently applied) reside mainly on the top of the canopy. Thus these two formulation types should produce differing patterns of pheromone dispersion because of the number of point sources used and their position in the canopy.

Similar issues of foliage density are found in numerous crops. Early in the season there may be few leaves in orchards and vineyards and consequently disruptant emanating from point-source formulations may be transported downwind (and possibly out of the treated area) in relatively discrete plumes. Foliage present later in the season will fragment and stir the plumes, so that their "active space" achieves a larger envelope close to the source, but may not extend as far downwind. The increased volume of foliage also may make it difficult to avoid creating pockets that contain insufficient disruptant to interfere with mate finding. Another issue interacting with foliage is that early in the season, comparatively lower temperatures may alter the performance of the formulation.

Another phenomenon of interest is adsorption of pheromone onto foliage and its subsequent re-entrainment into the atmosphere. This effect was first uncovered by Wall *et al.* (1981) with the pheromone of the pea moth

(*Cydia nigricana*). Leaves very near pheromone-baited traps adsorbed sufficient amounts of pheromone to become somewhat attractive on their own. Karg *et al.* (1994) later demonstrated in apple orchards treated with disruptant that adsorption onto foliage followed by re-entrainment modulated the distribution of disruptant in the orchards. The overall effect of such re-entrainment may be to even out the distribution of pheromone in treated areas, particularly when high dose, point-source dispensers are used. However, the contribution (if any) of the adsorption re-entrainment phenomenon in augmenting the efficacy of mating disruption has not been determined.

5.4 Mechanisms of communication disruption

5.4.1 *Normal patterns of mate finding*

Before considering how to interfere with mate finding, it will be useful to consider how insects normally follow a plume of pheromone in wind to its source. A flying male moth may orient toward a pheromone emitter over distances of tens to perhaps hundreds of meters by heading upwind whenever it detects pheromone. A walking insect similarly may find an upwind pheromone emitter, but typically over much smaller expanses such as in the range of only meters. To gauge wind direction, a flying organism uses visual feedback of how wind alters its trajectory, whereas a walking organism simply uses mechanoreceptive feedback (Cardé, 2003). Both flying and walking insects use the wind's direction (and not the pheromone's concentration) as a guide to the probable direction toward the source of pheromone, but simply heading upwind while sensing pheromone will not always bring a responder to the pheromone emitter. Two meteorological constraints make the task of tracing a plume to its origin quite challenging. First, as a plume of pheromone is transported downwind, turbulent forces generate pheromone-free gaps within the plume that can interrupt an insect's detection of pheromone. If the gaps grow large enough, insects simply may wander beyond the plume's overall boundary. (Molecular diffusion also dilutes the plume, but the magnitude of its effect is negligible compared to the forces of turbulent diffusion.) Second, changes in the wind field can cause the instantaneous wind direction within the plume to vary and often to be an unreliable indicator of the direction toward the source of the plume. Notwithstanding these two navigational difficulties, an upwind strategy while in contact with pheromone generally aids progress toward the source of pheromone and it is this orientation mechanism that moths and other organisms use to locate upwind odor sources. There remains a widespread belief that insects locate distant, upwind odor sources by using "concentration gradient" information. Reliable differences in concentration correlated with distance to the source only exist very close to a point source, and so this form of navigation is not available (Murlis *et al.*, 1992).

5.4.2 *Classification of mechanisms*

Mating disruption and communication disruption are two terms applied to interfering with mate location and mating by application of behavior-modifying chemicals. These labels were coined to avoid implying that a reduction in mating was due to a particular behavioral mechanism, as would be suggested, for example, when the method is termed either "confusion" or "masking". There are few cases in which we have sufficient evidence from field experimentation to verify how much a particular mechanism contributes to mating disruption and it is unlikely in any case that a single mechanism is entirely responsible for the efficacy achieved with any disruptant formulation.

Unraveling which mechanisms cause mating disruption is a daunting task. Laboratory bioassays, even those conducted in a wind tunnel, cannot duplicate the range of conditions found in the field: foliage; variable light levels; wind speed and turbulence; position of the calling female; etc. Field wind tunnels set over plants that have been treated with a disruptant formulation offer a testing milieu which approaches natural circumstances and night-time observations can be facilitated by video with infrared light (Cardé *et al.*, 1998), but males still must be released at the tunnel's downwind end and it remains difficult to regulate pre-release exposure to pheromone. Direct observations of male behavior are obscured by the simple fact that most moths seek mates at night. Nonetheless, there is convincing evidence for several mechanisms.

Three principal mechanisms that could diminish response to calling females have been identified (reviewed by Bartell, 1982; Cardé and Minks, 1995; Sanders, 1996). These are described in the following subsections.

Sensory adaptation

Sensory adaptation of the pheromone receptors and/or habituation of the response in the central nervous system. These two mechanisms do not differ in their effect on behavioral output: either the threshold of response is elevated (an increased concentration of pheromone is required to evoke a response) (Cardé and Minks, 1995; Mafra-Neto and Baker, 1996), or responsiveness is eliminated entirely. These collectively sometimes have been called "sensory impairment", "sensory interference", or "sensory fatigue". There is a large body of early literature showing that exposure to pheromone readily induces adaptation/habituation (reviewed by Bartell, 1982), and it is reasonable to conclude that some form of such sensory impairment generally contributes to disruption. Unless specific evidence for sensory adaptation is available, however, the term habituation is used in this review to refer to any reduction in responsiveness/elevation of threshold that is a consequence of prior exposure to pheromone. A strict demarcation between sensory adaptation and central nervous system habituation may in some cases be artificial. The level of peripheral sensory adaptation has been proposed to modulate the extent of habituation, with species with prolonged peripheral adaptation to pheromone being less apt to habituate under continuous exposure (Stelinski *et al.*, 2003*a*; *b*).

The extent to which disruptant-induced habituation mediates disruption of mate finding in the field remains enigmatic. Novak and Roelofs (1983), for example, found that males of the redbanded leafroller released into pheromone-permeated plots remained responsive to pheromone. However, laboratory bioassays by Bartell and Roelofs (1973) showed that brief exposures greatly inhibited subsequent responsiveness. Similarly, Figueredo and Baker (1992) demonstrated a persistent reduction in attraction of oriental fruit moth, *Grapholita (= Cydia) molesta*, to pheromone sources in a wind tunnel, following prior exposure to pheromone. Conversely, Rumbo and Vickers (1997) were unable to show in field plots treated with disruptant that oriental fruit moths were habituated to attraction. Some of the evident discrepancy between laboratory demonstrations of habituation after repetitive exposures to pheromone and the failure of long-term exposure to pheromone in field trials to account for disorientation of attraction may be attributable to the lower concentrations of pheromone that occur in field disruption plots and possibly a differing temporal pattern of encounter with pheromone (Rumbo and Vickers, 1997). Habituation is thus a mechanism that is readily demonstrated in a laboratory bioassay, but difficult to establish as contributing to mating disruption in the field.

Competition

Competition is simply males orienting along plumes from point-source dispensers. Such artificial sources of pheromone compete with females, and males attempting to locate these dispensers may do so at the expense of finding females. This mechanism also has been termed "false trail following". Many kinds of pheromone formulations, such as open-ended hollow plastic fibers (Scentry's NoMate®), sealed hollow plastic tubes (Shin-Etsu's Isomate®), plastic ampoules (BASF's RAK®), and plastic laminate flakes (Hercon's Disrupt®), emit pheromone at rates comparable to or well above those of calling females and such point-source dispensers can be as or more attractive than an individual calling female. The application density of these point sources can be quite high (e.g. about 1 per m^2 for hollow fibers) and time males spend orienting to these sources is unavailable to orient to females. The contribution of this mechanism to mating disruption is contingent on the ratio of point sources of formulation to calling females, the relative attractiveness of individual dispensers versus females, and possibly the position in the plant canopy of females and of point sources. There is direct evidence of pink bollworm, *Pectinophora gossypiella*, males orienting to and contacting Scentry hollow tube dispensers (Haynes *et al.*, 1986; Miller *et al.*, 1990). Not all point-source dispensers, however, are attractive to males.

Some systems release very high quantities of pheromone from dispensers that are widely dispersed (e.g. MSTRS® plastic bags, deployed at $\approx 20\,\mathrm{ha}^{-1}$) and, because of the high concentrations of pheromone they generate, these devices would not be expected to bring males to the source. Such devices might cause orientation (competition) well downwind, however, where the concentration of pheromone in filaments should be diluted sufficiently to induce upwind flight. Orientation to point sources of pheromone releasing very high concentrations of pheromone also may be fostered by the omnipresence of disruptant, which can elevate the male's threshold. This has been documented for the pink bollworm and codling moth (*Cydia pomonella*): males in disruption plots are attracted to high-dose lures that otherwise would be less attractive than low-dose lures (see section 5.8). Establishing the role of competition in mating disruption thus can be complicated by the effects a background of pheromone has on the upper and lower thresholds for orientation to pheromone plumes.

Camouflage

Camouflage is a mechanism that supposes the plume from a calling female becomes imperceptible as it is carried downwind and mixed by the forces of turbulent diffusion into the background of pheromone that is released from the formulation. The distance downwind at which the boundaries of the plume from the female become indistinguishable from the background of pheromone from the formulation will be contingent on the female's release rate and on the concentration and homogeneity in the environment of pheromone released from the formulation.

In other words, a female's plume might be camouflaged 10 m downwind, but not at 1 m. "Masking" is another term that seems to apply to this mechanism, but it may also refer to habituation.

Other mechanisms

Besides these three major mechanisms, other ways in which the formulation could alter male behavior have been proposed (Cardé and Minks, 1995). One effect is arrestment of upwind flight at concentrations of pheromone that are higher than found in a plume emanating from a calling female. This phenomenon is well exemplified in the oriental fruit moth, a species that seems particularly attuned to cease upwind progress when it reaches a concentration of pheromone that is just above that occurring downwind of a calling female (Baker and Cardé, 1979; Baker *et al.*, 1988). This suggests that application of sufficient pheromone in the environment would cause males to cease upwind plume following. Other species, such as the pink bollworm, do not have a similar threshold for arrestment of upwind progress. Pink bollworm males will orient to plumes from artificial sources releasing as much as 50 000 times as much pheromone as an individual female (Cardé and Mafra-Neto, unpublished).

Another effect is that the omnipresence of pheromone from the disruptant formulation can advance the time of male sexual activity several hours before females commence calling (Cardé *et al.*, 1998), as has been demonstrated in the pink bollworm. In this species males would then engage in plume-following to point-source dispensers, where they may encounter relatively high concentrations of pheromone. In turn, such exposure may diminish their subsequent ability to orient to plumes from females that have lower concentrations of pheromone.

This scenario illustrates another important issue; that several mechanisms can interact, with additive and possibly synergistic effects on efficacy. Moreover, the oriental fruit moth and pink bollworm examples also show how mechanisms can be contingent on both the response characteristics of the pest species and on the type of formulation that is used.

A microdispersible formulation (such as microcapsules) may not engage males of either species in plume following: high-dose, point-source formulations may not permit competition with oriental fruit moth females because of arrestment of upwind displacement in high concentration plumes, whereas with the pink bollworm, competition between high-dose artificial sources ought to be a significant factor in disruption efficacy.

We have no information on how mating disruption works for most species that have commercial disruption products. Studies with the pink bollworm and the oriental fruit moth, to be summarized in sections 5.7.1 and 5.7.3, have provided the most insights to date, pointing to multiple, interacting mechanisms.

5.4.3 *Recommending formulations*

An important but perhaps not plainly evident principle is that when it comes to efficacy of population regulation, formulations are not interchangeable. The first issue outlined above is that differing methods of dispensing disruptant promote different mechanisms. A second point is that formulations are applied at differing rates and achieve different atmospheric concentrations. A third issue is that some formulations are highly effective in protecting the disruptant from degradation and are efficient emitters; other formulations, because of deficiencies in either their protective or release characteristics, may be ineffective in promoting mating disruption.

The widely held assumption that any formulation containing the same AI (active ingredient) for disruption ought to be equivalent in efficacy stems from the experience of growers and extension personnel with conventional pesticides, where this supposition is apt to be correct. Those making recommendations for growers ought not to assume that registered, commercially available formulations for a given pest are interchangeable, because it is unlikely that they will have identical longevities in the field, emit pheromone at the same rate or in the same dispersion pattern, or have equivalent levels of efficacy in crop protection. In the United States, the Environmental Protection Agency (EPA) registers pesticides, including mating disruptants. This process dwells on the product's safety to the consumer, farm worker, and environment, but currently does not require the submission of efficacy data. In some cases, such as when the AI has not previously been registered, they may request data showing crop protection. The EPA registers a product based on a risk assessment that the benefit of the product outweighs possible risks to human health and the environment. The assumption in the United States is that the marketplace eventually will cull ineffective products. (However, North America, California, Canada, and Mexico currently require efficacy data.)

5.4.4 *Ersatz mixtures: incomplete and off-ratio blends for disruption*

The behavior-modifying ingredients in most commercial disruptants are synthetic copies of the natural pheromone: for most moth species the pheromone usually is comprised of a blend of two or three components; rarely is it just a single compound. A logical assumption is that the most efficacious disruptant would be an exact duplicate of the natural pheromone, because the complete pheromone blend has the lowest threshold for response and, in some species, it alone evokes attraction (Minks and Cardé, 1988). Indeed, most disruptants used commercially have the complete blend as the disruptant, but there are exceptions.

Successful disruption has been achieved using only a part of the pheromone blend in mating disruption of the smaller tea tortrix, *Adoxophyes honmai*, in Japan. Its pheromone is a blend of four components: Z11–14:Ac, E11–14:Ac, Z9–14:Ac, and 10-methyldodecyl acetate in a 63:31:4:2 ratio, but the commercial disruptant used until recently was comprised of Z11–14:Ac alone (Mochizuki *et al.*, 2002). Again, false trail following cannot be involved, because Z11–14:Ac by itself is unattractive to the smaller tea tortrix moth. The implications for evolution of resistance to this technique when an incomplete blend is used for mating disruption will be considered in section 5.10.

Another example of use of a partial blend is the European grape moth, *Eupoecilia ambiguella*, another tortricid. Its pheromone is comprised mainly of Z9–12:Ac, with several other minor components augmenting attractiveness of the main component (Arn *et al.*, 1986). The Z9–12:Ac disruptant used commercially, however, is contaminated with several percent of the E-isomer. (The presence of the E-isomer impurity in the technical product is a by-product of large-batch synthesis.) The Z9–12:Ac component alone is attractive, but its E-isomer is a strong antagonist of attraction. Nonetheless, this mixture is an efficacious disruptant. How it interferes with normal mate finding is not known, but competition between females and point sources of formulated pheromone cannot contribute to mating disruption, because the Z-E blend used is unattractive (Arn and Louis, 1996).

Another example of successful use of a disruptant that is not the natural pheromone is with the gypsy moth, *Lymantria dispar*. Its pheromone is (7R, 8S)-cis-epoxy-2-methyloctadecane, also known as (+)-disparlure. The racemate, (±)-disparlure, is inexpensive to synthesize, whereas synthesis of the pure (+)-enantiomer is quite costly. Only (±)-disparlure would be practical for mating disruption. In this species the (−)-enantiomer acts as a behavioral antagonist, reducing attractancy, so that traps baited with (±)-disparlure attract about a tenth as many males as those baited with (+)-disparlure. In the small-scale field tests that have been conducted, both (+)- and (±)-disparlure seem to be equally effective in disrupting mate finding (Plimmer *et al.*, 1982), and so all commercial formulations employ (±)-disparlure as the active ingredient.

The pea moth provides another exception to the generalization that the pheromone is the most efficacious disruptant. A combination of the pheromone (E8, E10–12:Ac) and 8% of two antagonists of attraction (E8, Z10–12:Ac and Z8, E10–12:Ac) appear to offer a greater disruptive effect than using the pheromone alone (Bengtsson *et al.*, 1994). The two antagonists are positional isomers of the pheromone and their co-occurrence is a result of isomerization of the isomerically pure pheromone after its field application in Isomate sealed polyethylene tube dispensers. In the first few days following application, males are attracted to the dispensers; after isomerization exceeds 5%, males are not attracted to these dispensers, and they were observed to fly out of the treated plots. The "repellent" effect evidently enhanced disruption. Bengtsson *et al.* (1994) reviewed several other cases in which a mixture of an attractant pheromone with an antagonist of attraction may offer superior disruption. Witzgall *et al.* (1999) described field tests of this approach with the codling moth, using E8, E10–12:Ac as an antagonist of the codling moth pheromone.

Several of these examples and others will be discussed fully in section 5.7, but they are mentioned here to emphasize that the mechanisms which disrupt mating will differ if the active ingredient of the disruptant formulation is not as attractive as the natural pheromone. The issue of whether the most efficacious disruptant would be the complete pheromone blend, partial blends, off ratios of the natural blend, or a mixture of the pheromone and a behavioral antagonist of attraction remains unsettled (Minks and Cardé, 1988) and probably can only be resolved by empirical tests with each moth species and method of formulation. Bengtsson *et al.* (1994) have suggested that a mixture of pheromone and an antagonist could be more disruptive than pheromone alone. The answer to the question of whether the best disruptant is the complete natural pheromone, a partial blend, an off ratio, or a mixture of one of the foregoing plus an antagonist of attraction, is likely to depend on the behavioral characteristics of the pest species, its pheromone system, and the type of formulation employed.

5.4.5 *Use of pheromone analogs for disruption*

There have been innumerable efforts to replace pheromones used for mating disruption with pheromone analogs (sometimes called "parapheromones") that evoke the same behavioral reaction (such as attraction), or, conversely, analogs (termed "behavioral antagonists" or "inhibitors") that interfere with mate finding (see review by Renou and Guerrero, 2000). The motivation for such studies has been varied. Analogs have been tested as replacements for pheromones that are either expensive to synthesize, or that are too labile to protect in formulations. For example, (Z)-9-tetradecen-1-ol formate can substitute for the less stable Z11–16:Ald in the two-component version of pheromone blend of the tobacco budworm, *Heliothis virescens* (Mitchell *et al.*, 1978). The formate also is effective by itself in interrupting the finding of females by males in small plots (Mitchell *et al.*, 1975). However, the viability of this formate for actual crop protection has not been evaluated, probably because it is not sufficiently disruptive. Other approaches have explored whether compounds (usually close analogs of the pheromone and labeled behavioral antagonists) that interfere with attraction when emitted with the pheromone from the same point source, might also interrupt mate finding when emitted from a formulation used for disruption. However, such antagonists tested alone for their efficacy in interfering with mate location typically have proved to be far less effective than broadcast application of formulated pheromone. To date, efforts to use such antagonists of attraction as disruptants have met with limited success in experimental field plots, and there is no example of commercial application. It is likely that non-pheromone analogs will remain more difficult to register than pheromones, unless a convincing argument can be made to regulatory authorities that analogs are qualified for the same "biopesticide" exemptions as synthetically equivalent pheromones.

5.4.6 *Use of lure and kill or "attracticide" formulations*

Pheromone can be released from point sources at rates that are attractive to males, and so it was logical to investigate if such formulations coupled with insecticide would enhance crop protection. In 1979 Conlee and Staten observed pink bollworm males contacting and attempting copulation with hollow fiber dispensers and theorized that the addition of insecticide to the formulation would enhance control (Conlee and Staten, 1987). Such an approach was first tried commercially with the pink bollworm using Scentry® hollow fiber dispensers (Doane and Brooks, 1981). This formulation is applied aerially with specialized equipment and the individual fibers adhere to cotton foliage with a droplet of polybutene sticker (Miller *et al.*, 1990) at a density of about 10 000 fibers ha^{-1}. Although this system disrupts mating on its own, in the 1980s Arizona cotton growers attempted to enhance efficacy by adding a small amount of insecticide to the sticker. Haynes *et al.* (1986) verified in wind tunnel experiments that pink bollworm males which contact poison-laced fibers received either a lethal dose or a sublethal exposure that rendered males much less capable of subsequent orientation to pheromone. In this case, efficacy of crop protection seems to rely, in part, on conventional mating disruption, because the insecticide-free fiber formulation achieved good crop protection. The addition of insecticide, therefore, provided an additive action of unknown magnitude, contributed by either male annihilation or an impairment of orientation capabilities.

An analogous method is being widely tested for control of many moth pests (Lösel *et al.*, 2000; 2002; Suckling, 2001; Ioratti and Angeli, 2002; Krupke *et al.*, 2002; Mitchell, 2002). Pheromone is mixed with a contact insecticide and applied as small ($\approx$ 50–100 μl) droplets of a viscous paste (Appeal®, Last Call®, Sirene®) to the foliage. Suckling and Brockerhoff (1999) have evaluated the potential of this method with the light brown apple moth, *Epiphyas postvittana*. Applying 450 droplets per ha of apple orchards reduced male catch by 96% in pheromone-baited traps deployed in the center of test plots, compared to experimental control traps. Droplets that were caged so that males could not contact the insecticide produced a 48% reduction. This innovative method of comparison indicates that about a half of the effect of this formulation could be attributed to competition between the droplets and the pheromone source (and therefore conventional mating disruption), with the remainder attributable to males being eliminated by contact with the formulation.

Charmillot *et al.* (2000) have used this system in Switzerland to control codling moths in isolated apple orchards. The formulation is delivered from hand-held applicators at rates of 20 to 80 droplets per tree and contains 0.16% pheromone and 6% permethrin. The amount of pheromone applied is less than $1\,g\,h^{-1}$ and it is active for 5–7 weeks. Males are attracted to these point sources and, upon contact, receive a lethal dose of insecticide. Provided that the initial density of codling moths is low, this method is effective, resulting in less than 1% fruit infestation. Its efficacy is not reliant in part on direct interference with mate finding via mating disruption (Charmillot *et al.*, 2000), unlike the case of the light brown apple moth (Suckling and Brockerhoff, 1999).

This method can be labor-intensive, necessitating 5 to 10 hours per ha for a single application in apple orchards (Charmillot *et al.*, 2000); recent improvements in techniques for applying this formulation, however, have reduced application time to 2 to 5 hours per ha. Efficacy of this method is highly contingent on having point sources of attractant that are very competitive with females. Using a copy of the female's complete pheromone blend in such formulations would seem compulsory. Any repulsive effect of the insecticide on the attractivity of the droplet would of course diminish efficacy, and the possibility of this system interacting with any resistance background to the pyrethoids or other contact insecticides may need to be considered in designing a management program (Poullot *et al.*, 2001).

5.5 *Effects of population size and patterns of adult movement*

One of the principal ecological factors that can affect the probability of successful mating disruption is the initial size of the population to be regulated. Because application of a mating disruptant may not prevent mating of all females resident in the treated plot, clearly the more females present, the higher the likelihood of crop damage exceeding acceptable levels. In the case of some species, however, such as the oriental fruit moth and the tomato pinworm, *Keiferia lycopersicella*, initial populations can be quite high and mating disruption nonetheless will be effective; conversely, in the case of the gypsy moth, mating disruption works only when populations are very low.

A second ecological consideration is the movement patterns of adult moths. If mated females emigrate in sufficient numbers from areas that are untreated to areas that are treated with pheromone, then mating disruption will be ineffective. Among the factors that determine how relevant such movement is to crop protection is the distance to be traversed from the source of mated females to the treated area and the size of the source population. The propensity of individual females to migrate can vary with a variety of factors, including time of year and quality of the habitat and availability of host material. Because it is typical that moths immigrating into a crop first settle near its outer edge, the larger the plot the less likely that immigration would be a factor in efficacy. Some application protocols explicitly account for this "edge effect" by treating plot edges with conventional insecticides or a higher rate of application of disruptant. Some moths, such as heliothine pests, are considered highly migratory, whereas other groups appear to have relatively limited adult movement or, in the case of the gypsy moth in North America, no significant migration because females are flightless.

There have been several suggestions based mainly on laboratory observations that female-released pheromone might also mediate female movement and calling behavior (e.g. Palanaswamy and Seabrook, 1978; 1985). Females of some moths are known to perceive their own pheromone (e.g. Cook *et al.*, 1978; see review by Schneider *et al.*, 1998). The adaptive value of such "autodetection" in natural situations could be to cause females to depart an area of high population density before calling, shift their calling position locally to an area where little or no pheromone was detected so that their pheromone plume would be distinct, or synchronize their times of calling (in the sense of a lek). As Schneider *et al.* pointed out, the adaptive value of autodetection in natural populations remains speculative.

In the case of crops treated with a disruptant formulation of pheromone, so far such supposed effects have not been documented to influence the efficacy of mating disruption, but there appear to be few field experiments that have addressed this possibility. What might be expected is that females that have departed a treated plot could mate outside of the treated area. Once mated, those females could re-enter the treated area, thereby decreasing the efficacy of mating disruption. Alternatively, females could move locally to a position where pheromone is not detected (or is at very low concentration, such as the top of the plant canopy). This theoretical possibility has been considered explicitly by Pearson and Schal (1999) for the sesiid moth *Vitacea polistiformis*, the grape root borer; as measured with electroantennograms, both males and females sense the female's pheromone.

The presence of pheromone released from the disruptant formulation in a treated plot could have a profound influence on the pattern of male movement. At one extreme, males exposed to pheromone could become quiescent and unresponsive to pheromone; at the other, males could move upwind continuously and, in a small plot, reach its upwind edge. Upwind displacement thus could aggregate males at certain portions of the plot and perhaps diminish efficacy of disruption in those sectors of the treated areas.

The importance of initial population density, plot size, and the timing and extent of adult migration will be considered further under specific Case Histories (section 5.7).

5.6 Effects of a delay in mating on fertility and fecundity

One underappreciated factor in the success of a mating disruption program may be its effect on the age at which a female first mates and in the number of matings that she procures over her lifetime (Vickers, 1997). Both of these factors can affect a female's fecundity (the number of eggs laid) and fertility (the proportion of laid eggs that are fertile). With the European corn borer, *Ostrinia nubilalis*, a delay of three days in mating decreased the number of eggs deposited by about one-quarter; a further delay of seven days decreased the number of eggs laid, but, more importantly, virtually eliminated egg fertility (Fadamiro and Baker, 1999). The same study also recorded a higher fecundity and fertility for multiply-mated females over those with a single coupling.

In field tests of mating disruption of the European corn borer (Fadamiro *et al.*, 1999) evaluated two formulations: Isomate® ropes set out on a 2-m spacing and meter spray timed-release (MSTRS®) devices deployed 35 m apart. With both formulations, males were on average ≈ 97% disrupted from locating pheromone-baited traps. In the first flight, ropes and MSTRS reduced the percentage of females mating by 17% and 10%, respectively; in the second flight the percent reductions were similar but statistically insignificant. In both flights the number of matings per female were reduced by roughly one-quarter. This study demonstrates that the overall efficacy of the mating disruption technique could be contingent on more subtle effects than simply whether or not a female has mated.

5.7 Case histories

5.7.1 *Pink bollworm (Pectinophora gossypiella)*

The pink bollworm is a major pest of cotton in most cotton-growing regions of the world. All formulations rely on the pheromone (a 1:1 mix of Z7Z11–16:Ac and Z7E11–16:Ac) for disruption. Formulation types in wide use include microcapsules, laminate flakes, hollow fibers, and sealed tubes (Campion *et al.*, 1989). This pest has multiple generations per year and it has become resistant to many conventional pesticides. The adult is also highly dispersive (Stern, 1979).

The first large-scale test demonstrating crop protection by disruption of communication for any pest insect was conducted by Harry Shorey and his colleagues at the University of California at Riverside (Gaston *et al.*, 1977) with the pink bollworm in the Coachella Valley of southern California, where this moth is a key pest of cotton. Pheromone was dispensed from open-ended hollow plastic fibers (Scentry) which were hand placed at three-week intervals on the top of the cotton canopy in a 1 by 1 m grid pattern. Individual fibers emitted 10 μg of pheromone per day. Because of concern that mated females would migrate into pheromone-treated plots, plot sizes (three replicates comprised of 5, 6, and 12 ha) were large compared with those typically used in efficacy tests of conventional pesticides. Efficacy of mating disruption was assessed by comparing the season-long rates of larval infestation of cotton bolls in pheromone-treated fields with fields under the conventional practice of insecticide treatment. Overall rates of boll damage throughout the season were comparable in the pheromone-treated and conventional practice fields, despite a ninefold reduction in use of insecticides in the pheromone-treated fields. In 1978 the United States EPA registered the Scentry® hollow fiber formulation, the first pheromone to be registered for areawide crop protection in the United States. At the same time, development of specialized equipment for application of fibers from aircraft permitted treatment of large areas, comparable to aerial spraying of insecticides, and this technical development led to widespread use of this formulation in Arizona and elsewhere.

An instructive example of successful mating disruption on a regional basis was the Parker Valley, Arizona, Project, summarized by Staten *et al.* (1996). Immediately prior to this program, boll infestation in Parker Valley generally exceeded 25%, despite intensive chemical treatments. In 1989, the last year of conventional insecticide application, the 11 250 ha of cotton in the Parker Valley had boll infestation rates of 18–36% in weekly samples taken from early August to mid-September. In the following four years, fields were treated with pheromone with two types of formulations. Most treatments were aerial applications of hollow fibers spaced approximately 10 days apart; sealed tubes (called "ropes") were hand applied to some fields early in the season (at the first blossom or "pin square" stage of cotton in May). Insecticide treatments were applied with the fibers in the first years of the program; dual applications of pheromone with an insecticide overspray were reduced to 1089 ha in 1993.

Boll infestations declined from the 18–36% weekly levels in 1989 under conventional insecticide practice to less than half those levels in 1990, less than 2% in 1991 and 1992, and a remarkable zero percent in 1993, when 22 282 bolls were sampled season-long, none of which had a pink bollworm larva. Parker Valley, a cotton-growing region along the Colorado River Valley, is somewhat isolated from other cotton areas in Arizona and California. In AWM programs such as this one, it is not feasible to have replicates or strictly comparable experimental controls. It is nonetheless informative to compare infestation levels in fields under conventional practice in central Arizona to the east in 1992 and the Palo Verde Valley to the south in 1991. Weekly samples of boll infestation in these two regions under conventional practice indeed were all much higher (five- to ten-fold or more) than those found in Parker Valley in the same year. This program and other similar highly successful efforts in Egypt and elsewhere (Campion *et al.*, 1989) clearly demonstrate the efficacy of mating disruption. Mating disruption programs were adopted in the 1990s mainly because of dissatisfaction with the ability of conventional, insecticide-based programs to provide a satisfactory level of pest control. More recently, pheromone applications for control of pink bollworm in Arizona have declined markedly. *Bt*-engineered cotton emerged as the favored cultivar because of its ability to control all pestiferous Lepidoptera on cotton, although in 2003 there has been some resurgence in use of conventional cotton varieties because of their adaptability to local growing conditions. Pheromone, along with other tactics, currently is being used as a major component of an AWM program to drive the pink bollworm to extinction from Texas to California and southward into Mexico.

5.7.2 *Tomato pinworm* (Keiferia lycopersicella)

The tomato pinworm is a major pest of tomatoes in Florida, Texas, and southern California and, in recent years, it has become a devastating pest in Mexico because of high levels of resistance to insecticides. Larvae of the tomato pinworm bore within stems and fruit of tomato. Unchecked, populations can proliferate to epidemic levels, with larvae boring through stems until the plant collapses on the ground; at these population levels, direct feeding on the fruit renders it unmarketable. The pheromone consists of one component, E4–13:Ac, and there are several registered formulations, including hollow fibers (Jenkins *et al.*, 1990).

This pest has been a particular problem in the Mexican state of Sinaloa. The four yearly plantings of tomatoes each received 25–40+ applications of insecticide cocktails (mixtures of two or more insecticides). Nonetheless pinworm damage levels were very high, sometimes causing the crop to be abandoned before harvest. An exceptional degree of resistance had rendered conventional insecticides impotent against the tomato pinworm in Sinaloa. Growers had little choice but to try alternative approaches to control this pest. The IPM program (Trumble and Alvarado-Rodriguez, 1993) adopted was based on disruption for direct control of the pinworm (Jenkins *et al.*, 1990), release of *Trichogramma* parasitoids for control of noctuids, particularly *Spodoptera exigua*, and some selective application of the pesticide abamectin for some additional suppression of the pinworm. This program has been evaluated not only for efficacy of control but also for its economic value. Growers on this program realized considerably higher profits, produced tomatoes with a greatly reduced pesticide input, and achieved a far safer environment for farm workers (Trumble and Alvarado-Rodriguez, 1993). There are no studies that explain how mating disruption works in this gelechiid.

5.7.3 *Oriental fruit moth* (Grapholita (=Cydia) molesta)

This tortricid is a worldwide pest of stone fruits, particularly peaches and nectarines. The pheromone of the oriental fruit moth consists of three main components, Z8–12:Ac and E8–12:Ac in about a 95:5 ratio, with several added percent of Z8–12:OH. Early field trials used the two acetates in their natural ratio as a disruptant, but subsequent field studies suggested that addition of the alcohol component lowered the amount of pheromone required for disrupting attraction to pheromone-baited and female-baited traps (Charlton and Cardé, 1981). Current commercial formulations utilize the three-component blend as a disruptant. One key aspect of successful control appears to be the requirement of formulations to release at least 5 mg of pheromone per hour per ha (Rothschild and Vickers, 1991). Some commercial formulations achieve longevity of 150 days in the field.

Successful control of this species in Australia and North America by mating disruption has been reviewed by Vickers (1990); Rothschild and Vickers (1991) and Sexton and Il'ichev (2000). Trimble *et al.* (2001) have shown in Ontario, Canada, that equivalent levels of protection of peaches from the oriental fruit moth are achieved with a conventional insecticide program and with control of the first generation with chlorpyrifos and mating disruption for the second and third generations. Of particular interest is the adoption of AWM programs in some regions of Australia (Il'ichev *et al.*, 2002). In the first year of a program in northern Victoria, disruptant was applied to 800 ha in 18 orchards; in the second year the program was expanded to 1100 ha in 40 orchards. Dispensers were applied to every fruit tree in the area, principally peaches and nectarines, but also to pears, apples, apricots, and plums. Efficacy was determined by damage to shoot tips and fruit. In the first year of the program conventional insecticide sprays were halved, and in the second year most growers did not apply any insecticide. In one peach orchard with a 25% fruit damage level before the AWM program, damage fell to 15% in the first year of treatment and nearly zero in the second year (Il'ichev *et al.*, 2002). These sharp reductions in damage indicate that this tortricid is very vulnerable to mating disruption and give support to the AWM approach to mating disruption.

Direct control of the oriental fruit moth using only disruptant was demonstrated in the Tulbagh Valley of South Africa (Barnes and Blomefield, 1997). The oriental fruit moth was first detected in South Africa in 1990 and rapidly spread to all stonefruit growing areas. Despite an intensive spray program of up to 13 applications of organophosphate insecticides, damage in canning peach orchards of Tulbagh Valley was extensive and prompted an AWM program of mating disruption in 1991/1992. Pheromone was applied to the entire 1200 ha crop of peaches and nectarines, and to the five border rows of all plum, prune, apricot, almond, and pear orchards. Shoot tip damage in peaches dropped from an average of 49% to 0% in one year, except one orchard with a late application of pheromone, which experienced no fruit damage but 6% tip damage. Orchards in Breeriver Valley 15 km away served as the experimental control. Despite rigorous spray programs, the sampled orchards in Breeriver experienced 12% and 15% fruit damage and high levels of shoot tip damage. This case is particularly impressive inasmuch as no insecticide sprays were applied to the pheromone-treated blocks of peaches.

The mechanisms responsible for mating disruption in this species remain speculative, but several experiments point to habituation being a factor in disruption. Males exposed continuously to a synthetic blend of the pheromone cease most orientation behavior within four minutes (Cardé, 1976). Males given two daily exposures to pheromone in a wind tunnel had reduced levels of attraction to pheromone on subsequent days (Figueredo and Baker, 1992): similar experiments by Sanders and Lucuik (1996) also demonstrated habituation of attraction with prior exposure. In addition, Rumbo and Vickers (1997) found habituation of the sequence of attraction to a pheromone source in a wind tunnel, with the degree of habituation being contingent on the duration and intensity of the pre-exposure stimulus. Rumbo and Vickers concluded, however, that the levels of pre-exposure necessary to achieve significant levels of habituation in wind tunnel trials and in orchard releases of males were above those that would be expected in field applications.

Males generally fail to proceed upwind in concentrations of pheromone that are approximately 10 to 30 times higher than those from calling females (Baker and Cardé, 1979). Such cessation of upwind flight seems attributable to two processes: complete adaptation of antennal neurons and attenuation of fluctuations in burst activity of these neurons (Baker *et al.*, 1988). Whether such "arrestment" of upwind flight occurs in plots treated with disruptant has not been verified.

Wind tunnel experiments (Valeur and Löfstedt, 1996) comparing orientation of males in single and overlapping plumes of varying concentrations of pheromone support both habituation/adaptation and competition as potential mechanisms for disruption in the field. These tests also suggested that the complete three-component blend (Z8–12:Ac, E8–12:OH, and Z8–12:OH) would be a more effective disruptant in the field than Z8–12:Ac and E8–12:Ac, as found in early field trials (Charlton and Cardé, 1981). Together, these behavioral tests lead to somewhat contradictory interpretations of the mechanisms responsible for efficacy of mating disruption in field applications.

5.7.4 *Codling moth* (Cydia pomonella)

This tortricid is a key pest of apple worldwide (except in Japan where this moth does not occur) and also of pears and walnuts. Its pheromone is E8,

E10-12:OH; other compounds related to the pheromone are emitted, but their role as possible minor pheromone components remains equivocal. Mating disruption has been extensively investigated as a control technique in apples, pears, and walnuts (Cardé and Minks, 1995). The most widespread use of mating disruption for management of the codling moth is in northwestern North America. In Washington, Oregon, and British Columbia, codling moth control has been achieved with several formulation types, but the most widely used formulation is Isomate ropes applied at a density of 1000 per ha (Gut and Brunner, 1998). In California, mating disruption is used predominantly on pears, followed by apples, and walnuts. In most walnut orchards pheromone is released as an aerosol from pressurized cans spaced 44 m apart along the borders of large orchard blocks (Shorey and Gerber, 1996).

Achieving acceptable levels of control (generally around 1% or less fruit or nut damage) is dependent on the population level of the codling moth being at an appropriately low level before application of mating disruptant. In cases where population density exceeds this minimum, conventional insecticide needs to be applied before treatment with mating disruptant. A second factor in attaining a satisfactory control is isolation of the treatment area from the influx of mated females. A number of field tests support the conclusion that the larger the area under mating disruption, the more efficacious mating disruption becomes (Charmillot, 1990; Cardé and Minks, 1995; Minks, 1996).

The number of hectares in the western United States under this technique rose from 600 in 1991 to 6500 in 1995 (Gut and Brunner, 1998). By 1999 the area under mating disruption reached 38 000 ha, although it has since declined to 32 000 ha, principally because of less favorable economic conditions for apple production (Brunner, 2003). Commercial IPM programs in the western USA typically add one to three applications of insecticide to supplement season-long pheromone applications. Pheromone-treated blocks seem more vulnerable to damage by leafroller moths and may require remedial insecticide applications. However, in pheromone blocks the densities of some natural predators and parasitoids were increased, although it is not clear whether these natural enemies contributed significantly to biological control of orchard pests (Gut and Brunner, 1998).

Studies of mating disruption of the codling moth over more than 25 years have established the viability of this technique, provided two conditions are met: first, the initial population of codling moths must be suitably low, and second, the treated orchards must be either quite isolated from sources of immigration, or, if such isolation is not feasible, the area of treatment must be quite large (100s of ha). In regions where there are many contiguous growers, an AWM program can coordinate the required population monitoring activities and grower compliance with application requirements. Supplementary insecticides are almost always required to maintain codling moth populations at manageable levels.

5.7.5 *Gypsy moth* (Lymantria dispar)

The gypsy moth's natural range spans from Japan and the Russian Far East to Europe, but it is a pest principally in North America, where it is an introduced species, and an important defoliator of deciduous forests. Its pheromone is (*7R,8S*)-*cis*-epoxy-2-methyloctadecane, called (+)-disparlure. Because of the high expense of synthesizing enantiomerically pure (+)-disparlure, the relatively inexpensive racemate, (±)-disparlure, is the only practical option for disruption. Various formulations of racemic disparlure have been tested extensively for efficacy in mating disruption since the 1970s.

The ecology of disruption of the gypsy moth mating is unique in two ways. First, there is a need to prevent mating over a large vertical expanse, because mating occurs throughout the tree canopy, although the majority of mating occurs within several meters of the ground. Pupation occurs in the duff on the forest floor to high on the tree trunks. A female typically calls, mates, and lays its egg mass very close to its pupation site. The height of the forest canopy (up to tens of meters in mature deciduous forests) poses a challenge to aerial application. For disruptant to be effective in mating disruption, it must be dispersed throughout the canopy.

A second distinction is that in North America migration of mated females into treated plots is not a consideration because females do not fly. Natural dispersal of gypsy moths occurs mainly by aerial transport of first instar larvae, but the usual distance of movement is only in the range of 10s to 100s of meters. Because application of pheromone for mating disruption is only over very large areas in AWM programs, movement of larvae into treated areas is not a consideration. This is an advantage for mating disruption.

Because the gypsy moth is univoltine, the objective of mating disruption is to reduce the population level in the year following treatment. This is most accurately assessed by changes in the density of egg masses, although accurate measurements are impossible below levels of 10 masses ha^{-1}. Other measures, such as larval density in the year following treatment, are subject to effects of natural enemies and diseases, and obviously these factors alone can influence population levels and damage. Because it is so difficult to establish in sparse populations of the gypsy moth that mating disruption changes population density, field tests which compare the effects of different treatment protocols, formulations, and rate of application, typically evaluate success by two measures: reductions in the mating frequency of laboratory-reared females held in mating stations and reduction in the capture of males in (+)-disparlure-baited traps.

Many field tests conducted in the 1970s and 1980s (Plimmer *et al.*, 1982) sought to demonstrate efficacy in population densities that, in retrospect, were far too high for this method to work. The reasons for these failures obviously are attributable to the ability of males to locate females when densities are high. In high densities a low proportion of males can be observed walking on tree trunks (Cardé and Hagaman, 1984). Copulation is induced by tactile cues that are perceived tarsally (Charlton and Cardé, 1990). Thus, mating may ensue without plume-following and the rate of such tactile encounters would increase with population density.

In low-density populations, however, tests have established the efficacy of mating disruption (Leonhardt *et al.*, 1996; Thorpe *et al.*, 2000). Disruptant in plastic laminate flakes was applied aerially at rates that produced release rates of 0.48 to 0.72 g of disparlure per day and success was measured by suppression of catch in pheromone-baited traps and mating of sentinel females, and in the following year, whether population densities remained low compared to densities in control plots. The effect of the disruptant extends up to 250 m beyond the boundaries of application (Sharov *et al.*, 2002), suggesting the aerial application need not provide completely thorough coverage. The success of these tests has led to an AWM program to suppress gypsy moths along the leading edge of its population expansion into the southeastern and midwestern United States. The recommended criteria for application are: (1) pheromone-baited traps deployed at 9 traps per 3 km^2 capture on average <15 males in the year prior to treatment and <30 in the year of treatment; (2) the treated area is >8 km from a source of many males; and (3) the area is sufficiently large to offset any migration of males into the treated area (Reardon *et al.*, 1998). In 2000 approximately 38 000 ha were treated in a band from North Carolina to Wisconsin, with the expectation of slowing the spread of the gypsy moth from an historical average of 21 to 8 km a year.

5.7.6 *Leafrollers on apple* (Tortricidae)

In some cropping systems there are a number of moth pests that require control. In apples in the eastern and midwestern United States, five tortricid leafrollers are damaging to both foliage and fruit in apple orchards (Table 5.2). These species use distinctive pheromone blends, but with some components in common. These leafrollers have either Δ11-14:Ac or

Table 5.2. *Pheromone components of pest tortricid leafrollers found in commercial apple orchards in the eastern and midwestern United States, eastern Canada, or in the western United States and Canada*

	Pheromone blends[a]					
Species	Z11–14:Ac	E11–14:Ac	Z11–14:OH	E11–14:OH	Z9–14:Ac	12:Ac
Agyrotaenia velutinana (redbanded leafroller)	88	12				+125%
Choristoneura rosaceana (obliquebanded leafroller)	95	5	+0.5–5%			
Pandemis limitata (threelined leafroller)	91				9	
Platynota flavedana (variegated leafroller)			10	90		
Platynota idaeusalis (tufted apple leafroller)		50		50		

[a]The pheromone blends listed are the mixtures that are attractive in the field and which are typically used in pheromone-baited monitoring traps. In western Canada, *Choristoneura rosaceana* uses Z11–14:Ald as a fourth component. References to identifications are in Arn *et al.* (2000).

Δ11–14:OH components in their pheromone, and so it was logical to examine the use of mixtures of these components in a "generic" blend that might prove efficacious for several leafroller species. Another leafroller, the smaller tea tortrix, had already been shown to be effectively disrupted in commercial use by use of only one of its components, Z11–14:Ac (Mochizuki *et al.*, 2002).

Gronning *et al.* (2000) compared the disruptive effect of generic pheromone blends on the redbanded leafroller and the oblique-banded leafroller (which share components, Table 5.2), and on the tufted apple bud moth and the variegated leafroller (which share components, Table 5.2). The "generic" blends contained either one component from each species group (Z11–14:Ac and E11–14:OH in a 1:1 mix), or three components from both groups (E11–14:Ac, Z11–14:Ac and E11–14:OH in either a 3:4:3 or a 4:2:4 ratio). The pheromone blends for the tufted apple budmoth and the variegated leafroller also were evaluated. These tests measured the ability of these blends to interfere with male capture in pheromone-baited traps for each species in mid-Atlantic apple orchards in small replicated plots. All three generic blends to some extent disrupted male orientation to traps in all four species and therefore showed the potential of this multispecies approach; actual crop protection remains to be demonstrated. None of the generic blends tested would have been attractive to these leafrollers, and so the mechanism of disruption cannot rely on competition of disruptant dispensers with calling females.

In the apple-growing region of northwestern North America, the obliquebanded and threelined are among the pestiferous tortricid leafrollers. Evenden *et al.* (1999*b*) have compared the four-component blend of the obliquebanded leafroller, the two-component blend of the threelined leafroller (Table 5.2), and a 1:1 mixture of Z9–14:Ac and the obliquebanded leafroller pheromone for disruptive effect on the threelined and obliquebanded leafrollers. Z9–14:Ac is a component of the threelined leafroller pheromone and it is an antagonist of attraction to the obliquebanded leafroller when admixed with its pheromone. In small (0.1 ha) plot field tests of disorientation to tethered females, the four-component obliquebanded leafroller pheromone seemed as effective in suppression of female mating as a Z11–14:Ac alone or this component in combination with Z9–14:Ac (Evenden *et al.*, 1999*b*). Similar tests exploring complete versus incomplete blends with the obliquebanded leafroller also failed to demonstrate that the complete blend was more efficacious than a two-component blend (Evenden *et al.*, 1999*a*). These findings provide evidence that a generic blend approach to leafroller disruption may be an effective strategy, and they further suggest that competition does not contribute significantly to disruption. In wind tunnel trials, habituation did not seem to be a major contributor to disruption (Evenden *et al.*, 2000). The obliquebanded leafroller, however, seems to be very difficult to disrupt (see section 5.9).

5.7.7 *Clearwing borers (Sesiidae)*

Sesiid larvae bore into stems, bark, and roots of plants in more than 40 families, and they are important pests on fruit trees and grape and berry vines. The structure of the pheromones in this family appears to be remarkably undiversified, usually comprised of one or two components that are usually a Z3, Z13–18 or E3, Z13–18 chain motif, with an Ac or OH terminal moiety, or very similar analogs. Tests with the lesser peach tree borer, *Synanthedon pictipes*, in Virginia peach orchards used 500 point source dispensers containing 40 mg of the pheromone, E3, Z13–18:Ac, set out at the beginning of each generation; in Pennsylvania 70 mg dispensers were used season-long. Damage was assessed by the presence of pupal exuviae on tree trucks. Compared to blocks treated with a conventional insecticide spray, damage was reduced by 19–97%, so that in this application, pheromones were superior to the application of insecticides to trunks with handguns (Pfeiffer *et al.*, 1991). Snow (1990) reported similar findings with the lesser peachtree borer, and also showed similar levels of control with the peach tree borer, *Synanthedon exitosa*, when both components of its pheromone, Z3, Z13–18:Ac and E3, Z13–18:Ac were evaporated from dispensers hung on peach trees.

5.7.8 *Worldwide use of pheromones for mating disruption*

The foregoing examples consider the majority of cases of successful application in terms of area under treatment. The worldwide use of mating disruption in 2002 across all crops has been estimated at 433 000 ha (Table 5.3) and includes a number of additional species and cropping systems not considered in this review. Some of these examples are considered in Cardé and Minks (1995).

5.8 Methods of testing formulations

Most of the methods used to test efficacy of disruption have changed little since they were reviewed by Roelofs (1979). Only replicated field tests can establish whether a given formulation is capable of disrupting mate finding and ultimately protecting a crop. Principal concerns in such field tests of crop protection are isolation of the treated plots from outside sources of mated females and ensuring that initial population levels are appropriate for this technique. Although crop protection is the decisive gauge of success, there are many other measures as to whether a tested system is efficacious and which can be used to guide its development. Because the goal of mating disruption is interruption of mate location, the most widely used method to evaluate a formulation is to assess how well it disrupts male attraction to either pheromone- or female-baited traps, or suppresses mating of tethered or wing-clipped females, compared to similar traps or females in untreated (experimental control) plots. Typically, disruption of attraction or of mating relative to the experimental checks ought to be >95% for the method to warrant further evaluation in large-scale field trials. Because of the difficulty of finding a series of plots that are of the appropriate size and isolation, and which harbor comparable populations, for initial field trials it typically is necessary to use small replicated plots (on the order of 100 m^2) embedded in a test area under

Table 5.3. *Estimates of worldwide use of mating disruption for control of moth pests in 2002*[a]

Species	Principal crop	Area (ha)	Region
Pectinophora gossypiella	Cotton	55 000	USA, Israel
Cydia pomonella	Apple, pear, almond, walnut	120 000	USA, Australia, EU, South America
Grapholita molesta	Peach, nectarine, apple, pear	50 000	USA, Australia, EU, South America
Lobesia botrana	Grape	41 000	EU
Eupoecilia ambiguella	Grape	32 000	EU
Endopiza viteana	Grape	1000	Canada, USA
Chilo suppressalis	Rice	4000	EU
Leafrollers (Tortricidae)[b]	Tea, pear, apple, peach, grape	24 000	Japan, Australia, USA, New Zealand
Synanthedon spp.[c]	Apricot, black currant, peach	5000	Japan, USA, New Zealand
Zeuzerina pyrina	Pear, olive	2000	EU
Plutella xylostella	Cabbage	2000	Japan
Keiferia lycopersicella	Tomato	10 000	Mexico, USA
Lymantria dispar	Deciduous forests	60 000	USA
Others[d]	Vegetables, apple, peach, golf turf	27 000	Japan, USA
TOTAL		433 000	

[a]Figures courtesy of Shin-Etzu Corporation.
[b]*Adoxophyes, Archips, Epiphyas, Platynota stultana.*
[c]*Synanthedon exitosa, S. hector, S. pictipes, S. tipuliformis.*
[d]*Spodoptera exigua, S. litura, Phyllonorycter ringoniella, Carposina niponensis, Lyonetia clerkella, Anarsi lineatella, Platyptilia carduidactyla.*

conventional control practices (e.g. Charlton and Cardé, 1981; Roelofs and Novak, 1981). This method allows direct comparison of several formulation types, rates of application, etc., before undertaking large-scale evaluations of crop protection. Small-plot comparisons, however, may limit expression of some disruption mechanisms such as habituation. Males in small plots may not be exposed to a given formulation for as prolonged a period as would occur in large plots.

In some evaluations, females have been captured in either light traps or traps baited with odors of adult-food or its host. These females can then be assessed for whether or not they have been mated; analogous procedures for males (Evenden *et al.*, 2003) may as well be useful. The females most attracted to host odors, however, may be those that have already mated, and so this method may not be an accurate measure of the proportion of females that have been prevented from mating.

Male and female codling moths are attracted to an ester found in pear fruits (Light *et al.*, 2001). Efforts to use a synthetic copy of this host odor in a trap to sample codling moths in mating disruption plots (especially in apple and walnut orchards) have not yielded a consistent pattern of insight into whether mating disruption has been efficacious in a given orchard (although clearly mating disruption does protect against the codling moth, provided the initial population is suitably low). The potential value of this approach is that males should be attracted to ester-baited traps in plots under mating disruption: of course, their presence in such plots would be unreliably indicated in pheromone-baited traps. The capture of many males in ester traps, but much less so in pheromone traps, could signify that mating disruption is efficacious. Females also should be attracted; their number could provide a useful measure of whether females are in the treated area, and possibly whether they have been mated. Such information could be valuable for pest management specialists in guiding their decisions on timing the application of disruptant (or if needed, remedial insecticide sprays). In experimental plots, efficacy of mating disruption could be assessed directly by dissecting females and determining their mating status.

Such approaches eventually may provide insights into efficacy, but to date, the pear ester attractant has not proven to be a consistent sampling tool for assessing mating disruption of the codling moth. The probability of capture of both sexes of the codling moth in ester-baited traps in disruptant-treated orchards has been unpredictable among various hosts (apple and walnut), generation, year, and region where the tests are conducted.

A second approach to determining efficacy in mating disruption plots is the use of "high-dose" lures. Doane and Brooks (1981) were the first to note that a high-dose lure, tenfold higher than used in a conventional monitoring trap, attracted pink bollworm males in a cotton field that had received a protective application of pheromone, whereas catch in the lower-dose monitoring trap was suppressed. Similarly, a high dose for the codling moth can be used to determine the presence of codling moths in pheromone-treated orchards (Barrett, 1995). The value of using traps baited with low and high does of pheromone as a device to monitor efficacy of disruption may not extend to other moth species.

Another factor to determine is the emission performance of the formulation. The kinds of evaluations that can be undertaken include: gravimetric analysis, residual analysis, volume changes, and volatile trapping. Dispensers can be field-aged or field conditions can be simulated in the laboratory. The advantages and limitations of these approaches have been summarized elsewhere (e.g. Weatherston, 1990) and are beyond the scope of this review. Emission rates determined in the laboratory can be a useful starting point, but a natural plant canopy and meteorological conditions in the field are very difficult to duplicate in the laboratory. One field method is to collect airborne disruptant with an adsorptive trap, extract this material with solvent, and then quantify it by gas chromatography (GC) or GC-mass spectrometry. This method has been used, for example, to show that the Isomate® rope formulation of the pink bollworm produces a concentration near the top of the cotton canopy of about 1.4 to 2.0 ng m^{-3} (Flint *et al.*, 1990). The limitation of this method is twofold: the quantification procedure is time consuming and it generally requires sampling times of many hours (12 hours in the foregoing case). It thus yields a time-averaged concentration that cannot provide insights into the moment-to-moment distribution of disruptant in the crop.

There are several methods that use an insect antenna as a pheromone detector, relying either on antennal (Karg *et al.*, 1994; Färbert *et al.*, 1996; 1997; Suckling and Karg, 1996) or single sensillum responses to odor levels (van der Pers and Minks, 1996). These methods have the great advantage of allowing real-time comparisons (relative concentrations over several seconds), facilitating many comparisons of density, for example at different positions within or above the canopy or as the wind profile changes. It is also possible to calibrate these detection systems (e.g. Koch *et al.*, 2002), so that airborne concentrations (e.g. ng m^{-3}) can be provided, rather than having concentrations expressed as "relative units".

5.9 Do species differ in their intrinsic susceptibility to mating disruption?

It is enormously problematic to define as a species characteristic its susceptibility to the technique of mating disruption. There is variation in density, patterns of movement, and conventional control practices that accompany mating disruption, not only between species but among populations of a given species. Therefore, it is difficult to disentangle the behavioral basis of disruption from these ecological and management factors. Moreover, the mechanistic basis of disruption is not fully understood for any species, making susceptibility definable only on the basis of end-effects, such as how much the catch by pheromone-baited traps is suppressed, or how well the crop is protected.

Nonetheless, there appear to be some species that can be disrupted even when their populations are initially at very high levels and others that can be disrupted only when their populations are sparse. The oriental fruit moth and tomato pinworm are two exemplars of easily disruptable moths. In the former species, habituation may be an important mechanism contributing to disruption. Both species may be subject to arrestment of upwind plume following at high concentrations of pheromone. In contrast, the pink bollworm may orient to plumes from dispensers with very high release rates and this species can be difficult to habituate (Cardé *et al.*, 1998). This species nonetheless has been controlled at initially high densities in an AWM program (Staten *et al.*, 1996). The codling moth and the gypsy moth, conversely, appear to be species that are successfully controlled by mating disruption only when initial populations are, by comparison, relatively low.

The obliquebanded leafroller is a species that so far has proven difficult to control in apple orchards, even when initial populations are intrinsically low (Agnello *et al.*, 1996; Lawson *et al.*, 1996 and references therein), and this tortricid has been suggested to be a species that is relatively difficult to disrupt (Stelinski *et al.*, 2003*a*; *b*). The explanation for the relative intractability of this leafroller to disruption has been hypothesized to be related to its susceptibility to long-term adaptation of its peripheral chemoreceptors. Evidence considered earlier (Evenden *et al.*, 1999*a*;*b*; 2000), points to competition between point source formulations and females, and to habituation contributing little to disruption in the obliquebanded leafroller.

5.10 Prospects for evolution of resistance

The use of a disruptant in a given area over many years provides selective pressure for the evolution of traits that could bypass mating disruption. The proposed mechanisms for resistance entail a variety of adaptations: an enhancement of the specificity of the male's response to pheromone; an increase in the rate of the female's emission of pheromone; a shift in the pheromone system, involving a coordinated alteration in male response and female production; the tendency for males to be habituated by pheromone; female calling at the top of the plant or other positions in the canopy where there may be less pheromone; and female movement out of the disruptant-treated area before mating (Cardé, 1976; 1990; Cardé and Haynes, 2004). These are reviewed here briefly.

When the disruptant is not a complete copy of the natural blend, males that can distinguish a natural plume amidst the background of disruptant would be favored. This explanation of resistance relies only on the male "improving" an extant behavior. The salutary lesson is that use of a partial or incomplete copy of the natural pheromone blend for disruption carries the risk of promoting a resistant strain based only on improving a male's ability to distinguish the complete blend. (Another reason to use the complete blend, of course, is that a partial blend may be less efficacious (Minks and Cardé, 1988).)

This mechanism likely applies to the only documented case of the evolution of mating resistance. In Japan, mating disruption of the smaller tea tortrix relied on using the most prevalent component of the four-component blend (Mochizuki *et al.*, 2002). After some 15 years of application, efficacy declined in one region, as evidenced by both damage levels and capture of males in pheromone-baited traps. Effective control was restored by changing the disruptant formulation to add the three missing components. The use of a partial blend of a pheromone should place strong selective pressure on the treated populations to shift the male's ability to detect the correct blend amidst a background of disruptant, but other potential mechanisms, such as a reduction in habituation, could be at play. Without detailed information on how mating disruption works in the smaller tea tortrix, understanding how resistance evolved will remain speculative. It is important to emphasize that in this example, resistance is to the disruptant, a component of the pheromone, and not to the complete pheromone.

Another potential strategy is for females to increase emission of pheromone, thereby becoming more apparent amongst the background of disruptant (Cardé, 1976; Shani and Clearwater, 2001). The extent to which this trait could confer resistance is speculative, but, in disruptant-treated fields, pheromone traps baited with high release-rate dispensers catch males, whereas those with a reduced dose do not (Doane and Brooks, 1981; see also Barrett, 1995). Directional selection experiments with the pink bollworm for increased emission suggested that this trait could evolve more rapidly than an alteration in component ratio (Collins *et al.*, 1990). Field populations of pink bollworm in southern California treated with mating disruptant for 10 years had a 20% elevation in pheromone emission, but no alteration in component ratio (Haynes and Baker, 1988). This change would be consistent with predictions of resistance mechanisms, but whether a 20% increase in pheromone would increase mating success of females in a field under mating disruption is untested.

Another proposed mechanism involves a coordinated shift by both males and females in their pheromone channel, so that the new attractant is apparent amongst the background of disruptant. Such a change could involve a shift in the ratio of components or the addition of a new component. Given that female production and male response traits generally do not appear to be linked genetically, such changes may not evolve quickly. One of the formulations that ought to foster such a shift is "lure and kill". Any deviation in an individual female's signal from the formulation would be of selective advantage for a male to detect. As well, odorants emitted from the sticker or insecticide could serve to signify that the point source is not a female, and therefore serve as an independent basis of behavioral resistance.

Resistance to disruptant also might entail altering a male's ability to be habituated by the omnipresence of pheromone, a trait which seems to vary among individual males of the pink bollworm moth (Cardé *et al.*, 1998). Those males that are less readily habituated presumably will continue to search for females. Females also might migrate out of a pheromone-treated plot, mate, and then re-enter the plot. Females of some moths, as noted earlier, appear capable of detecting their own pheromone, and so an ability to detect the presence of pheromone from the formulation could serve as a preadaptation for this change in female behavior.

A parallel strategy for a female "escaping" the effect of a disruptant treatment simply could involve shifting the place of calling within the canopy to positions that render the plume more apparent (Cardé, 1976). Pink bollworm females, for example, seem to call preferentially at the top of the cotton canopy and, depending on the type of formulation employed, there may be less pheromone from the disruptant at the top than within the canopy. Similarly, in the case of the codling moth, it appears that pheromone-baited traps placed near the top of the tree in a disruptant plot are more likely to trap males (Barrett, 1995; Weissling and Knight, 1995), implying that females so positioned also would have a higher probability of luring a mate than females calling from within the canopy. A more complicated variant on this strategy would be to depart a treated field for mating, and to return for oviposition.

Although the smaller tea tortrix remains the only documented case of resistance to mating disruption (not to the complete pheromone but to a component of the pheromone), some loss of efficacy over time has been suggested in a few other species in some localities. Such changes, however, may be the result of changes in grower practices, such as cutting the rate of application of disruptant. Other factors that can contribute to the loss of efficacy are of immigration of mated females from untreated plots, changes in the regime of conventional insecticides, and variation in quality control in commercial disruption products.

5.11 Future directions and prospects

Mating disruption in pest management programs has been implemented to date, not for its safety to consumers, workers, or the environment, but rather because other control techniques, such as conventional pesticides, are ineffective. The first major successes with mating disruption occurred because of an inability to control pink bollworm, oriental fruit moth, and tomato pinworm with conventional insecticides due to insecticide resistance. Legislation in the United States, the European Union, and elsewhere, that is aimed at reducing use of neuroactive pesticides and lowering the total amount of pesticides applied, also may promote future development and acceptance of the mating disruption programs.

A further impetus for adoption of mating disruption comes from increasing grower and government acceptance of areawide approaches to insect management. Three ambitious programs illustrate the potential of AWM. In the case of the gypsy moth, mating disruption is used across an advancing front, several thousand kilometers in extent, of this species in the southern and midwestern United States in an effort to "slow the spread" of this species to the south and west. Such a program is only feasible with a sophisticated management structure and the ability to aerially apply disruptant to low-density populations over wide expanses.

Another AWM program underway in the southwestern United States, is attempting to drive the pink bollworm to "extinction" in cotton-growing areas from Texas to California. The program is geographically sequential, having started in Texas; it is now moving westward and into New Mexico and southward into Chihuahua in northern Mexico. Control measures for pink bollworm rely on a range of tactics, mainly *Bt*-cotton, or where non *Bt*-cotton is to be protected, application of either several kinds of disruptant formulation, conventional insecticides, or aerial inundation with sterile moths. The long-term success of this effort remains to be established. Two concerns are that the pink bollworm is an extremely migratory species. First, males are found in substantial numbers in pheromone-baited traps stationed in the desert of southern California 100s of kilometers from cotton, particularly late in the season (Stern, 1979); presumably females move comparable distances. Second, this moth is present at low levels elsewhere in some of the southeastern portions of the United States. Following eradication, the possibility of recolonization thus remains, and therefore a continuing program of surveillance with pheromone-baited traps will be necessary.

The AWM programs for codling moth in Washington and Oregon and southeastern British Columbia similarly owe much of their success to their use in large, usually contiguous portions of this apple-growing region. For many moth species, the AWM approach seems to enhance disruption efficacy over a patchwork of pheromone-treated and insecticide-treated plots.

Future advances in techniques of disrupting mating could involve having genetically modified plants that synthesize the pheromone of their moth pests, which then presumably would be released from the plants in sufficient quantities to duplicate the effect of applying disruptant formulations. Biosynthesis pathways are well established for many moth pheromones and, in a few cases, the genes responsible for steps in the biosynthesis have been sequenced and cloned (e.g. Roelofs *et al.*, 2002). Although this approach appears technically feasible, its implementation would require considerable innovation and perhaps changes in governmental policies and consumer acceptance of genetically modified crops.

Improvements in the mating disruption techniques that are now readily practicable include development of new release technologies that have increased field longevity and do not require tailor-made, specialized application equipment. Testing release characteristics and longevity in the field in real time is possible with field instrumentation that uses the moth's antenna as a detection device (Karg *et al.*, 1994; Färbert *et al.*, 1996; 1997; Suckling and Karg, 1996). This innovation means that pheromone density can be measured directly (Koch *et al.*, 2002), rather than relying on an indirect sentinel of loss of efficacy, the failure of a formulation to suppress catch in pheromone-baited monitoring traps. This methodology could speed up testing of release characteristics and longevity of formulations.

Another area worthy of continuing study is whether the complete pheromone blend is really the best disruptant. Evidence for most species tends to point to the complete pheromone being more disruptive than partial blends, off ratios, and blends containing antagonists of attraction, but there are few comparative studies, and at least a few cases that seem to support the converse conclusion. Understanding how mating disruption works in the field might suggest new avenues for improving formulations, deployment patterns, and selection of the most effective chemicals.

Finally, it is worth re-emphasizing that for mating disruption to be adopted, it must be more than simply cost effective and efficacious for the target pest in question; it must be compatible with the economics of management of the entire pest complex that affects the crop. Compared to conventional insecticides, mating disruption can be more expensive because of the high costs of pheromone synthesis and formulation, yet it is only active against the target species. On the other hand, as a behavioral control tactic, it is inherently fragile, because its effective use is typically contingent on having good information on a pest's density. Finally, mating disruption is rarely a "stand alone" technology; in most applications some insecticide treatments will need to be applied to the crop to reduce the population to a density appropriate for mating disruption, and to help maintain populations at these levels. There have been notable exceptions to this generalization, namely direct control of initially substantial populations of oriental fruit moth, tomato pinworm, and pink bollworm without a contribution from conventional insecticides. Mating disruption will continue to be used when it is the best available and the most economic control tactic within an integrated pest management program.

Acknowledgments

Numerous colleagues over many years have helped me consider the many issues and examples raised in this review. In particular I am indebted to the late Charles Doane for stimulating my thoughts on how mating disruption works and to Albert Minks for collaborating on two previous reviews on mating disruption. Jack Jenkins, Philipp Kirsch, and Iain Weatherston were kind enough to comment on this manuscript and offer numerous important insights. Many colleagues generously provided updated information and reprints. I am most grateful to Shin-Etsu Corporation for allowing reproduction of their summary of worldwide use of pheromones, presented as Table 5.3.

References

Agnello, A.M., Reissig, W.H., Spangler, S.M., Charlton, R.E. and Kain, D.P. (1996). Trap response and fruit damage by obliquebanded leafroller (Lepidoptera: Tortricidae) in pheromone-treated apple orchards. *Environmental Entomology*, **25**, 268–82.

Alpizar, D., Fallas, M., Oehlschlager, A.C. *et al.* (2002). Pheromone mass trapping of the West Indian sugarcane weevil and the American palm weevil (Coleoptera: Curculionidae) in palmito palm. *Florida Entomologist*, **85**, 426–30.

Arn, H. and Louis, F. (1996). Mating disruption in European vineyards. In R. T. Cardé and A. K. Minks (eds.), *Insect Pheromone Research. New Directions*. New York: Chapman & Hall. pp. 377–82.

Arn, H., Rauscher, S., Buser, H. R. and Guerin, P. M. (1986). Sex pheromone of *Eupoecilia ambiguella* female: analysis and male response to ternary blend. *Journal of Chemical Ecology*, **12**, 1417–29.

Arn, H., Tóth, M. and Priesner, E. (2000). *The Pherolist*. www.nyaes.cornell.edu/pheronet (last updated on 2 September 2000, accessed on September 1, 2003). Superseded by P. Witzgall, T. Lindblom, M. Bengtsson and M. Tóth (2004), *The Pherolist*. www-pherolist.slu.se

Atterholt, C. A., Delwiche, M. J., Rice, R. E. and Krochta, J. M. (1999). Controlled release of insect pheromones from paraffin wax and emulsions. *Journal of Controlled Release*, **57**, 233–47.

Baker, T. C. and Cardé, R. T. (1979). Analysis of pheromone-mediated behaviors in male *Grapholitha molesta*, the oriental fruit moth (Lepidoptera: Tortricidae). *Environmental Entomology*, **8**, 956–68.

Baker, T. C., Hansson, B. S., Löfstedt, C. and Löfqvist, J. (1988). Adaptation of antennal neurons is associated with cessation of pheromone-mediated flight. *Proceedings of the National Academy of Sciences USA*, **85**, 9826–30.

Barnes, B. N. and Blomefield, T. L. (1997). Goading growers towards mating disruption: the South African experience with *Grapholita molesta* and *Cydia pomonella* (Lepidoptera: Tortricidae). In P. Witzgall and H. Arn (eds.), *Technology Transfer in Mating Disruption*. IOBC/wprs Bulletin 20(1), pp. 45–56.

Barrett, B. A. (1995). Effect of synthetic pheromone permeation on captures of male codling moth (Lepidoptera: Tortricidae) in pheromone and virgin female-baited traps at different tree heights in small orchard blocks. *Environmental Entomology*, **24**, 1201–15.

Bartell, R. J. (1982). Mechanisms of communication disruption by pheromone in the control of Lepidoptera: a review. *Physiological Entomology*, **7**, 353–64.

Bartell, R. J. and Roelofs, W. L. (1973). Inhibition of sexual response in males of the moth *Argyrotaenia velutinana* by brief exposures to synthetic pheromone or its geometrical isomer. *Journal of Insect Physiology*, **19**, 655–61.

Bengstsson, M., Karg, G., Kirsch, P. A. *et al.* (1994). Mating disruption of pea moth *Cydia nigricana* F. (Lepidoptera: Tortricidae) by a repellent blend of sex pheromone and attractant inhibitors. *Journal of Chemical Ecology*, **20**, 871–87.

Borden, J. H. (1996). Disruption of semiochemical-mediated aggregation in bark beetles. In R. T. Cardé and A. K. Minks (eds.), *Insect Pheromone Research. New Directions*. New York: Chapman & Hall. pp. 421–38.

Brunner, J. F. (2003). *Building a pheromone-based pest management system for apple production in Washington State*. 3rd International Pheromone Symposium. Bäckaskog, Sweden. Abstracts, p. 51.

Campion, D. G., Critchley, B. R. and McVeigh, L. J. (1989). Mating disruption. In A. R. Justum and R. F. S. Gordon (eds.), *Insect Pheromones in Plant Protection*. Chichester, UK: John Wiley & Sons. pp. 89–119.

Cardé, R. T. (1976). Utilization of pheromones in the population management of moth pests. *Environmental Health Perspectives*, **14**, 133–44.

Cardé, R. T. (1990). Principles of mating disruption. In R. L. Ridway, R. M. Silverstein and M. Inscoe (eds.), *Behavior Modifying Chemicals for Insect Management: Applications of Pheromones and Other Attractants*. New York: Marcel Dekker. pp. 47–91.

Cardé, R. T. (2003). Orientation. In V. H. Resh and R. T. Cardé (eds.), *Encyclopedia of Insects*. San Diego, CA: Academic Press. pp. 823–7.

Cardé, R. T. and Hagaman, T. E. (1984). Mate location strategies of gypsy moths in dense populations. *Journal of Chemical Ecology*, **10**, 25–31.

Cardé, R. T. and Haynes, K. F. (2004). Structure of the pheromone communication channel in moths. In R. T. Cardé and J. G. Millar (eds.), *Advances in the Chemical Ecology of Insects*. Cambridge, UK: Cambridge University Press. pp. 283–332.

Cardé, R. T. and Minks, A. K. (1995). Control of moth pests by mating disruption: successes and constraints. *Annual Review of Entomology*, **40**, 559–85.

Cardé, R. T., Staten, R. T. and Mafra-Neto, A. (1998). Behaviour of pink bollworm males near high-dose, point sources of pheromone in field wind tunnels: insights into mechanisms of mating disruption. *Entomologia Experimentalis et Applicata*, **89**, 35–46.

Cardé, R. T., Trammel, K. and Roelofs, W. L. (1975). Disruption of sex attraction of the redbanded leafroller (*Argyrotaenia velutinana*) with microencapsulated pheromone components. *Environmental Entomology*, **4**, 448–50.

Charlton, R. E. and Cardé, R. T. (1981). Comparing the effectiveness of sexual communication disruption in the oriental fruit moth (*Grapholitha molesta*) using different combinations and dosages of its pheromone blend. *Journal of Chemical Ecology*, **7**, 501–8.

Charlton, R. E. and Cardé, R. T. (1990). Factors mediating copulatory behavior and close-range mate recognition in the male gypsy moth, *Lymantria dispar* (L.). *Canadian Journal of Zoology*, **68**, 1995–2004.

Charmillot, P. J. (1990). Mating disruption technique to control codling moth in western Switzerland. In R. L. Ridway, R. M. Silverstein and M. Inscoe (eds.), *Behavior Modifying Chemicals for Insect Management: Applications of Pheromones and Other Attractants*. New York: Marcel Dekker. pp. 165–82.

Charmillot, P. J., Hofer, D. and Pasquier, D. (2000). Attract and kill: a new method for control of the codling moth *Cydia pomonella*. *Entomologia Experimentalis et Applicata*, **94**, 211–16.

Charmillot, P. J., Pasquier, D., Scalco, A. and Hofer, D. (1996). Essais de lutte contre le carpocapse *Cydia pomonella* L. par un procédé attracticide. *Mitteilungen der Schwiezerischen Entomologischen Gesellschaft*, **69**, 431–9.

Collins, R., Rosenblum, S. and Cardé, R. T. (1990). Selection for increased pheromone titre in the pink bollworm moth, *Pectinophora gossypiella* (Lepidoptera: Gelechiidae). *Physiological Entomology*, **15**, 141–7.

Conlee, J. K. and Staten, R. T. (1987). *Device for insect control*. U.S. Patent No. 4,671,010.

Cook, B. J., Shelton, W. D. and Staten, R. T. (1978). Antennal responses of the pink bollworm to gossyplure. *The Southwestern Entomologist*, **3**, 141–6.

Cork, A., Alam, S. N., Rouf, F. M. A. and Talekar, N. S. (2003). Female sex pheromone of the brinjal fruit and shoot borer, *Leucinodes orbonalis* (Lepidoptera: Pyralidae): trap optimization and application in IMP trials. *Bulletin of Entomological Research*, **93**, 107–13.

Doane, C. C. and Brooks, T. W. (1981). Research and development of pheromones for insect control with emphasis on the pink bollworm. In E. R. Mitchell (ed.), *Plenum, Management of Insect Pests with Semiochemicals. Concepts and Practice*. New York. pp. 285–303.

Evenden, M. L., Delury, L. E., Judd, G. J. R. and Borden, J. H. (2003). Assessing the mating status of male obliquebanded leafrollers *Choristoneura rosaceana* (Lepidoptera: Tortricidae) by dissection of male and female moths. *Annals of the Entomological Society of America*, **96**, 217–24.

Evenden, M. L., Judd, G. J. R. and Borden, J. H. (1999*a*). Pheromone-mediated mating disruption of *Choristoneura rosaceana*: is the most attractive blend really the most effective? *Entomologia Experimentalis et Applicata*, **90**, 37–47.

Evenden, M. L., Judd, G. J. R. and Borden, J. H. (1999*b*). Simultaneous disruption of pheromone communication in *Choristoneura rosaceana* and *Pandemis limitata* with pheromone and antagonist blends. *Journal of Chemical Ecology*, **25**, 501–17.

Evenden, M. L., Judd, G. J. R. and Borden, J. H. (2000). Investigations of mechanisms of pheromone communication disruption of *Choristoneura rosaceana* (Harris) in a wind tunnel. *Journal of Insect Behavior*, **13**, 499–510.

Fadamiro, H. Y. and Baker, T. C. (1999). Reproductive performance and longevity of female European corn borer, *Ostrinia nubilalis*: effects of multiple mating, delay in mating, and adult feeding. *Journal of Insect Physiology*, **45**, 385–92.

Fadamiro, H. Y., Cossé, A. A. and Baker, T. C. (1999). Mating disruption of European corn borer, *Ostrinia nubilalis* by using two types of sex pheromone dispensers deployed in grassy aggregation sites in Iowa cornfields. *Journal of Asia-Pacific Entomology*, **2**, 121–32.

Färbert, P., Koch, U. T., Färbert, A. and Staten, R. T. (1996). Measuring pheromone concentration in cotton fields with the EAG method. In R. T. Cardé and A. K. Minks (eds.), *Insect Pheromone Research. New Directions*. New York: Chapman & Hall. pp. 347–58.

Färbert, P., Koch, U. T., Färbert, A., Staten, R. T. and Cardé, R. T. (1997). Pheromone concentration measured with EAG in cotton fields treated for mating disruption of *Pectinophora gossypiella* (Lepidoptera: Gelechiidae). *Environmental Entomology*, **26**, 1105–16.

Figueredo, A. J. and Baker, T. C. (1992). Reduction of the response to sex pheromone in the oriental fruit moth, *Grapholita molesta* (Lepidoptera: Tortricidae) following successive pheromone exposures. *Journal of Insect Behavior*, **5**, 347–63.

Flint, H. M., Yamamoto, A., Parks, N. J. and Nyomura, K. (1990). Aerial concentration of gossyplure, the sex pheromone of the pink bollworm (Lepidoptera: Gelechiidae), in cotton fields treated with long-lasting dispensers. *Environmental Entomology*, **19**, 1845–51.

Gaston, L.K., Shorey, H.H. and Saario, S.A. (1967). Insect population control by the use of sex pheromones to inhibit orientation between the sexes. *Nature*, **213**, 1155.

Gaston, L.K., Kaae, R.S., Shorey, H.H. and Seller, D. (1977). Controlling the pink bollworm by disrupting sex pheromone communication between adult moths. *Science*, **196**, 904–5.

Gronning, E.K., Borchert, D.M., Pfeiffer, D.G. *et al.* (2000). Effect of specific and generic sex attractant blends on pheromone trap captures of four leafroller species in mid-Atlantic apple orchards. *Journal of Economic Entomology*, **93**, 157–64.

Gut, L.J. and Brunner, J.F. (1998). Pheromone-based management of codling moth (Lepidoptera: Tortricidae) in Washington apple orchards. *Journal of Agricultural Entomology*, **15**, 387–406.

Haynes, K.F., Li, W.G. and Baker, T.C. (1986). Control of pink bollworm moth (Lepidoptera: Gelechiidae) with insecticide and pheromones (attracticide): lethal and sublethal effects. *Journal of Economic Entomology*, **79**, 1466–71.

Haynes, K.F. and Baker, T.C. (1988). Potential for evolution of resistance to pheromones. Worldwide and local variation in chemical communication system of the pink bollworm moth, *Pectinophora gossypiella*. *Journal of Chemical Ecology*, **14**, 1547–560.

Howse, P., Stevens, I. and Jones, O. (1998). *Insect Pheromones and their Use in Pest Management*. London: Chapman & Hall.

Il'ichev, A.L., Gut, L.J., Williams, D.G., Hossain, M.S. and Jerie, P.H. (2002). Area-wide approach for improved control of oriental fruit moth *Grapholita molesta* (Busck) (Lepidoptera: Torticidae) by mating disruption. *General and Applied Entomology*, **31**, 7–15.

Ioriatti, C. and Angeli, G. (2002). Control of codling moth by attract and kill. In P. Witzgall, B. Mazomenos and M. Konstantopoulou (eds.), *Pheromones and Other Biological Techniques for Insect Control in Orchards and Vineyards*. International Organization for Biological Control, Montpelier, France, Bulletin 25 (9). pp. 129–36.

Issacs, R., Ulczynski, M., Wright, B., Gut, L.J. and Miller, J.R. (1999). Performance of the microsprayer, with application for pheromone-mediated control of insect pests. *Journal of Economic Entomology*, **92**, 1157–64.

Jenkins, J.W., Doane, C.C., Schuster, D.J., McLaughlin, J.R. and Jimenez, M.H. (1990). Development and commercial application of sex pheromone for control of the tomato pinworm. In R.L. Ridway, R.M. Silverstein and M. Inscroe (eds.), *Behavior Modifying Chemicals for Insect Management: Applications of Pheromones and Other Attractants*. New York: Marcel Dekker. pp. 269–80.

Justum, A.R. and Gordon, R.F.S. (eds.) (1989). *Insect Pheromones in Plant Protection*. Chichester, UK: John Wiley and Sons.

Kaae, R.S. and Shorey, H.H. (1973). Sex pheromones of Lepidoptera. 44. Influence of environmental conditions on the location of pheromones communication and mating in *Pectinophora gossypiella*. *Environmental Entomology*, **2**, 1081–4.

Karg, G., Suckling, D. M. and Bradley, S. J. (1994). Adsorption and release of pheromone of *Epiphyas postvittana* (Lepidoptera: Tortricidae) by apple leaves. *Journal of Chemical Ecology*, **20**, 1825–41.

Koch, U. T., Cardé, A. M. and Cardé, R. T. (2002). Calibration of an EAG system to measure airborne concentration of pheromone formulated for mating disruption of the pink bollworm moth, *Pectinophora gossypiella* (Saunders) (Lepidoptera: Gelechiidae). *Journal of Applied Entomology*, **126**, 431–5.

Krupke, C. H., Roitberg, B. D. and Judd, G. J. R. (2002). Field and laboratory responses of male codling moth (Lepidoptera: Tortricidae) to a pheromone-based attractant-and-kill strategy. *Environmental Entomology*, **31**, 190–7.

Kydonieus, A. F. and Beroza, M. (1982). *Insect Suppression with Controlled Release Pheromone Systems*. Vols. I and II. Boca Raton, FL: CRC Press.

Lawson, D. S., Reissig, W. H., Agnello, A. M., Nyrop, J. P. and Roelofs, W. L. (1996). Interference with mate-finding communication system of the obliquebanded leafroller (Lepidoptera: Tortricidae) with sex pheromones. *Environmental Entomology*, **25**, 895–905.

Leonhardt, B. A., Mastro, V. C., Leonard, D. S. *et al.* (1996). Control of low-density gypsy moth (Lepidoptera: Lymatriidae) populations by mating disruption with pheromone. *Journal of Chemical Ecology*, **22**, 1255–72.

Light, D. M., Knight, A. L., Henrick, C. A. *et al.* (2001). A pear-derived kairomone with pheromonal potency that attracts male and female codling moth, *Cydia pomonella* (L.). *Naturwissenschaften*, **88**, 333–8.

Lösel, P. M., Penners, G., Potting, R. P. J. *et al.* (2000). Laboratory and field experiments towards the development of an attract and kill strategy for control of the codling moth, *Cydia pomonella*. *Entomologia Experimentalis et Applicata*, **95**, 39–46.

Lösel, P. M., Potting, R. P. J., Ebbinghaus, D., Elbert, A. and Scherkenbeck, J. (2002). Factors affecting the performance of an attracticide against of the codling moth *Cydia pomonella*. *Pest Management Science*, **58**, 1029–37.

Mafra-Neto, A. and Baker, T. C. (1996). Elevation of pheromone response threshold in almond moth males pre-exposed to pheromone spray. *Physiological Entomology*, **21**, 217–22.

Mafra-Neto, A. and Habib, M. (1996). Evidence that mass trapping suppresses pink bollworm populations in cotton fields. *Entomologia Exerimentalis et Applicata*, **81**, 315–23.

Mayer, M. S. and McLaughlin, J. R. (1991). *Handbook of Insect Pheromones and Sex Attractants*. Boca Raton, FL: CRC Press.

Miller, E., Staten, R. T., Nowell, C. and Gourd, J. (1990). Pink bollworm (Lepidoptera: Gelechiidae): point source density and its relationship to efficacy in attracticide formulations of gossyplure. *Journal of Economic Entomology*, **83**, 1321–5.

Minks, A. K. (1996). Mating disruption of the codling moth. In R. T. Cardé and A. K. Minks (eds.) *Insect Pheromone Research. New Directions*. New York: Chapman and Hall. pp. 372–82.

Minks, A. K. and Cardé, R. T. (1988). Disruption of pheromone communication in moths: is the natural blend really most efficacious? *Entomologia Exerimentalis et Applicata*, **49**, 25–36.

Mitchell, E. R. (2002). Promising new technology for managing diamondback moth (Lepidoptera: Plutellidae) in cabbage with pheromone. *Journal of Environmental Science and Health*, **B37**, 277–90.

Mitchell, E. R. (ed.) (1981). *Management of Insect Pests with Semiochemicals. Concepts and Practice*. New York: Plenum.

Mitchell, E. R., Jacobson, M. and Baumhover, A. H. (1975). *Heliothis* spp.: disruption of pheromonal communication with (*Z*)-9-tetradecen-1-ol formate. *Environmental Entomology*, **4**, 577–9.

Mitchell, E. R., Tumlinson, J. H. and Baumhover, A. H. (1978). *Heliothis virescens*: attraction of males to blends of (*Z*)-9-tetradecen-1-ol formate and (*Z*)-9-tetradecenal. *Journal of Chemical Ecology*, **4**, 709–16.

Mochizuki, F., Fukumoto, T., Noguchi, H. *et al.* (2002). Resistance to a mating disruptant composed of (*Z*)-11-tetraecenyl acetate in the smaller tea tortrix, *Adoxophyes honmai* (Yasda). *Applied Entomology and Zoology*, **37**, 299–304.

Murlis, J., Elkinton, J. S. and Cardé, R. T. (1992). Odor plumes and how insects use them. *Annual Review of Entomology*, **37**, 505–32.

Novak, M. A. and Roelofs, W. L. (1983). Behavior of male redbanded leafroller moths, *Arygrotaenia velutinana* (Lepidoptera: Tortricidae) in small disruption plots. *Environmental Entomology*, **14**, 12–16.

Oehlschlager, A. C., Chinchilla, C., Castillo, G. and Gonzalez, L. (2002). Control of red ring disease by mass trapping of *Rhynchophorus palmarum* (Coleoptera: Curculionidae). *Florida Entomologist*, **85**, 507–13.

Ogawa, K. (2000). Pest control by pheromone mating disruption and the role of natural enemies. *Journal of Pesticide Science*, **25**, 456–61.

Palanaswamy, P. and Seabrook, W. D. (1978). Behavioral responses of the female eastern spruce budworm *Choristoneura fumiferana* (Lepidoptera: Tortricidae) to the sex pheromone of her own species. *Journal of Chemical Ecology*, **4**, 649–56.

Palanaswamy, P. and Seabrook, W. D. (1985). The alteration of calling behavior by female *Choristoneura fumiferana* when exposed to synthetic sex pheromone. *Entomologia Experimentalis et Applicata*, **37**, 13–16.

Pearson, G. A. and Schal, C. (1999). Electroantennogram responses of both sexes of grape root borer (Lepidoptera:Sesiidae) to synthetic female sex pheromone. *Environmental Entomology*, **28**, 943–6.

Pfeiffer, D. G., Killian, J. C., Rajotee, E. G., Hull, L. A. and Snow, J. W. (1991). Mating disruption for reduction of damage by lesser peach tree borer (Lepidoptera: Sesiidae) in Virginia and Pennsylvania peach orchards. *Journal of Economic Entomology*, **84**, 218–23.

Plimmer, J. R., Leonhardt, B. A., Webb, R. E. and Schwalbe, C. P. (1982). Management of the gypsy moth with its sex attractant pheromone. In B. A. Leonhardt and M. Beroza (eds.), *Insect Pheromone Technology: Chemistry and Applications*. American Chemical Society Sym. Ser. 190. pp. 232–42.

Polavarapu, S., Wicki, M., Vogel, K., Lonergan, G. and Nielsen, K. (2002). Disruption of sexual communication of oriental beetles (Coleoptera: Scarabaeidae) with a microencapsulated formulation of sex pheromone components in blueberries and ornamental nurseries. *Environmental Entomology*, **31**, 1268–75.

Poullot, D., Beslay, D., Bouvier, J.C. and Sauphanor, B. (2001). Is attract-and-kill technology potent against insecticide-resistant Lepidoptera? *Pest Management Science*, **57**, 729–36.

Reardon, R.C., Leonard, D.S., Mastro, V.C. *et al.* (1998). *Using Mating Disruption to Manage Gypsy Moth: A Review*. USDA Forest Service FHTET-98-01, Morgantown, West Virginia, USA.

Renou, M. and Guerrero, A. (2000). Insect parapheromones in olfaction research and semiochemical-based pest control strategies. *Annual Review of Entomology*, **45**, 605–30.

Ridway, R.L., Silverstein, R.M. and Inscoe, M. (eds.) (1990). *Behavior Modifying Chemicals for Insect Management: Applications of Pheromones and Other Attractants*. New York: Marcel Dekker.

Roelofs, W.L. (ed.) (1979). *Establishing Efficacy of Sex Attractants and Disruptants for Insect Control*. Lanham, Maryland: Entomological Society of America.

Roelofs, W.L. and Novak, M.A. (1981). Small-plot disorientation for screening potential disruptants. In E.R. Mitchell (ed.), *Management of Insects Pests with Semiochemicals*. Concepts and Practice. New York: Plenum. pp. 229–42.

Roelofs, W.L., Liu, W., Hao, G. *et al.* (2002). Evolution of moth sex pheromones via ancestral genes. *Proceedings of the National Academy of Sciences USA*, **99**, 13621–6.

Rothschild, G.H.L. and Vickers, R.A. (1991). Biology, ecology and control of the oriental fruit moth. In L.P.S. Van Der Geest and H.S. Evenhuis (eds.), *Tortricid Pests. Their Biology, Ecology and Control*. Amsterdam: Elsevier. pp. 389–412.

Rumbo, E.R. and Vickers, R.A. (1997). Prolonged adaptation as possible mating disruption mechanism in the oriental fruit moth, *Cydia (=Grapholita) molesta*. *Journal of Chemical Ecology*, **23**, 445–57.

Sanders, C.J. (1996). Mechanisms of mating disruption in moths. In R.T. Cardé and A.K. Minks (eds.), *Insect Pheromone Research. New Directions*. New York: Chapman and Hall. pp. 333–46.

Sanders, C.J. and Lucuik, G.S. (1996). Disruption of oriental fruit moth to calling females in a wind tunnel by different concentrations of synthetic pheromone. *Journal of Chemical Ecology*, **22**, 1971–86.

Schneider, D., Schulz, S., Priesner, E., Ziesmann, J. and Francke, W. (1998). Autodetection and chemistry of female and male sex pheromone in both sexes of the tiger moth *Panaxia quadripunctaria*. *Journal of Comparative Physiology A*, **182**, 153–61.

Sciarappa, W.J., Polavarapu, S., Holdcraft, R.J. and Barry, J.D. (2005). Disruption of sexual communication of oriental beetles (Coleoptera: Scarabaeidae) in highbush blueberries with retrievable pheromone sources. *Environmental Entomology*, **34**, 54–8.

Sexton, S.B. and Il'ichev, A.L. (2000). Pheromone mating disruption with reference to oriental fruit moth *Grapholita molesta* Busck (Lepidoptera: Tortricidae) by mating disruption. *General and Applied Entomology*, **29**, 63–8.

Shani, A. and Clearwater, J. (2001). Evasion of mating disruption in *Ephestia cautella* (Walker) by increased pheromone production relative to that of undisrupted populations. *Journal of Stored Products Research*, **37**, 237–52.

Sharov, A.A., Thorpe, K.W. and Tcheslavskaia, K. (2002). Effect of synthetic pheromone on gypsy moth (Lepidoptera: Lymantriidae) trap catch and mating success beyond treated areas. *Environmental Entomology*, **31**, 1119–27.

Shorey, H.H. and Gerber, R.G. (1996). Use of puffers for disruption of sex pheromone communication of codling moths (Lepidoptera: Tortricidae) in walnut orchards. *Environmental Entomology*, **25**, 1398–400.

Snow, J.W. (1990). Peachtree borer and lesser peachtree borer control in the United States. In R.L. Ridway, R.M. Silverstein and M. Inscoe (eds.), *Behavior Modifying Chemicals for Insect Management: Applications of Pheromones and Other Attractants*. New York: Marcel Dekker. pp. 241–53.

Staten, R.T., El-Lissy, O. and Antilla, L. (1996). Successful area-wide program to control pink bollworm by mating disruption. In R.T. Cardé and A.K. Minks (eds.), *Insect Pheromone Research. New Directions*. New York: Chapman and Hall. pp. 383–96.

Stelinski, L.L., Miller, J.R. and Gut, L.J. (2003*a*). Presence of long-lasting peripheral adaptation in oblique-banded leafroller, *Choristoneura rosaceana* and absence of such adaptation redbanded leafroller, *Argyrotaenia velutinana*. *Journal of Chemical Ecology*, **29**, 405–23.

Stelinski, L.L., Miller, J.R. and Gut, L.J. (2003*b*). Concentration of air-borne pheromone required for long-lasting peripheral adaptation in the obliquebanded leafroller, *Choristoneura rosaceana*. *Physiological Entomology*, **28**, 97–107.

Stern, V.M. (1979). Long and short range dispersal of the pink bollworm *Pectinophora gossypiella* over southern California. *Environmental Entomology*, **8**, 524–7.

Suckling, D.M. (2001). Issues affecting the use of pheromones and other semiochemicals in orchards. *Crop Protection*, **19**, 677–83.

Suckling, D.M. and Karg, G. (1996). Mating disruption of the light brown apple moth: portable electroantennogram equipment and other aspects. In R.T. Cardé and A.K. Minks (eds.), *Insect Pheromone Research. New Directions*. New York: Chapman and Hall. pp. 411–20.

Suckling, D.M. and Brockerhoff, E.G. (1999). Control of the light brown apple moth (Lepidoptera: Tortricidae) using an atracticide. *Journal of Economic Entomology*, **92**, 367–72.

Thorpe, K.W., Leonard, D.S., Mastro, V.C. *et al.* (2000). Effectiveness of gypsy moth mating disruption from aerial applications of plastic laminate flakes with and without a sticking agent. *Agricultural and Forest Entomology*, **2**, 1–7.

Trammel, K., Roelofs, W.L. and Glass, E.H. (1974). Sex-pheromone trapping of males for control of redbanded leafroller in apple orchards. *Journal of Economic Entomology*, **67**, 159–64.

Trimble, R.M., Pree, D.J. and Carter, N.J. (2001). Integrated control of oriental fruit moth (Lepidoptera: Tortricidae) in peach orchards using insecticide and mating disruption. *Journal of Economic Entomology*, **94**, 476–85.

Trumble, J.T. and Alvarado-Rodriguez, B. (1993). Development and economic evaluation of an IPM program for fresh market tomato production in Mexico. *Agriculture, Ecosystems and Environment*, **43**, 267–84.

Valeur, P.G. and Löfstedt, C. (1996). Behaviour of male oriental fruit moth, *Grapholita molesta*, in overlapping sex pheromone plumes in a wind tunnel. *Entomologia Experimentalis et Applicata*, **79**, 51–9.

Van der Pers, J. and Minks, A.K. (1996). Measuring pheromone dispersion in the field with single sensillum recording technique. In R.T. Cardé and A.K. Minks (eds.), *Insect Pheromone Research. New Directions*. New York: Chapman and Hall. pp. 359–71.

Vickers, R.A. (1990). Oriental fruit moth in Australia and Canada. In R.L. Ridway, R.M. Silverstein and M. Inscoe (eds.), *Behavior Modifying Chemicals for Insect Management: Applications of Pheromones and Other Attractants*. New York: Marcel Dekker. pp. 183–92.

Vickers, R.A. (1997). Effect of a delayed mating on oviposition pattern, fecundity and fertility in codling moth, *Cydia pomonella* (L) Lepidoptera, Torticidae). *Australian Journal of Entomology*, **36**, 179–82.

Waldstein, D.E. and Gut, L.J. (2003). Comparison of microcapsules density with various apple tissues and formulations of oriental fruit moth (Lepidoptera: Tortricidae) sprayable pheromone. *Journal of Economic Entomology*, **96**, 58–63.

Wall, C., Sturgeon, D.M., Greenway, A.R. and Perry, J.N. (1981). Contamination of vegetation with synthetic sex attractant released from traps for pea moths, *Cydia nigricana*. *Entomologia Experimentalis et Applicata*, **30**, 111–15.

Weatherston, I. (1990). Principles of design of controlled-release formulation. In R.L. Ridway, R.M. Silverstein and M. Inscoe (eds.), *Behavior Modifying Chemicals for Insect Management: Applications of Pheromones and Other Attractants*. New York: Marcel Dekker. pp. 93–112.

Weissling, T.J. and Knight, A.L. (1995). Vertical distribution of codling moth adults in pheromone-treated and untreated plots. *Entomologia Experimentalis et Applicata*, **77**, 271–5.

Witzgall, P., Bäckman, A.C., Svensson, M. *et al.* (1999). Behavioral observations of codling moth, *Cydia pomonella*, in orchards permeated with synthetic pheromone. *BioControl*, **44**, 211–37.

Zhang, G.-F., Meng, A.Z., Han, Y. and Sheng, C.F. (2002). Chinese tortix (*Cydia trasias*) (Lepidoptera: Olethreutidae): suppression on street-planting trees by mass trapping with sex pheromone traps. *Environmental Entomology*, **31**, 602–7.

6

Nutritional ecology of plant feeding arthropods and IPM

ANTÔNIO R. PANIZZI

6.1 Introduction

Insect nutritional ecology has been defined as an area of entomology that involves the integration of biochemical, physiological, and behavioral information, within the context of ecology and evolution (Slansky and Rodriguez, 1987*a*). Such a broad view suggests the need for basic studies essential to understanding the different life styles of insects.

Considering the damage inflicted to plant structures by feeding arthropods, it is possible to identify several feeding guilds, from the more conspicuous foliage and fruit chewers to the less noticeable seed-suckers, fruit-borers, and root-feeders. All of these, and many others, have been studied and reviewed under the paradigm of insect nutritional ecology (see chapters in Slansky and Rodriguez, 1987*b*). In general, these reviews using the insect nutritional ecology model, have focused primarily on basic aspects of the different insects (feeding guild biology), and only secondarily have dealt with applied aspects, despite the enormous importance of insects in these guilds as pests of major crops worldwide.

Within the context of integrated pest management (IPM) systems, several tactics taking into account the nutritional ecology model sensu lato can be considered. They include host plant resistance, trap crops, asynchrony of foods and pests phenology, crop consortiums, and functional allelochemicals. These tactics, although considered in several IPM text books (e.g. Pimentel, 1981; Kogan, 1986*a*; Rechcigl and Rechcigl, 2000; Flint and Gouveia, 2001; Pedigo, 2002; Pimentel, 2002; Norris *et al.*, 2003), still must be further explored within the context of the insect nutritional ecology paradigm, considering each of the major feeding guilds associated with plants.

One attempt to stress IPM under the scope of nutritional ecology is the book *Insect Nutritional Ecology and its Implication on Pest Management* (in Portuguese) (Panizzi and Parra, 1991*a*), in which greater attention was given to management tactics considering insect feeding activity, host plant preferences, host plant impact on pest populations, and feeding behavior.

In this chapter, I will touch upon basic information for holistic IPM programs, including insect–plant interactions, plant diversity and stability, and IPM tactics in the context of insect nutritional ecology. As a case study, I will present in greater detail a system with soybean as the major commodity and the guild of seed-sucking insects associated with it. This guild includes many severe pests of several crops worldwide (Schaefer and Panizzi, 2000), and it is the most important pest complex of soybean in the neotropics, a region that hosts the second largest soybean production area in the world. Using this system, it will be shown how basic information on interactions of these pests with their entire host plant range may be used to mitigate impact on the main crop plant.

6.2 Nutritional ecology of phytophagous arthropods and IPM

6.2.1 *Insect–plant interactions*

Insect–plant interactions have been explored in many ways, and the literature covering this subject has exploded in the last 20+ years (Ahmad, 1983; Crawley, 1983; Bernays, 1989–1994; Bernays and Chapman, 1994; Brackenbury, 1995; Jolivet, 1998). Phytophagous insects plus the plants they feed upon make up ca. 50% of all living species; members of Lepidoptera, Hemiptera, and Orthoptera are mostly phytophages (Strong *et al.*, 1984).

Despite the gigantic biomass formed by plants, only nine orders of insects utilize plants as their main food, which suggests that plants may not be an ideal food. Due to many physical (pilosity, toughness of tissues, thorns, etc.) and chemical (non-nutritional compounds, imbalance of nutrients, lack of water, etc.) attributes, insects cannot or do not explore plants fully as food sources (Edwards and Wratten, 1980).

Because of the diversity of plant defenses, and insect adaptations to feed on the defended plants, studies on co-evolution have proliferated during the past 30–40 years, since the publication of the paper by Erlich and Raven (1964) on the co-evolution of butterflies and plants. Despite this and many other studies that followed, several authors do not consider co-evolution to be the general mechanism driving insect–plant interactions or regard it as the mechanism responsible for the structure of phytophagous insect communities (e.g. Janzen, 1980; Fox, 1981; Futuyma, 1983; Jermy, 1984; Strong *et al.*, 1984). Insects and plants co-exist, and considering the integrated management of pests on crops, the theoretical bases of insect–plant interactions provide subsidies for research on host plant resistance and the practice of IPM (Kogan, 1986*a*).

6.2.2 *Plant diversity and stability*

The issue of species diversity and stability of biotic communities has been the object of considerable interest and debate (see also Chapters 3, 8 and 15). Although this seems to be the case for natural ecosystems, when considering agroecosystems, ecologists and pest management experts and practitioners still argue whether the 'diversity–stability hypothesis' holds true.

In general, studies suggest that, with the increase of biodiversity, i.e. all species of plants, animals, and microorganisms interacting in an ecosystem, it is possible to stabilize the community of insects, and to enhance the management of pests (Altieri and Letourneau, 1984; Andow, 1991). With expansion of monocultures plant biodiversity is reduced, with consequent habitat destruction, decrease in resource availability, and reduction in numbers of arthropod species; this leads to changes in the functioning of the ecosystem, affecting its productivity and sustainability (Altieri and Nicholls, 1999).

Southwood and Way (1970) considered factors influencing the degree of biodiversity in agroecosystems: (a) the diversity of vegetation within and around the agroecosystem; (b) the temporal and spatial permanence of the various crops within the agroecosystem; (c) the intensity of management practices, such as tillage and pesticide applications; and (d) the degree of isolation of the agroecosystem from natural vegetation. The role of uncultivated land in the biology of crop pests and their natural enemies has been recognized (van Emden, 1965).

In agroecosystems, biodiversity can be planned or associated, as suggested by Vandermeer (1995). In the first case, biodiversity consists of cultivated crops, livestock, and associated organisms, which are introduced into the system on purpose, for economic or aesthetic reasons, and are managed intensively. In the second case, biodiversity includes all organisms, from plants and higher animals to microbes, which naturally existed or moved into the system from surrounding areas. This associated biodiversity is important to maintain or mitigate the unbalance that usually is associated with the planned biodiversity.

It may be stated that the stability of ecosystems, in general, is a result of the addition of all interactions among the living organisms. Therefore, the more structured the agroecosystem the greater the stability. Altieri (1994) reported that cropping systems with taller plants (e.g. corn) mixed with shorter plants (squash or beans) provide more niches, enhancing species biodiversity. In southern Brazil, small growers cultivate beans, cassava, and small grains, in areas surrounded by taller plants such as corn or pigeon pea. These latter plants not only provide increased species diversity, but also function as barriers to insects' dispersion, preventing pest outbreaks. Producers of organic soybean plant the beans in relatively small areas surrounded by natural vegetation or corn to reduce the attack of pests (see section 6.3.2.3: Managing mixed crops to mitigate the heteropterans' impact to soybean).

6.2.3 *IPM tactics in the context of nutritional ecology*

The IPM tactics of host plant resistance, trap cropping, and mixed cropping and the allelochemicals associated with those systems, can be profitably analyzed under the context of nutritional ecology.

Host plant resistance

The use of cultivars resistant to pests is one of the most effective, economical, and environmentally safe management tactics (Pedigo, 2002) and should be a key component of any IPM system.

The development of host plant resistance within the context of the nutritional ecology model includes the interrelationships of food attributes, with the insect consumption and utilization of the food, and its consequences to the insect performance and fitness. These interrelationships between insect nutritional ecology and host plant resistance were illustrated by Slansky (1990) (Figure 6.1). In this diagram, studies on insect nutritional ecology focus on the understanding of the effect of food on the insect biology, while host plant resistance attempts to manipulate food attributes to manage insect pests. Therefore, the basic insect nutritional ecology approach supports the applied approach of host plant resistance, and the convergence of the two disciplines results in a better understanding of the whole process.

Of the three fundamental modalities of host plant resistance, i.e. antibiosis, antixenosis, and plant tolerance, stated over 50 years ago by Painter (1951), the first component – antibiosis – relates to the context of insect nutritional ecology. Plant attributes comprising nutrients, non-nutrients, and morphological features, will dictate the extent of the food's impact on the insect's biology. This impact may result in death of immatures,

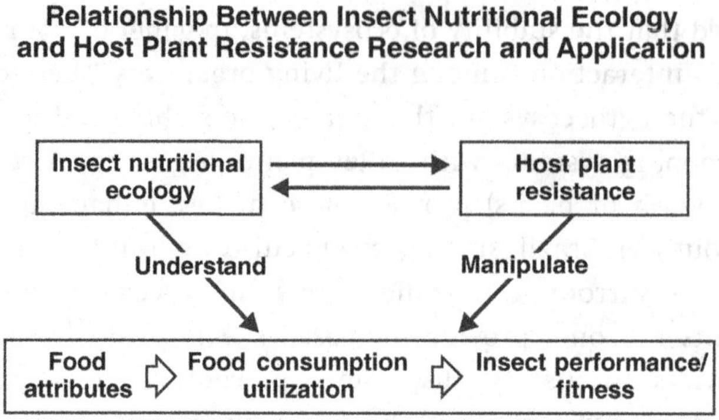

Figure 6.1. Schematic representation of the interrelationships between insect nutritional ecology and host plant resistance (source: Slansky, 1990).

reduced growth rates, increased mortality of pupae, small adults with reduced fecundity, shortened adult life span, morphological malformations, restlessness, and other abnormal behaviors (Pedigo, 2002).

With the introduction of genetically modified (GM) crops carrying toxins, host plant resistance is gaining a new momentum. This approach using modern biotechnology is considered a new technological breakthrough in agriculture, comparable to the green revolution of the early 1970s. For example, transgenic plants expressing the bacterium *Bacillus thuringiensis* Berliner, which produces toxins that confer pest resistance to plants, has been introduced in at least 18 crops; corn, cotton, and potato GM cultivars are already commercially available (Shelton *et al.*, 2002). In 2001, ca. 13 million hectares were cultivated with *Bt*-corn and *Bt*-cotton, mainly in the USA and Canada (James, 2001). *Bt*-crops are also being cultivated in China, India, South Africa, and Argentina (Carpenter *et al.*, 2002). Other toxins, such as inhibitors of digestive enzymes (proteinases, amylases) and lectins, are also being introduced to plants to give them protective effects (Gatehouse and Gatehouse, 2000). These and other toxins, being introduced in cultivars of many crops, will certainly make the host plant resistance strategy a major component of IPM programs worldwide. However, concerns about the possible environmental effects of genetically modified crops resistant to insects have risen, and this issue has been extensively debated (see review by Fontes *et al.*, 2002).

Trap crops

Trap crops are plants, usually preferred hosts, planted to attract insects and, in consequence, to divert their attack from the crops. This can be reached by deviating the pests from attacking the target crop, and concentrating them in great numbers in restricted areas, where control measures can be taken, usually much more economically compared with conventional control methods, such as the use of pesticides (see review by Hokkanen, 1991).

This tactic (trap crop) has strong components in the context of the nutritional ecology model. These components include, first, the feeding preference. Although most insects are polyphagous or olygophagous, they tend to show preferences for certain plant taxa, and one can use this preference to attract the insects. This preference will be dictated, at least in part, by the nutritional value of the plants. Apparently, insects can predict or evaluate the nutritional value of plants, and "choose" them for oviposition. Although less-preferred host plants also have an important role in the insect's biology (see section 6.3.3.4: The role of less-preferred plant food sources), the preferred hosts usually contribute more to the insect's fitness.

A second component of the trap crop tactic, considering the nutritional ecology paradigm, has to do with the impact of the trap crop on the performance of larvae/nymphs and adults. Usually, on these preferred hosts, the maximum potential contribution to the next generation is expected to be achieved, with production of the fittest individual. Survivorship of immatures and reproduction of adults are the greatest. Therefore, populations will tend to explode on these preferred hosts and exhaust them; an accurate estimate of the holding capacity of the trap crop should be determined so pest insects do not leave the trap crop because of interspecific competition and lack of food. Thus it is important to determine when to interfere with control measures to avoid dispersion of populations to the target crop.

A third component of trap crops, considering the nutritional ecology model, is that preferred, and therefore, highly nutritional host plants, may allow pest species to store energy in their bodies to overcome unfavorable periods of food scarcity. Although not considered widespread, this is a very important event in the biology of those insects that accumulate energy. Feeding on a rich nutritional food source, such as trap crops, in particular at times that precede the winter, might be crucial for these insects' survival.

Cultivation of mixed crops

The cultivation of mixed crops is another tactic of IPM that fits in the context of nutritional ecology. In general, we may say that as the diversification of cropping systems is increased, or, as the number of cultivated plant species is increased in a particular system (polycultures), outbreaks of herbivores populations are decreased (Andow, 1991; Altieri, 1994; see also section 6.2.2).

There are several reasons why polycultures are less susceptible to pest attack. First, different species of plants intercropped may provide mutual protection by acting as a physical barrier, each for the other; second, one species of plant may camouflage another species of plant, forming a mosaic that will confound the behavior of pests; and third, the odor produced by a particular plant may repel or disrupt the searching ability of pest species (Altieri, 1994).

Another major point that makes polycultures less susceptible to pest outbreaks is the greater occurrence of natural enemies (predators and parasitoids) in such a system than in monocultures. An extensive body of literature demonstrates this fact (references in Altieri and Nicholls, 1999; Horn, 2002; Norris *et al.*, 2003). Also, the dispersal of insects in response to vegetation diversity is greatly affected. These authors stated that the establishment of a system of corridors of natural vegetation linking crop fields may serve multiple purposes in implementing IPM at the landscape level. For example, it may block plant inocula dispersion, it may block pest movement, and it may produce biomass for soil fertility, among other effects. The fact is that by making cropping systems more diverse, we make them more sustainable with greater conservation of resources (Vandermeer, 1995).

To function properly as a management tactic, the cultivation of mixed crop demands a very accurate study of the local conditions. In general, there is a need to get information on the population trends of the different pests locally before deciding upon any type of polycultivation. Once the decision of establishing a system using several crops is taken, one should decide which crops and what percentage of the total area should be dedicated to each of them. As mentioned before, a certain amount of the area should be allocated to host the natural vegetation, to provide refuges and corridors linking the system to allow the balance of pests with their natural enemies. A strong program of monitoring these insects, i.e. pests and their natural enemies, during the cropping season and after harvest is crucial to understand the flows of insects from one crop to the other and to the natural vegetation. There is a need to assess each agricultural system separately, to understand the many interactions of pests and natural enemies, which will depend on the size of the field, location, plant composition, surrounding vegetation, and cultural management (Altieri and Nicholls, 1999).

Functional allelochemicals

Allelochemicals are compounds that mediate behavioral or physiological interactions among organisms of different species. There are thousands of compounds mediating a myriad of interactions, within the three classical categories of allelochemicals: kairomones (i.e. allelochemicals that provide an adaptive advantage to the perceiver), allomones (i.e. allelochemicals that provide an adaptive advantage to the emitter), and synomones (i.e. allelochemicals that provide an adaptive advantage to both the perceiver and the emitter). For the purpose of IPM, the classification proposed by Kogan (1986*b*) is a good example of how these compounds function: as kairomones they may function as attractants, driving the insects toward the host plant; as arrestants, slowing or stopping movement; and as feeding or oviposition excitants, provoking biting/piercing or oviposition. In the second case, as allomones, they may function as antixenotics, by orienting insects away from the plant (repellents), speeding up movement (locomotory excitants), inhibiting biting/piercing (suppressants), and preventing maintenance of feeding/oviposition (deterrents); or as antibiotic, causing intoxication (toxins) or reducing food utilization processes (digestibility reducing factors).

Most plants synthesize toxins that affect herbivores. Those toxins that increase the fitness of the plants have a metabolic cost. Studies indicate that there is a balance between the cost and the various ecological effects (Karban and Baldwin, 1997), although it is often difficult to measure either the costs or the benefits associated with defensive substances.

Plant toxins have played an important role in agricultural plants, and most crops contain one or more types of toxins (see Seigler, 2002 for important groups of toxins in major crops). Some plants produce toxins in their roots with toxic and/or repellent effects to root-feeders, such as nematodes. These plants are called antagonistic plants (see review by Owino, 2002).

Resistance has been managed in crops either by using traditional plant breeding or new molecular techniques (Karban and Baldwin, 1997). Despite the many successful examples of breeding for changes in secondary metabolite chemistry to enhance resistance in crop plants, undesirable side-effects have been observed. For example, cotton lines with high contents of gossypol, a sesquiterpene toxin, show resistance to bollworm larvae and other herbivores, but also show detrimental effects to humans and livestock that use cotton products (Gershenzon and Croteau, 1991). On the other hand, the elimination of cyanogenic glycosides from the tuber roots of cassava, mitigating the poisoning effects to humans, is highly desirable, but this might increase herbivory and fungal attacks on plants free of these compounds (Moeller and Siegler, 1999). Therefore, there is a need to balance the cost/benefits of manipulating plant toxins.

Many studies report a wide range of allelochemical interactions. Borden (2002) exemplifies these interactions among terrestrial plants, arthropods, and vertebrates.

Despite these many studies and examples in the literature, the adoption of allelochemicals as pest management tools has been limited, for several reasons, some of them discussed above. Clearly, much remains to be done and there is no doubt that the management of insect pests through the many possible uses of allelochemicals will play a major role in IPM programs in the future.

6.3 Case study: nutritional ecology and management of heteropterans pests of soybean

As an example within the context of the nutritional ecology of plant feeding arthropods and IPM, the feeding guild of the heteropterans pests of soybean will be analyzed. Soybean, *Glycine max*, is a legume crop cultivated worldwide. The crop is colonized by many insects, including heteropterans that feed mostly on seeds, which are particularly important as pests in the neotropics (Kogan and Turnipseed, 1987; Panizzi and Corrêa-Ferreira,1997).

At the beginning of soybean cultivation in the neotropics, particularly in Brazil (today the world's second largest producer) few species of heteropterans were reported to cause damage to the crop. Lists of hemipterans pests contained fewer than 13 species during the 1970s, and included species not necessarily associated with soybean but with weeds present in soybean fields (Bertels and Ferreira, 1973; Corseuil *et al.*, 1974; Smith, 1978). More recently, in the mid-1980s, 32 species were listed as present on soybean fields in Brazil and in other countries of South America (Panizzi and Slansky, 1985*a*). The number of hemipterans has certainly increased even more during the period 1990–2007, due to the expansion of the crop towards new areas in Brazil close to the Equator.

Not only has the number of heteropteran species associated with soybean increased, but their pest status has also changed. For instance, the small green stink bug, *Piezodorus guildinii* (Westwood), rarely found on soybean in the early 1970s, today is very abundant and is found everywhere that soybean is cultivated in Brazil, from the south (32° S latitude) to the northeast areas close to the Equator (4° S latitude) (Panizzi and Corrêa-Ferreira, 1997).

Another example is the neotropical brown stink bug, *Euschistus heros* (Fabricius), also formerly rare and today the most abundant stink bug on soybean in Brazil. Beyond its adaptation to feed and reproduce on soybean, its ability to survive in partial hibernation under fallen dead leaves (Panizzi and Niva, 1994), living during this time on stored lipids (Panizzi and Hirose, 1995*a*; Panizzi and Vivan, 1997), and escaping the action of natural enemies (Panizzi and Oliveira, 1999), allows the bug's populations to increase and surpass in numbers the others species.

Finally, the alydid *Neomegalotomus parvus* (Westwood) is another example of a bug that has adapted to feed and reproduce on soybean, and dramatically increased in numbers. Initially known as a pest of common bean, *Phaseolus vulgaris*, and with a restricted distribution (Chandler, 1989), it is now present on soybean throughout the country, particularly in the warmer regions (A. R. Panizzi, unpublished). Although not so harmful to the plant's yield as stink bugs, *N. parvus* is able to cause significant damage to soybean seeds, particularly affecting the seed quality (Santos and Panizzi, 1998*a*).

6.3.1 *Life history of heteropterans*

Most pest species of heteropterans spend only a third of their lifetimes feeding on spring/summer crops, usually their preferred hosts. The rest of the time they feed and breed on alternate hosts or occupy overwintering sites. Most of the studies on the field biology of heteropterans concentrate on crops, paying less attention or completely overlooking the role of the wild, usually less-preferred, food plants. Although heteropterans may not breed on these plants (at least on some of these plants), they provide nutrients and water. Some alternate hosts may not be even recognized as potential toxic plants, despite their polyphagy and wide capacity to overcome toxic allelochemicals or lack of essential nutrients.

Usually the bugs move out to alternate hosts as the crops mature or are harvested, and continue to breed and develop, particularly in tropical or subtropical areas, where the bugs are active during the entire year. Considering the species associated with soybean, several studies report on alternate hosts in many parts of the world (e.g. Kiritani *et al.*, 1965; Miner, 1966; Singh, 1972; Jones, 1979; Todd and Herzog, 1980; Jones and Sullivan, 1981, 1982; Velasco *et al.*, 1995; Panizzi, 1997). The use of alternate host plant sequences by bugs and their manipulation seeking to manage them will be discussed (see section 6.3.3.1). Despite these many studies, much remains to be done, particularly with regard to the role of, among the alternate hosts, the less preferred ones used only occasionally, as pointed out (Panizzi, 2000 – see section 6.3.3.4: The role of less-preferred plant food sources, for details).

Feeding

Hemipterans (heteropterans) that feed on plants insert their stylets into plant tissues. According to Hori (2000), the bugs will feed in one of the following ways: stylet-sheath feeding, lacerate-and-flush feeding, macerate-and-flush feeding, and osmotic pump feeding. In stylet-sheath feeding, the bugs insert their stylets in the tissue, mostly in the phloem, destroying few cells; and a stylet sheath is produced, which remains in the plant tissues and can be used to estimate feeding frequency of these insects (Bowling, 1979, 1980). The resulting damage is a minor mechanical damage (Miles and Taylor, 1994). The external part of the stylet sheath is actually seen and recorded, and was called "flange" by Nault and Gyrisco (1966), who worked with other plant-sucking insects (aphids).

In lacerate-and-flush feeding, the bugs move their stylets vigorously back and forth, and several cells are lacerated. In the macerate-and-flush feeding type the cells are macerated by the action of salivary pectinase. In both cases, the cells' contents are injected with saliva, damaging several cells. Finally, osmotic pump feeding occurs through the secretion of salivary sucrase injected in the plant tissue to increase osmotic concentration of intercellular fluids containing sugars and amino acids, which are then sucked, leaving empty cells around the stylets (Hori, 2000).

The damage to plant tissues, including seeds and fruits, results from the frequency of stylet penetration and feeding duration, associated with salivary secretions that can be toxic and cause tissue necrosis. Slansky and Panizzi (1987) reviewed the nutritional ecology of phytophagous hemipterans specializing on seeds/fruits, and provided details on their feeding behaviors. More recently, Hori (2000) reviewed the salivary secretions and tissue damage.

In the case of soybean, feeding of heteropterans and their resulting damage has been studied by several authors. The reviews on the feeding damage caused by pentatomids in general are included in DeWitt and Godfrey (1972), Todd and Herzog (1980), Panizzi and Slansky (1985*a*), and Todd (1989). Significant reductions in yield, quality, and germination can result from feeding by these pests (e.g. Todd and Turnipseed, 1974). The oil and protein content of soybean seeds apparently are unaffected by moderate to heavy levels of feeding (Thomas *et al.*, 1974), but the chemical composition of the fatty acids can change (Todd *et al.*, 1973). Oil and protein can affect nymphal developmental time and adult weight (Calhoun *et al.*, 1988).

Also, soybean seeds damaged by pentatomids are more susceptible to attack by stored product pests (Todd and Womack, 1973); damaged seeds can have an increased incidence of pathogenic organisms (Ragsdale *et al.*, 1979; Panizzi *et al.*, 1979). The feeding punctures (flanges) on seeds cause minute darkish spots, and generally chalky-colored air spaces are produced when the cell contents are withdrawn; later, dark discoloration may surround the punctures and the inner membrane of the seedcoat may be abnormally fused to the cotyledons (Miner, 1966).

The southern green stink bug, *N. viridula*, apparently feeds preferentially on soybean seeds in the upper half of the plant until high infestations develop, and damage-free seeds compensate for damaged seeds with increases in 100-seed weight (Russin *et al.*, 1987). Feeding damage potential is similar for adults and fifth instars; less damage is caused by third and fourth instars (McPherson *et al.*, 1979). The damage caused by adults also varies with age and physiological condition. For instance, the number of feeding punctures produced by adults increased from day 1 to day 4 and leveled thereafter; adults previously deprived of food for 24 hours produced the same number of punctures as those fed continuously, but starved adults fed longer than non-starved adults (Panizzi, 1995). Also, *N. viridula* feed more frequently on seeds closest to the pedicel compared to other seeds (Panizzi *et al.*, 1995).

Other species of pentatomids associated with soybean may feed on stems, like *Edessa meditabunda* (Fabricius) (Galileo and Heinrichs, 1979; Panizzi and Machado-Neto, 1992). The bug's preference for stems is unique among the species of stink bugs that colonize soybean worldwide (Kogan and Turnipseed, 1987) and may represent the use of an unfilled niche. This preference for vegetative plant tissue somewhat decreases its effect on seed yield, as reported for soybean (Costa and Link, 1977).

Mating and egg production

Pre-mating and mating behavior of heteropterans may involve several cues, including production of odors and sounds. Among species associated with soybean, the southern green stink bug, *N. viridula*, has been the most investigated in this regard. Males produce sex pheromones, which are important for mate finding, and have an important ecological impact, by having a concomitant attraction to parasitic flies (tachinids) (Mitchell and Mau, 1971; Harris and Todd, 1980; Borges *et al.*, 1987; Borges, 1995). For this species, sound production is an important component of mating, and several authors have studied different aspects of this communication (Harris *et al.*, 1982; Čokl, 1983; Ota and Čokl, 1991; Čokl *et al.*, 1999, 2000).

During copulation, the pair may feed and males may guard females via prolonged copulation to protect their sperm precedence after multiple matings (McLain, 1981; Carroll, 1988).

Egg production of heteropterans is highly variable and depends, basically, on the quality of the food ingested. Among hemipteran pests of soybean in Brazil, for example, the fecundity (eggs/female) of the neotropical brown stink bug, *Euschistus heros* (Fabricius), varied from zero when feeding on star bristle, *Acanthospermum hispidum*, to 286.2 when feeding on soybean, with intermediate values when feeding on other host plants (Table 6.1). The fecundity of another pentatomid, *Loxa deducta* (Walker), varied from 26.0 eggs/female on soybean to almost 10× more (236.0 eggs/female) on privet, *Ligustrum lucidum*. This last host plant also yields much higher fecundity to *N. viridula* (256.5 eggs/female) than do most other host plants (Panizzi *et al.*, 1996). This species of privet is known to host over 10 species of pentatomids in subtropical Brazil (Panizzi and Grazia, 2001).

The highly polyphagous southern green stink bug, *N. viridula*, also has a variable fecundity, from zero eggs/female when fed on temporary food plants to as much as 296.9 eggs/female when fed on sesame, *Sesamum indicum* (Table 6.1). High fecundity is also observed for females fed on hemp sesbania, *Sesbania emerus* and privet, *L. lucidum*. For most foods, fecundity varied from 50 to 100 eggs/female.

The less polyphagous *Piezodorus guildinii* (Westwood) will produce from 11.0 eggs/female when fed on pods of pigeon pea, *Cajanus cajan*, up to almost 50× more eggs on another host, *Indigofera truxillensis* (506.7 eggs/female). This tremendous variability in the fecundity of this bug illustrates the importance and the variability of the quality of the food ingested.

Data on the fecundity of another seed-sucker, the alydid *Neomegalotomus parvus* (Westwood), also demonstrate the importance of food affecting fecundity. For this bug, greater fecundity is observed when females feed on mature seeds/pods, rather than on immature seeds/pods (Table 6.1). This means that not only the species of food plant will affect fecundity, but the degree of maturity of their seeds/fruits, as well.

Nymphal biology

As the nymphs emerge from the eggs, they normally stay on or near the eggshells. A mixture of visual, olfactory, and tactile cues keeps them grouped together. For example, the southern green stink bug *N. viridula* uses tactile stimuli to stay aggregated during the first two days of the first stadium. Beyond this period, chemical cues (n-tridecane) are used to maintain

Table 6.1. *Fecundity of heteropterans pests of soybean in the neotropics, feeding on different host plants (wild and cultivated)*[a]

Adult species/host species[b]	Eggs/female	References
Euschistus heros		
Acanthospermum hispidum plants[c]	0	Panizzi and Rossi, 1991
Cajanus cajan	119.4	Panizzi and Oliveira, 1998
Euphorbia heterophylla	61.7	Pinto and Panizzi, 1994
Glycine max	197.8	Panizzi and Oliveira, 1998
	287.2	Villas Bôas and Panizzi, 1980
Loxa deducta		
Glycine max	27.0	Panizzi and Rossi, 1991
Leucaena leucocephala	65.6	
Ligustrum lucidum	236.0	Panizzi *et al*., 1998
Neomegalotomus parvus		
Cajanus cajan immature pods	55.8	Santos and Panizzi, 1998*b*
C. cajan mature pods	117.9	
C. cajan mature seeds	99.8	
Dolichos lablab mature seeds	68.1	
Glycine max immature pods	23.2	
	63.5	Panizzi, 1988
G. max mature pods	68.1	
G. max mature seeds	69.1	
	107.6	Panizzi, 1988
Lupinus luteus immature pods	12.0	
Phaseolus vulgaris pods	13.7	
Nezara viridula		
Acanthospermum hispidum plants[c]	0	Panizzi and Rossi, 1991
Brassica kaber	107.4	Panizzi and Meneguim, 1989
Crotalaria lanceolata	29.0	Panizzi and Slansky, 1991
Datura stramonium plants[c]	30.8[d]	Velasco and Walter, 1992
Desmodium tortuosum	61.0	Panizzi and Slansky, 1991
Glycine max	67.7	Panizzi *et al*., 1996
G. max	99.3	Panizzi and Alves, 1993
G. max	110.0	Panizzi and Slansky, 1991
G. max	139.7	Panizzi and Hirose, 1995*b*
G. max	203.7	Panizzi and Saraiva, 1993
G. max plants[c]	124.8[c]	Velasco and Walter, 1992
Leonurus sibiricus	91.7	Panizzi and Meneguim, 1989
Ligustrum lucidum[e]	256.5	Panizzi *et al*., 1996
Macroptilium lathyroides plants[c]	0[d]	Velasco and Walter, 1992
Physalis virginiana plants[c]	56.8[d]	

Table 6.1. *(cont.)*

Adult species/host species[b]	Eggs/female	References
Raphanus raphanistrum	68.8	Panizzi and Saraiva, 1993
Rapistrum rugosum plants[c]	94.3	Velasco and Walter, 1992
Ricinus communis	0	Panizzi and Meneguim, 1989
R. communis plants[c]	95.0	Velasco and Walter, 1992
Sesamum indicum	297.9	Panizzi and Hirose, 1995*b*
Sesbania emerus	273.9	Panizzi and Slansky, 1991
S. vesicaria	40.0	
Trifolium repens plants[c]	0	Velasco and Walter, 1992
Triticum aestivum	0	Panizzi, 1997
Piezodorus guildinii		
Cajanus cajan	11.0	Panizzi *et al.*, 2000a
Crotalaria lanceolata	36.3	Panizzi *et al.*, 2002
C. lanceolata	58.2	Panizzi and Slansky, 1985*b*
Glycine max	28.0	
G. max plants[c]	31.1	Panizzi and Smith, 1977
G. max	72.2	Panizzi *et al.*, 2000a
G. max	78.7	Panizzi *et al.*, 2002
Indigofera endecaphylla	315.5	Panizzi, 1992
I. hirsuta	115.2	
I. hirsuta	204.8	Panizzi and Slansky, 1985*b*
I. suffruticosa	196.7	Panizzi, 1992
I. truxillensis	507.7	
Sesbania aculeata	205.1	Panizzi, 1987

[a]Adapted from Panizzi, 1997.
[b]Unless otherwise indicated, all hosts are fruit.
[c]Fruiting plants.
[d]Nymphs/female.
[e]Erroneously referred to as *L. japonicum*.

the individuals together. However, depending on the concentration, this chemical may also act as a dispersant of the colony (Lockwood and Story, 1985).

At this age (first instar), nymphs aggregated will not feed. There has been speculation about possible ingestion of egg shell residues or intake of microorganisms (symbionts) or of water during this period, but so far, nothing has been proven. Apparently, the colony functions as an organism, humidity being the key factor. Observations conducted in our laboratory indicate that a gradient of humidity keeps the colony as a whole. With a decrease in humidity, nymphs move from the colony. When nymphs were put in contact with filter paper oversaturated with water, all died, indicating that lack or excess of water is harmful to the colony's survivorship (Hirose *et al.*, 2006).

In general, because first instars from egg masses remain on or near the eggshells, they may be more susceptible to predators. During this time they hardly move, whereas nymphs from isolated eggs are much more restless and have a greater propensity to walk, which facilitates escape from predators or parasites. On the other hand, however, groups of nymphs may discourage the action of natural enemies. Lockwood and Story (1986) reported that aggregations of *N. viridula* nymphs suffer less predation by *Podisus maculiventris* (Say) than isolated nymphs. Also ants will feed on egg masses of *N. viridula*, but will not feed on newly hatched nymphs (A. R. Panizzi, unpublished). Perhaps their odor, movements, or appearance avoid the attack of ants.

In contrast to nymphs originating from egg masses, nymphs of other species emerge from single eggs. These nymphs abandon the eggshells soon after emergence, and, usually, do not aggregate (Panizzi, 2004). There is no complete evidence that first instars from eggs laid singly will always feed, but in some cases different foods offered to first instars will have a different impact on their ability to survive. For example, when first instars of the alydid *N. parvus* were offered mature soybean seeds, no mortality was observed; with immature soybean seeds 16.7% of the nymphs died. With immature pods of soybean and green beans, *Phaseolus vulgaris*, and mature seeds of lupine, *Lupinus luteus*, nymphal mortality was less than 1.7%; and with vegetative soybean tissue (stems and leaves), nymph mortality was 2.5% and 5.0%, respectively (Panizzi, 1988). These data suggest that, for these first instars, few nutrients plus water may be used, and that on most foods, even those unpreferred such as vegetative materials, nymphs developed well. With no food at all, nymphs of *N. parvus* will pass to the second instar (A. R. Panizzi, unpublished). In another species of alydid, *Megalotomus quinquespinosus* (Say), first instars are reported not to feed (Yonke and Medler, 1965). First instars of the rhopalid *J. choprai* will feed on mature seeds of the balloon vine, *Cardiospermum halicacabum* (Sapindaceae), a weed plant of soybean fields in south Brazil (A. R. Panizzi, unpublished).

Depending on the food, nymph developmental time and survivorship will vary greatly, as we shall see by examining the data related to hemipterans pests of soybean in Brazil. For example, nymph developmental time (2nd to 5th stadia) of the alydid *N. parvus* ranged from 16.3 to 34.1 days, and nymph mortality varied from 12.5% to 93.3% (Table 6.2). These two parameters are affected not only by the species of plant fed upon, but also by the degree of maturity of the fruits, and whether the seeds are exposed or protected

Table 6.2. *Developmental time and mortality of various nymphs of heteropterans pests of soybean in the neotropics, feeding on different host plants (wild and cultivated)*[a]

Adult species/host species[b]	Days[c]	Mortality (%)	References
Euschistus heros			
Cajanus cajan	22.1–22.8	25.0	Panizzi and Oliveira, 1998
Euphorbia heterophylla	20.8–21.3	21.5	Pinto and Panizzi, 1994
Glycine max	23.1–23.9	28.6	
G. max	23.9	16.5	Villas Bôas and Panizzi, 1980
G. max	21.1–22.0	17.5	Panizzi and Oliveira, 1998
Loxa deducta			
Glycine max	35.8	68.8	Panizzi and Rossi, 1991
Leucaena leucocephala	56.0–56.6	82.6	
Ligustrum lucidum	49.1	17.1	Panizzi *et al.*, 1998
Neomegalotomus parvus			
Cajanus cajan immature pods	18.8–19.4	17.1	Santos and Panizzi, 1998*b*
C. cajan mature pods	19.4–20.1	25.8	
C. cajan seeds	17.4–19.3	12.5	
Dolichos lablab mature seeds	20.0–21.3	13.3	
Glycine max immature pods	20.4–23.3	77.5	
	17.3–18.3	78.3	Panizzi, 1988
G. max mature pods	21.0–24.5	77.5	
G. max mature seeds	19.0–22.3	12.5	
	23.1–24.9	51.7	Panizzi, 1988
G. max immature seeds	18.8–19.9	65.0	
Lupinus luteus immature pods	27.7	93.3	
Phaseolus vulgaris immature pods	31.0–34.1	81.7	
Nezara viridula			
Brassica kaber	26.1–27.5	25.0	Panizzi and Slansky, 1991
Cassia fasciculata	29.4	42.0	Jones, 1979
	–	100.0	Panizzi and Slansky, 1991
C. occidentalis	26.7	0.0	Jones, 1979
Crotalaria lanceolata	27.2–33.9	85.0	Panizzi and Slansky, 1991
C. spectabilis	37.3	26.0	Jones, 1979
Croton glandulosus	43.5	80.0	
Datura stramonium plants[c]	38.7	59.5	Velasco and Walter, 1992
Desmodium canun	–	100.0	Panizzi and Rossini, 1987
D. tortuosum	22.0–23.5	65.0	Panizzi and Slansky, 1991
Ebelmoschus esculentus	33.5	10.0	Jones, 1979
Glycine max	26.2–26.3	60.0	Panizzi and Alves, 1993
G. max	25.2–27.8	28.9	Panizzi and Rossini, 1987

Table 6.2. (*cont.*)

Adult species/host species[b]	Days[c]	Mortality (%)	References
G. max	25.9–26.0	15.0	Panizzi and Saraiva, 1993
G. max	23.0	2.0	Jones, 1979
G. max	22.9–23.2	22.5	Panizzi and Slansky, 1991
G. max	22.3–23.4	20.0	Panizzi *et al.*, 1996
G. max plants[c]	32.8	25.5	Velasco and Walter, 1992
G. wightii	25.0–27.5	93.3	Panizzi and Rossini, 1987
Indigofera hirsuta	–	100.0	Panizzi and Slansky, 1991
Leonurus sibiricus	30.4–31.9	25.0	Panizzi and Meneguim, 1989
Lepidium virginicum	–	100.0	Jones, 1979
Ligustrum lucidum[e]	26.9–30.1	38.7	Panizzi *et al.*, 1996
Macroptilium lathyroides plants	33.5	61.7	Velasco and Walter, 1992
Melilotus indica plants[d]	47.6	63.7	
Physalis virginiana plants[d]	47.5	65.0	
Prunus serotina	29.3	78.0	Jones, 1979
Raphanus raphanistrum	35.4–39.3	56.2	Panizzi and Saraiva, 1993
R. raphanistrum	27.5	2.0	Jones, 1979
Rapistrum rugosum plants[d]	44.1	65.2	Velasco and Walter, 1992
Ricinus communis	42.3–42.6	60.2	Panizzi and Meneguim, 1989
R. communis plants[d]	50.2	86.5	Velasco and Walter, 1992
Sesbania aculeata	–	100.0	Panizzi and Rossini, 1987
S. emerus	20.3–20.8	10.0	Panizzi and Slansky, 1991
S. vesicaria seeds (immature)	20.5–22.2	40.0	
S. vesicaria pods	–	100.0	
Trifolium repens plants[d]	64.0	98.4	Velasco and Walter, 1992
Piezodorus guildinii			
Cajanus cajan	22.5–25.3	94.4	Panizzi *et al.*, 2000a
Crotalaria lanceolata	18.2–18.7	64.0	Panizzi *et al.*, 2002
Glycine max	19.7	–	Panizzi and Smith, 1977
G. max	22.4–23.0	57.7	Panizzi *et al.*, 2000a
G. max	20.0–23.0	88.0	Panizzi *et al.*, 2002
Indigofera endecaphylla	21.9–22.0	12.5	Panizzi, 1992
I. Hirsuta	24.9–25.9	58.3	
I. suffruticosa	28.5–30.3	84.2	
I. truxillensis	22.0–22.3	26.7	
Sesbania aculeata	22.2–22.6	25.0	Panizzi, 1987

[a]Adapted from Panizzi, 1997.

[b]Unless otherwise indicated, all hosts are fruit.

[c]From second stadium to adult.

[d]Fruiting plants.

[e]Erroneously referred to as *L. japonicum*.

by pod walls. In general, on mature exposed seeds, nymphs performed better than on immature seeds/pods.

For the southern green stink bug, *N. viridula*, nymphal developmental time varied from 22.0 to 50.2 days, and nymphal mortality from 0 to 100%, most values falling in the range of 22–26 days and 15–30% mortality on the preferred food, soybean (Table 6.2).

For the neotropical pentatomid, *P. guildinii*, nymphal developmental time ranged from 18.2 to 30.3 days, and nymph mortality from 12.5% to 94.4%. Best results were observed on indigo (*Indigofera endecaphylla* and *I. truxillensis*) and sesbania (*S. aculeata*) legumes (Table 6.2).

For the neotropical pentatomid, *Loxa deducta*, nymphal developmental time ranged 35.8–56.6 days, and nymph mortality 16.1–82.6%, with a better performance on fruits of privet, *Ligustrum lucidum* (Table 6.2).

The neotropical brown stink bug, *E. heros*, also showed variability on nymph developmental time and mortality, but to a lesser extent, compared to the former species of hemipterans.

Nymph/adult dispersion and host choice

As nymphs reach adulthood, the bugs are ready to begin dispersal to new areas. Although nymphs of heteropterans may disperse from the oviposition site by walking, they can only move limited distances. For instance, nymphs of the southern green stink bug, *N. viridula*, and of the neotropical small green stink bug, *P. guildinii*, disperse as far as 12 m from the release site during their development in soybean fields. Nymphs will also move longer distances along the row of soybean plants than across the rows (Figure 6.2) (Panizzi *et al.*, 1980). The main distance is covered by late (4th and 5th) instars, when the gregarious behavior mitigates.

Adult heteropterans are responsible for most dispersion, and several species are known to migrate, such as the classical pest of cereals in the Middle East – "Sunn pests" or "Soun pests" – pentatomids of the genus *Aelia* (references in Panizzi *et al.*, 2000), and scutellerids of the genus *Eurygaster* (references in Javahery *et al.*, 2000). Other species disperse by flight among host trees, such as the pentatomid *Bathycoelia thalassina* (Herrich-Schaeffer), pest of cocoa in Africa (Owusu-Manu, 1977).

Although there are only a few studies of dispersal of adult heteropterans in soybean, Costa and Link (1982) found that *P. guildinii* adults have greater mobility than *N. viridula*, and females of both species disperse farther than males.

As adults reach other areas, they start looking for suitable host plants. Although in general polyphagous, a close association of heteropterans of

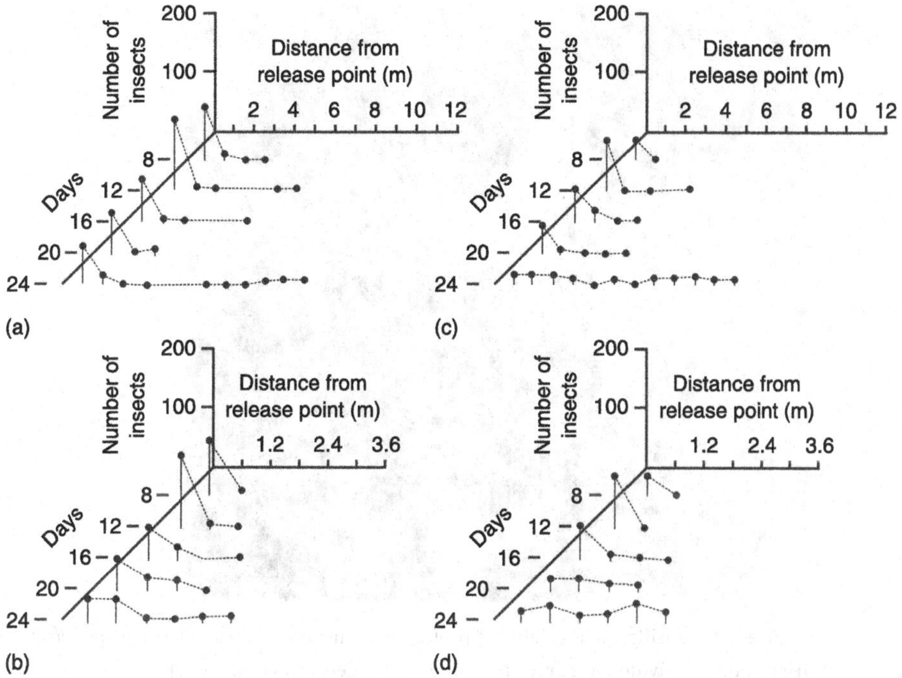

Figure 6.2. Number of nymphs re-captured 8, 12, 16, 20 and 24 days after release in soybean fields. A, B = *Piezodorus guildinii* along and across rows, respectively. C, D = *Nezara viridula* along and across rows, respectively (source: Panizzi *et al*., 1980).

certain taxa with particular plant taxa is observed. For instance, among Alydidae, the Leptocorisinae are primarily grass feeders and the Alydinae are primarily legume feeders (Schaefer and Mitchell, 1983); the pentatomid *N. viridula* prefer legumes and brassicas (Todd and Herzog, 1980); another pentatomid, *Edessa meditabunda* (Fabricius), prefers legumes and solanaceous plants (Silva *et al*., 1968); and pentatomids of the genera *Acrosternum* and *Euschistus* tend to be associated with legumes, whereas those of the genera *Aelia, Mormidea*, and *Oebalus* prefer to feed on graminaceous plants (references in Panizzi *et al*., 2000b).

In the process of host choice, several organs are used, such as eyes, antennae and palps. Chemical and physical characteristics of the plant make it acceptable or not to the pest. A series of behaviors of the bugs is associated with host acceptance, and the intensity of these behaviors will vary, depending on the suitability of the food involved. For example, time of dabbing/antennation of the legume feeder, *Neomegalotomus parvus* (Westwood) (Alydidae), varies from 137 seconds on soybean pod, to 102 seconds on common bean pod, down to 81 seconds on pigeon pea pod, which means that

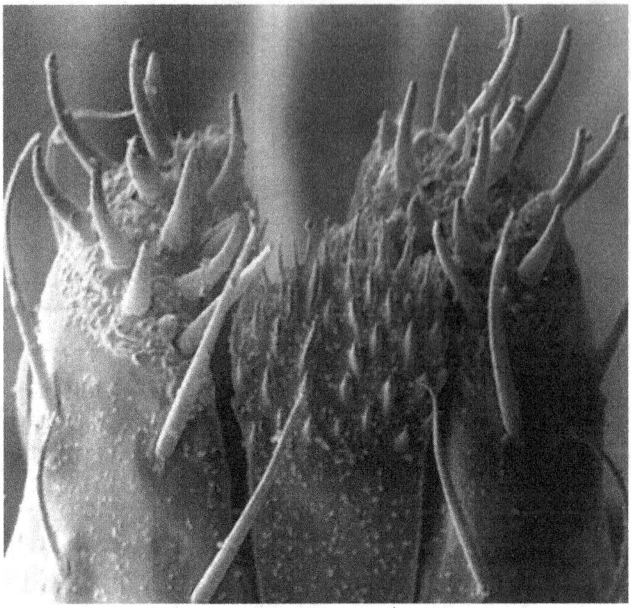

Figure 6.3. Sensilla on the labial tip of second instars of *Neomegalotomus parvus* (Heteroptera: Alydidae) (bar = 10 μm) (source: Ventura *et al.*, 2000).

this last food is accepted faster. Also, probing frequency (%) will vary from 46 to 71.8 to 100 on these food plants, respectively, again indicating that pigeon pea pod is preferred (Ventura *et al.*, 2000). Sensilla on the labium (Figure 6.3) are involved in the process of food acceptance by functioning as taste receptors.

Impact on adult performance by switching food from nymph to adult

Adults of hemipterans quite often disperse from their nymphal hosts to feed and reproduce on other species of plants. By doing so, their progeny will frequently feed on different food sources. This switching in food (host plant) from nymph to adult, although a common event in the biology of bugs, has been, in general, little investigated.

The switch in food from nymph to adult will usually have a variable (positive or negative) impact on adult performance or may not affect the performance at all. There are not many examples in the literature; perhaps the southern green stink bug, *N. viridula*, has been the most studied heteropteran in this respect.

For instance, when *N. viridula* nymphs and adults are fed the same food, poor nymphal performance on *Crotalaria lanceolata* pods and *Glycine max*

Table 6.3. *Reproductive performance of* Nezara viridula *females feeding on different legume foods (immature pods, unless indicated otherwise) as affected by switching or not switching foods from nymph to adult*[a]

Food		Number		Number/♀ (X ± SEM)[c]		Egg hatch, [d](X ± SEM)
Nymphs	Adult	Pairs	% ♀ ovip.,[b]	Egg masses	Eggs	
P. vulgaris	*P. vulgaris*	21	57.1	3.1 (0.5) ab	185.3 (33.0) ab	75.6 (5.8) a
S. emerus	*S. emerus*	13	84.6	3.7 (0.5) a	273.9 (36.1) a	55.8 (6.4) ab
P. vulgaris	*S. emerus*	10	80.0	2.4 (0.6) B	172.1 (50.1) B	64.4 (6.8) A
A. hypogaea[a]	*A. hypogaea*[e]	5	60.0	3.0 (1.0) ab[f]	99.7 (50.4) bc*	26.1 (11.2) b*
P. vulgaris	*A. hypogaea*[e]	10	100.0	5.7 (1.0) A	446.4 (93.7) a[e]*	62.3 (10.3) A*
G. max	*G. max*	17	76.5	1.9 (0.2) bc	110.0 (11.8) bc	61.5 (10.2) a
P. vulgaris	*G. max*	10	90.0	2.4 (0.4) B	149.1 (20.4) B	70.0 (9.6) A
D. tortuosum	*D. tortuosum*	16	56.2	1.3 (0.2) c[f]	61.0 (15.0) c*	59.5 (11.3) ab
P. vulgaris	*D. tortuosum*	10	70.0	2,7 (0.3) B*	153.1 (17.8) B*	43.8 (16.2) A
G. max[e]	*G. max*[a]	10	10.0[a]	1.0	23.0	0
P. vulgaris	*G. max*[e]	10	70.0	2.6 (0.4) B	204.4 (28.3) B	72.5 (9.9) A
C. lanceolata	*C. lanceolata*	8	12.5[g]	1.0	29.0	0
P. vulgaris	*C. lanceolata*	10	30.0	2.3 (0.3) B	122.7 (30.7) B	68.2 (18.5) A
S. vesicaria[h]	*S. vesicaria*	12	8.3[g]	1.0	40.0	15.0
P. vulgaris	*S. vesicaria*	10	40.0	1.0 (0.0) B	87.5 (4.3) B	82.2 (6.6) A

[a]Source: Panizzi and Slansky, 1991.

[b]For both series, % females ovipositing was dependent of food (nymphs and adults, same food: $G = 54.48$; df = 7; $P < 0.001$; adults switched to different food: $G = 12.24$; df = 5; $P < 0.05$).

[c]Means in each column followed by the same lowercase letter (nymphs and adult same food), and upper case letter (adults reared as nymphs on *P. vulgaris* and then switched to the various foods) are not significantly different ($P = 0.05$), Duncan's multiple range test. Asterisk indicates significant difference between the two series within each food ($P = 0.05$; *t* test).

[d]Data transformed to arcsine for analysis.

[e]Mature seeds.

[f]Data were included in the analysis although the residuals were not normally distributed.

[g]Because only one female laid one egg mass in each of these treatments, data for these were excluded from the statistic analyses.

[h]Immature seeds.

mature seeds carried over to poor performance of adults (Table 6.3). On these two foods only one female oviposited. It took more than twice as long (49 days) to produce the single egg mass on *C. lanceolata* as on the other foods (22 to 24 days), and none of the eggs hatched. Mean longevity of females on both of these foods, and of males on *C. lanceolata*, was shortened considerably (Figure 6.4) (Panizzi and Slansky, 1991).

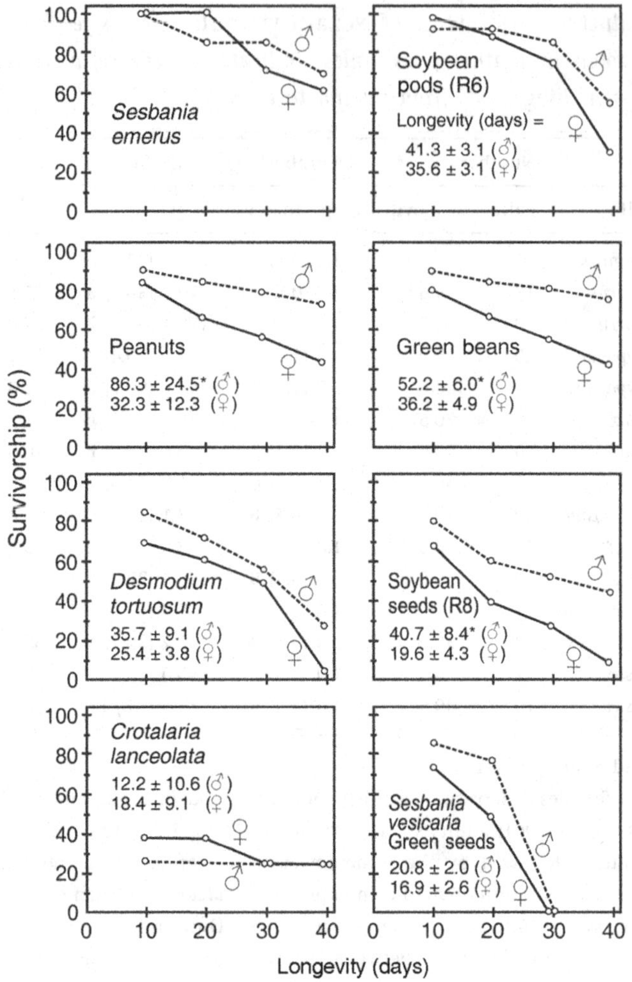

Figure 6.4. Survivorship (%) up to 40 days and longevity of adult *Nezara viridula* feeding on eight foods (nymphs and adults were fed the same food). No significant differences in female longevity among foods; for males, peanuts (a), green beans (b), soybean pods, *D. tortuosum*, and mature soybean seeds (bc), *S. vesicaria* (cd), and *C. lanceolata* (d) (Duncan's multiple range test, P = 0.05). Asterisk indicates significant differences (*t* test, P = 0.05) in mean longevity between the sexes in each food. Longevity of adults on *S. emerus* not available (source: Panizzi and Slansky, 1991).

Performance of *Sesbania vesicaria*-fed adults also reared as nymphs on this food was poor (Table 6.3), none surviving beyond 30 days (Figure 6.5). Seven of 12 females were observed copulating, but only one female laid one egg mass, and few eggs hatched (Table 6.3). In contrast, the high performance

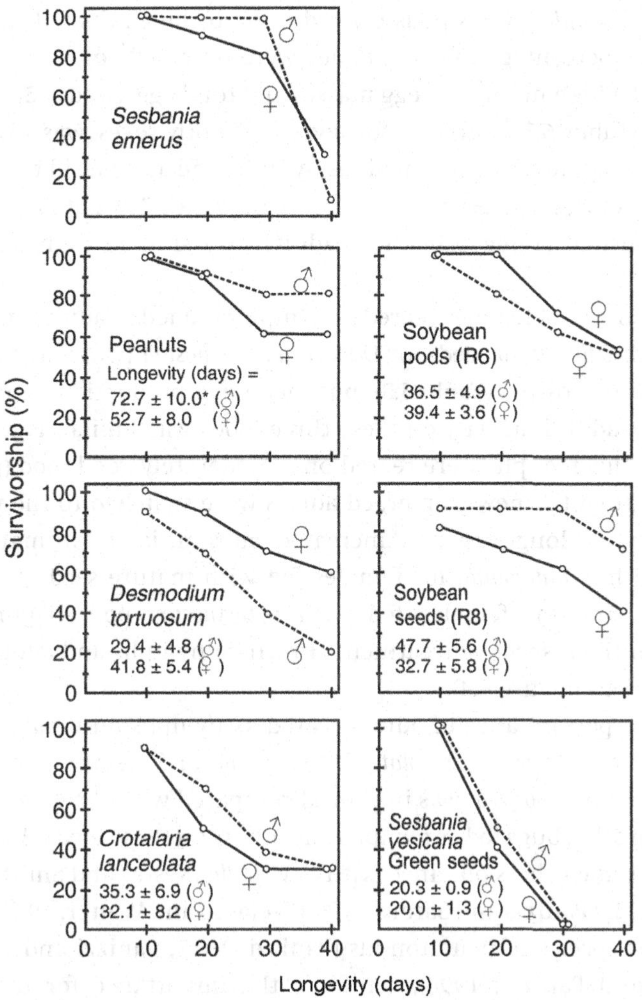

Figure 6.5. Survivorship (%) up to 40 days and longevity of adult *Nezara viridula* feeding on seven foods (nymphs in all treatments were reared on green beans, *Phaseolus vulgaris*). Significant differences in female longevity are peanuts and *D. tortuosum* (ab), soybean pods, soybean seeds and *C. lanceolata* (bc), and *S. vesicaria* (c); for males, peanuts (a), mature soybean seeds (b), soybean pods, *D. tortuosum* and *C. lanceolata* (bc), and *S. vesicaria* (c) (Duncan's multiple range test, P = 0.05). Asterisk indicates significant differences (*t* test, P = 0.05) in mean longevity between the sexes in each food. Longevity of adults on *S. emerus* not available (source: Panizzi and Slansky, 1991).

of nymphs on *Sesbania emerus* pods carried over to the adults fed with this legume: a large percentage of the females were observed copulating (85%), all of these laid a high number of egg masses and total eggs, and $>50\%$ of the eggs hatched (Table 6.3). Survival to day 40 of both sexes was also high (Figure 6.4). Females fed *P. vulgaris* pods as nymphs and adults laid the second largest number of egg masses and total eggs, with about 75% hatch, although only 57% oviposited (Table 6.3); these adults also showed high longevity (Figure 6.4).

A high percentage of females reared as nymphs and fed as adults on *Glycine max* pods, *Arachys hypogaea* seeds, or *Desmodium tortuosum* pods was observed copulating and oviposited (Table 6.3), but mean egg production/female was low. In general, adult longevity on these three foods was similar (Figure 6.4).

When *N. viridula* nymphs were reared on a moderately good food (pods of *P. vulgaris*) and then the newly emerged adults were switched to the various other foods, adult longevity was increased substantially for males and females fed with *C. lanceolata*, for females fed with mature seeds of *G. max* or *A. hypogaea*, and for females fed with *D. tortuosum* pods (Figure 6.5), compared with the respective treatments in which nymphs and adults were fed the same foods (Figure 6.4).

Reproductive performance of adults reared as nymphs on *P. vulgaris* and then switched to pods of *C. lanceolata*, *D. tortuosum*, or *S. vesicaria*, or mature seeds of *G. max* or *A. hypogaea*, was improved compared with the general poor performance of stink bugs fed as nymphs and adults on these foods (Table 6.3).

These results and others obtained with *N. viridula* (Kester and Smith, 1984; Panizzi *et al.*, 1989; Panizzi and Saraiva, 1993; Velasco and Walter, 1992; 1993), and with other species of pentatomids (Panizzi, 1987; Panizzi and Slansky, 1985*b*; Pinto and Panizzi, 1994), reinforce the importance for the adult performance of heteropterans of the switch in food from nymph to adult.

6.3.2 *Managing heteropterans on soybean*

Host plant resistance

As indicated previously (see section 6.2.3.1: Host plant resistance), host plant resistance is an important IPM tactic in the context of insect nutritional ecology. In the case of heteropterans, many studies have been conducted over the years, including evaluation of commercial cultivars, evaluation of genotypes from germplasm banks, and development of new cultivars.

Early studies by McPherson *et al.* (1979) with soybean suggested the commercial cultivar "Lee 68" might possess some mechanism of tolerance to stink bug feeding. Similarly, Link *et al.* (1971; 1973) found a lower percentage of damaged seeds for the cultivar "Bienville", compared with cv. "Santa Rosa" and "Industrial", and that cv. "Serrana" were less affected by stink bugs than "Bienville". Jones and Sullivan (1978) found cv. "Essex", which matured earlier than other cultivars, to escape severe damage by stink bugs. This observation of early maturity cultivars avoiding stink bug damage, which was also observed by other researchers elsewhere, was the basis for the development of massive breeding programs that led to the development of early maturity cultivars which escape the damage by heteropterans, particularly, in Brazil.

The evaluation of germplasm led to the discovery of several plant introductions (PIs) with variable degrees of resistance to several insect pests, including stink bugs. For instance, Turnipseed and Sullivan (1975) reported adverse effects of PIs 229358, 227687, and 171451, and of the line ED 73–371 to *N. viridula* nymphs. Jones and Sullivan (1979) showed that PI 229358 was the most consistently resistant PI to *N. viridula* nymphs. Another germplasm, PI 171444 was also shown to be resistant to *N. viridula* due to antibiosis and antixenosis (Gilman *et al.*, 1982; Kester *et al.*, 1984). The lines IAC 74–2832 and Chi-Kei No. 1B showed less damage by stink bugs than did many cultivars and lines evaluated in the field (Panizzi *et al.*, 1981).

Despite these many years of research on host (soybean) plant resistance to stink bugs, it was only in 1989 that the first variety was released, which presented antibiosis and tolerance types of resistance (Rossetto, 1989). This variety was named IAC-100, which name stands for the "Instituto Agronômico de Campinas" (IAC) in São Paulo, Brazil, marking its 100-year anniversary in 1989. It was cultivated by certain growers after its release, but was soon replaced by other cultivars with higher seed yield, despite their susceptibility to heteropterans.

As pointed out by Boethel (1999), soybean breeders and entomologists have discovered many obstacles in their attempts to develop soybean insect-resistant cultivars. The incorporation of the *Bt* gene into soybean against chewing insects, raised hopes for the revitalization of host plant resistance. But, so far, the discovery of an effective toxin against heteropterans that might be incorporated into soybean by those working on traditional breeding and in biotechnology remains a challenge.

Use of trap crops

The use of trap crops, in the context of using same species of host plant in different phenological stage of development to attract pest species that prefer to feed on plants during a certain time of plant development, has been used with some success on soybean to manage heteropterans in different parts of the world.

Apparently, the first study carried out on soybean was by Newsom and Herzog (1977), who reported on the attractiveness of soybean planted early to stink bugs in Louisiana, USA. The bugs concentrated in small areas of early planted soybean which, because planted earlier, were already in the reproductive stage, with pods filled with seeds that attracted the bugs. In the remaining area, which was planted later, the plants were still in the vegetative stage, and less attractive. Chemical control was applied to the plants in reproduction, avoiding the dispersal of the bugs to the surrounding soybean crop. Similar results were reported later on soybean in the USA by Ragsdale *et al.* (1981) and McPherson and Newsom (1984).

In Brazil, early maturity and early planted soybean, occupying ca. 10% of soybean fields (Figure 6.6), were reported to attract several species of stink bugs, reducing the degree of colonization of the main area (Panizzi, 1980). This tactic to control stink bugs on soybean fields was used together with the release of egg parasitoids, such as *Trissolcus basalis* (Wollaston),

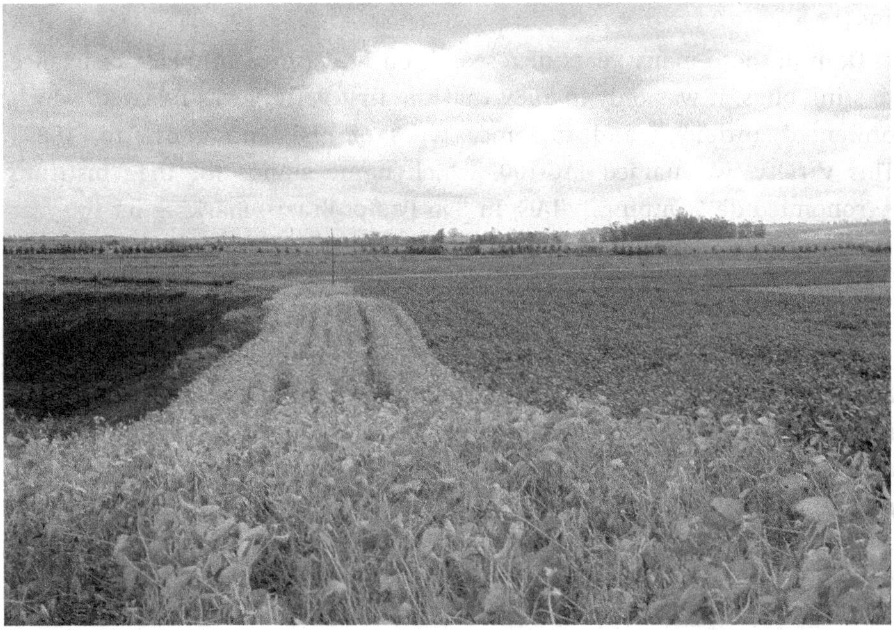

Figure 6.6. Early maturity soybean variety cultivated on the perimeter of a soybean field (ca. 10% of the total area) as a trap crop to attract stink bugs in southern Brazil (photo: A.R. Panizzi).

in the trap area early in the season; this was effective in managing pest populations of stink bugs (Corrêa-Ferreira, 1987). Additional studies on the use of the trap crop technique for stink bugs were also conducted in the expanding area of soybean cultivation in central Brazil (Kobayashi and Cosenza, 1987).

Despite these many studies and favorable results, the trap crop technique for management of stink bugs on soybean and other crops has been limited to special situations, such as small isolated fields or organic fields where the use of pesticides is prohibited. There are several reasons why this technique is not widely adopted by growers: the polyphagous feeding habits of heteropterans which increase the difficulty in attracting the bugs to a specific trap crop more effectively; the limited knowledge of the host plants'/bugs' interactions; and the lack of interest by growers who, in general, prefer more conventional methods of pest control (e.g. chemical control), which are considered dependable and easier to use.

Managing mixed crops to mitigate the heteropterans' impact to soybean

Soybean is usually cultivated in large-size areas worldwide. However, in some regions of the world where the crop is expanding, such as in the tropics, a growing percentage of the total acreage is in small fields. These fields are usually exploited by small farmers with specific purposes, such as the production of organic soybean or vegetable type soybean to be used for human consumption.

In these small fields, mixing crops in the same area or cultivation of several crops in adjacent areas are common events. For example, in some areas of southern Brazil, the landscape with relatively small soybean fields is surrounded by natural vegetation and other crops (Figure 6.7). In this scenario, soybean usually escapes the damage caused by stink bugs. It is known that in the tropics soybean cultivated near other legume fields is much less damaged by heteropterans than when cultivated alone (Jackai, 1984; Naito, 1996).

For over 15 years we have cultivated soybean for research in small fields surrounded by several other crops. We observed a much slower rate of plant colonization by heteropterans than in soybean fields in the so-called "open areas". Even in these latter areas, which are usually flatter and larger than the small fields surrounded by other crops, additions of different crops tend to mitigate the impact of heteropterans.

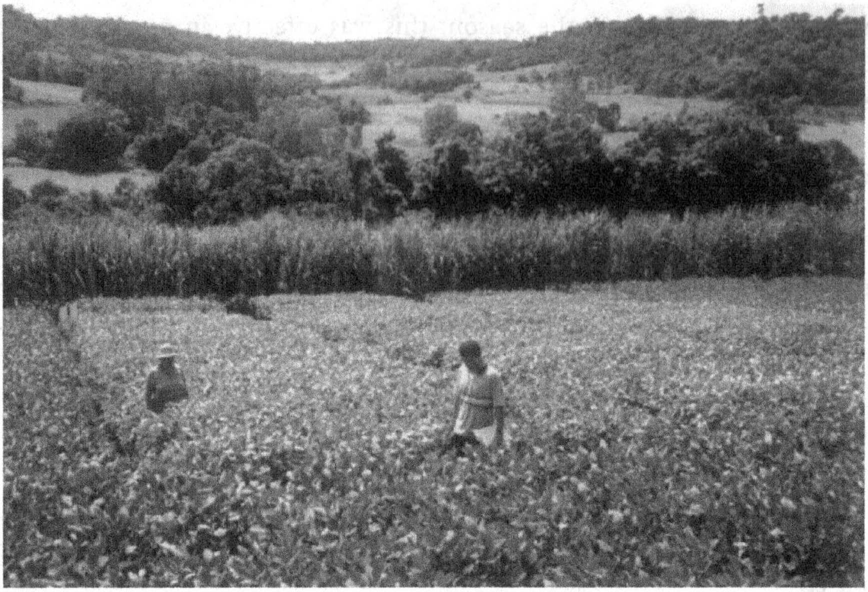

Figure 6.7. Landscape with relatively small soybean fields surrounded by natural vegetation and other crops. In this scenario soybean usually escape the damage caused by stink bugs (photo: Marcos Lena – GaMa Comercial Importadora e Exportadora Ltda).

Use of substances that interfere with the feeding process to reduce the heteropterans' impact on soybean

The use of secondary compounds or allelochemicals that interfere with the feeding process of insects on plants has been discussed briefly (see section 6.2.3.4: Functional allelochemicals). With regard to heteropterans that feed on soybean, an example of a substance that interferes with their feeding behavior and is being used to manage these pests is presented in detail below.

Field observations in soybean fields in southern Brazil, of an apparent attraction of stink bugs to clothes or handles of tools, caused growers and extension entomologists to speculate that human sweat was attracting the bugs. Field trials were set up using sodium chloride, mixed with water and sprayed over soybean plants. Initial studies, in the greenhouse, with potted soybean plants in cages, indicated that the southern green stink bug, *Nezara viridula* (L.), preferred plants sprayed with NaCl over plants sprayed with water only (Corso, 1989). Results of additional tests, using a mixture of NaCl (0.5%) with half the recommended dosage of conventional pesticides to control stink bugs, indicated a similar efficacy in control. The reduced dosage was promptly adopted by growers for economic reasons (Corso, 1990).

Additional field studies were conducted at the research station of the National Soybean Research Center of Embrapa (Empresa Brasileira de Pesquisa Agropecuária), in Londrina, Paraná state, by Panizzi and Oliveira (1993), to test the "attraction" of NaCl to stink bugs. They selected field plots (32 m × 7 m) in nine different locations of soybean fields, and sprayed half the plots with NaCl (0.5%). The other half was sprayed with water only. The bugs were sampled ca. twice a week during 11 weeks (15 sampling dates), using the beat cloth method; and the number of nymphs and adults of the three major species of pentatomids (i.e. *Nezara viridula*, *Piezodorus guildinii*, and *Euschistus heros*) were recorded. The results indicated that nymphs and adults of all three species were consistently more abundant in areas treated with salt than in the untreated areas (Figure 6.8).

Because laboratory bioassays indicated that NaCl did not have a synergistic effect when mixed with insecticides (Sosa-Gómez *et al.*, 1993), additional investigations were conducted by Niva and Panizzi (1996) to test the hypothesis that common salt was interfering in the stink bug feeding behavior. They compared the feeding behavior of adults of the southern green stink bug, *Nezara viridula*, on soybean pods treated with NaCl (0.5%) and soybean pods treated with water only (control). They offered soybean pods with and without salt to bugs confined in restricted arenas (Petri dish 14 cm × 2 cm), and recorded the time spent on food-touching with

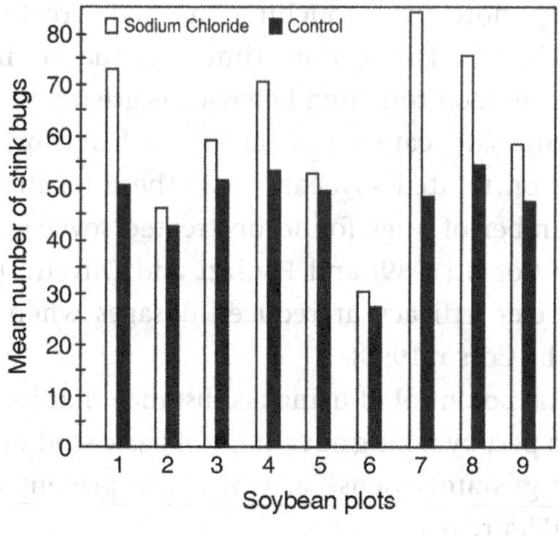

Figure 6.8. Mean number of nymphs + adults of three major species of pentatomids (*Nezara viridula*, *Euschistus heros*, and *Piezodorus guildinii*) captured on nine different soybean plots. Each mean corresponds to 15 sampling dates. Plots were half-sprayed with a solution of sodium chloride (0.5%) and half-sprayed with water only. Londrina, PR, 1990 (source: Panizzi and Oliveira, 1993).

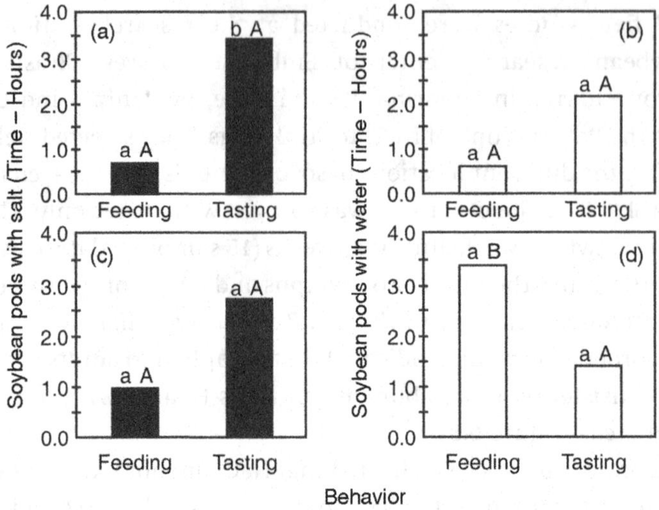

Figure 6.9. Total duration (hours) of the behaviors associated with the feeding process of the southern green stink bug, *Nezara virdula*, on soybean pods treated with a solution of sodium chloride (0.5%) (A and C) compared to control pods (treated with water only) (B and D). Different lower case letters (between behaviors within each treatment) and different upper case letter (between treatments within each behavior) indicates significant differences using the aleatoric test ($P < 0.05$) (source: Niva and Panizzi, 1996).

mouthparts and feeding (i.e. insertion of the stylets into the soybean pods). The bugs spent considerably more time touching the pods treated with salt than the untreated pods, and the feeding time was similar in both treatments (Figure 6.9). Because food touching behavior is greatly increased on soybean pods treated with salt, causing an arrestant behavior, which means the bugs stay longer on treated soybean pods. These results might explain both the greater number of bugs found on treated soybean plants than on control plants (see Corso (1989) and Panizzi and Oliveira (1993)), and also the greater insecticide efficacy at reduced dosages when mixed with pesticides, as reported by Corso (1990).

This example illustrates the potential of using a substance that interferes with the feeding to manage a pest by taking advantage of basic studies on the feeding behavior driven by gustatory sensilla, which are present on the labial tips of heteropterans (Figure 6.3).

6.3.3 *Managing heteropterans on host plants*

Most species of heteropterans spend only a third of their lifetimes feeding and breeding on cultivated spring/summer crops. The rest of the time these bugs are found feeding and reproducing on alternate (wild) host plants, or occupying overwintering sites provided by these hosts, such as under the trees' bark or underneath fallen dead leaves. Therefore, it is important to monitor the bugs' population while living on these wild plants or underneath debris, and to devise tactics to manipulate these pests before they colonize cultivated plants. This is, perhaps, one of the greatest challenges faced by entomologists, because much knowledge is needed regarding the biology, the ecology, the behavior, etc., and because little has been investigated compared with what is known about the activities of these bugs in damaging plants cultivated on an economic scale.

To design strategies to manage pest species, one must know which wild host plants are used by heteropterans, how suitable they are for nymphal development and adult reproduction, which sequences of plants are used by sequential generations, and when dispersal occurs from crop plants to wild plants and vice versa (Panizzi, 1997).

Host plant sequences

In general, heteropterans explore a variety of host plants within and between generations. Nymphs and adults move among the same or different plant species, which may be colonized in sequence. There are several examples of sequences of plants used by different species of heteropterans (references in Panizzi, 1997) (Figures 6.10–6.12).

In Paraná state (Brazil), the highly polyphagous southern green stink bug, *Nezara viridula*, colonizes soybean during late spring and summer, completing three generations on this crop, before it moves to alternate hosts such as *Crotalaria lanceolata*, where a fourth generation is completed (Figure 6.10). During this time it may feed on the weed star bristle, *Acanthospermum hispidum*, but no reproduction occurs on this plant. A fifth generation is completed during the mild winter of northern Paraná, on host plants such as wild radish, *Raphanus raphanistrum*; mustard, *Brassica campestris*; and pigeon pea, *Cajanus cajan*. During winter, *N. viridula* may feed on wheat, *Triticum aestivum*, but does not reproduce on this plant. A sixth generation is completed, during spring, on Siberian motherworth, *Leonurus sibiricus*. During the entire year, the southern green stink bug is observed on castor bean, *Ricinus communis*, on which plant it may feed but not reproduce.

The less polyphagous small green stink bug, *Piezodorus guildinii*, also completes three generations on soybean (Figure 6.11). A fourth generation is completed on legumes such as crotalaria, pigeon pea, and several indigo species (*Indigofera hirsuta*, *I. truxillensis*, and *I. suffruticosa*). During the winter it feeds on indigo legumes, but, in contrast to *N. viridula*, it does not

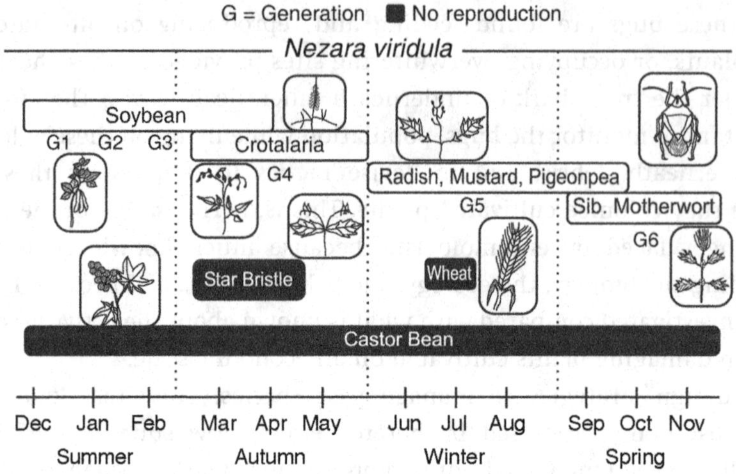

Figure 6.10. Wild and cultivated plant species of the host plant sequences used by successive generations of the southern green stink bug, *Nezara virdula*, in northern Paraná state, Brazil (blank boxes indicate occurrence of reproduction) (source: Panizzi, 1997, with adaptations).

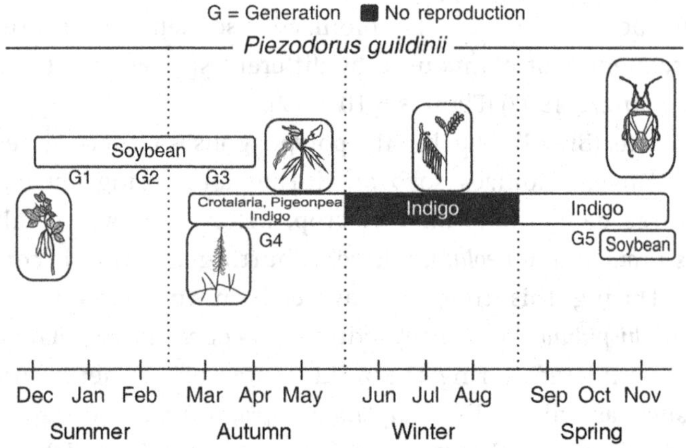

Figure 6.11. Wild and cultivated plant species of the host plant sequences used by successive generations of the small green stink bug, *Piezodorus guildinii*, in northern Paraná state, Brazil (blank boxes indicate occurrence of reproduction) (source: Panizzi, 1997, with adaptations).

reproduce at this time. A neotropical species, *P. guildinii* is less adapted to the somewhat cooler temperatures of the winter mouths. A fifth generation is completed on indigo legumes, before the bug starts colonizing soybean again during late spring.

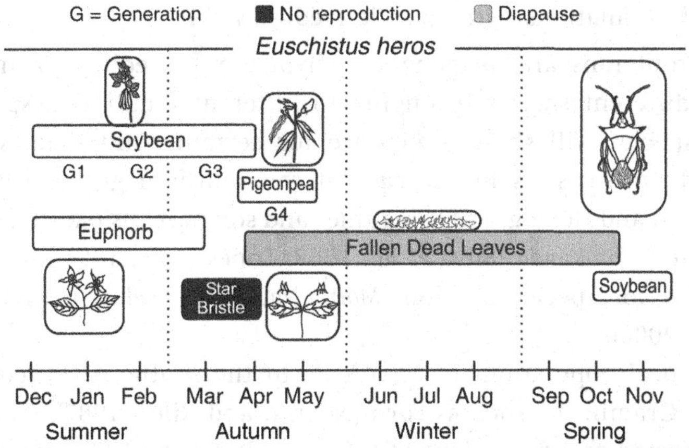

Figure 6.12. Wild and cultivated plant species of the host plant sequences used by successive generations of the neotropical brown stink bug, *Euschistus heros*, in northern Paraná state, Brazil (blank boxes indicate occurrence of reproduction) (source: Panizzi, 1997, with adaptations).

The neotropical brown stink bug, *Euschistus heros*, like the previous two species, completes three generations on soybean. During late summer/early fall, a fourth generation is completed on pigeon pea, *Cajanus cajan*. During the summer it may be found feeding on the euphorb *Euphorbia heterophylla*, but reproduction on this plant was observed to occur only under laboratory conditions (Pinto and Panizzi, 1994). *E. heros* may feed, but will not reproduce, on the weed plant star bristle, *Acanthospermum hispidum*. It is interesting to note that on this plant, this typical seed-sucker feeds on the stems of the plant (see section 6.3.3.4: The role of less-preferred plant food sources, for details). At this time, *E. heros* starts moving to shelters underneath leaf litter, where it remains until the next summer (Figure 6.12). During this time this bug accumulates lipids and does not feed, remaining in a state of partial hibernation (Panizzi and Niva, 1994; Panizzi and Hirose, 1995*a*; Panizzi and Vivan, 1997). Despite completing fewer generations than the former two species, *E. heros* is the most abundant species, particularly in the warmer regions. Its long time in shelters enables it to avoid the attack of natural enemies, thereby increasing its survivorship (Panizzi and Oliveira, 1999).

In the study of host plant sequences used by heteropterans it is important to determine which plants are used in sequence and how suitable they are to nymphal development and adult reproduction. By doing so, one will know which host plants are the most important to the bug's biology, and on which plants studies should be concentrated in order to devise management tactics to mitigate their impact to crops.

Local populations with specific feeding habits

Heteropterans are, in general, polyphagous, feeding on an array of plant of different species belonging to different families. Despite this polyphagy, species will show preference for certain plant taxa, such as legumes and brassicas, as in the case of the southern green stink bug, *N. viridula* (Todd and Herzog, 1980), legumes and solanaceous plants, as in the case of *Edessa meditabunda* (Silva *et al.*, 1968; Lopes *et al.*, 1974), or grasses in general, as for species of *Aelia*, *Mormidea*, and *Oebalus* (references in Panizzi *et al.*, 2000b).

However, local populations of *N. viridula* in the southern United States may feed on Gramineae, such as corn (Negron and Riley, 1987), which has not been reported as a food plant of this bug elsewhere. Local populations of the neotropical brown stink bug, *E. heros*, will feed on the euphorb, *E. heterophylla*, but, in general, this bug does not explore this plant as host (Pinto and Panizzi, 1994). These and several other examples demonstrate that, depending on time of exposure to restricted hosts and their availability, polyphagous species will act as monophagous or olygophagous (Fox and Morrow, 1981).

This phenomenon of local populations with specific feeding habits makes clear the complexity of the biology of phytophagous heteropterans. What may be valid information in one place may not apply in another. This indicates that to devise management tactics which involve manipulation of host plants, studies should be done locally. The host plant sequences used by each species at each place should be determined and fully understood, considering such biotic factors as characteristics of the bug species, of the host plant species and host plant sequences; and such abiotic factors as rain regime, range of favorable temperatures that allow the bug reproduction, and photoperiod.

Manipulation of preferred host plants as traps

There have been several studies concerning preferred host plants as traps, as a tool to manage pest species (see references in Hokkanen, 1991). In the case of heteropterans, several studies have been conducted, manipulating plant phenology, as in the case of using the preference of stink bugs to feed on soybean plants with pods/seeds compared with plants in the vegetative period (see section 6.3.2.2: Use of trap crops).

Because heteropterans are, in general, polyphagous, this makes the use of the trap crop technique, in the context of attraction by different plant species, a more complicated issue.

Despite their polyphagy, several attempts have been made to use the classical trap crop concept to manage heteropterans. An early report is by Watson (1924), who referred to the use of legumes (*Crotalaria*) to attract populations of the southern green stink bug, *N. viridula*, in citrus orchards in Florida, USA. The bugs concentrated on the legume plants and were killed before colonizing the citrus plants. More recently, Ludwig and Kok (1998) evaluated trap crops to manage the harlequin bug, *Murgantia histrionica* (Hahn) (Pentatomidae), on broccoli.

In the tropics, Jackai (1984) reported on the attraction of cowpea, *Vigna unguiculata*, to stink bugs, mitigating their damage on soybean in Africa. And in Indonesia, the alydid *Riptortus linearis* (L.), also a soybean pest, was controlled using the legume *Sesbania rostrata* as a trap crop (Naito, 1996).

In southern Brazil, there is good potential to use some legume host plants as traps for the heteropterans that feed on soybean. For instance, the close association of the small green stink bug, *P. guildinii*, with wild legumes (indigo, genus *Indigofera*), can be used to attract the bugs and concentrate them in particular areas where they can be eliminated. Similarly, pigeon pea, *Cajanus cajan*, which is used on contour lines as a wind barrier, can be used as a trap crop for this and other pentatomids, such as *N. viridula* and *E. heros*, and for alydids, such as *Neomegalotomus parvus* (Westwood). This legume produces pods almost all year round, and is attractive to bugs when they leave their preferred host, soybean.

The role of less-preferred plant food sources

In general, during their lifetime, insects are faced with less-preferred food sources and must adapt to explore alternate food sources when more preferred species are unavailable.

Most species of hemipterans spend only a third of their lifetime feeding on spring/summer crops, usually their preferred hosts. The rest of the time they spend feeding and breeding on alternate hosts, some of them of low nutritional quality, or occupying overwintering sites. Therefore, the less-preferred food plants are usually overlooked, and their roles in the life history of hemipterans are, in general, underestimated.

Although hemipterans do not breed on these plants (at least on some of these plants), they provide nutrients, to some extent, and water, as well. However, because bugs are not used to them, sometimes they may not recognize these "host plants" as potential toxic plants, despite their polyphagy and wide capacity to overcome toxic allelochemicals or lack of essential nutrients.

Among the less-preferred host plants of hemipterans, some are cultivated and some are wild, uncultivated plants. Usually they occur near cultivated fields where the preferred hosts were harvested or ended their cycle and became mature. In some cases, weeds that remain green between mature plants of a certain crop are temporarily used as a source of nutrients and water. This situation is common in tropical or subtropical areas, where most bugs are active during the entire year (some species, however, enter diapause, underneath debris, without feeding, such as the neotropical brown stink bug, *Euschistus heros* (F.) [Panizzi and Vivan, 1997]).

When phytophagous hemipterans are faced with a scarcity of preferred host plants, and environmental conditions are favorable, i.e. temperatures and humidity are relatively high and photoperiod adequate, bugs will feed and remain active on less-preferred plant food sources. However, they may be forced to change their feeding habits used on preferred plants. This may happen for several reasons: the less-preferred plants possess seeds or fruits the bugs are not accustomed to feed upon; the less-preferred plants may be at a vegetative stage and lacking seeds and fruits; or the less-preferred plants may produce fruits and seeds suitable but inaccessible (out of reach, as e.g. seeds protected by thick pod walls, or by an empty space between the pod walls and the seeds). Faced with one of these conditions or others, bugs must change their feeding habits and feed on other plant structures, usually not explored as food sources.

For instance, the southern green stink bug, *N. viridula*, will feed on less-preferred host plants in northern Paraná state, Brazil, such as star bristle, *Acanthospermum hispidum*. Nymph mortality on this plant is high in the laboratory (in the field nymphs may not even feed on this plant); adults will not reproduce on it, and their longevity is reduced: although a seed-feeder, this bug strongly prefers feeding on stems of this plant (Panizzi and Rossi, 1991). The stems are mostly filled with an aqueous tissue and the insects apparently detect this abundant source of water.

On castor bean, *Ricinus communis*, late instars and adult *N. viridula* show an atypical feeding behavior by feeding on the leaf veins (Panizzi, 2000). Eggs are not laid by females on castor bean leaves, unless accidentally. On wheat, *Triticum aestivum*, *N. viridula* adults have been observed feeding on reproductive plants during mild winters. Adults will feed on seedheads, but will not lay eggs on plants. Attempts to raise nymphs in the laboratory using seedheads or mature seeds did not succeed.

N. viridula, although extremely polyphagous, does not breed on graminaceous plants. There are reports of its damage to wheat in Brazil (Maia, 1973), and to wheat and corn in the United States (Viator *et al.*, 1983;

Negron and Riley, 1987). However, these may be cases of local populations with specific feeding habits, as previously discussed. In northern Paraná state, Brazil, *N. viridula* may eventually feed on corn, but on the stems, not on the ears, of seedling corn grown under the no-tillage cultivation system. Bugs that stay in areas with weed plants, or with scattered cultivated host plants, will eventually feed on corn seedlings that are established in these areas. However, these events are uncommon.

Other species of hemipterans, like *N. viridula*, will also feed on less-preferred food plants. For instance, the neotropical brown stink bug *E. heros*, a typical seed-sucker, will feed on star bristle stems (Panizzi, 2000).

Another pentatomid, *Dichelops melacanthus* (Dallas), previously reported as a pest of soybean, and feeding on pods (Galileo *et al.*, 1977), has been observed feeding on corn, and on wheat. It is interesting that on these two graminaceous plants, the bugs feed on stems of young plants, causing substantial damage. This change in feeding habits, from reproductive structures of more preferred hosts, such as legumes (soybean), to vegetative tissues of less-preferred hosts (graminaceous), is attributed to the low availability of preferred hosts. After soybean harvest, *D. melacanthus*, stays on the ground underneath debris, and will feed on corn or wheat plants growing on areas under conservation tillage. In these areas, bugs found shelter (straw) and food (dried seeds dropped on the ground) and will thrive, unlike what occurs on areas under conventional cultivation systems, where bugs are dislodged from their shelters and killed by plowing.

A similar situation occurs with the alydid *Neomegalotomus parvus* Westwood, a typical seed-sucker that feeds on mature seeds of legumes. On areas under conservation tillage, this bug will feed on soybean seedlings. In areas under conservation tillage, it stays on the ground feeding on its preferred food (mature seeds) and will complement its diet with young plants (Panizzi and Chocorosqui, 1999).

In conclusion, although many aspects of the biology of hemipterans have been investigated, perhaps an aspect least studied is this subject of hemipterans on less-preferred plant food sources. If we are to develop holistic integrated pest management systems, more attention must be devoted to this subject.

6.3.4 *Managing heteropterans in overwintering sites and host plants*

After colonizing spring/summer crops, heteropterans disperse to overwintering sites or, especially in the tropics, to alternate hosts. In general, bugs begin to disperse even before the crop they are feeding on completes maturation. For example, pentatomids that feed on soybean will start leaving

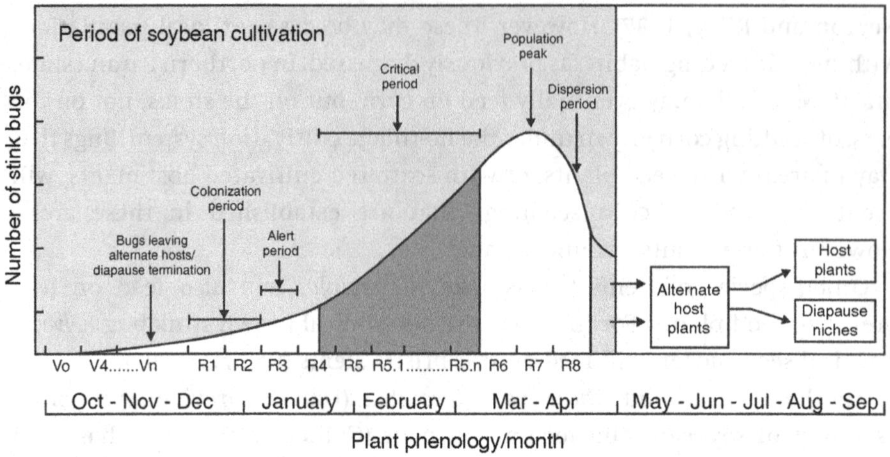

Figure 6.13. Phenology of the population of pentatomids associated with the main host plant (soybean) in southern Brazil. Notice that, after colonizing soybean, the bugs will either colonize alternate host plants, on which they may complete further generations, or look for hidden sites (underneath dead leaves, debris, etc.) to pass the unfavorable season in a state of partial diapause (dormancy) (source: Corrêa-Ferreira and Panizzi, 1999).

the crop after reaching the population peak, during the time plants begin to senesce. This process of crop abandonment increases in intensity as the plants dry out completely and become mature (Figure 6.13).

In general, after leaving the summer crops, heteropterans feed on alternate hosts and may complete an extra generation before moving to diapause sites, or may continue to breed on these alternate hosts (Figure 6.13). This will depend not only on the favorability of abiotic factors, such as temperature and photoperiod, but also on the capability of certain species to reproduce on these alternate host plants. These bugs will emerge from diapause sites or alternate hosts to start colonizing spring/summer preferred hosts, such as soybean (Figure 6.13). In soybean, colonization begins during the vegetative period (V0–Vn), increases with reproduction during blooming/early pod set (R1 to R3), and the numbers increase to reach the so-called critical period at pod-filling (R4 to R5.n); at this stage the damage to the crop is crucial. At the end of the pod-filling period (R6), the bugs' population reaches its peak, and dispersal begins again (Figure 6.13).

Managing crop residues

The management of crop residues in order to mitigate the impact of pest species is becoming increasingly important, due to the adoption of no-tillage or minimum-tillage cropping systems in many regions of the world.

These systems provide favorable conditions for soil-inhabiting insects or those that live in or under debris.

At least three species of soybean heteropterans have been favored by no-tillage or minimum-tillage cropping systems. These are the neotropical brown stink bug, *Euschistus heros* (Fabricius), the neotropical green-belly stink bug, *Dichelops melacanthus* (Dallas) (Pentatomidae), and the brownish root bug, *Scaptocoris castanea* Perty (Cydnidae).

E. heros, because of its habit of hiding underneath crop residues during more than six months of the year, particularly during late fall/winter, and early spring (see Figure 6.12), has increased in abundance dramatically. Considered a secondary pest in the 1970s, today it is the most common stink bug pest of soybean in Brazil.

D. melacanthus has also increased its abundance, probably because of the adoption of no-tillage cultivation systems. Once considered a minor pest of soybean, together with *D. furcatus* (Fabricius), *D. melacanthus* is now a major pest of corn and wheat. It also remains on the ground in partial hibernation (oligopause) and when corn or wheat are sown directly over the crop residues during the fall, in southern Brazil, it attacks the seedlings and the resulting plants show severe damage (Chocorosqui and Panizzi, 2001). Similar damage by stink bugs to seedling corn has also been observed in the United States (Sedlacek and Townsend,1988; Apriyanto *et al.*, 1989).

The third species, *S. castanea*, attacks the roots of many economically important plants, such as corn, cotton, rice, groundnuts, sugarcane, potato, peas, tomatoes, pimentos, and lucerne, as well as wild uncultivated plants in the neotropics. It can also be devastating to soybean (Lis *et al.*, 2000 and references therein). A root-feeder, it spends most of its life underground.

To control these three species and others that live in the soil at least part of their life, management of crop residues is mandatory. Plowing or burning the residues is recommended.

Monitoring the bugs in overwintering niches to determine crop colonization

Perhaps one of the most important steps toward implementing holistic IPM programs is to monitor overwintering niches and host plants to determine the abundance of pest populations and likely time of crop invasion. This "preventive" step is, in general, overlooked and its importance underestimated.

How can one estimate the intensity of stink bugs colonizing a soybean field by monitoring the bugs during the overwintering period? This depends on several factors. For example, temperature and humidity are crucial. If, after soybean harvest, the temperature falls below 5 °C during the fall/winter and remains low for a certain period, a high mortality of bugs is expected. Similarly, if strong spring rains precede the cultivation period, the population of bugs on alternate host plants or in the soil under crop residues will be heavily affected. These two factors might mitigate the impact of bugs during the following soybean season.

Another important factor influencing the population dynamics of the bugs during overwintering is the cultivation system. As mentioned above, the no-tillage or minimum tillage cultivation systems may promote a greater than expected population of bugs, particularly of *E. heros* and *D. melacanthus*. These two species overwinter on the soil under crop residues. Plowing eliminates a great portion of these two bugs' populations.

Finally, the presence of host plants may allow predicting which species of stink bugs are likely to predominate in the following soybean season. For example, the presence of indigo legumes as overwintering host plants will increase the population of the small green stink bug, *P. guildinii*. Similarly, the weed plant Siberian motherwort, *Leonurus sibiricus*, which grows before soybean during early spring, is a preferred host of the southern green stink bug, *N. viridula*, allowing reproduction and, therefore, population build-up. These and other examples illustrate that it is possible, to a certain degree, to predict both qualitatively and quantitatively the populations of stink bugs that colonize soybean.

6.4 Conclusions

As stated at the beginning of this chapter, research on insect nutritional ecology has concentrated on basic aspects, relating food characteristics to food intake and utilization by insects, and its consequences on their performance. Insect nutritional ecology as a more applied field, as a support to pest management programs, has been, in general, overlooked. An exception is the paper by Slansky (1990), which relates insect nutritional ecology to host plant resistance.

Several decades ago, during the 1960s and 1970s, several authors concentrated on pest management strategies, considering insect nutritional aspects in a broad context (references in Panizzi and Parra, 1991*b*). Today, after more than 30 years, these pest management strategies based on insect nutrition, such as host plant resistance, trap crops, polycultivation, and use of allelochemicals, remain a challenge to be fully implemented in IPM programs.

As new areas in biology gain momentum, like the development of genetically modified crops resistant to insects, insect nutritional ecology becomes a very important area of research in entomology, now within a more applied context. These cultivars of many important crops are being widely adopted by growers all over the world, and will certainly impact the insect pests causing dramatic changes. Many questions will arise, such as how these genetically modified plants will fit into current IPM programs. Clearly, much research will be needed to change the traditional IPM programs to accommodate this new technological tool.

To conclude, it is reasonable to assume that as we develop new IPM programs, more efficient and more ecologically sound, the tactics taking into account the interactions of insects with their food will play a growing role to achieve our goals.

Acknowledgments

I thank the editors, Marcos Kogan and Paul Jepson, for the invitation to write this chapter. The valuable comments and suggestions made by Carl W. Schaefer are deeply appreciated. This work is a contribution of the Laboratory of Insect Bioecology of the National Soybean Research Center of Embrapa, Londrina, PR, Brazil. It was supported by funds from Embrapa and from the Conselho Nacional de Desenvolvimento Científico e Tecnológico (CNPq).

References

Ahmad, S. (1983). *Herbivorous Insects: Host-Seeking Behaviour and Mechanisms*. New York: Academic Press.

Altieri, M.A. (1994). *Biodiversity and Pest Management in Agroecosystems*. New York, Food Products Press.

Altieri, M.A. and Letourneau, D.L. (1984). Vegetation diversity and insect pest outbreaks. *Critical Review of Plant Science*, **2**, 131–69.

Altieri, M.A. and Nicholls, C.I. (1999). Biodiversity, ecosystem function, and insect pest management in agricultural systems. In W.W. Collins and C.O. Qualset (eds.), *Biodiversity in agroecosystems*. Boca Raton, FL: CRC Press. pp. 69–84.

Andow, D.A. (1991). Vegetational diversity and arthropod population response. *Annual Review of Entomology*, **36**, 561–86.

Apriyanto, D., Townsend, L.H. and Sedlacek, J.D. (1989). Yield reduction from feeding by *Euschistus servus* and *E. variolarius* (Heteroptera: Pentatomidae) on stage V2 field corn. *Journal of Economic Entomology*, **82**, 445–8.

Bernays, E.A. (ed.). (1989–1994). *Insect–Plant Interactions*. Vols. I–V, Boca Raton, FL: CRC Press.

Bernays, E.A. and Chapman, R.F. (1994). *Host-Plant Selection By Phytophagous Insects*. New York: Chapman and Hall.

Bertels, A. and Ferreira, E. (1973). Levantamento atualizado dos insetos que vivem nas culturas de campo no Rio Grande do Sul. Univ. Católica Pelotas, *Série Publicação Científica*, **1**, 17 pp.

Boethel, D.J. (1999). Assessment of soybean germplasm for multiple insect resistance. In S.L. Clement and S.S. Quisenberry (eds.), *Global Plant Genetic Resources for Insect-Resistant Crops*. Boca Raton, FL: CRC Press. pp. 101–29.

Borden, J.H. (2002). Allelochemics. In D. Pimentel (ed.), *Encyclopedia of Pest Management*. New York: Marcel Dekker, Inc. pp. 14–16.

Borges, M. (1995). Attractant compounds of the southern green stink bug, *Nezara viridula* (L.) (Heteroptera: Pentatomidae). *Anais da Sociedade Entomológica do Brasil*, **24**, 215–25.

Borges, M., Jepson, P.C. and House, P.E. (1987). Long range mate location and close range courtship behaviour of the green stink bug, *Nezara viridula*, and its mediation by sex pheromones. *Entomologia Experimentalis et Applicata*, **44**, 205–12.

Bowling, C.C. (1979). The stylet sheath as an indicator of feeding activity of the rice stink bug. *Journal of Economic Entomology*, **72**, 259–60.

Bowling, C.C. (1980). The stylet sheath as an indicator of feeding activity by the southern green stink bug on soybeans. *Journal of Economic Entomology*, **73**, 1–3.

Brackenbury, J. (1995). *Insects and Flowers: A Biological Partnership*. London: Cassell Brandford.

Calhoun, D.S., Funderburk, J.E. and Teare, I.D. (1988). Soybean seed crude protein and oil levels in relation to weight, developmental time, and survival of southern green stink bug (Hemiptera: Pentatomidae). *Environmental Entomology*, **17**, 727–29.

Carpenter, J., Felsot, A., Goode, T. *et al.* (2002). Comparative environmental impacts of biotechnology-derived and traditional soybean, corn, and cotton crops. Iowa, USA: Council for Agricultural Science and Technology.

Carroll, S.P. (1988). Contrasts in reproductive ecology between temperate and tropical populations of *Jadera haematoloma*, a mate-guarding hemipteran (Rhopalidae). *Annals of the Entomological Society of America*, **81**, 54–63.

Chandler, L. (1989). The broad-headed bug, *Megalotomus parvus* (Westwood) (Hemiptera: Alydidae), a dry season pest of beans in Brazil. *Annual Report Bean Improvement Cooperation*, **32**, 84–5.

Chocorosqui, V.R. and Panizzi, A.R. (2001). Evolução dos danos causados por *Dichelops melacanthus* (Dallas) (Heteroptera: Pentatomidae) a plântulas de milho e trigo. Resumos XIX Congresso Brasileiro de Entomologia, Manaus, Amazonas. p. 276.

Čokl, A. (1983). Functional properties of vibroreceptors in the legs of *Nezara viridula* (L.) (Heteroptera, Pentatomidae). *Journal of Comparative Physiology*, **150**, 261–9.

Čokl, A., Virant-Doberlet, M. and McDowell, A. (1999). Vibrational directionality in the southern green stink bug, *Nezara viridula* (L.), is mediated by female song. *Animal Behavior*, **58**, 1277–83.

Čokl, A., Virant-Doberlet, M. and Stritih, N. (2000). The structure and function of songs emitted by southern green stink bugs from Brazil, Florida, Italy and Slovenia. *Physiological Entomology*, **25**, 196–205.

Corrêa-Ferreira, B. S. (1987). Liberação do parasitóide *Trissolcus basalis* em cultivar armadilha e seu efeito na população de percevejos da soja. *Resultados de Pesquisa de Soja*, **20**, 142–3.

Corrêa-Ferreira, B. S. and Panizzi, A. R. (1999). Percevejos da soja e seu manejo. Embrapa Soja, *Circular Técnica*, **24**, 45 pp.

Corseuil, E., Cruz, F. Z. and Meyer, L. M. C. (1974). *Insetos Nocivos à Soja no Rio Grande do Sul*. Porto Alegre, RS, Brasil: UFRGS Faculdade de Agronomia, Gráfica da Escola Profissional Champagnat.

Corso, I. C. (1989). Atratividade do sal de cozinha para espécimes de *Nezara viridula* (L. 1758). Resultados de Pesquisa de Soja 1988/1989. *Documentos*, **45**, 78–9.

Corso, I. C. (1990). Uso de sal de cozinha na redução da dose de inseticida para controle de percevejos da soja. *Centro Nacional de Pesquisa de Soja, Londrina, PR, Comunicado Técnico*, **45**, 1–6.

Costa, E. C. and Link, D. (1977). Danos causados por algumas espécies de Pentatomidae em duas variedades de soja. *Revista do Centro de Ciências Rurais*, **7**, 199–206.

Costa, E. C. and Link, D. (1982). Dispersão de adultos de *Piezodorus guildinii* e *Nezara viridula* (Hemiptera: Pentatomidae) em soja. *Revista do Centro de Ciências Ruais*, **12**, 51–6.

Crawley, M. J. (1983). *The Herbivory: The Dynamics of Animal-Plant Interactions*. Berkeley, CA: University of California Press.

DeWitt, N. B. and Godfrey, G. L. (1972). The literature of arthropods associated with soybeans. II. A bibliography of the southern green stink bug *Nezara viridula* (Linneaus) (Hemiptera: Pentatomidae). *Illinois Natural History Survey Bulletin Notes*, **78**, 1–23.

Edwards, P. J. and Wratten, S. D. (1980). *Ecology of Insect–Plant Interactions*. London: Edward Arnold.

Erlich, P. R. and Raven, P. H. (1964). Butterflies and plants: a study in coevolution. *Evolution*, **18**, 586–608.

Flint, M. L. and Gouveia, P. (2001). *IPM in Practice – Principles and Methods of Integrated Pest Management*. Oakland, CA: University of California.

Fontes, E. M. G., Pires, C. S. S., Sujii, E. R. and Panizzi, A. R. (2002). The environmental effects of genetically modified crops resistant to insects. *Neotropical Entomology*, **31**, 497–513.

Fox, L. R. (1981). Defense and dynamics in plant–herbivore systems. *American Zoologist*, **21**, 853–64.

Fox, L. R. and Morrow, P. A. (1981). Specialization: species property of local phenomenon? *Science*, **211**, 887–93.

Futuyma, D. J. (1983). Evolutionary interactions among herbivorous insects and plants. In D. G. Futuyma and M. Slatkin (eds.), *Coevolution*. Sunderland, MA: Sinauer Assoc. Inc. pp. 207–31.

Galileo, M.H.M. and Heinrichs, E.A. (1979). Danos causados à soja em diferentes níveis e épocas de infestação durante o crescimento. *Pesquisa Agropecuária Brasileira*, **14**, 279–82.

Galileo, M.H.M, Gastal, H.A.O. and Grazia, J. (1977). Levantamento populacional de Pentatomidae (Hemiptera) em cultura de soja (*Glycine max* (L.) Merr.) no município de Guaíba, Rio Grande do Sul. *Revista Brasileira de Biologia*, **37**, 111–20.

Gatehouse, J.A. and Gatehouse, A.M.R. (2000). Genetic engineering of plants for insect resistance. In J.E. Rechcigl and N.A. Rechcigl (eds.), *Biological and Biotechnological Control of Insect Pests*. Boca Raton, FL: Lewis Publishers. pp. 211–41.

Gershenzon, J. and Croteau, R. (1991). Terpenoids. In G.A. Rosenthal and M.R. Berenbaum (eds.), *Herbivores: Their Interactions with Secondary Plant Metabolites*. San Diego, CA: Academic Press. pp. 165–219.

Gilman, D.F., McPherson, R.M., Newsom, L.D. *et al.* (1982). Resistance in soybeans to the southern green stink bug. *Crop Science*, **22**, 573–6.

Harris, V.E. and Todd, J.W. (1980). Male-mediated aggregation of male, female and 5th instar southern green stink bug and concomitant attraction of a tachinid parasite, *Trichopoda pennipes*. *Entomologia Experimentalis et Applicata*, **27**, 117–26.

Harris, V.E., Todd, J.W., Webb, J.C. and Benner, J.C. (1982). Acoustical and behavioral analysis of the songs of the southern green stink bug, *Nezara viridula*. *Annals of the Entomological Society of America*, **75**, 234–49.

Hirose, E., Panizzi, A.R. and Cattelan, A.J. (2006). Effect of relative humidity on emergence and on dispersal and regrouping of first instar *Nezara viridula* (L.) (Hemiptera: Pentatomidae). *Neotropical Entomology*, **35**, 757–61.

Hokkanen, H.M.T. (1991). Trap cropping in pest management. *Annual Review of Entomology*, **36**, 119–38.

Hori, K. (2000). Possible causes of disease symptons resulting from the feeding of phytophagous Heteroptera. In C.W. Schaefer and A.R. Panizzi (eds.), *Heteroptera of Economic Importance*. Boca Raton, FL: CRC Press. pp. 11–35.

Horn, D.J. (2002). Ecological aspects of pest management In D. Pimentel (ed.), *Encyclopedia of Pest Management*. New York: Marcel Dekker, Inc. pp. 211–13.

Jackai, L.E.N. (1984). Using trap plants in the control of insect pests of tropical legumes. In P.C. Matteson (ed.), *Proceedings of the International Workshop in Integrated Pest Control of Grain Legumes*. Brasília, Brazil: Embrapa. pp. 101–12.

James, C.A. (2001). *Global Review of Commercialized Transgenic Crops, 2001*. ISAAA Briefs Number 24. Ithaca, New York: ISAAA.

Janzen, D.H. (1980). When is it coevolution? *Evolution*, **84**, 611–12.

Javahery, M., Schaefer, C.W. and Lattin, J.D. (2000). Shield bugs (Scutelleridae). In C.W. Schaefer and A.R. Panizzi (eds.), *Heteroptera of Economic Importance*. Boca Raton, FL: CRC Press. pp. 475–503.

Jermy, T. (1984). Evolution of insect/host plant relationships. *American Naturalist*, **124**, 609–30.

Jolivet, P. (1998). *Interrelationship between Insects and Plants*. Boca Raton, FL: CRC Press.

Jones, W.A., Jr. (1979). *The Distribution and Ecology of Pentatomid Pests of Soybeans in South Carolina*. Ph.D. thesis, Clemson University, Clemson, SC.

Jones, W. A., Jr. and Sullivan, M. J. (1978). Susceptibility of certain soybean cultivars to damage by stink bugs. *Journal of Economic Entomology*, **71**, 534–6.

Jones, W. A., Jr. and Sullivan, M. J. (1979). Soybean resistance to the southern green stink bug, *Nezara viridula*. *Journal of Economic Entomology*, **72**, 628–32.

Jones, W. A., Jr. and Sullivan, M. J. (1981). Overwintering habitats, spring emergence patterns, and winter mortality of some South Carolina Hemiptera. *Environmental Entomology*, **10**, 409–14.

Jones, W. A., Jr. and Sullivan, M. J. (1982). Role of host plants in population dynamics of stink bug pests of soybean in South Carolina Hemiptera. *Environmental Entomology*, **11**, 867–75.

Karban, R. and Baldwin, I. T. (1997). *Induced Responses to Herbivory*. Chicago, IL: University of Chicago Press.

Kester, K. M. and Smith, C. M. (1984). Effects of diet on growth, fecundity and duration of tethered flight of *Nezara viridula*. *Entomologia Experimentalis et Applicata*, **35**, 75–81.

Kester, K. M., Smith, C. M. and Gilman, D. F. (1984). Mechanisms of resistance in soybean *Glycine max* (L.) Merrill genotype PI 171444 to the southern green stink bug *Nezara viridula* (L.). *Environmental Entomology*, **13**, 1208–15.

Kiritani, K., Hokyo, N., Kimura, K. and Nakasuji, F. (1965). Imaginal dispersal of the southern green stink bug, *Nezara viridula* L., in relation to feeding and oviposition. Jap. *Japanese Journal of Applied Entomology and Zoology*, **9**, 291–96.

Kobayashi, T. and Cosenza, G. W. (1987). Integrated control of soybean stink bugs in the Cerrados. *Japanese Agriculture Research Quarterly*, **20**, 229–36.

Kogan, M. (ed.). (1986*a*). *Ecological Theory and Integrated Pest Management Practice*. New York: Wiley.

Kogan, M. (1986*b*). Plant defense strategies and host-plant resistance. In M. Kogan (ed.), *Ecological Theory and Integrated Pest Management Practice*. New York: Wiley. pp. 83–134.

Kogan, M. and Turnipseed, S. G. (1987). Ecology and management of soybean arthropods. *Annual Review of Entomology*, **32**, 507–38.

Link, D., Estefanel, V. and Santos, O. S. (1971). Danos causados por percevejos fitófagos em grãos de soja. *Revista do Centro de Ciências Rurais*, **1**, 9–13.

Link, D., Estefanel, V., Santos, O. S., Mezzomo, M. C. and Abreu, L. E. V. (1973). Influência do ataque de pentatomídeos nas características agronômicas do grão de soja, *Glycine max* (L.) Mer. *Anais da Sociedade Entomológica do Brasil*, **2**, 59–65.

Lis, J. A., Becker, M. and Schaefer, C. W. (2000). Burrower bugs (Cydnidae). In C. W. Schaefer and A. R. Panizzi (eds.), *Heteroptera of Economic Importance*. Boca Raton, FL: CRC Press. pp. 405–19.

Lockwood, J. A. and Story, R. N. (1985). Bifunctional pheromone in the first instar of the southern green stink bug, *Nezara viridula* (L.) (Hemiptera: Pentatomidae): its characterization and interaction with other stimuli. *Annals of the Entomological Society of America*, **78**, 474–79.

Lockwood, J. A. and Story, R. N. (1986). Adaptive functions of nymphal aggregation in the southern green stink bug, *Nezara viridula* (L.) (Hemiptera: Pentatomidae). *Environmental Entomology*, **15**, 739–9.

Lopes, O.J., Link, D. and Basso, I.V. (1974). Pentatomídeos de Santa Maria – lista preliminar de plantas hospedeiras. *Revista do Centro de Ciências Rurais*, **4**, 317–22.

Ludwig, S.W. and Kok, L.T. (1998). Evaluation of trap crops to manage harlequin bugs, *Murgantia histrionica* (Hahn) (Hemiptera: Pentatomidae) on broccoli. *Crop Protection*, **17**, 123–8.

Maia, N.G. (1973). Ocorrência do percevejo da soja – *Nezara viridula* (L.) em espigas de trigo no Rio Grande do Sul. *Agronomia Sulriograndense*, **9**, 241–3.

McLain, D.K. (1981). Sperm precedence and prolonged copulation in the southern green stink bug, *Nezara viridula*. *Journal of the Georgia Entomological Society*, **16**, 70–6.

McPherson, R.M. and Newsom, L.D. (1984). Trap crops for control of stink bugs in soybean. *Journal of the Georgia Entomological Society*, **19**, 470–80.

McPherson, R.M., Newsom, L.D. and Farthing, B.F. (1979). Evaluation of four stink bug species from three genera affecting soybean yield and quality in Louisiana. *Journal of Economic Entomology*, **72**, 188–94.

Miles, P.W. and Taylor, G.S. (1994). 'Osmotic pump' feeding by coreids. *Entomologia Experimentalis et Applicata*, **73**, 163–73.

Miner, F.D. (1966). Biology and control of stink bugs on soybeans. Arkansas. *Arkansas Agricultural Experiment Station Bulletin*, **708**, 1–40.

Mitchell, W.C. and Mau, R.F.L. (1971). Response of the southern green stink bug and its parasite, *Trichopoda pennipes*, to male stink bug pheromones. *Journal of Economic Entomology*, **64**, 856–9.

Moeller, B.L. and Seigler, D.S. (1999). Biosynthesis of cyanogenic glycosides, cyanolipids, and related compounds. In B.J. Singh (ed.), *Plant Amino Acids*. New York: Marcel Dekker, Inc. pp. 563–609.

Naito, A. (1996). Insect pest control through use of trap crops. *Agrochemicals Japan*, **68**, 9–11.

Nault, L.R. and Gyrisco, G.G. (1966). Relation of the feeding process of the pea aphid to the inoculation of pea enation mosaic virus. *Annals of the Entomological Society of America*, **59**, 1185–96.

Negron, J.F. and Riley, T.J. (1987). Southern green stink bug, *Nezara viridula* (Heteroptera: Pentatomidae) feeding on corn. *Journal of Economic Entomology*, **80**, 666–9.

Newsom, L.D. and Herzog, D.C. (1977). Trap crops for control of soybean pests. *Louisiana Agriculture*, **20**, 14–15.

Niva, C.C. and Panizzi, A.R. (1996). Efeitos do cloreto de sódio no comportamento de *Nezara viridula* (L.) (Heteroptera: Pentatomidae) em vagem de soja. *Anais da Sociedade Entomológica do Brasil*, **25**, 251–256.

Norris, R.F., Caswell-Chen, E.P. and Kogan, M. (2003). *Concepts in Integrated Pest Management*. Upper Saddle River, NJ: Prentice Hall.

Ota, D. and Čokl, A. (1991). Mate location in the southern green stink bug, *Nezara viridula* (Heteroptera: Pentatomidae), mediated through substrate-borne signals on ivy. *Journal of Insect Behavior*, **4**, 441–6.

Owino, P.O. (2002). Antagonistic plants. In D. Pimentel (ed.), *Encyclopedia of Pest Management*. New York: Marcel Dekker, Inc. pp. 21–3.

Owusu-Manu, E. (1977). Flight activity and dispersal of *Bathycoelia thalassina* (Herrich-Schaeffer) Hemiptera: Pentatomidae. *Ghana Journal of Agricultural Science*, **10**, 23–6.

Painter, R. H. (1951). *Insect Resistance in Crop Plants*. New York: Macmillan.

Panizzi, A. R. (1980). Uso de cultivar armadilha no controle de percevejos em soja. *Trigo e Soja*, **47**, 11–14.

Panizzi, A. R. (1987). Impacto de leguminosas na biologia de ninfas e efeito da troca de alimento no desempenho de adultos de *Piezodorus guildinii* (Hemiptera: Pentatomidae). *Revista Brasileira de Biologia*, **47**, 585–91.

Panizzi, A. R. (1988). Biology of *Megalotomus parvus* (Westwood) (Heteroptera: Alydidae) on selected leguminous food plants. *Insect Science and its Application*, **9**, 279–85.

Panizzi, A. R. (1992). Performance of *Piezodorus guildinii* on four species of *Indigofera* legumes. *Entomologia Experimentalis et Applicata*, **63**, 221–8.

Panizzi, A. R. (1995). Feeding frequency, duration and preference of the southern green stink bug (Heteroptera: Pentatomidae) as affected by stage of development, age, and physiological condition. *Anais da Sociedade Entomológica do Brasil*, **24**, 437–44.

Panizzi, A. R. (1997). Wild hosts of pentatomids: ecological significance and role in their pest status on crops. *Annual Review of Entomology*, **42**, 99–122.

Panizzi, A. R. (2000). Suboptimal nutrition and feeding behavior of hemipterans on less preferred plant food sources. *Anais da Sociedade Entomológica do Brasil*, **29**, 1–12.

Panizzi, A. R. (2004). Adaptive advantages for egg and nymph survivorship by egg deposition in masses or singly in seed-sucking Heteroptera. In G. T. Gujar (ed.), *Contemporary Trends in Insect Science*. New Delhi: Campus Books International. pp. 60–73.

Panizzi, A. R. and Alves, R. M. L. (1993). Performance of nymphs and adults of the southern green stink bug (Heteroptera: Pentatomidae) exposed to soybean pods at different phenological stages of development. *Journal of Economic Entomology*, **86**, 1088–93.

Panizzi, A. R., Cardoso, S. R. and Chocorosqui, V. R. (2002). Nymph and adult performance of the small green stink bug, *Piezodorus guildinii* (Westwood) on lanceleaf crotalaria and soybean. *Brazilian Archives of Biology and Technology*, **45**, 53–8.

Panizzi, A. R., Cardoso, S. R. and Oliveira, E. D. M. (2000a). Status of pigeon pea as an alternative host of *Piezodorus guildinii* (Hemiptera: Pentatomidae), a pest of soybean. *Florida Entomologist*, **83**, 334–42.

Panizzi, A. R. and Chocorosqui, V. R. (1999). Pragas - elas vieram com tudo. *Cultivar*, **11**, 8–10.

Panizzi, A. R. and Corrêa-Ferreira, B. S. (1997). Dynamics in the insect fauna adaptation to soybean in the tropics. *Trends in Entomology*, **1**, 71–88.

Panizzi, A. R., Galileo, M. H. M., Gastal, H. A. O., Toledo, J. F. F. and Wild, C. H. (1980). Dispersal of *Nezara viridula* and *Piezodorus guildinii* nymphs in soybeans. *Environmental Entomology*, **9**, 293–6.

Panizzi, A. R. and Grazia, J. (2001). Stink bugs (Pentatomidae) and a unique host plant in the Brazilian subtropics. *Iheringia, Série Zoologia*, **90**, 21–35.

Panizzi, A. R. and Hirose, E. (1995*a*). Seasonal body weight, lipid content, and impact of starvation and water stress on adult survivorship and longevity of *Nezara viridula* and *Euschistus heros*. *Entomologia Experimentalis et Applicata*, **76**, 247–53.

Panizzi, A. R. and Hirose, E. (1995*b*). Survival, reproduction, and starvation resistance of adult southern green stink bug (Heteroptera: Pentatomidae) reared on sesame or soybean. *Annals of the Entomological Society of America*, **88**, 661–5.

Panizzi, A. R. and Machado-Neto, E. (1992). Development of nymphs and feeding habits of nymphal and adult *Edessa meditabunda* (Heteroptera: Pentatomidae) on soybean and sunflower. *Annals of the Entomological Society of America*, **85**, 477–81.

Panizzi, A. R., McPherson, J. E., James, D. G., Javahery, M. and McPherson, R. M. (2000b). Economic importance of stink bugs (Pentatomidae). In C. W. Schaefer and A. R. Panizzi (eds.), *Heteroptera of Economic Importance*. Boca Raton, FL: CRC Press. pp. 421–74.

Panizzi, A. R. and Meneguim, A. M. (1989). Performance of nymphal and adult *Nezara viridula* on selected alternate host plants. *Entomologia Experimentalis et Applicata*, **50**, 215–23.

Panizzi, A. R., Meneguim, A. M. and Rossini, M. C. (1989). Impacto da troca de alimento da fase ninfal para a fase adulta e do estresse nutricional na fase adulta na biologia de *Nezara viridula* (Hemiptera: Pentatomidae). *Pesquisa Agropecuária Brasileira*, **24**, 945–54.

Panizzi, A. R., Mourão, A. P. M. and Oliveira, E. D. M. (1998). Nymph and adult biology and seasonal abundance of *Loxa deducta* (Walker) on privet, *Ligustrum lucidum*. *Anais da Sociedade Entomológica do Brasil*, **27**, 199–206.

Panizzi, A. R. and Niva, C. C. (1994). Overwintering strategy of the brown stink bug in northern Paraná. *Pesquisa Agropecuária Brasileira*, **29**, 509–11.

Panizzi, A. R., Niva, C. C. and Hirose, E. (1995). Feeding preference by stink bugs (Heteroptera: Pentatomidae) for seeds within soybean pods. *Journal of Entomological Science*, **30**, 333–41.

Panizzi, A. R. and Oliveira, E. D. M. (1998). Performance and seasonal abundance of the neotropical brown stink bug, *Euschistus heros* nymphs and adults on a novel food plant (pigeon pea) and soybean. *Entomologia Experimentalis et Applicata*, **88**, 169–75.

Panizzi, A. R. and Oliveira, E. D. M. (1999). Seasonal occurrence of tachinid parasitism on stink bugs with different overwintering strategies. *Anais da Sociedade Entomológica do Brasil*, **28**, 169–72.

Panizzi, A. R. and Oliveira, N. (1993). Atração do cloreto de sódio (sal de cozinha) aos percevejos-pragas da soja. *Resultados de Pesquisa de Soja 1989/1990, Embrapa Soja, Documentos*, **58**, 71–6.

Panizzi, A. R. and Parra, J. R. P. (eds.). (1991*a*). *Ecologia Nutricional de Insetos e suas Implicações no Manejo de Pragas*. São Paulo, SP: Ed. Manole.

Panizzi, A. R. and Parra, J. R. P. (1991*b*). A ecologia nutricional e o manejo integrado de pragas. In A. R. Panizzi and J. R. P. Parra (eds.), *Ecologia Nutricional de Insetos e suas Implicações no Manejo de Pragas*. São Paulo, SP, Ed. Manole. pp. 313–36.

Panizzi, A.R. and Rossi, C.E. (1991). The role of *Acanthospermum hispidum* in the phenology of *Euschistus heros* and of *Nezara viridula*. *Entomologia Experimentalis et Applicata*, **59**, 67–74.

Panizzi, A.R. and Rossini, M.C. (1987). Impacto de várias leguminosas na biologia de ninfas de *Nezara viridula* (Hemiptera:Pentatomidae). *Revista Brasileira de Biologia*, **47**, 507–12.

Panizzi, A.R. and Saraiva, S.I. (1993). Performance of nymphal and adult southern green stink bug on an overwintering host and impact of nymph to adult food-switch. *Entomologia Experimentalis et Applicata*, **68**, 109–15.

Panizzi, A.R. and Slansky, F., Jr. (1985a). Review of phytophagous pentatomids (Hemiptera: Pentatomidae) associated with soybean in the Americas. *Florida Entomologist*, **68**, 184–214.

Panizzi, A.R. and Slansky, F., Jr. (1985*b*). Legume host impact on performance of adult *Piezodorus guildinii* (Westwood) (Hemiptera: Pentatomidae). *Environmental Entomology*, **14**, 237–42.

Panizzi, A.R. and Slansky, F., Jr. (1991). Suitability of selected legumes and the effect of nymphal and adult nutrition in the southern green stink bug (Hemiptera: Heteroptera: Pentatomidae). *Journal of Economic Entomology*, **84**, 103–13.

Panizzi, A.R. and Smith, J.G. (1977). Biology of *Piezodorus guildinii*: oviposition, development time, adult sex ratio, and longevity. *Annals of the Entomological Society of America*, **70**, 35–39.

Panizzi, A.R., Smith, J.G., Pereira, L.A.G. and Yamashita, J. (1979). Efeitos dos danos de *Piezodorus guildinii* (Westwood, 1837) no rendimento e qualidade da soja. *Anais do I Seminário Nacional de Pesquisa de Soja*. Vol. II, pp. 59–78.

Panizzi, A.R. and Vivan, L.M. (1997). Seasonal abundance of the neotropical brown stink bug, *Euschistus heros* in overwintering sites and the breaking of dormancy. *Entomologia Experimentalis et Applicata*, **82**, 213–16.

Panizzi, A.R., Vivan, L.M., Corrêa-Ferreira, B.S. and Foerster, L.A. (1996). Performance of southern green stink bug (Heteroptera: Pentatomidae) nymphs and adults on a novel food plant (Japanese privet) and other hosts. *Annals of the Entomological Society of America*, **89**, 822–6.

Panizzi, M.C.C., Bays, I.A., Kiihl, R.A.S. and Porto, M.P. (1981). Identificação de genótipos fontes de resistência a percevejos-pragas da soja. *Pesquisa Agropecuária Brasileira*, **16**, 33–6.

Pedigo, L.P. (2002). *Entomology and Pest Management*, 4th edn. Upper Saddle River, NJ: Prentice Hall.

Pimentel, D. (ed.). (1981). *Handbook of Pest Management in Agriculture*. Vols. I–III. Boca Raton, FL: CRC Press.

Pimentel, D. (ed.). (2002). *Encyclopedia of Pest Management*. New York: Marcel Dekker, Inc.

Pinto, S.B. and Panizzi, A.R. (1994). Performance of nymphal and adult *Euschistus heros* (F.) on milkweed and on soybean and effect of food switch on adult survivorship, reproduction and weight gain. *Anais da Sociedade Entomológica do Brasil*, **23**, 549–55.

Ragsdale, D. W., Larson, A. D. and Newsom, L. D. (1979). Microorganisms associated with feeding and from various organs of *Nezara viridula*. *Journal of Economic Entomology*, **72**, 725–6.

Ragsdale, D. W., Larson, A. D. and Newsom, L. D. (1981). Quantitative assessment of the predators of *Nezara viridula* eggs and nymphs within a soybean agroecosystem using an ELISA. *Environmental Entomology*, **10**, 402–405.

Rechcigl, J. E. and Rechcigl, N. A. (eds.). (2000). *Insect Pest Management – Techniques for Environmental Protection*. Boca Raton, FL: Lewis Publishers.

Rossetto, C. J. (1989). Breeding for resistance to stink bugs. In A. J. Pascale (ed.), *Proceedings of the World Soybean Research Conference IV*. Buenos Aires: Impresiones Amawald S.A. p. 2046.

Russin, J. S., Layton, M. B., Orr, D. B. and Boethel, D. J. (1987). Within-plant distribution of, and partial compensation for, stink bug (Heteroptera: Pentatomidae) damage to soybean seeds. *Journal of Economic Entomology*, **80**, 215–20.

Santos, C. H. and Panizzi, A. R. (1998*a*). Danos qualitativos causados por *Neomegalotomus parvus* (Westwood) em sementes de soja. *Anais da Sociedade Entomológica do Brasil*, **27**, 387–93.

Santos, C. H. and Panizzi, A. R. (1998*b*). Nymphal and adult performance of *Neomegalotomus parvus* (Hemiptera: Alydidae) on wild and cultivated legumes. *Annals of the Entomological Society of America*, **91**, 445–51.

Schaefer, C. W. and Panizzi, A. R. (eds.). (2000). *Heteroptera of Economic Importance*. Boca Raton, FL: CRC Press.

Schaefer, C. W. and Mitchell, P. L. (1983). Food plants of the Coreoidea (Hemiptera: Heteroptera). *Annals of the Entomological Society of America*, **76**, 591–615.

Sedlacek, J. D. and Townsend, L. H. (1988). Impact of *Euschistus servus* and *E. variolarius* (Heteroptera: Pentatomidae) feeding on early growth stages of corn. *Journal of Economic Entomology*, **81**, 840–4.

Seigler, D. S. (2002). Toxins in plants. In D. Pimentel (ed.), *Encyclopedia of Pest Management*. New York: Marcel Dekker, Inc. pp. 840–2.

Shelton, A. M., Zhao, J. Z. and Roush, R. T. (2002). Economic, ecological, food safety, and social consequences of the deployment of *Bt* transgenic plants. *Annual Review of Entomology*, **47**, 845–81.

Silva, A. G. D.'A., Gonçalves, C. R., Galvão, D. M. *et al.* (1968). *Quarto Catálogo dos Insetos que Vivem nas Plantas do Brasil – Seus Parasitas e Predadores*. Rio de Janeiro: Ministério de Agricultura. *Parte II, vol. I.*

Singh, Z. (1972). Bionomics of the Southern Green Stink Bug, *Nezara Viridula*. (Linn.) (Hemiptera: Pentatomidae) in Central India. Ph.D. thesis, University of Illinois, Urbana-Champaign.

Slansky, F., Jr. (1990). Insect nutritional ecology as a basis for studying host plant resistance. *Florida Entomologist*, **73**, 359–78.

Slansky, F., Jr. and Rodriguez, J. G., (1987*a*). Nutritional ecology of insects, mites, spiders, and related invertebrates. In F. Slansky, Jr. and J. G. Rodriguez (eds.), *Nutritional Ecology of Insects, Mites, Spiders and Related Invertebrates*. New York: Wiley. pp. 1–69.

Slansky, F., Jr. and Rodriguez, J. G. (eds.), (1987*b*). *Nutritional Ecology of Insects, Mites, Spiders and Related Invertebrates*. New York: Wiley.

Slansky, F., Jr. and Panizzi, A. R. (1987). Nutritional ecology of seed-sucking insects. In F. Slansky, Jr. and J. G. Rodriguez (eds.), *Nutritional Ecology of Insects, Mites, Spiders and Related Invertebrates*. New York: Wiley. pp. 283–320.

Smith, J. G. (1978). Pests of soybean in Brazil. In S. R. Singh, H. F. van Emden and T. A. Taylor (eds.), *Pests of Grain Legumes: Ecology and Control*. London: Academic Press, pp. 167–76.

Sosa-Gómez, D. R., Takachi, C. Y. and Moscardi, F. (1993). Determinação de sinergismo e suscetibilidade diferencial de *Nezara viridula* (L.) e *Euschistus heros* (F.) (Hemiptera: Pentatomidae) à inseticidas em mistura com cloreto de sódio. *Anais da Sociedade Entomológica do Brasil*, **22**, 569–76.

Southwood, T. R. E. and Way, M. J. (1970). Ecological background to pest management. In R. L. Rabb and F. E. Guthrie (eds.), *Concepts of Pest Management*. Raleigh, NC: North Carolina State University Press. pp. 6–29.

Strong, D. R., Lawton, J. H. and Southwood, T. R. E. (1984). *Insects on Plants. Community Patterns and Mechanisms*. Cambridge, MA: Harvard University Press.

Thomas, G. D., Ignoffo, C. M., Morgan, C. E. and Dickerson, W. A. (1974). Southern green stink bug: influence on yield and quality of soybeans. *Journal of Economic Entomology*, **67**, 501–3.

Todd, J. W. (1989). Ecology and behavior of *Nezara viridula*. *Annual Review of Entomology*, **34**, 273–92.

Todd, J. W. and Herzog, D. C. (1980). Sampling phytophagous Pentatomidae on soybean. In M. Kogan and D. C. Herzog (eds.), *Sampling Methods in Soybean Entomology*. New York: Springer-Verlag. pp. 438–78.

Todd, J. W., Jellum, M. D. and Leuck, D. B. (1973). Effects of southern green stink bug damage on fatty acid composition of soybean oil. *Environmental Entomology*, **2**, 685–89.

Turnipseed, S. G. and Sullivan, M. J. (1975). Plant resistance in soybean insect management. In L. D. Hill (ed.), *Proceedings of the World Soybean Research Conference I*. Urbana-Champaign, IL: Interstate Printers and Publishers. pp. 549–60.

Todd, J. W. and Turnipseed, S. G. (1974). Effects of southern green stink bug damage on yield and quality of soybeans. *Journal of Economic Entomology*, **67**, 421–26.

Todd, J. W. and Womack, H. (1973). Secondary infestations of cigarette beetle in soybean seed damaged by southern green stink bug. *Environmental Entomology*, **2**, 720.

Vandermeer, J. (1995). The ecological basis of alternative agriculture. *Annual Review of Ecology and Systematics*, **26**, 201–24.

van Emden, H. F. (1965). The role of uncultivated land in the biology of crop pests and beneficial insects. *Scientific Horticulture*, **17**, 121–26.

Velasco, L. R. I. and Walter, G. H. (1992). Availability of different host plant species and changing abundance of the polyphagous bug *Nezara viridula* (Hemiptera: Pentatomidae). *Environmental Entomology*, **21**, 751–9.

Velasco, L. R. I. and Walter, G. H. (1993). Potential of host-switching in *Nezara viridula* (Hemiptera: Pentatomidae) to enhance survival and reproduction. *Environmental Entomology*, **22**, 326–33.

Velasco, L. R. I., Walter, G. H. and Harris, V. E. (1995). Voltinism and host plant use by *Nezara viridula* (L.) (Hemiptera: Pentatomidae) in Southeastern Queensland. *Journal of the Australian Entomological Society*, **34**, 193–203.

Ventura, M. U., Montalván, R. and Panizzi, A. R. (2000). Feeding preferences and related types of behaviour of *Neomegalotomus parvus*. *Entomologia Experimentalis et Applicata*, **97**, 309–15.

Viator, H. P., Pantoja, A. and Smith, C. M. (1983). Damage to wheat seed quality and yield by the rice stink bug and southern green stink bug (Hemiptera: Pentatomidae). *Journal of Economic Entomology*, **76**, 1410–13.

Villas Bôas, G. L. and Panizzi, A. R. (1980). Biologia de *Euschistus heros* (Fabricius1789) em soja (*Glycine max* L. Merrill). *Anais da Sociedade Entomológica do Brasil*, **9**, 105–13.

Watson, J. R. (1924). *Crotalaria* as a trap crop for pumpkin bugs in citrus groves. *The Florida Grower*, **29**, 6–7.

Yonke, T. R. and Medler, J. T. (1965). Biology of *Megalotomus quinquespinosus* (Hemiptera: Alydidae). *Annals of the Entomological Society of America*, **58**, 222–24.

7

Conservation, biodiversity, and integrated pest management

S. D. WRATTEN, D. F. HOCHULI, G. M. GURR, J. TYLIANAKIS AND S. L. SCARRATT

7.1 Introduction

Conservation biology has been described as a "mission oriented discipline" (Soule and Wilcox, 1980), while Samways (1994) goes further, describing it as a "crisis science", in recognition of the immediate and adverse impacts facing our biosphere. Similarly, the development of integrated pest management (IPM) over the past 50 years has been driven by "real world" pressures, in this case, the failure of unilateral pesticidal management to provide effective control of pests. A further similarity between the disciplines of conservation biology – the science of preserving biodiversity (Pullin, 2002) – and IPM is that both are ecological disciplines. Despite this common ground, however, the two broad disciplines have developed in relative isolation from each other. One reflection of this is that conservation biologists have tended to use ecological theory to a considerably greater extent than have those involved in IPM. Indeed, the growth of the discipline of conservation biology and the conceptual framework that has developed around it is one of the most prominent advances in ecology in recent years (Caughley, 1994; Dobson *et al.*, 1997). Recognizing the theoretical maturity of conservation biology (at least compared with IPM), one aim of this chapter is to explore the "common ground" of these disciplines, with the intention of identifying research directions that may further the discipline of IPM. Associated with this objective, we aim to avoid the reductionist tendency of scientific exploration and consider the extent to which the objectives of IPM and conservation biology may be compatible in agricultural landscapes. The need for such compatibility is acute. Not only do pests continue to cause severe losses to agricultural production

(Oerke *et al.*, 1994), but so much of the Earth's land surface is used for agriculture – approximately 35% (Gerard, 1995) – that conservation cannot be confined to reserves. Increasingly it will need to be integrated with agriculture on farmlands. "The struggle to maintain biodiversity is going to be won or lost in agricultural ecosystems" (McIntyre *et al.*, 1992).

Despite the background outlined above, the goals of conservation biology are sometimes seen as incompatible with many modern agricultural practices. Hill *et al.* (1995) tabulated 20 such practices with clearly identifiable ecological impacts. Attempts to manage pests are responsible for some of these effects, including on- and off-farm impact of pesticides on non-target species, the sometimes over-use of tillage to control soil pests, as well as habitat destruction and fragmentation. Thus, advances in IPM that reduce such impacts offer scope to make agroecosystems, and the habitats affected by them, better able to contribute to the conservation of biodiversity.

The objective of ecological sustainability has been incorporated into planning and management by many farmers and agencies (see Robertson and Harwood, 2001). This objective is not, however, always easy given the pressure to produce cheap, temporally reliable food (Passioura, 1999). This is reflected also in the "pesticide paradox" (Gurr *et al.*, 1996), whereby consumers demand cosmetically perfect produce yet abhor the use of pesticides which, in many cases, are used to prevent such damage. Reducing reliance on pesticides by widespread implementation of IPM is a critical objective for twenty-first century agriculture. This chapter will explore the extent to which that objective is compatible with conservation of biodiversity in farming landscapes and also the lessons that can be drawn from the field of conservation biology.

By integrating issues in conservation biology, biodiversity, and IPM, we aim to identify some new directions for IPM research. These will need to recognize the increasing significance of farmlands as the setting for conservation of animals and plants, many of which play no direct role in agricultural production but need to be conserved there because of loss of habitat elsewhere. In doing this we address a gap in current research, linking ecological theory to managed systems (Brown *et al.*, 2001), suggesting applications of conservation biology to the implementation of IPM. We also identify the many parallels between applied ecological research contributing to conservation biology and that supporting IPM, arguing that practical integration of their conceptual frameworks and ultimate goals is an inevitable consequence of the pressures for conservation outcomes and ecologically sustainable production.

7.2 Conservation biology and biodiversity

The status of conservation biology among ecologists has grown immensely over the past decade (Caughley and Gunn, 1996; Burgman and Lindenmayer, 1998; New, 2000), despite a broadly perceived crisis in its rigor and direction in its formative years (Caughley, 1994). Goals for conservation biology are often set at multiple spatial, temporal, and biological scales, with the latter imposing a conceptual breadth through contributions from biologists from diverse backgrounds. Conceptual advances attributed to the discipline revolve mainly around research on landscapes and genes (Burgman and Lindenmayer, 1998), although advocates for species-centered research and its effectiveness make a compelling case for not abandoning that approach (Caughley and Gunn, 1996).

The goal of conservation biology is essentially simple: to preserve, through effective management, biodiversity (at all its scales, see below) while ensuring its long-term viability. The emergence of conservation biology as a discipline has been driven by a recognition of the need for targeted research of highly specific applied questions, identified and often funded by those responsible for managing conservation issues. Considerable effort has recently been directed at reconciling the gamut of system- and species-specific literature in conservation biology so that general themes can be identified, despite the difficulties of doing so unequivocally (Beck, 1997; Lawton, 1999). Generalizations in conservation biology focus on responses to perceived threats and impacts, appropriate mechanisms for the management of landscapes, species, and populations as well as methods for prioritizing conservation actions at all of its relevant scales (Burgman and Lindenmayer, 1998). The ongoing and active discourse about the strength and validity of the generalizations outlined below highlights the youth of the discipline, as well as the difficulty of making generalizations at multiple biological and spatial scales. Nevertheless, the inherent scope of conservation biology dictates that general principles must operate across these scales to be effective.

Much current research focuses on the development of a greater understanding of mechanisms to aid in conservation of biodiversity at multiple scales and the testing of different tools to do this. Among the many arguments used to make the case for conserving biodiversity, the search for an understanding of the ecological roles of biodiversity and their significance in natural systems is one of the most prominent (Johnson *et al.*, 1996; Chapin *et al.*, 2000). This chapter differs from such studies in addressing agricultural systems.

7.3 Tenets and themes of conservation biology and biodiversity

The uncertainty pervading attempts to generalize in conservation biology reinforces the need for unambiguous operational definitions in any critical assessment of conservation biology and biodiversity (Simberloff, 1988; Peters, 1995; Ghilarov, 1996; Lawton, 1999). In all cases below, the generalizations are scale-dependent and goal-dependent; the themes and conclusions presented reflect consensus despite considerable debate about the strength with which they are held.

7.3.1 *The theory of island biogeography*

One of the key pieces of theory to have influenced conservation biology is the theory of island biogeography (MacArthur and Wilson, 1967). This predicts that the numbers of species found on a given island is the product of the rates of immigration and extinction. Thus, the numbers of species on small, remote islands are low because relatively few colonizers arrive and rates of extinction among established species are high. Conversely, larger islands sited closer to large landmasses receive more colonizers and these are able to establish large populations, which by virtue of the types of factors outlined below, are less likely to become extinct within a given time period. The theory of island biogeography was extended to conservation biology by Diamond (1975), who considered optimal-sized and shaped designs for nature reserves. This exercise took the pragmatic view that even if nature reserves may, at the time of their creation, be distinct from the surrounding habitat only in terms of their declared anthropogenic status, they tend to become floristically and faunistically distinct over time as adjoining areas are subject to changes in use, such as conversion to farmland or housing. Thus, nature reserves mimic islands, being defined by their size and level of remoteness from other areas with which they may exchange organisms. They do, however, differ from islands in being subject to relatively severe edge effects arising from adverse impacts of neighboring land use practices, such as the agricultural ones outlined above. One potential advantage of such reserves over true islands is that the level of isolation may be mitigated by connectivity (in the form of a vegetated corridor) or small "stepping stones" of habitat which, though sub-optimal, may suffice to allow movement of organisms between reserves separated by otherwise inhospitable habitats.

7.3.2 *Connectivity and metapopulations*

The importance of connectivity to invertebrates has been demonstrated in an eloquent study in which the moss covering stones was manipulated into patterns with differing levels of connectivity (Gilbert *et al.*, 1998). Patterns ranged from undisturbed (control) to isolated patches and included broken or unbroken corridors as intermediate treatments. Species richness declined with isolation and, of particular relevance for IPM, this effect was reflected for predatory species.

The importance of isolation will vary from species to species according to its vagility. For species that are able to move readily between habitat patches in a highly disturbed environment, survival may be possible even if the size of individual patches is such that the chances of extinction within each is high. The term "metapopulation" was coined by Levins (1969) (see Chapter 3) to describe such an arrangement. The sustainability of such systems demands that the factors that may cause extinction do not apply synchronously to all patches (Hanski, 1997) for this would not allow recolonization from occupied patches (Figure 7.1). Metapopulations are acknowledged to be important in the dynamics of natural enemy (e.g. Bonsall and Hassell, 2000) and pest (e.g. Jervis, 1997) systems.

7.4 SLOSS and IPM

The considerations sketched out above developed into a controversy over the optimal design of nature reserves that became known as the SLOSS (Single Large or Several Small) debate (Dobson and Rodriguez, 2001). From a biodiversity conservation perspective, the SLOSS debate essentially revolves

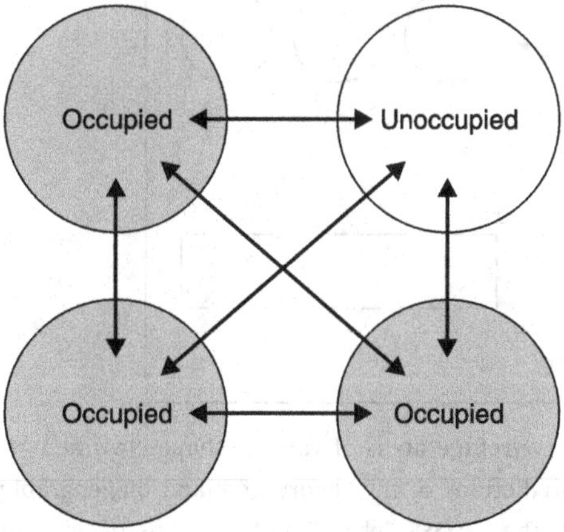

Figure 7.1. Schematic representation of the metapopulation model in which a species is present in only some of the potentially suitable patches of habitat between which individuals are able to move. (Adapted from Pullin, A.S., 2002. *Conservation Biology*. Cambridge University Press, Cambridge.)

around whether a single large reserve with robust populations is better than having the same total land area divided into several small reserves (Figure 7.2a). The latter scenario may support fewer species and the population of each may be at a relatively high risk of extinction but it can "capture" a wider range of habitat types and decreases the likelihood of extinction from a catastrophic event (such as fire or epidemic) which could affect a species confined to a single large reserve. The relevance of the SLOSS debate has not, to the best of our knowledge, been considered from the perspective of IPM but at least some ecological studies indicate its relevance.

For vertebrate pests, which require relatively large areas, theory suggests scope to effect control by reducing the size of patches of suitable habitat ("reserves") located in farming landscapes to below the size that would support a viable population. This concept is supported by empirical data (Newmark, 1987) but the minimum sizes implied by this relationship are relatively large in relation to the typical size of remnant vegetation patches found in many agricultural landscapes. Reducing the size of such habitat may also be incompatible with the broader objective of maintaining biodiversity in agricultural landscapes. An alternative is to site vulnerable enterprises in locations most remote from habitat associated with risk. An analysis of the

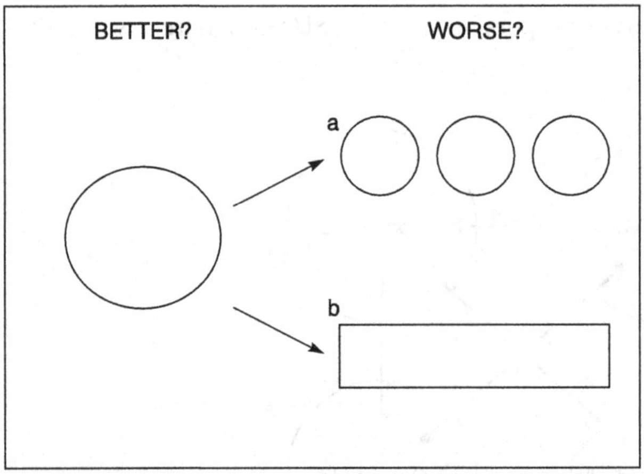

Figure 7.2. Examples of reserve/refuge areas of varying shape classified "better" and "worse" according to predictions of the theory of island biogeography but which may be challenged by the "SLOSS debate" and demands of on-farm IPM (see text for explanation). (Adapted from Diamond, J. M., 1975. The island dilemma: lessons of modern biogeographic studies for the design of natural reserves. In: *Biological Conservation*, Volume 7. Elsevier Science, Amsterdam. pp. 129–146.)

factors affecting lynx predation on sheep in the French Jura, for example, showed the importance of forested areas (Stahl *et al.*, 2002).

Achieving adequate protection from vertebrate pest damage can, however, be difficult, especially for highly mobile species such as birds, because short-term movement over large distances is possible, demanding the clearing of suitable habitat from large areas. The same challenge applies also to some arthropod pest species that may undertake long-range dispersal (e.g. *Helicoverpa* spp [Fitt, 1989]).

The SLOSS debate is less relevant to considerations of species that are undesired, such as pests as considered above, than to desired species where conservation is the goal. For the natural enemies of pests, conservation is a component of IPM for which there is growing interest (Barbosa, 1998; Pickett and Bugg, 1998). An important difference, however, between conservation biology and conservation of natural enemies in IPM is that the aim of the latter is to maximize the impact of predators and parasitoids on pests within crops, rather than simply conserving populations of natural enemies in separate reserves/refuges. Many species of natural enemies have small body size and short individual longevity and some are flightless. These factors demand that viable populations are maintained close to crops if they are to effect control of its pests. The limited powers of dispersal of many natural enemies may also require a departure from the idealized, circular shaped reserve with minimal edge to area ratio. Indeed, linear features with a high edge:area ratio (Figure 7.2b) may be demanded also by the practicalities of such features being accommodated on farms. Ways in which this may be accomplished are further explored below in relation to two further aspects of conservation biology theory; the declining population and small population paradigms.

7.5 Declining population and small population paradigms

The likelihood of local (and ultimately global) extinctions of populations can be predicted by examining the factors causing their declines and assessing the extent to which their size is likely to contribute to their loss (Caughley, 1994; Caughley and Gunn, 1996). The "small population paradigm" highlights how stochastic factors affect the persistence of populations, whereas the "declining population paradigm" focuses on how deterministic factors influence their trajectories. Caughley (1994) argues strongly that the small population paradigm has little to contribute to conservation biology in practice outside of a general conceptual underpinning, and that the effective diagnosis of declines and their treatment through management "cures" (i.e. the declining population paradigm) was the major mechanism by which conservation biologists could contribute to the conservation of species.

The declining population paradigm requires that we identify deterministic causes of declines (negative growth) in populations to identify systemic pressures on the persistence of populations. The identification of key threats is critical to the effective management of the declining populations and relies on clear diagnoses of all factors contributing to the declines through experimental approaches and predictive modeling (Caughley and Gunn, 1996; Brook *et al.*, 2000). In an agricultural system, for example, densities of natural enemy species may decline in response to factors such as use of monocultures, intensive tillage practices, use of fire, large field sizes, efficient weed control, and impoverished non-crop vegetation. Some factors will cause obvious direct mortality of natural enemies while others may act in less immediate fashion by, for example, decreasing availability of key resources such as nectar, pollen, alternative prey or hosts, as well as habitats with moderated microclimates.

Assessments of declines in systems reveal a willingness to accept "intuitive, common sense" explanations for possible declines, even in the absence of data supporting them (Caughley and Gunn, 1996). For instance, management options for conserving the declining populations of the large blue butterfly, *Maculinea arion*, in the UK focused on apparent anthropogenic impacts at coarse scales (e.g. grazing, collecting) rather than the complex obligate relationship the lycaenid butterfly had with ants (Elmes and Thomas, 1992). Inappropriate management responses contributed to the ongoing decline of populations and the ultimate extinction of the species in the UK. Further case studies (outlined in Caughley and Gunn, 1996) highlight the importance of clear, experimental approaches to identifying factors that influence the trajectory of declining populations so that appropriate management actions can be taken. This call for an increased emphasis on experimental approaches in conservation biology is made by numerous proponents (Thomas, 1996; Debinski and Holt, 2000) and has recently been echoed for conservation biological control (Gurr *et al.*, 2003*a*).

In the case of impacts of natural enemies of crop pests, there is good evidence that use of conservation biological control approaches as simple as providing a strip of nectar-producing flowers in the field margin can improve local densities (Landis *et al.*, 2000). There are, however, examples of habitat manipulation attempts failing to bring about any increase in pest suppression or reduction in crop damage (Gurr *et al.*, 2000). These illustrate the applicability to IPM of the need for experimental studies to identify the causes of declining populations, as discussed for conservation biology in the above paragraph. In the case of parasitoids of *Pseudaletia unipuncta* in the USA, for example, studies showed that percent parasitism was significantly higher in "complex" landscapes containing woodlots than in a simple landscape but was unaffected by proximity to hedgerows at the field scale within each type of landscape (Marino and Landis, 1996). Work on a large spatial scale by Roland and Taylor (1997) has shown that the response of four species of parasitoid of the forest-tent caterpillar *Malacosoma disstria* to habitat fragmentation differs markedly in relation to parasitoid body size. The species studied ranged in weight, an expression of their relative sizes, from 34 mg for *Carcelia malacosomae*, 41 mg for *Patelloa pachypyga*, 58 mg for *Arachnidomyia aldrichi*, and 68 mg for *Leschenaultia exul*. The larger three species are considered most important in control of the host insect and it was for these species that areas with relatively large blocks of contiguous forests (212 to 850 meters square) were required for optimal performance.

Studies of this type offer clear scope to identify systems in which landscape-level, as opposed to field-level, manipulation of the habitat is required to enhance IPM and, therefore, the extent to which a natural enemy taxon or crop systems is amenable to manipulation.

The risk of populations becoming extinct through a range of stochastic and deterministic processes increases immensely if they are small (Caughley and Gunn, 1996; Burgman and Lindenmayer, 1998; New, 2000). The range of mechanisms that increase the chance of extinction for small populations include random catastrophes, variation in environmental characteristics, and demographic accidents, all of which conspire to make small size one of the attributes that most threatens endangered populations. Agricultural systems tend to have higher levels of disturbance than do natural systems. This takes the form of tillage; applications of insecticides, fungicides and herbicides; and harvest operations. In the case of annual crops the latter can result in complete clearance of all vegetation from the field, though even mowing or fruit picking in hay and perennial fruit systems, respectively, may disrupt populations of natural enemies (e.g. Hossain *et al.*, 2002).

7.6 Population viability analysis (PVA)

The importance of the factors explored above is highlighted in the development of approaches to population viability analysis (PVA) which predicts risks of extinction for populations using the gamut of factors to which populations are exposed (Brook, 2000; Brook *et al.*, 2000). Although many options for PVA are now available, the process involves the development of a deterministic model for describing the population biology of the species of interest using ecologically meaningful data (e.g. age structure, predation, competition, density dependence) followed by the introduction of the effects of uncertainty, for demographic and environmental parameters (Burgman and Lindenmayer, 1998). The main use of PVA is to identify priorities for management and guide future research (Boyce, 1992).

Though PVA has been applied to a variety of natural systems, its utility in agricultural IPM is yet to be explored. It has been applied most often to mammals and plants, though Baguette *et al.* (2000) have employed PVA to explore butterfly ecology. Their study showed that even within a relatively discrete taxon ("butterflies") there may be significant differences between species and that recommendations made on the basis of two of the species would not be expected to lead to the persistence of the third. This has considerable, though yet to be explored, consequences for natural enemies.

Data from studies of re-introductions and translocations carried out for conservation purposes suggest that establishment is most successful when the numbers of animals released is large but the relationship is not strong (Fischer and Lindenmayer, 2000; Figure 7.3). In the case of biological control, it is generally accepted that releases of large numbers of individuals increases the probability of establishment (Gurr *et al.*, 2000) with larger introductions avoiding Allee effects (Dobson and Rodriguez, 2001) resulting from dispersal and reduced mate finding (Hopper and Roush, 1993). Some experimental work has been done on optimal release numbers for *Sericothrips staphylinus*,

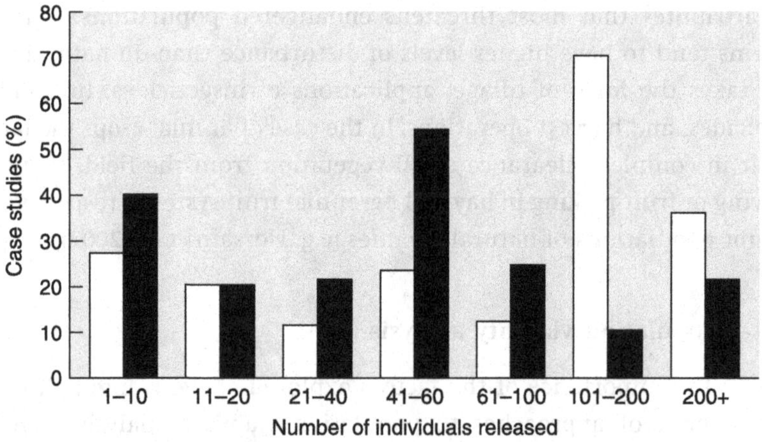

Figure 7.3. Effect of release size on success (white) and failure (black) of reintroduction attempts. (From: Pullin, A. S., 2002. *Conservation Biology*. Cambridge University Press, Cambridge.)

a biological control agent of the environmental weed, gorse (Memmott *et al.*, 1996). That study showed that the success rate of agent establishment rose in an approximately linear pattern from zero, when eight or fewer thrips were released per site, to 100% when 810 were released. PVA offers scope to refine such work, so making release strategies more efficient and enhancing the success rate of classical biological control.

In the case of biological control exerted by native agents, a "super-guild" of predators and parasitoids from many animal classes is involved and this complicates decision making on the optimal way of minimizing local extinctions. PVA of natural enemy populations may help by identifying those factors most likely to lead to local extinction of key biological control species, thus directing habitat manipulation efforts to interventions most likely to reduce extinction. Studies could include decisions on which natural enemy species or guild to focus upon (e.g. soil-dwelling predators versus parasitoid flies) or upon which type of intervention to use for a given agent (e.g. providing alternative hosts for immature parasitoids versus nectar for adults), as well as the spatial and temporal arrangement of habitat features such as flower strips or hedgerow facsimiles.

7.7 Landscape ecology and conservation

Consensus among conservation biologists is that the major anthropogenic pressures on terrestrial organisms come from the destruction and fragmentation of habitats. Thus, it is not surprising that recent approaches to conservation biology have shifted from species-specific approaches to the examination of landscapes and their ecology. The emphasis on studies outlining responses to habitat fragmentation has been on mensurative approaches, with most detailing shifts in the composition of species in habitat fragments and some studies outlining the ecological consequences of changes in species composition in fragments, often relating the extents of changes to the sizes of remnants (Robinson *et al.*, 1992; Simberloff, 1993; Abensperg *et al.*, 1996; Major *et al.*, 1999; Gibb and Hochuli, 2002).

As predicted by the theory of island biogeography, small habitat fragments generally support fewer species than do larger fragments (Debinski and Holt, 2000), with those species often represented by relatively low numbers of individuals (Connor *et al.*, 2000). Smaller fragments are disproportionately affected by a range of edge effects (Turner, 1996; Ozanne *et al.*, 1997) which may facilitate colonization by invasive species. A recent overview of experimental approaches for examining habitat fragmentation revealed little consistency in results across studies, especially with regard to species richness and abundance relative to fragment size (Debinski and Holt, 2000). Highly mobile taxa and early successional plant species, long-lived species, and generalist predators did not respond in the "expected" manner, most likely due to inconsistencies in edge effects, competitive release in the habitat fragments, and the spatial scale of the experiments (Debinski and Holt, 2000). One of the more consistently supported hypotheses was that movement and species richness were positively affected by corridors and connectivity, respectively. The review also identified that strong patterns revealed in three long-term (>14 years) studies would have been missed in shorter-term investigations, indicating the need for judicious interpretation of currently available evidence.

7.8 Invasive pest species and changes in distributions

Key threats to biodiversity often emerge when invasive and introduced species play roles as dominant competitors or predators in natural systems, thus excluding native flora and fauna. Mounting concern over the impact of exotic biological control agents on non-target species (Howarth, 2000) has driven the development of testing protocols and codes of conduct to avoid cases of "biological pollution" such as the invasive cane toad (*Bufo marinus*) in Australia. This species was introduced into Queensland in the 1930s for control of sugarcane beetles yet it failed to provide control and has become a pest of ever-widening distribution. Even in cases where agents have exerted useful control over the intended target, scope exists for unwanted impact on other species. Insects released in the USA for control of pest thistles are known to attack endangered endemic thistles (Strong, 1997). Such effects have to be balanced against the alternative options for control of exotic pests: to use pesticides, associated with well-known environmental impact; or to do nothing, which may leave natural systems exposed to invasion by weeds.

Despite the well-known successes of classical biological control, effective suppression of insect targets has been achieved in approximately 10% of cases only (Greathead and Greathead, 1992) with the success rate relatively constant from the 1880s to 2000 (Gurr *et al.*, 2000). Other than uncertainties relating to optimal release strategies (identified above), it has been suggested that an important factor in the failure of most classical biological control attempts is the hostility of the environment into which agents are released (Gurr and Wratten, 1999). Conservation biology suggests that the simplicity of agricultural systems with associated high levels of disturbance and fragmentation may be especially detrimental to organisms such as biological control agents that occupy high trophic levels. Habitat manipulation offers scope to enhance the performance of classical biological control agents (Gurr and Wratten, 1999; Baggen *et al.*, 1998)

7.9 Biodiversity in natural systems

The search for general principles relating biodiversity to the functioning of natural systems has become a dominant theme in recent ecological research (Cameron, 2002), driven by the potential impacts of species loss from natural systems (Naeem *et al.*, 1994; Moffat, 1996; Wardle *et al.*, 2000). Predictions about the relationship between ecosystem function and biodiversity focus on general hypotheses, usually supported by experimental studies in micro- and mesocosms (Naeem *et al.*, 1995; Lawton, 1996; Hodgson *et al.*, 1998), the scale of which is often criticized (Carpenter, 1996; Drake *et al.*, 1996). Many of the general themes explored the focus on productivity (Tilman *et al.*, 1996; Chapin *et al.*, 1998; Lawton *et al.*, 1998; Wardle *et al.*, 2000), which is of limited relevance to usefulness to IPM. Other factors linked to biodiversity include community invasibility (Tilman, 1997), stability (Johnson *et al.*, 1996; Tilman, 1996; McCann, 2000), predictability (McGrady-Steed *et al.*, 1997), resilience (Grimm and Wissel, 1997), and fragility (Nilsson and Grelsson, 1995). Criticisms of biodiversity as an ecological concept focus on the multiple, all-encompassing definitions ascribed to it for many levels of biological organization (Ghilarov, 1996). Despite attempts to incorporate measurements of ecosystem, phylogenetic, functional, and genetic diversity into its assessment (Andersen, 1995; Humphries *et al.*, 1995; Purvis and Hector, 2000), species are still the most commonly used unit for the assessment of biodiversity and are used here as the default unit for discussion.

The major relevance of biodiversity to IPM relates to how it may affect the extent, strength, and breadth of ecological interactions. The diversity/stability relationship is of particular interest to IPM as it integrates diversity, trophic relationships, and temporal persistence in assemblages.

7.9.1 Biodiversity in agricultural landscapes

The potential for approaches that enhance the conservation value and biodiversity of agricultural landscapes to benefit agricultural production is recognized (Stinner *et al.*, 1997) although evidence supporting this is scarce (Wood and Lenne, 1997). Examples do, however, exist to support the often-voiced view that diversity is in some, though often poorly defined way, good for IPM. A study of the effects of botanical diversity in Swiss vineyards reported by Hani *et al.* (1998) showed that faunistic diversity of beneficial and "indifferent" (i.e. neutral) organisms increased as botanical diversity rose while faunistic diversity of pest species was not affected. Other studies illustrate that a greater faunistic diversity can lead to lower densities of key pest species. One of the earliest of these showed that densities of cereal aphids were negatively correlated with non-aphid diversity of arthropods (Potts and Vickerman, 1974). Andow (1991) performed a comprehensive analysis of studies of the effect of diversity on pests and concluded that herbivorous arthropods were generally less abundant on plants in polycultures, though many species (20.2%) responded in a variable fashion, and others (15.3%) were more abundant in polycultures.

Explanations for the inconsistent effect of diversity on pest control are explored by Wilby and Thomas (2002). An important component of this is the range of possible relationships between species richness and the rate of any ecological process (e.g. predation or parasitism). Systems may exhibit "perfect complementarity" such that the presence of each additional species of natural enemy provides an incremental increase in process rate. In contrast, other systems, or even some pest species within the same system, may respond differently to species richness (with redundancy or an idiosyncratic response) with the latter relationship making it especially hard to predict the effects of altered species richness. It is, therefore, important to develop an understanding of which aspects of biodiversity may benefit IPM and the mechanisms by which this may operate.

Farmlands are typically hostile for predators and parasitoids because of the agronomic factors described above. The areas in which natural enemies may have refuge from such factors include conspicuous ones such as forest, and less conspicuous features such as weedy borders. Both can have a large effect on densities of natural enemies in adjacent crops (Altieri and Todd, 1981; Figure 7.4). Landscape features such as forest are increasingly rare and hard to

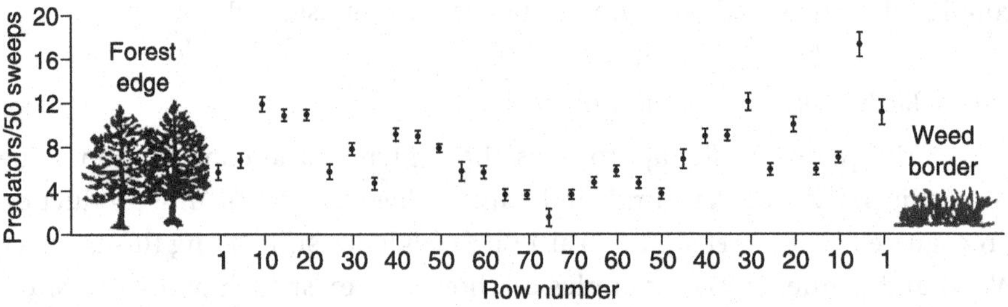

Figure 7.4. Response of predator densities in a soybean field to adjacent vegetation. (From: Altieri, M.A. and Todd, J.W., 1981. Some influences of vegetational diversity on insect communities of Georgia soybean fields. *Protection Ecology* 3:333–337. Reprinted with permission from Elsevier.)

accommodate on farms but simpler structures such as hedgerows may be as beneficial for communities of natural enemies (Wratten *et al.*, 1998).

Large-scale clearance of hedgerows to accommodate highly efficient, mechanized production has adversely affected the ability of certain natural enemies to suppress aphid pests. Hedgerows may often be confined to the margins of very large fields or, still worse, have been replaced by simpler structures (such as post and wire fences), over large areas of land. These factors have resulted in a situation where the large field sizes common to industrialized agriculture may have the predator densities shown in the center of Figure 7.4 over much of their area. Scope to address this problem using hedgerow facsimiles comes from studies in which raised earth banks sown to perennial, tussocky grasses have been established within cereal fields (Thomas *et al.*, 1991). These "beetle banks" harbor high densities of predatory beetles and spiders during winter which disperse into the crop in spring (Figure 7.5), affording effective management of pests. Unlike hedges, beetle

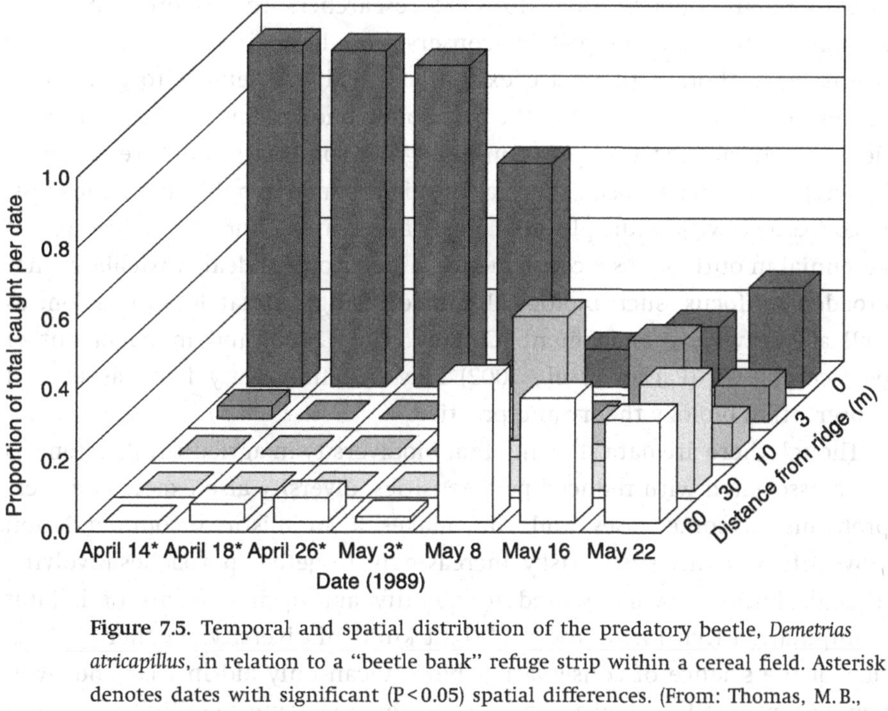

Figure 7.5. Temporal and spatial distribution of the predatory beetle, *Demetrias atricapillus*, in relation to a "beetle bank" refuge strip within a cereal field. Asterisk denotes dates with significant ($P < 0.05$) spatial differences. (From: Thomas, M. B., Wratten S. D. and Sotherton N. W. (1991) Creation of 'island habitats' in farmland to manipulate populations of beneficial arthropods: predator densities and emigration. *Journal of Applied Ecology*, **28**, 906–917. Reprinted with permission of Blackwell Publishing Ltd.)

banks are cheap, can be established, and, if necessary, removed rapidly, and allow passage of large farm machinery so they are compatible with modern farming practice in addition to their functional role in IPM. Beetle banks also have conservation utility for species of no direct significant to agricultural production. An example of such conservation is the elevated harvest mouse, *Micromys minutus*, densities found on beetle banks (Bence *et al.*, 1999). Similarly, field margins and non-crop flora have been shown to lead to elevated densities or even new populations of endangered birds (see examples in Landis *et al.*, 2000). Functional elements of ecological landscapes such as shelterbelts, field margins, watercourses, and roadsides can become nurseries for other taxa of indigenous biodiversity (Meurk and Swaffield, 2000).

7.10 Conclusions

Pressures are mounting for a less reductionist approach to be taken in the applied ecological research domains of biodiversity conservation and integrated pest management. Experimental approaches are required for both and there are opportunities for IPM researchers to explore aspects of ecological theory developed by conservation biologists and biodiversity researchers. Though there are exceptions, IPM has tended to have been narrow in outlook, focusing on the pests (often one or a few species) in a single field in a given year. Conservation biology in contrast (though represented by single charismatic species such as the giant panda in public consciousness) is concerned with multiple taxa over larger areas and, by definition, is perennial in outlook. As a consequence, aspects of IPM dealt with above that broaden its focus, such biological control using habitat manipulation, as well as areawide management (Calkins *et al.*, 2000) and management of pest complexes (Parker *et al.*, 2002) may benefit greatly from aspects of conservation biology theory and practice.

Though there are data showing that biodiversity in agricultural systems is often associated with reduced pest densities, diversity also exacerbates pest problems in some cases and this makes a simplistic recommendation "diversify to reduce pests" risky. Increasingly, targeted approaches involving ecological theory will be used to identify appropriate forms of habitat manipulation that introduce the "right kind" of diversity. Further exploration of the science of conservation biology can only aid in that endeavor. Population viability analysis has scope for analyzing viability of natural enemy populations in agricultural landscapes. Increasingly, those practicing IPM (especially systems involving conservation biological control) are using habitat manipulation practices that can simultaneously conserve endangered flora and fauna. Cross-disciplinary studies are required to maximize the scope for synergy. The cause of conservation biology in agricultural landscapes – where it must work (Gillespie and New, 1998) – will be furthered greatly if the methods it employs are aligned as closely as possible with the habitat manipulation approaches being developed for IPM. On a still broader perspective, IPM and on-farm conservation of flora and fauna are just two examples of a larger hierarchy of "ecosystem services" provided by on-farm biodiversity (Gurr *et al.*, 2003*b*). Further research in such directions will help agriculture to maximize economic, social, and ecological sustainability; minimize adverse impacts on the environment; and contribute to the conservation of biodiversity.

References

Abensperg, T.M., Smith, G.T., Arnold, G.W. and Steven, D.E. (1996). The effects of habitat fragmentation and livestock-grazing on animal communities in remnants of gimlet *Eucalyptus salubris* woodland in the Western Australian wheatbelt. I. Arthropods. *Journal of Applied Ecology*, **33**, 1281–301.

Altieri, M.A. and Todd, J.W. (1981). Some influences of vegetational diversity on insect communities of Georgia soybean fields. *Protection Ecology*, **3**, 333–7.

Andersen, A.N. (1995). Measuring more of biodiversity: genus richness as a surrogate for species richness in Australian ant faunas. *Biological Conservation*, **73**, 39–43.

Andow, D.A. (1991). Vegetational diversity and arthropod population response. *Annual Review of Entomology*, **36**, 561–86.

Baggen, L.R. and Gurr, G.M. (1998). The influence of food on *Copidosoma koehleri*, and the use of flowering plants as a habitat management tool to enhance biological control of potato moth, *Phthorimaea operculella*. *Biological Control*, **11**, 9–17.

Baguette, M., Petit, S. and Queva, F. (2000). Population spatial structure and migration of three butterfly species within the same habitat network: consequences for conservation. *Journal of Applied Ecology*, **37**, 100–7.

Barbosa, P. (1997). *Conservation Biological Control*. San Diego, CA: Academic Press.

Beck, M.W. (1997). Inference and generality in ecology: current problems and an experimental solution. *Oikos*, **78**, 265–73.

Bence, S., Standen, K. and Griffiths, M. (1999). Nest site selection by the harvest mouse (*Micromys minutus*) on arable farmland. *Aspects of Applied Biology*, **54**, 197–202.

Bonsall, M.B. and Hassell, M.P. (2000). The effects of metapopulation structure on indirect interactions in host–parasitoid assemblages. *Proceedings of the Royal Society of London, Series B: Biological Sciences*, **267**, 2207–12.

Boyce, M.S. (1992). Population viability analysis. *Annual Review of Ecology and Systematics*, **23**, 481–506.

Brook, B.W. (2000). Pessimistic and optimistic bias in population viability analysis. *Conservation Biology*, **14**, 564–6.

Brook, B. W., O'Grady, J. J., Chapman, A. P. *et al.* (2000). Predictive accuracy of population viability analysis in conservation biology. *Nature*, **404**, 385–7.

Brown, S., Scatena, F. N. and Ewel, J. J. (2001). Managed ecosystems deserve greater attention. *Bulletin of the Ecological Society of America*, January, 91–3.

Burgman, M. A. and Lindenmayer, D. A. (1997). *Conservation Biology for the Australian Environment*. Chipping Norton: Surrey Beatty.

Calkins, C. O., Knight, A. L., Richardson, G. and Bloem, K. A. (2000). *Areawide Population Suppression of Codling Moth. Areawide Control of Fruit Flies and Other Insect Pests*. In Joint Proceedings of the International Conference on Area-Wide Control of Insect Pests, 28 May–2 June, 1998 and the Fifth International Symposium on Fruit Flies of Economic Importance, Penang, Malaysia, 1–5 June, 1997. Penerbit Pulau Pinang, Malaysia: Universiti Sains Malaysia, pp. 215–19.

Cameron, T. (2002). The year of the 'diversity–ecosystem function' debate. *Trends in Ecology and Evolution*, **17**, 495–6.

Carpenter, S. R. (1996). Microcosm experiments have limited relevance for community and ecosystem ecology. *Ecology*, **77**, 677–80.

Caughley, G. (1994). Directions in conservation biology. *Journal of Animal Ecology*, **63**, 215–44.

Caughley, G. and Gunn, A. (1996). *Conservation Biology In Theory and Practice*. Carlton, VA, Australia: Blackwell Science Pty Ltd.

Chapin, F. S. I., Sala, O. E., Burke, I. C. *et al.* (1997). Ecosystem consequences of changing biodiversity. *Bioscience*, **48**, 45–52.

Chapin, F. S., Zavaleta, E. S., Eviner, V. T. *et al.* (2000). Consequences of changing biodiversity. *Nature*, **405**, 234–42.

Connor, E. F., Courtney, A. C. and Yoder, J. M. (2000). Individuals–area relationships: The relationship between animal population density and area. *Ecology*, **81**, 734–47.

Debinski, D. M. and Holt, R. D. (2000). A survey and overview of habitat fragmentation experiments. *Conservation Biology*, **14**, 342–55.

Diamond, J. M. (1975). The island dilemma: lessons of modern biogeographic studies for the design of natural reserves. In *Biological Conservation, Volume 7*. Amsterdam: Elsevier Science. pp. 129–46.

Dobson, A. P. and Rodriguez, J. P. (2001). Conservation biology, discipline of. In *Encyclopedia of Biodiversity, Volume 1*. New York: Academic Press. pp. 855–64.

Dobson, A. P., Bradshaw, A. D. and Baker, A. J. M. (1997). Hopes for the future: restoration ecology and conservation biology. *Science*, **277**, 515–22.

Drake, J. A., Huxel, G. R. and Hewitt, C. L. (1996). Microcosms as models for generating and testing community theory. *Ecology*, **77**, 670–7.

Elmes, G. W. and Thomas, J. A. (1992). Complexity of species conservation in managed habitats: interaction between *Maculinea* butterflies and their ant hosts. *Biodiversity and Conservation*, **1**, 155–69.

Fischer, J. and Lindenmayer, D. B. (2000). An assessment of the results of animal relocations. *Biological Conservation*, **96**, 1–11.

Fitt, G. P. (1989). The ecology of Heliothis species in relation to agroecosystems. *Annual Review of Entomology*, **34**, 17–52.

Gerard, P.W. (1995). Agricultural Practices, Farm Policy and the Conservation of Biological Diversity. USDI Biological Sciences Report No. 4, Washington, DC.

Ghilarov, A. (1996). What does 'biodiversity' mean: Scientific problem or convenient myth? *Trends in Ecology and Evolution*, **11**, 304–6.

Gibb, H. and Hochuli, D.F. (2002). Habitat fragmentation in an urban environment: large and small fragments support different anthropod assemblages. *Biological Conservation*, **106**, 91–100.

Gilbert, F., Gonzales, A. and Evans Freke, I. (1997). Corridors maintain species richness in fragmented landscapes of a microcosm. *Proceedings of the Royal Society of London B*, **265**, 577–83.

Gillespie, R.G. and New, T.R. (1997). Conservation and pest management: irresolvable clash or complementary fields of research? In M.P. Zaluki, R.A.I. Drew and G.G. White (eds.), *Pest Management – Future Challenges: Proceedings of the Sixth Australian Applied Entomological Research Conference*, 29 September–2 October 1998. Volume 2. The University of Queensland, Australia. pp. 195–207.

Greathead, D.J. and Greathead, A.H. (1992). Biological control of insect pests by insect parasitoids and predators: the BIOCAT database. *Biocontrol News and Information*, **13**, 61–8.

Grimm, V. and Wissel, C. (1997). Babel, or the ecological stability discussions: an inventory and analysis of terminology and a guide for avoiding confusion. *Oecologia*, **109**, 323–34.

Gurr, G.M., Valentine, B.J., Azam, M.N.G. and Thwaite, W.G. (1996). Evolution of arthropod pest management in apples. *Agricultural Zoology Reviews*, **7**, 35–69.

Gurr, G.M. and Wratten, S.D. (1999). 'Integrated biological control': a proposal for enhancing success in biological control. *International Journal of Pest Management*, **45**, 81–4.

Gurr, G.M., Barlow, N., Memmott, J., Wratten, S.D. and Greathead, D.J. (2000). A history of methodological, theoretical and empirical approaches to biological control. In G.M. Gurr and S.D. Wratten (eds.), *Biological Control: Measures of Success*. Kluwer: Dordrecht. pp. 3–37.

Gurr, G.M., Wratten, S.D., Tylianakis, J., Kean, J. and Keller, M. (2003*a*). Providing plant foods for insect natural enemies in farming systems: balancing practicalities and theory. In F.L. Wackers, p.C.J. van Rijn and J. Bruin (eds.), *Plant-Derived Food and Plant-Carnivore Mutualism*. Cambridge, UK: Cambridge University Press (in press).

Gurr, G.M., Wratten, S.D. and Luna, J. (2003*b*). Multi-function agricultural biodiversity: pest management and other benefits. *Basic and Applied Ecology*, **4**, 107–16.

Hani, F.J., Boller, E.F. and Keller, S. (1997). Natural regulation at the Farm Level. In C.H. Pickett and R.L. Bugg (eds.), *Enhancing Biological Control Habitat Management to Promote Natural Enemies of Agricultural Pests*. Berkeley, CA: University of California Press. pp. 161–210.

Hanski, I. (1997). Metapopulation dynamics: from concepts to observations and models. In I.A. Hanski and M.E. Gilpin (eds.), *Metapopulation Biology: Ecology Genetics and Evolution*. San Diego: Academic Press. pp. 69–91.

Hill, D., Andrews, J., Sotherton, N. and Hawkins, J. (1995). Farmlands. In W.J. Sutherland and D.A. Hill (eds.), *Managing Habitats for Conservation*. Cambridge, UK: Cambridge University Press. pp. 231–66.

Hodgson, J.G., Thompson, K., Wilson P.J. and Bogaard, A. (1997). Does biodiversity determine ecosystem function? The Ecotron experiment reconsidered. *Functional Ecology*, **12**, 843–56.

Hopper, K.R. and Roush, R.T. (1993). Mate finding, dispersal, number released, and the success of biological-control introductions. *Ecological Entomology*, **18**, 321–31.

Hossain, Z., Gurr, G.M., Wratten, S.D. and Raman, A. (2002). Habitat manipulation in lucerne (*Medicago sativa* L.): arthropod population dynamics in harvested and 'refuge' crop strips. *Journal of Applied Ecology*, **39**, 445–54.

Howarth, F.G. (2000). Non-target effects of biological control agents. In G.M. Gurr and S.D. Wratten (eds.), *Biological Control: Measures of Success*. Kluwer: Dordrecht. pp. 369–403.

Humphries, C.J., Williams, P.H. and Vane-Wright, R.I. (1995). Measuring biodiversity value for conservation. *Annual Review of Ecology and Systematics*, **26**, 93–111.

Jervis, M.A. (1997). Metapopulation dynamics and the control of mobile agricultural pests: fresh insights. *International Journal of Pest Management*, **43**, 251–2.

Johnson, K.H., Vogt, K.A., Clark, H.J., Schmitz, O.J. and Vogt, D.J. (1996). Biodiversity and the productivity and stability of ecosystems. *Trends in Ecology and Evolution*, **11**, 372–7.

Landis, D., Wratten, S.D. and Gurr, G.M. (2000). Habitat Management for Natural Enemies. *Annual Review of Entomology*, **45**, 175–201.

Lawton, J.H. (1996). The Ecotron facility at Silwood Park: The value of 'big bottle' experiments. *Ecology*, **77**, 665–9.

Lawton, J. (1999). Are there general laws in ecology? *Oikos*, **84**, 177–92.

Lawton, J.H., Naeem, S., Thompson, L.J., Hector, A. and Crawley, M.J. (1997). Biodiversity and ecosystem function: getting the Ecotron experiment in its correct context. *Functional Ecology*, **12**, 848–52.

Levins, R. (1969). Some demographic and genetic consequences of environmental heterogeneity for biological control. *Bulletin of the Entomological Society of America*, **15**, 237–40.

MacArthur, R.H. and Wilson, E.O. (1967). *The Theory of Island Biogeography*. Princeton: Princeton University Press.

Major, R.E., Christie, F.J., Gowing, G. and Ivison, T.J. (1999). Age structure and density of red-capped robin populations vary with habitat size and shape. *Journal of Applied Ecology*, **36**, 901–7.

Marino, P.C. and Landis, D.A. (1996). Effect of landscape structure on parasitoid diversity and parasitism in agroecosystems. *Ecological Applications*, **6**, 276–84.

McCann, K.S. (2000). The diversity-stability debate. *Nature*, **405**, 228–33.

McGrady-Steed, J., Harris, P.M. and Morin, P.J. (1997). Biodiversity regulates ecosystem predictability. *Nature*, **390**, 162–5.

McIntyre, S., Barrett, G.W., Kitching, R.L. and Recher, H.F. (1992). Species triage – Seeing beyond wounded rhinos. *Conservation Biology*, **6**, 604–6.

Memmott, J., Fowler, S.V., Harman, H.M. and Hayes, L.M. (1966). How best to release a biological control agent. In *Proceedings of the 9th International Symposium on Biological Control of Weeds, Stellenbosch, South Africa, 19–26 January 1996*. Rondebosch, South Africa: University of Cape Town. pp. 291–6.

Meurk, C.D. and Swaffield, S.R. (2000). A landscape ecological framework for indigenous regeneration in rural New Zealand–Aotearoa. *Landscape and Urban Planning*, **50**, 129–44.

Moffat, A.S. (1996). Biodiversity is a boon to ecosystems, not species. *Science*, **271**, 1497.

Naeem, S., Lawton, J.H., Thompson, L.J., Lawler, S.P. and Woodfin, R.M. (1995). Biotic diversity and ecosystem processes: Using the Ecotron to study a complex relationship. *Endeavour* (Cambridge), **19**, 58–63.

Naeem, S., Thompson, L.J., Lawler, S.P., Lawton, J.H. and Woodfin, R.M. (1994). Declining biodiversity can alter the performance of ecosystems. *Nature*, **368**, 734–7.

New, T.R. (2000). *Conservation Biology: An Introduction for Southern Australia*. Melbourne: Oxford University Press.

Newmark, W.D. (1987). The land-bridge island perspective on mammalian extinctions in North American parks. *Nature*, **325**, 430–2.

Nilsson, C. and Grelsson, G. (1995). The fragility of ecosystems: A review. *Journal of Applied Ecology*, **32**, 677–92.

Oerke, E.-C., Dehne, H.-W., Schönbeck, F. and Weber, A. (1994). *Crop Production and Crop Protection: Estimated Losses in Major Food and Cash Crops*. Amsterdam: Elsevier.

Ozanne, C.M.P., Hambler, C., Foggo, A. and Speight, M.R. (1997). The significance of edge effects in the management of forests for invertebrate biodiversity. In N.E. Stork, J. Adis and R.K. Didham (eds.), *Canopy Arthropods*. London: Chapman and Hall. pp. 534–50.

Passioura, J.B. (1999). Can we bring about a perennially peopled and productive countryside? *Agroforestry Systems*, **45**, 411–21.

Parker, W.E., Collier, R.H., Ellis, P.R. *et al.* (2002). Matching control options to a pest complex: the integrated pest management of aphids in sequentially-planted crops of outdoor lettuce. *Crop Protection*, **21**, 235–47.

Peters, R.H. (1995). *A Critique for Ecology*. Melbourne, Australia: Press Syndicate of the University of Cambridge.

Pickett, C.H. and Bugg, R.L. (eds.) (1997). *Enhancing Biological Control Habitat Management to Promote Natural Enemies of Agricultural Pests*. Berkeley, CA: University of California Press.

Potts, G.R. and Vickerman, G.P. (1974). Studies on the cereal ecosystem. *Advances in Ecological Research*, **8**, 107–97.

Pullin, A.S. (2002). *Conservation Biology*. Cambridge, UK: Cambridge University Press.

Purvis, A. and Hector, A. (2000). Getting the measure of biodiversity. *Nature*, **405**, 212–19.

Robertson, G. P. and Harwood, R. R. (2001). Agriculture, sustainable. In *Encyclopedia of Biodiversity, Volume 1*. New York: Academic Press. pp. 99–107.

Robinson, G. R., Holt, R. D., Gaines, M. S. *et al.* (1992). Diverse and contrasting effects of habitat fragmentation. *Science*, 257.

Roland, J. and Taylor, P. D. (1997). Insect parasitoid species respond to forest structure at different spatial scales. *Nature*, **386**, 710–13.

Samways, M. W. (1994). *Insect Conservation Biology*. London: Chapman and Hall.

Samways, M. J. (1997). Classical biological control and biodiversity conservation: What risks are we prepared to accept? *Biodiversity and Conservation*, **6**, 1309–16.

Simberloff, D. (1987). The contribution of population and community biology to conservation science. *Annual Review of Ecology and Systematics*, **19**, 473–511.

Simberloff, D. (1993). Fragmentation, corridors, and the Longleaf Pine community. In C. Moritz and J. Kikkawa (eds.), *Conservation Biology in Australia and Oceania*. Chipping Norton: Surrey Beatty and Sons. pp. 47–56.

Soule, M. E. and Wilcox, B. A. (1980). *Conservation Biology: an Evolutionary-Ecological Perspective*. Sunderland, MA: Sinauer Associates.

Stahl, P., Vandel, J. M., Ruette, S. *et al.* (2002). Factors affecting lynx predation on sheep in the French Jura. *Journal of Applied Ecology*, **39**, 204–16.

Steffan-Dewenter, I. and Tscharntke, T. (2002). Insect communities and biotic interactions on fragmented calcareous grasslands – a mini review. *Biological Conservation*, **104**, 275–84.

Stinner, D. H., Stinner, B. R. and Martsolf, E. (1997). Biodiversity as an organizing principle in agroecosystem management: case studies of holistic resource management practitioners in the USA. *Agriculture Ecosystems and Environment*, **62**, 199–213.

Strong, D. R., Ecology – fear no weevil? *Science*, **277**, 1058–9.

Thomas, J. A. (1996). The case for a science-based strategy for conserving threatened butterfly populations in the UK and north Europe. In J. Settele, C. Margules, P. Poschold and K. Henle (eds.), *Species Survival in Fragmented Landscapes*. Dordrecht, Netherlands: Kluwer Academic Publishers. pp. 1–6.

Thomas, M. B., Wratten, S. D. and Southerton, N. W. (1991). Creation of 'island' habitats in farmland to manipulate populations of beneficial arthropods: predator densities and emigration. *Journal of Applied Ecology*, **28**, 906–17.

Tilman, D. (1996). Biodiversity: population versus ecosystem stability. *Ecology*, **77**, 350–63.

Tilman, D. (1997). Community invasibility, recruitment limitation, and grassland biodiversity. *Ecology*, **78**, 81–92.

Tilman, D., Wedin, D. and Knops, J. (1996). Productivity and sustainability influenced by biodiversity in grassland ecosystems. *Nature*, **379**, 718–20.

Turner, I. M. (1996). Species loss in fragments of tropical rain forest: a review of the evidence. *Journal of Applied Ecology*, **33**, 200–9.

Wardle, D. A., Huston, M. A., Grime, J. P. *et al.* (2000). Biodiversity and ecosystem function: an issue in ecology. *Bulletin of the Ecological Society of America*, **81**, 235–9.

Wilby, A. and Thomas, M.B. (2002). Are the ecological concepts of assembly and function of biodiversity useful frameworks for understanding natural pest control? *Agricultural and Forest Entomology*, **4**, 237–43.

Wood, D. and Lenne, J.M. (1997). The conservation of agrobiodiversity on-farm: questioning the emerging paradigm. *Biodiversity and Conservation*, **6**, 109–29.

Wratten, S.D., van Emden, H.F. and Thomas, M.B. (1998). Within-field and border refugia for the enhancement of natural enemies. In C.H. Pickett and R.L. Bugg (eds.), *Enhancing Biological Control*. Berkeley, CA: University of California Press. pp. 375–403.

8

Ecological risks of biological control agents: impacts on IPM

H.M.T. HOKKANEN, J.C. VAN LENTEREN AND I. MENZLER-HOKKANEN

8.1 Ecological concerns of using biological control agents

In the more than 100 years of biological control, hundreds of species of exotic natural enemies have been imported, mass-reared and released as biological control agents, resulting in successful control of many species of pests and considerable reductions in pesticide use (e.g. Greathead, 1995; van Lenteren, 2000*a*; 2003; Gurr and Wratten, 2000). Negative environmental effects of these releases have rarely been reported (Hokkanen and Lynch, 1995; Follett *et al.*, 2000; Louda *et al.*, 2003; Lynch and Thomas, 2000; Lynch *et al.*, 2001). An important difference between biological control and the use of chemical pesticides is that natural enemies are often self-perpetuating and self-dispersing, and, as a result, biological control is regularly irreversible, although this is not always the case in inundative types of biological control. Inundative biological control is the release of large numbers of mass-produced biological control agents to reduce a pest population without necessarily achieving continuing impact or establishment; classical biological control is the intentional introduction and permanent establishment of an exotic biological agent for long-term pest management. It is exactly the self-perpetuation, self-dispersal and irreversibility that is so highly valued in properly executed classical biological control programs: it makes them sustainable and highly economic compared with any other control method (Bellows and Fisher, 1999; van Lenteren, 2001). The current popularity of biological control may, however, result in problems, as an increasing number of activities will be executed by persons not trained in identification, evaluation, and release of biological control agents (Howarth, 2000). Therefore, methodologies for risk assessment are now being developed as a basis for regulation of import and release of exotic natural enemies (e.g. Barratt *et al.*, 1999; van Lenteren *et al.*, 2003; OECD, 2004). These regulations vary from very strict forms of registration making biological control almost impossible, to the simple need to notify the wish to import a natural enemy.

8.1.1 *Effects of invasions*

The effects of insect invasions are documented poorly, except for economically important organisms like pest species and natural enemies. Insect invasions in natural ecosystems have received hardly any attention. It is estimated that 5% of all unintentionally introduced organisms (plants and animals) establish, and that 7% of these become pest species (Di Castri, 1989; van Lenteren, 1995). Pimentel *et al.* (1989) and Pimentel (1995) discussed the effects of intentional introductions into the USA of agricultural and ornamental plants, and agricultural, sport, and pet animals. Of these introductions (some 7800 species) 160 species became pests; 2.2% of the plant species and 1.4% of the animal species, a rather high risk rate. The percentage establishment is much higher when species are purposefully introduced. From the literature on insect biological control it is known that 25–34% of the introduced arthropod species are able to establish (e.g. Hall and Ehler, 1979; van Lenteren, 1983). In weed control the success of establishment of natural enemies is still higher: 65% (Crawley, 1986). Pimentel *et al.* (1989) remarked that insects introduced for biological control have generally had minimal or no environmental impacts and relate that to greater ecological knowledge and established regulations.

In general, two types of negative effects of invasions are distinguished: (a) colonizing species becoming a pest, or (b) colonizing species leading to extinction of native species. Data on extinctions due to colonization are, among others, provided by Simberloff (1981). He evaluated 10 papers covering 850 plant and animal species introductions, and concluded that less than 10% of the introductions caused species extinctions through habitat alteration, predation, or competition (71 out of 850 introductions): introductions apparently tend to add species to a community, rather than causing extinctions, but there is a considerable risk that changes in the composition of biota occur after an invasion. Predation was the principle cause of extinctions, followed by habitat change, and competition caused only few extinctions. Interpretation of the effects of invasions lead to very different and conflicting conclusions, e.g. "There is no invasion of natural communities without disturbance" (Fox and Fox, 1986) and "Sufficient examples can now be assembled to indicate that invasion can proceed without continuing disturbance" (Mack, 1985). Extinction in 8.3% (71 out of 850) of the cases can be considered as a very high risk rate.

8.1.2 *Effects of exotic natural enemies*

What kind of effects can we expect from exotic natural enemies? This question is not easy to answer, because rather unexpected events may occur after release, and the devil is often in the details. An example is as follows. In the 1950s, the myxomatosis virus was introduced in the UK to reduce the rabbit population (Crawley, 1983). After the death of the rabbits, the turf increased in height and there was a spectacular increase in the abundance of flowers, among other species those of orchids. But in the absence of rabbit grazing, turf was losing its density, and several plant species decreased in quantity. An increase in woody plants occurred. Brambles and gorse bushes developed. Longer-term effects of the virus were a reduction in plant-species richness, increased dominance by a small number of rank grasses and hastened succession towards woodland. Non-target effects on the fauna were also observed (although many effects were not perceived): as a result of changes in ground cover, numbers of *Myrmica* ant species decreased, and this in its turn lead to a strong reduction and local disappearance of myrmecophilous species of *Maculinea* butterflies.

As direct effects of introductions and releases, extinction or reduction in numbers of native non-target species may occur. Several ecologists have stated that, because insects have more restricted diets than vertebrates and that particular categories of insects like parasitoids are extremely specific in host use, cases of introduced insects endangering native fauna would be rare compared with those involving vertebrate introductions (e.g. Harris, 1990). Polyphagous insect predators may, however, attack non-target organisms and be harmful. In a rather controversial article about the environmental impacts as a result of purposeful introductions in classical biological control, Howarth (1991) states that two zygaenid moth species (a pest and a non-pest species) may have gone extinct in Fiji as a result of the introduction of a polyphagous tachinid parasitoid, *Bessa remota*. He further believes that the introduction of generalist predators and parasitoids of Lepidoptera have caused the extinction of at least 15 species (including 5 pest species) of larger native moths of Hawaii. Most of the other examples of insect introductions he gives relate to declines in numbers rather than extinctions. The examples of non-target effects put forward by Howarth (1985, 1991) have been seriously criticized by e.g. Follett *et al.* (2000), but recently

Louda *et al.* (2003) provided 10 cases making clear that serious negative ecological effects have been observed after the release of exotic natural enemies for control of weeds and insects. It remains, however, very difficult to show that in an ecosystem with a rare plant-eating insect species which is attacked by many different natural enemy species, introduction of a new natural enemy further reduces the density of the herbivore, or merely replaces some other mortality factor. However, extinction of pest or non-target organisms as a result of biological control is expected to be unlikely. Rather, a low population level of both pest and natural enemy will develop. The search behavior of natural enemies and the ways in which herbivores can defend themselves against natural enemies prevent extinction. The most critical ecological issues are to estimate the probabilities of attack of non-target organisms, and the dispersal and establishment capacities of the biological control agent. Only a few natural enemies are strictly monophagous (Zwölfer, 1971), but many are oligophagous and thus have a restricted host/prey choice. Sometimes the biological control industry favors the release of polyphagous natural enemies, in order to be able to apply them for the control of various, taxonomically unrelated pest species. These natural enemies in particular are expected to cause non-target effects.

Indirect effects arising from this are (1) preying on, or parasitizing of, by the introduced organism on native natural enemies may cause a reduction in numbers of these species; (2) competition for host or prey with local organisms may negatively affect non-target native natural enemies; and (3) the habitat may be modified. It has been suggested that, in some cases, releases of polyphagous predators and parasitoids have not only led to a decimation of pest caterpillars, but also to a reduction of non-target caterpillars, resulting in a decline in native predaceous wasps and native bird populations (Simberloff, 1992). Myriad indirect effects are possible. Species can interact through shared prey or hosts, shared predators, parasitoids or pathogens. Natural historic information on species involved in such indirect effects can be used (or collected) for the prediction of such indirect effects.

Effects on other trophic levels include intraguild predation, omnivory, vectoring, and hybridization. When the released biological control agent is able to attack not only herbivores but also species that are feeding on these herbivores themselves ("intraguild predation" (Rosenheim *et al.*, 1995) or "facultative hyperparasitism" (Sullivan, 1987)), this is of specific interest, as these other natural enemies might be important in the regulation of some of these herbivores (see indirect effects). Some natural enemies also feed on plant material during part of their life cycle, and could damage plants as a result (omnivory, Coll and Guershon, 2002). Further effects are: (1) the attack of the released biological control agents by other organisms in the ecosystem; (2) vectoring of pathogens by the natural enemy (Bjørnson and Schütte, 2003); (3) hybridization between the biological control agent and indigenous biotypes of the same or very closely related natural enemy species (for more information, see van Lenteren *et al.*, 2003).

Negative effects of insect biological control might be prevented when, as in modern biological weed control programs (e.g. Wapshere, 1974; Blossey, 1995; Lonsdale *et al.*, 2001), the effect is not only determined on the target species, but also on indigenous non-target species (van Lenteren, 1995; Blossey, 1995). Until relatively recently, testing of indigenous non-target species has rarely been applied as part of pre-release evaluation programs for arthropod natural enemies (van Lenteren and Woets, 1988; Waage, 1997; Barratt *et al.*, 1999). However, in classical biological control, there are several cases where non-target species testing has been applied properly. Examples are the programs for control of *Sitona* beetles in lucerne in New Zealand (e.g. Barratt *et al.*, 1999) and that for control of cassava mealybugs in Africa (Neuenschwander and Markham, 2001). Implementation of regulation is considered by many countries, and is expected to significantly increase during the coming decade as a result of the agreements reached during the sixth United Nations meeting on the Convention on Biological Diversity (Convention on Biological Diversity (CBD), 2002) as an approach to prevent the spread of invasive alien species. Later in this chapter, a risk assessment procedure for natural enemies is discussed (van Lenteren *et al.*, 2003), as well as a guideline for regulation of import and release of exotic natural enemies (OECD, 2004).

Another way to reduce the risks associated with the release of exotic natural enemies would be to limit the number of releases themselves by increasing the use of native natural enemies. Although this seems a logical approach, in the last four decades many exotic biological control agents were imported and released without thorough evaluation of the properties of native natural enemies that could have been candidates for inundative biological control (e.g. van Lenteren 2000*a*; 2003).

8.2 Historical record of actual hazards in biocontrol

Several reviews have addressed the issue of non-target impacts in biocontrol (e.g. Howarth, 1991; Stiling and Simberloff, 1999; Lynch *et al.*, 2001; Louda *et al.*, 2003). While earlier papers raised questions regarding specific biocontrol introductions, and brought up conceptual issues regarding the complexity of such effects, recent papers have tackled the problem in a more quantitative manner.

Lynch *et al.* (2001) provide the only attempt to analyze in detail all the reported non-target effects worldwide. They used the BIOCAT database of CABI Bioscience, a compilation of all biological control introductions. An exhaustive literature search, and information extracted via direct contacts with biological control specialists, amended the existing database.

8.2.1 *Classical introductions*

BIOCAT contained data on 5279 insect introductions for classical biocontrol. Only about 1.5% of these had associated with them some data on the realized field-specificity of the agent, and on the existence of non-target effects of any level. These data can give us some information, but because they address only a small number of non-randomly selected cases, they only can be taken as indicative of the size of the problem. Clearly, cases associated with obvious and conspicuous non-target effects are known and reported, while those with subtle and less dramatic impacts, as well as those where no non-target impacts exist, remain undetected.

Analysis of the reported non-target studies, however, revealed some clear patterns. The quantitative study of mortality induced on non-targets showed that minor effects vastly outnumber the major effects. Thus, only 8% of the instances with some non-target mortality seem serious, in the sense that they may have led to population level consequences. However, as 63% of releases have involved polyphagous species, attacking an average of about two non-target host species each (Stiling and Simberloff, 1999), perhaps 10% of the introductions have had serious population level consequences. It therefore can be estimated that about 530 non-targets may have been affected at the population level over the history of biocontrol. Significantly, however, most of these impacts have occurred during the early days of the more than 100 years of biocontrol, and the proportion of introductions recorded to have led to non-target parasitism or feeding has decreased tenfold over the course of this history (Figure 8.1). Thus, while analyses of outcome patterns from biocontrol introductions from the pre-pesticide era can be illustrative, they offer little guidance to making predictions on the safety of current practices in biological control. At most, they set a bottom line, "worst case scenario", but these should not be used to depreciate the potential of using biocontrol as it is practiced today.

Other retrospective studies have revealed similar data, as provided by Lynch *et al.* (2001). Funasaki *et al.* (1988) report for Hawaii that native non-target species were attacked by 7% of the predators, and by 10% of

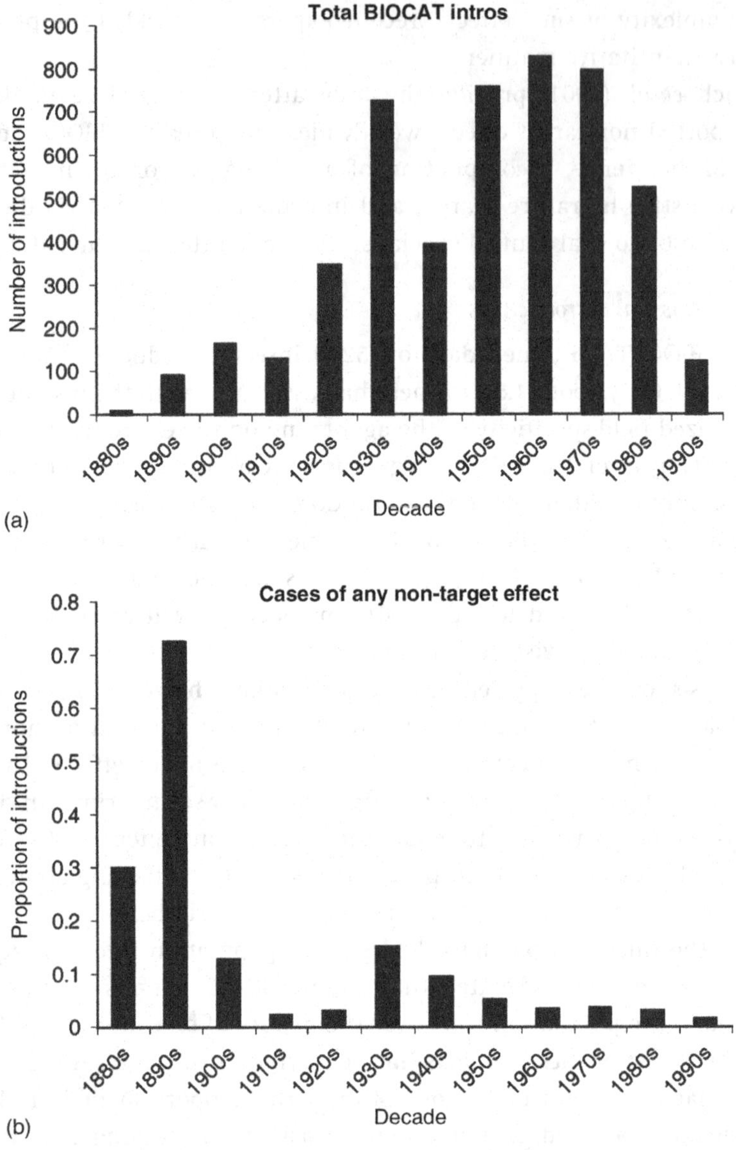

Figure 8.1. Patterns in recorded non-target effects of biocontrol introductions over decades: (a) total numbers of classical biocontrol introductions per decade; (b) proportion leading to any non-target effect. Based on data in BIOCAT. After Lynch *et al.* (2001).

parasitoids introduced for insect control. Hawkins and Marino (1997) estimate that 16% of exotic insect parasitoids in North America have colonized native non-target species. In most cases, however, the non-target feeding is likely to be minor, and it appears that the "tens-rule" of

Williamson (1996) roughly applies to these situations as well: 10% of (accidentally) introduced species establish, of these 10% attack non-target species, and of these, 10% cause serious population-level consequences.

A further analysis by Lynch *et al.* (2001) of the nature of the implicated problems found evidence mostly for only minor utilization of the non-target host or prey species, suggesting that many agents utilize non-target hosts at a low level, but do not generate sufficient mortality to imply some kind of population-level effect. Concerning the more severe cases, their data indicated that many of the occasional cases where initial suspicions of an effect at the population level existed, were found to be unsubstantiated. However, when such effects were detected, most of them implied >40% population level impacts at a local or global scale. This shows either that severe effects are more likely than minor ones, or that severe non-target suppression is necessary for the effect to be noticed. The latter explanation implies that minor population impacts might be much more frequent than what they appear to be.

There is strong suspicion about the role of biocontrol in the apparent extinction of certain species in island ecosystems, but little evidence singling it out as an overriding cause in particular cases. There is one relatively clear example for Fiji, however. This is the local extinction of the native moth *Heteropan dolens*, following the introduction of parasitoid *Bessa remota* against the levuana moth in 1925 (Tothill *et al.*, 1930; Howarth, 1985; Roberts, 1986). It seems unlikely that this was an isolated incident amongst the island ecosystems where biocontrol took place at that time.

Even less evidence exists for the role of biocontrol introductions in insect extinction in the post-DDT era. Biocontrol is implicated as a possible contributory factor in the extinction of *Murgantia histronica*, which occurred following two introductions into Hawaii in 1962 (Howarth, 1991). However, a recent review of this case (Follett *et al.*, 1999) suggests that the evidence is essentially circumstantial. Furthermore, *Murgantia* was considered a pest itself, and its role as a potential host of one of the agents used was well known. A more recent example of a possible extinction is that of the displacement of the native parasitoid, *Apanteles diatraeae*, by the introduced parasitoid, *Cotesia flavipes*, on the sugarcane borer *Diatraea saccharalis*. The native parasitoid could no longer be collected from the sugarcane borer in 1984–1985, implying severe population reduction and possible extinction (Bennett, 1993). Such cases of displacement are difficult to interpret, however. Firstly, extinction per se has not been shown (for example, the natural habitat and native hosts were not examined). Secondly, a native species that maintains itself only by parasitizing a pest on an agricultural crop could be said to be "living on borrowed time" in any case. Extinction in a natural habitat would be more meaningful, but less likely to be detected.

It can be concluded that very little is known for sure about extinctions caused by biocontrol introductions. During the early twentieth century these probably occurred on a number of occasions on island habitats, but by now the passage of time, presence of other factors and lack of detailed investigation makes it difficult to confirm the causes in many individual cases. From 1960 onwards, while there are some "borderline cases", there seems to be little evidence for extinction caused principally by insect introductions.

8.2.2 *Inundative releases*

Other strategies of biological control, such as inundative releases or conservation biological control, are also likely to produce non-target effects, as are any other methods of pest management. These have seldom been studied or documented, except for some field and microcosm trials involving inundative applications of some predators, parasitoids, and insect pathogens (see van Lenteren *et al.*, 2003; Hokkanen and Hajek, 2003). Lynch *et al.* (2001) found 35 non-target effect records (agent/target/non-target combinations) from publications on inundative methods. The quantification of these in terms of numbers of species affected is difficult, however, as soil-dwelling organisms were often grouped into broad classes in these studies. Many of these agents are generalists, but their use is considered safe due to their relative lack of persistence, and the consequent transience or non-existence of their non-target impacts. Field tests of actual host range are rare, but many studies where (local) population effects have been looked for, have found none. It appears that many releases are, in this sense, safe. A substantial number of low-intensity population level impacts and a few more serious ones were apparent in the review of Lynch *et al.* (2001). Overall, these authors estimate that about 49% of inundative releases outdoors may have some population consequences for local fauna, and that 8% produce >40% population reductions temporarily. Despite such impacts, the use of inundative methods may be justified because of the transience and locality of the impacts, particularly in agroecosystems distant from truly natural habitats, or with agents that disperse on only a small scale.

8.3 Minimizing ecological risks

As discussed above, negative environmental effects of biological control have seldom been reported, although over the past 100 years more than 6000 exotic natural enemies have been imported, mass-reared, and released as biological control agents. To ensure the continuing safety and positive public image of biological control, many countries are now requiring risk assessment for all biological control agents. This is new for macro-organisms in inundative use, although as classical agents these have been regulated for a long time already, in particular for weed biocontrol. Microbial agents have historically first been tested and applied on a small scale, and without many restrictions. Regulatory authorities first became interested in them, when larger-scale production and application became possible. The main worry in the beginning was the safety of workers and consumers: is the production and application safe for the personnel involved (e.g. toxic metabolites), and for the general public if used on edible crops? As a consequence, microbial control agents have been subject to regulatory approval (e.g. EU Directive 91/414). Macroscopic organisms used in inundative control, such as nematodes, mites and insect parasitoids, usually have been exempted from these regulations.

Several reasons for a more comprehensive regulatory procedure for biological control agents have been presented (see Blum *et al.*, 2003). These include potential human health hazards, environmental risks, and questions of efficacy (see Hokkanen and Hajek, 2003). Consideration of these has led to differing policies in different countries. For example, the UK regulates all imports of any organisms from elsewhere, but does not regulate the use of indigenous control agents. Norway has banned the import of all species that cannot be proven to be native to Norway, but allows other imports, and also regulates the use of native organisms. Switzerland, Austria, and Sweden also have implemented a formal registration procedure for all macro-organisms. Elsewhere, in some countries the procedures (e.g. those of Australia, New Zealand, and Hawaii; see articles in Lockwood *et al.*, 2001) are already so strict that import and release of exotic natural enemies is extremely difficult. Some other countries have no regulations at all, so any species can be imported and released. There is a general trend, however, towards more stringent regulatory requirements (e.g. Barratt *et al.*, 1999).

8.3.1 Components of an ecological risk assessment

The ecological risk assessment is the most critical and difficult part of the overall risk assessment procedure in biological control. A general framework developed for such an assessment (van Lenteren *et al.*, 2003) identifies the following seven basic steps. These consider different aspects of natural enemy biology and the environment of the system into which the natural enemy will be introduced, in order to evaluate the potential impacts on non-target species and ecological risks, as follows:

1. defining ecological context and the selection of appropriate non-target species
2. host specificity testing
3. natural enemy dispersal capability
4. potential for natural enemy establishment
5. direct effects on non-targets
6. indirect effects on non-targets
7. risk assessment.

Evaluation of risks related to releases of natural enemies demands integration of many aspects of their biology, as well as information on ecological interactions. For a full risk assessment, three steps are distinguished: (1) risk identification and evaluation procedure concerning the release of a natural enemy; (2) risk management plan dealing with risk reduction and risk mitigation; and (3) risk–benefit analysis of the proposed release of the natural enemy, together with risk–benefit analyses of current and alternative pest management methods.

The risk of adverse effects caused by the release of a biological control agent is the product of the impact of likelihood (probability) and the impact of magnitude (consequence) (see Hickson *et al.*, 2000; van Lenteren *et al.*, 2003). Five groups of risks are considered related to the release of exotic biological control agents. These are: establishment; dispersal; host specificity; direct effects; and indirect non-target effects.

$$\text{RISK} = (\text{LIKELIHOOD of adverse effect}) \times (\text{MAGNITUDE of effect})$$

In order to assess risks, first the likelihood and magnitude of adverse effects are established for each of the five risk groups. For details of this process, see van Lenteren *et al.*, 2003. However, without adding a numerical value to each evaluation criterion it remains a qualitative procedure, making comparison of natural enemies difficult. Therefore, the following values could be assigned to each criterion (van Lenteren *et al.*, 2003):

Likelihood	Magnitude
Very unlikely = 1	Minimal = 1
Unlikely = 2	Minor = 2
Possible = 3	Moderate = 3
Likely = 4	Major = 4
Very likely = 5	Massive = 5

Table 8.1. *Example of calculating the lowest possible risk index for a biological control agent*

Criterion		Likelihood	Magnitude	L × M
Establishment	1	1	1	
Dispersal		1	1	1
Host range		1	1	1
Direct effects		1	1	1
Indirect effects	1	1	1	
SUM = risk index 5				

The overall risk index for each natural enemy is then obtained by first multiplying the figures obtained for likelihood and magnitude, and then by adding the resulting figures obtained for dispersal, establishment, host specificity, and direct and indirect effects. The minimum score therefore is 5 (5 times 1 × 1, see Table 8.1), and the maximum score 125 (5 times 5 × 5).

8.3.2 *Application of ecological risk assessment*

In a first application of this methodology, van Lenteren *et al.* (2003) analyzed 31 cases of natural enemy introductions. The risk indices obtained ranged from 7 to 105, and revealed several interesting patterns. The lowest risk index (7) found was for *Thripobius semiluteus*, an inundative biological control agent used in greenhouses (no establishment, poor dispersal outside greenhouse, monophagous, no direct or indirect non-target effects), but rising to 12 when used in the field (some establishment, reasonable dispersal, monophagous, no direct or indirect non-target effects). Quite similarly, the lowest risk index found for a classical biological control agent was 13 (e.g. *Aphelinus mali* and *Polynema striaticorne*; good establishment and dispersal in target habitat but not in non-target habitats, monophagous, no direct or indirect non-target effects). In addition, low risk indices (below 35) were found for many parasitoids, several predatory mites, and one predatory insect.

Intermediate risk indices (between 35 and 70) were found for all guilds of arthropod natural enemies: parasitoids, predatory insects, predatory mites, parasitic nematodes, and entomopathogenic fungi. Entomopathogens (*Beauveria, Metarhizium* and *Steinernema*) all scored intermediate because of

Table 8.2. *Case: risk index for* Steinernema feltiae *in field crops in Finland*

	Criterion	Likelihood	Magnitude	L × M
Establishment	3	5	15	
Dispersal		1	1	
Host range		5	5	25
Direct effects		4	2	8
Indirect effects	4		1	4
SUM = risk index 53				

their broad host range, but their very limited dispersal capacities strongly reduces risk. For example, in the case of *S. feltiae* (see Table 8.2) it could be argued (and further research might show) that the actual realized host range in the agricultural fields should not score 5 × 5 but, e.g. 3 × 2, which then already would change the initial rating significantly.

The highest risk indices were found for predatory insects (*Harmonia axyridis* (101), *Hippodamia convergens* (105), *Podisus maculiventris* (88), *Orius insidiosus* (77)) and parasitoids (*Encarsia pergandiella* and *Trichogramma brassicae* (both 73) and *Cales noacki* (87)). This is not surprising, as they would all be classified by biological control experts in the high-risk category based on what is known of their biology.

The risk assessment methodology applied by van Lenteren *et al.* (2003) also pointed out clearly that different values for the same organism will be obtained when evaluated for different release areas. For example, the predatory mite *Phytoseiulus persimilis* had a higher risk index when released in the open field in Mediterranean Italy (24), than when released in greenhouses in temperate climate countries (10). Risk assessment methodologies may also sometimes lead to underestimating risk, particularly if biological knowledge of the ecosystem where the natural enemy will be released is poor. For instance, when an agent is reported monophagous in its area of origin and is subsequently released into a new region where the native host fauna has been poorly surveyed or examined, range expansion to non-target hosts might occur (e.g. Brower, 1991; Barratt *et al.*, 1997). Prior to performing proper host specificity testing, adequate knowledge of the potential non-target species and habitats in the area of release is necessary. In Europe and North America, where the arthropod fauna is relatively well known, this information is likely to be available. In some other regions, however, where the native fauna has been poorly investigated, care should be taken with regard to whether the state of knowledge at pre-release and early post-release is sufficient to predict the occurrence of host-range expansion and the magnitude of its ecological impact.

Risk assessment methodologies may also sometimes lead to overestimating risk. Based on information from the literature, the tachinid parasitoid *Trichopoda pennipes*, which was accidentally introduced in Italy more than 10 years ago, would receive an intermediate risk index rating of 61. This rating is mainly the result of its reported polyphagy in its area of origin. After extensive sampling in the field in Italy 10 years after its introduction, it appeared to have become established in central Italy, but it attacks only one host (the pest species *Nezara viridula*) and none of the other related native pentatomid species occurring in the same area (Salerno *et al.*, 2002).

It is clear that interpretation of risk indices such as those proposed by van Lenteren *et al.* (2003) should be performed with great care, and can only be carried out by biological control experts who know the biology of the natural enemy under consideration. Risk indices will also vary according to the region for which they were made. Therefore, risk indices should not be seen as absolute values, but as indicators to which a judgement can be connected when considering whether to release an agent or not.

The next step in the risk assessment process is to discuss risk management, including risk mitigation and risk reduction. For example, in some cases risks may be minimized by imposing label restrictions, such as concerning the types of crops on which the use of the organism is, or is not, allowed (e.g. treatment of flowering plants with a mycoinsecticide), or by requesting specific application techniques (e.g. soil incorporation only for insect pathogenic nematodes) (Menzler-Hokkanen and Hokkanen, 2005).

The final step in making a justified ecological risk analysis for a new biological control agent is to conduct a risk–benefit analysis, which should include a comparative performance of pest management methods, particularly based on environmental aspects.

8.4 Impacts on IPM

Biological control has always been considered as a cornerstone of integrated pest management programs. Any biocontrol strategy may play a key role in integrated pest management (IPM), and often more than one approach is utilized in concert in a particular crop ecosystem, such as in fruit orchards: for example, conservation biological control could be making use of generalist predators, inundative releases are carried out with *Trichogramma*, codling moth virus, or with entomopathogenic nematodes, and classical agents such as specific parasitoids may control potentially severe exotic pests. The enormous economic and social benefits from biological control have been in many cases well documented (e.g. Greathead, 1995; Lubulwa and McMeniman, 1998; Gutierrez *et al.*, 1999), and the production and sales of some 200 different species of beneficial organisms for inundative uses supports a flourishing industry of small and medium sized companies. IPM based on biological control has been adopted eagerly in many branches of speciality production, most notably in greenhouses, where in some crops almost 100% of crop protection needs are met by biocontrol. One of the best-functioning IPM programs is in the production of greenhouse vegetables in Europe, where some 15 species of natural enemies may be used together with climate control for plant disease management, combined with the use of resistant plant materials (van Lenteren, 2000*b*). Overall, over 100 species of natural enemies are commercially available for these systems. Also the rapidly growing sector of organic production solely relies on biocontrol-based IPM.

Interest in biological control solutions to pest problems has been steadily increasing, in particular as many of the chemical pesticides have been phased out largely because of their environmental risks. Urgent needs to tackle the growing number of exotic pest invaders in agriculture, forestry, and in amenity areas and wildlife refuges, further turns the attention to biological controls. In the USA alone the current costs of exotic invaders accrue to about $137 billion per year (Pimentel *et al.*, 2000), and the figures rise at an accelerating rate. The "do nothing" approach is usually not an option in agriculture and forestry, nor does it preserve wilderness areas from invasive species.

Recent concerns over the ecological safety of biological control introductions and the debate over non-target effects has polarized biologists. Many practitioners view biocontrol as a progressive, environmentally benign alternative to chemical pesticides, while reputable ecologists question its safety record and produce evidence of non-target attacks by biological control agents. This has prompted some to conclude that ecology, which is the basic science behind the practice of biological control, has done much to discredit biological control, but little to aid it (Syrett, 2002). Public attitudes toward biological control have now become more cautious, although the environmental community traditionally has been very supportive of it (Goldburg, 1996; Syrett, 2002). As the perceived risks have increased, regulatory authorities have started to demand much more detailed impact assessments and benefit–risk analyses. In many countries the process of obtaining permission to introduce and release new organisms has become complex, lengthy, and costly: some countries prohibit such introductions altogether (e.g. Norway, UK).

While classical introductions have been regulated to some extent for a long time, this has not been usually the case with inundative applications of beneficial organisms, for example in the greenhouse. However, some 25 countries currently require registration for all biological control agents, and a guidance document has been produced by the OECD (2004) to harmonize these requirements between the different countries, and to encourage also other countries to adopt such regulations. There is currently a heated debate about the value of such blanket regulations, which the industry (IBMA, International Biocontrol Manufacturer's Association) sees as a threat to sensible biological control. Registration of all biocontrol agents is considered to delay product development and to push costs so high that most companies will not be able to carry through the development of their products no matter how promising candidates they have in the pipeline. This, of course, would strongly affect the availability of IPM-compatible control agents for plant protection. These concerns are not without foundation, as registration processes can be notoriously lengthy. For example, in Finland it took nine years to get a registration for *Streptomyces griseoviridis*-based biofungicide (Mycostop), and currently *Pseudomonas chloroaphis* is in its eighth year in the registration process at the European Union, despite the fact that it has already been used without any observed problems for over six years on over 500 000 ha in Scandinavia with a preliminary registration (Ehlers, 2004, personal communication). As macro-organisms such as predatory mites, parasitic insects, and entomopathogenic nematodes have so far been exempted from registration in most countries, the biocontrol agent producers now fear that subjecting them to similar requirements as for micro-organisms and as proposed by the OECD, will all but suffocate their industry.

An alternative approach, advocated by the IBMA, is the one used by the European and Mediterranean Plant Protection Organization (EPPO), i.e. the positive list approach. Here a biological control agent can be placed on a positive list after appropriate safety testing, or as with current agents, after a long record of safe use. After placement on a positive list, no restrictions on its use in the specified geographical region exist.

The increased awareness of the possible non-target or other negative ecological impacts has already resulted in the slowing down of biocontrol introductions into the USA, for example (Syrett, 2002), and the increased regulatory hurdles are likely to seriously affect the availability of suitable biological control agents for inundative releases. Regulatory process may tip the balance to drive out of the market small products or enterprises by its sheer cost and delay in getting any returns from the investments into product development. For example, after Sweden decided to regulate the use of macrobial control agents, small companies withdrew many products for small applications (they were not worth the trouble any more) and currently Swedish growers have only some 20–25 biocontrol products available to them (Blum *et al.*, 2003).

To overcome some of the challenges concerning the safety and regulatory policy developments, an international workshop was held in Germany in 2003 (International Organization for Integrated and Biological Control, IOBC/WPRS, together with EU-COST Actions 850 and 832). Also the biological control industry was well represented at the meeting. The workshop reached a consensus on the following principles (summarized in Blum *et al.*, 2003).

A risk assessment procedure (some light form of "regulation") is necessary for the introduction and use of biological control agents not native to the area of intended inundative use. Such a procedure is specific to each ecological region, and should follow the principles laid out in the publication of van Lenteren *et al.* (2003). In practice, each country would regulate the releases into their country, but could copy the data generated for another country within the same ecological region.

As a rule, no risk assessment nor any regulation is necessary for native biological control agents. Potential environmental impacts are likely to be reversible, and relate to massive uses close to environmentally sensitive areas. These situations should be taken care of by other regulations instead of a blanket regulation of biological control agents; however, if problems arise, the use of native organisms could be regulated in these specific cases after the problem has been identified.

The risk assessment regulation for the exotic agents should lead to an establishment of a positive list of such agents for each ecological region, which would omit any further need for testing of these agents in countries within that region.

1. The implementation of the evaluation procedure should be administered by an international organization such as the EPPO.
2. The research which might be required for the risk assessment in addition to the research and development necessary for the product itself and its use, should be at least partly publicly funded (possibly with matching funds collectively from the industry, e.g. IBMA) and should result in a generic safety registration of each particular agent (freely available for any producer afterwards).

3. All stakeholders must be present to give their input into the development of any new regulations.

Concerning classical biocontrol introductions, a balance must be found between the risks of non-target impacts and stakeholder benefits. Cock (2003) illustrates this need in his account of several case studies from developing countries. He emphasizes that due to funding and capacity constraints, comprehensive risk assessments of biological control introductions are unlikely to take place in many cases in developing countries. Despite that, the national authorities have to make potentially irreversible decisions of such introductions in light of the available information, even if from a distance or with hindsight we may disagree with them. The case studies which Cock (2003) presents show that the impacts of not implementing biological control, and the benefits to be achieved by doing so successfully, are likely to be orders of magnitude greater than the non-target impact associated with introducing the biological control agents. In all cases, a full account of stakeholder benefits and risks, and the costs and risks of alternative actions, should be carried out when deciding about the future of biological control, and the role that it can play in executing IPM programs.

Overall, a compromise must be found between some form of relevant regulation and its costs, in order not to completely hamper the use of exotic natural enemies. Further, applications of naturally occurring native organisms seldom would require any regulation as regards their environmental safety. In addition, risks of biological control employing exotic natural enemies need to be compared with the risks caused by other methods of control, in particular with the current practice. This usually involves chemical control, which often interferes negatively with IPM, while biocontrol is an essential element of such programs.

Acknowledgements

Our work has been supported by the EU research grants FAIR5-CT97-3489 (ERBIC), QLK5-CT-2001-01447 (MASTER) and FP6-2004-SSP-4-022709 (REBECA).

References

Barratt, B.I.P., Evans, A.A., Ferguson, C.M. *et al.* (1997). Laboratory nontarget host range of the introduced parasitoids *Microctonus aethiopoides* and *Microctonus hyperodae* (Hymenoptera: Braconidae) compared with field parasitism in New Zealand. *Environmental Entomology*, **26**, 694–702.

Barratt, B. I. P., Ferguson, C. M., McNeil, M. R. and Goldson, S. L. (1999). Parasitoid host specificity testing to predict field host range. In T. M. Whithers, L. Barton Browne and J. N. Stanley, J. N. (eds.), *Host Specificity Testing in Australia: Towards Improved Assays for Biological Control*. Brisbane, Australia: CRC for Tropical Pest Management. pp. 70–83.

Bellows, T. S and Fisher, T. W. (eds.) (1999). *Handbook of Biological Control*. San Diego, CA: Academic Press.

Bennett, F. D. (1993). Do introduced parasitoids displace native ones? *Florida Entomologist*, **76**, 54–63.

Bjørnson, S. and Schütte, C. (2003). Pathogens of mass-produced natural enemies and pollinators. In J. C. van Lenteren (ed.), *Quality Control and Production of Biological Control Agents: Theory and Testing Procedures*. Wallingford, UK: CABI Publishing. pp. 133–65.

Blossey, B. (1995). Host specificity screening of insect biological control agents as part of an environmental risk assessment. In H. M. T. Hokkanen and J. M. Lynch (eds.), *Biological Control: Benefits and Risks*. Cambridge, UK: Cambridge University Press. pp. 84–8.

Blum, B., Ehlers, R., Haukeland-Salinas, S. *et al.* (2003). Biological control agents: safety and regulatory policy. *BioControl*, **48**, 477–84.

Brower, J. H. (1991). Potential host range and performance of a reportedly monophagous parasitoid, *Pteromalus cerealellae* (Hymenoptera: Pteromalidae). *Entomological News*, **102**, 231–5.

CBD (2002). Decision VI/23 - *Alien species that threaten ecosystems, habitats or species. Decisions adopted by the Conference of Parties to the Convention on Biological Diversity at its sixth meeting, The Hague, 7–19 April 2002*. Annex 1: pp. 155–77. http://www.biodiv.org/doc/decisions/COP-06-dec-en.pdf

Cock, M. J. W. (2003). Risks of non-target impact versus stakeholder benefits in classical biological control of arthropods: selected case studies from developing countries. In R. G. Van Driesche (ed.), *Proceedings of the International Symposium on Biological Control of Arthropods*. USDA Forest Service Publication FHTET-03-05. pp. 25–33.

Coll, M. and Guershon, M. (2002). Omnivory in terrestrial arthropods: mixing plant and prey diets. *Annual Review of Entomology*, **47**, 267–97.

Crawley, M. J. (1983). *Herbivory: the Dynamics of Animal–Plant Interactions*. Oxford, UK: Blackwell.

Crawley, M. J. (1986). The population biology of invaders. *Philosophical Transactions Royal Society London, B*, **314**, 711–31.

Di Castri, F. (1989). History of biological invasions with special emphasis on the old world. In J. A. Drake (ed.), *Biological Invasions: A Global Perspective*. Chichester, UK: Wiley. pp. 1–30.

Follett, P. A., Duan, J., Messing, R. H. and Jones, V. P. (2000). Parasitoid drift after biological control introductions: re-examining Pandora's box. *American Entomologist*, **46**, 82–94.

Follett, P. A., Johnson, T. M., and Jones, V. P. (1999). Parasitoid drift in Hawaiian Pentatomoids. In P. A. Follet and J. J. Duan (eds.), *Nontarget Effects of Biological Control*. Dordrecht, The Netherlands: Kluwer Academic Publishers. pp. 77–93.

Fox, M.D. and Fox, B.J. (1986). The susceptibility of natural communities to invasion. In R.H. Groves and J.J. Burdon (eds.), *Ecology of Biological Invasions: An Australian Perspective*. Canberra: Australian Academy of Science. pp. 57–66.

Funasaki, G.F., Po-Yung, L., Nakahara, L.M., Beardsley, J.W. and Ota, A.K. (1988). A review of biological control introductions in Hawaii: 1890–1985. *Proceedings of the Hawaiian Entomological Society*, **28**, 105–60.

Goldburg, R. (1996). *Is the public behind biological control?* Cornell Community Conference on Biological Control. www.nysaes.cornell.edu/ent/bcconf/talks/goldburg.html

Greathead, D.J. (1995). Benefits and risks of classical biological control. In H.M.T. Hokkanen and J.M. Lynch (eds.), *Biological Control: Benefits and Risks*. Cambridge, UK: Cambridge University Press. pp. 53–63.

Gurr, G. and Wratten, S. (eds.) (2000). *Measures of Success in Biological Control*. Dordrecht: Kluwer Academic Publishers.

Gutierrez, A.P., Caltagirone, L.E. and Meikle, W. (1999). Evaluation of results: economics of biological control. In T.S. Bellows, Jr. and T.W. Fisher (eds.), *Handbook of Biological Control*. San Diego, CA: Academic Press. pp. 243–52.

Hall, R.W. and Ehler, L.E. (1979). Rate of establishment of natural enemies in classical biological control. *Bulletin of the Entomological Society of America*, **25**, 280–2.

Harris, P. (1990). Environmental impact of introduced biological control agents. In M. Mackauer, L.E. Ehler and J. Roland (eds.), *Critical Issues in Biological Control*. Andover, UK: Intercept Ltd. 289–300.

Hawkins, B.A. and Marino, P.C. (1997). The colonization of native phytophangous insects in North America by exotic parasitoids. *Oecologia*, **112**, 566–71.

Hickson, R., Moeed, A. and Hannah, D. (2000). HSNO, ERMA and risk management. *New Zealand Science Review*, **57**, 72–7.

Hokkanen, H.M.T. and Lynch, J.M. (eds.) (1995). *Biological Control: Benefits and Risks*. Cambridge, UK: Cambridge University Press.

Hokkanen, H.M.T. and Hajek, A.E. (2003). *Environmental Impacts of Microbial Insecticides: Need and Methods for Risk Assessment*. Dordrecht, The Netherlands: Kluwer Academic Publishers.

Howarth, F.G. (1985). Impacts of alien land arthropods and molluscs on native plants and animals in Hawaii. In C.P. Stone and J.M. Scott (eds.), *Hawaii's Terrestrial Ecosystems: Preservation and Management*. Honolulu: University of Hawaii. pp. 149–78.

Howarth, F.G. (1991). Environmental impacts of classical biological control. *Annual Review of Entomology*, **36**, 485–508.

Howarth, F.G. (2000). Environmental issues concerning the importation of non-indigenous biological control agents. In J.A. Lockwood, F.G. Howarth and M.F. Purcell (eds.), *Balancing Nature: Assessing the Impact of Importing Non-Native Biological Control Agents*. Thomas Say Publications in Entomology: Proceedings. Lanham, Maryland: Entomological Society of America. pp. 70–98.

Lockwood, J. A., Howarth, F. G. and Purcell, M. F. (eds.) (2001). *Balancing Nature: Assessing the Impact of Importing Non-Native Biological Control Agents*. Thomas Say Publications in Entomology: Proceedings. Lanham, Maryland: Entomological Society of America.

Lonsdale, W. M., Briese, D. T. and Cullen, J. M. (2001). Risk analysis and weed biological control. In E. Wajnberg, J. K. Scott and P. C. Quimby (eds.), *Evaluating Indirect Ecological Effects of Biological Control*. Wallingford, UK: CABI Publishing. pp. 185–210.

Louda, S. M., Pemberton, R. W., Johnson, M. T. and Follett, P. A. (2003). Nontarget effects: the Achilles' Heel of biological control? Retrospective analysis to reduce risk associated with biocontrol introductions. *Annual Review of Entomology*, **48**, 365–8.

Lubulwa, G. and McMenimam, S. (1998). ACIAR-supported biological control projects in the South Pacific (1983–1996): an economic assessment. *Biocontrol News and Information*, **19**, 91N–98N.

Lynch, L. D. and Thomas, M. (2000). Nontarget effects in the biocontrol of insects with insects, nematodes and microbial agents: the evidence. *Biocontrol News and Information*, **21**, 117N–130N.

Lynch, L. D., Hokkanen, H. M. T., Babendreier, D. *et al.* (2001). Indirect effects in the biological control of arthropods with arthropods. In E. Wajnberg, J. C. Scott and P. C. Quimby (eds.), *Evaluating Indirect Ecological Effects of Biological Control*. Wallingford, UK: CABI Publishing. pp. 99–125.

Mack, R. N. (1985). Invading plants: their potential contribution to population biology. In J. White (ed.), *Studies in Plant Demography: A Festschrift for John L. Harper*. London: Academic Press. 127–41.

Menzler-Hokkanen, I. and Hokkanen, H. M. T. (2005). *Developing entomopathogenic nematode delivery systems for biological control of oilseed rape pests*. IOBC/WPRS Bulletin, **28**, 19–22.

Neuenschwander, P. and Markham, R. (2001). Biological control in Africa and its possible effects on biodiversity. In E. Wajnberg, J. C. Scott and P. C. Quimby (eds.), *Evaluating Indirect Ecological Effects of Biological Control*. Wallingford, UK: CABI Publishing. pp. 127–46.

OECD (2004). *Guidance for Information Requirements for Regulation of Invertebrates as Biological Control Agents (IBCAs)*. ENV/JM/MONO (2004)1. Paris: Organization for Economic Cooperation and Development (OECD).

Pimentel, D. (1986). Biological invasions of plants and animals in agriculture and forestry. In H. A. Mooney and J. A. Drake (eds.), *Ecology of Biological Invasions of North America and Hawaii*. New York: Springer. 149–62.

Pimentel, D. (1995). Biotechnology: environmental impacts of introducing crops and biocontrol agents in North American agriculture. In H. M. T. Hokkanen and J. M. Lynch (eds.), *Biological Control: Benefits and Risks*. Cambridge, UK: Cambridge University Press. pp. 13–28.

Pimentel, D., Hunter, M. S., LaGro, J. A. *et al.* (1989). Benefits and risks of genetic engineering in agriculture. *BioScience*, **39**, 606–14.

Pimentel, D., Lach, L., Zuniga, R. and Morrison, D. (2000). Environmental and economic costs of non-indigenous species in the United States. *BioScience*, **50**, 53–65.

Roberts, L. I. N. (1986). The practice of biological control – implications for conservation, science, and the community. *Weta News Bulletin of the Entomological Society of New Zealand*, **9**, 76–84.

Rosenheim, J. A., Kaya, H. K., Ehler, L. E., Marois, J. J. and Jafee, B. A. (1995). Intraguild predation among biological control agents: theory and evidence. *Biological Control*, **5**, 303–35.

Salerno, G., Colazza, S. and Bin, F. (2002). *Nezara viridula* (Heteroptera: Pentatomidae) parasitism by the tachinid fly *Trichopoda pennipes* ten years after its accidental introduction into Italy from the New World. *BioControl*, **47**, 617–24.

Simberloff, D. (1981). Community effects of introduced species. In M. H. Nitecki (ed.), *Biotic Crises in Ecological and Evolutionary Time*. New York: Academic Press. pp. 53–81.

Simberloff, D. (1992). Conservation of pristine habitats and unintended effects of biological control. In W. C. Kaufmann and R. J. Nichols (eds.), *Selection Criteria and Ecological Consequences of Importing Natural Enemies*. Maryland, USA: Entomological Society of America. pp. 103–17.

Stiling, P. and Simberloff, D. (1999). The frequency and strength of non-target effects of invertebrate biological control agents of plant pests and weeds. In P. A. Follet, and J. J. Duan (eds.), *Nontarget Effects of Biological Control*. Dordrecht, The Netherlands: Kluwer Academic Publishers. pp. 31–43.

Sullivan, D. J. (1987). Insect hyperparasitism. *Annual Review of Entomology*, **32**, 49–70.

Syrett, P. (2002). New restraints on biological control. In G. J. Hallman and C. P. Schwalbe (eds.), *Invasive Arthropods in Agriculture*. Enfield, NH, USA: Science Publishers, Inc. pp. 363–94.

Tothill, J. D., Taylor, T. H. C. and Paine, R. W. (1930). *The Coconut Moth in Fiji (A History of its Control by Means of Parasites)*. London: Imperial Institute of Entomology.

van Lenteren, J. C. and Woets, J. (1988). Biological and integrated pest control in greenhouses. *Annual Review of Entomology*, **33**, 239–68.

van Lenteren, J. C. (1983). The potential of entomophagous parasites for pest control. *Agriculture, Ecosystems and Environment*, **10**, 143–58.

van Lenteren, J. C. (1995). Frequency and consequences of insect invasions. In H. M. T. Hokkanen and J. M. Lynch (eds.), *Biological Control: Benefits and Risks*. Cambridge, UK: Cambridge University Press. pp. 30–43.

van Lenteren, J. C. (2000*a*). Measures of success in biological control of arthropods by augmentation of natural enemies. In G. Gurr and S. Wratten (eds.), *Measures of Success in Biological Control*. Dordrecht: Kluwer Academic Publishers. pp. 77–103.

van Lenteren, J. C. (2000*b*). A greenhouse without pesticides: fact of fantasy? *Crop Protection*, **19**, 375–84.

van Lenteren, J. C. (2001). Harvesting safely from biodiversity: natural enemies as sustainable and environmentally friendly solutions for pest control. In J. A. Lockwood, F. G. Howarth and M. F. Purcell (eds.), *Balancing Nature: Assessing the Impact of Importing Non-Native Biological Control Agents*. Thomas Say Publications in Entomology: Proceedings, Lanham, Maryland: Entomological Society of America. pp. 15–30.

van Lenteren, J. C. (ed.) (2003). *Quality Control of Natural Enemies Used in Biological Pest Management: Theoretical Background and Development of Testing Procedures*. Wallingford, UK: CABI Publishing.

van Lenteren, J. C., Babendreier, D., Bigler, F. *et al.* (2003). Environmental risk assessment of exotic natural enemies used in inundative biological control. *BioControl*, **48**, 3–38.

Waage, J. K. (1997). Global developments in biological control and the implications for Europe. In I. M. Smith (ed.), *EPPO/CABI Workshop on Safety and Efficacy of Biological Control in Europe*. Oxford: Blackwell Science Ltd. pp. 5–13.

Wapshere, A. J. (1974). A strategy for evaluating the safety of organisms for biological weed control. *Annals of Applied Biology*, **77**, 201–11.

Williamson, M. (1996). *Biological Invasions*. London: Chapman and Hall.

Zwölfer, H. (1971). The structure and effect of parasite complexes attacking phytophagous host insects. In P. J. de Boer and G. R. Gradwell (eds.), *Dynamics of Numbers in Populations*. Oosterbeek, The Netherlands: Advanced Study Institute. pp. 405–16.

9

Ecology of natural enemies and genetically engineered host plants

G. G. KENNEDY AND F. GOULD

9.1 Introduction

Plants play an important role in the interaction between phytophagous arthropods and the parasitoids, predators, and pathogens that attack them. They provide habitat for both phytophagous arthropods and their natural enemies, and they provide behavioral cues that are important in host/prey location by parasitoids and predators (Vet and Dicke, 1992). Plants also serve as the primary source of food for phytophagous species and, in the case of some parasitoids and predators, as a source of supplemental food. The nutritional quality and phytochemical content of this plant-supplied food is known to affect the vulnerability of phytophagous species to attack by parasitoids, predators, and pathogens, as well their suitability as hosts or prey following attack. The nutritional quality and phytochemical content of plant-supplied food can also affect parasitoids and predators that feed on plant tissues and plant products, such as pollen, nectar, and plant sap.

Plant effects on phytophagous arthropods and their natural enemies occur largely at the level of the individual but have consequences at the population level for both pests and their natural enemies. It is these population-level effects, which can be manifested in both ecological time and in evolutionary time, that are most important in the context of crop protection and integrated pest management (IPM).

Crop varieties developed through conventional plant breeding to be resistant to arthropod pests have long been used, with dramatic success, in crop protection. In general, the use of resistant varieties in pest management has been considered to be compatible with biological control in the context of IPM. Using a straightforward mathematical model to investigate the population level consequences of the combined effects of host plant resistance and biological control in relation to crop protection, van Emden (1966) demonstrated that a moderate level of host plant resistance interacting with natural enemies of pests could prevent pest outbreaks in some situations in which neither the resistance nor biological control operating alone was sufficient to effect control.

Although several studies have subsequently provided support for van Emden's conclusions, others have pointed out that, under some conditions, antagonistic interactions between host plant resistance and natural enemies have the potential to interfere with the effectiveness of biological control (e.g. Hare, 1992). While there is extensive evidence documenting that individuals of important parasitoid and predator species of major crop pests are adversely affected by some resistance traits, there is little documentation that these effects have had population-level consequences, which interfered significantly with pest management and crop protection. One possible reason for this is that most pest species are attacked by a complex of natural enemies. The individual species that make up this complex interact with the crop and pest differently. If only some species in the complex are affected, the impact of resistance on the overall natural enemy complex may be minimal (e.g. Kennedy, 2003). Another possible reason is that most natural enemies exploit host populations across an array of plant species, including multiple crops. Because different resistance mechanisms targeting different pest species have been used in different crops and the land area planted to crop varieties possessing the same resistance mechanisms has generally been limited, the opportunity for widespread impacts on natural enemy populations has been limited. Other possible reasons include the difficulty of measuring relatively small population-level effects and the possibility that effects seen as negative at the individual level are not negative at the population level.

The relatively recent commercialization and widespread adoption of transgenic crop varieties expressing the δ-endotoxin of *Bacillus thuringiensis* (Bt) and resistant to key pest species, has led to concern over the potential for transgenic, insecticidal crop varieties to affect natural enemies and to disrupt biological control (Hilbeck, 2002). Given that the effects on natural enemies of pest resistance traits are independent of the methods used to incorporate the traits into the plant, there is no reason to expect a priori that the effects of transgenic, pest-resistant plants on natural enemies will differ in kind from effects of pest-resistant plants developed using the methods of conventional plant breeding. However, the power of recombinant DNA technology creates the potential for transgenic, pest-resistant crop varieties to exert a greater impact on natural enemies and on biological control than has been observed for conventionally bred pest resistant varieties. This is because:

- this technology enables the expression in crop plants of foreign genes coding for completely novel traits that confer pest resistance;
- it facilitates the use of the same or closely related resistance traits in multiple crops (e.g. Bt δ-endotoxins in maize, cotton, rape, rice, and potato);
- it has the potential to make high levels of resistance to key pest species available in virtually all crops;
- transgenic, insect-resistant crop varieties expressing the same or similar resistance mechanisms (Bt δ-endotoxins) are being planted on an unprecedented scale (e.g. major crops, primarily cotton and corn, genetically transformed to express one or more Bt δ-endotoxins were planted and grown on 11.4 million hectares worldwide (Shelton *et al.*, 2002)).

In this chapter, we provide a brief overview of the kinds of interactions that occur between conventional plant resistance traits and natural enemies. We then summarize the known effects of Bt δ-endotoxin-expressing crops on natural enemies and attempt to address the question, "How should we evaluate the significance of potential and observed effects of genetically engineered traits on natural enemies?".

9.2 Plant/natural enemy interactions

A variety of plant characteristics are known to affect predaceous and parasitic arthropods. These have been summarized by Bottrell *et al.* (1998) and include morphological characteristics such as the size and shape of plants and plant parts, epicuticular waxes, pubescence, and color. They also include phytochemicals that provide behavioral cues used by natural enemies in locating their hosts/prey (Vet and Dicke, 1992; Dicke, 1999) or entrap small arthropods (Rabb and Bradley, 1968; Obrycki and Tauber, 1984). Phytochemicals may also act as toxins that have lethal or sublethal effects on natural enemies (e.g. Kennedy, 2003). Particular plant traits may favor natural enemies by increasing the vulnerability of their prey or hosts to attack or by providing behavioral cues that enhance the enemy's ability to locate its prey/hosts. Alternatively, plant traits may adversely affect the natural enemies by interfering with their ability to locate prey, or by causing detrimental physiological effects or mortality (Price, 1986; Hare, 1992; Kennedy, 2003).

9.2.1 *Individual-level effects*

The effects of plant traits conferring pest resistance may be mediated by direct contact between the natural enemy and the resistance factor. Alternatively, they may be mediated indirectly through prey/hosts that are affected by the resistance mechanisms or by the sequestration of plant defensive compounds by plant feeding hosts/prey (multitrophic level effects) (Bottrell *et al.*, 1998; Barbosa and Benrey, 1998). The type and magnitude of effects are determined by both the specific attributes of the resistance mechanism (e.g. toxin, semiochemical, plant trichomes) and by the specific details of the interaction between the natural enemies, their prey/host, and the plant (Kennedy, 2003).

Feeding on plant materials by parasitic and predaceous insects is common. Hagen (1987) reported that 122 of 163 families of parasitic and predaceous insects contained species that may feed on plant materials as both immatures and adults. Natural enemies that feed on plant sap (e.g. hemipteran predators *Podisus maculiventris, Geocoris* spp.), pollen (e.g. *Coleomegilla maculata*) or nectar (e.g. parasitic hymenoptera) are much more likely to be directly affected by plant toxins that are expressed in these plant products, or by reductions in their nutritional quality, than are natural enemies that do not feed on them.

Similarly, natural enemies that have an intimate association with the plant surface are more likely to be directly affected by plant toxins that are present in the plant cuticle or in glandular trichomes. For example, resistance to numerous insect pests in the wild tomato species, *Lycopersicon hirsutum f. glabratum*, is conferred by toxic methyl ketones in the tips of glandular trichomes on the foliage and stems. The tomato fruitworm, *Helicoverpa zea*, is able to detoxify the primary toxin, 2-tridecanone, and is minimally affected by it. The methyl ketone/glandular trichome mediated resistance differentially affects the tachinids *Archytas marmoratus* and *Eucelatoria bryani*, which are larval parasitoids of *H. zea*, because the two species differ in the way they interact with the food plant of their host. *A. marmoratus* larviposits on its host's food plant. The parasitoid larvae attach to passing host larvae and penetrate the cuticle. The parasitoid undergoes limited development within its host until the host pupates. It then develops rapidly and emerges as an adult from the host pupa. In contrast, *E. bryani* larviposits directly into host larvae and the parasitoid larvae develop rapidly, killing their host within a few days. Parasitized host larvae typically drop to the ground before they are killed and the parasitoid larvae emerge to pupate. Thus, unlike *A. marmoratus, E. bryani* has minimal direct contact with its host's food plant. Parasitism of *H. zea* larvae by *A. marmoratus* is reduced on tomato plants expressing high levels of 2-tridecanone and high trichome densities compared to levels on plants that do not express 2-tridecanone. This is due to the direct exposure of the parasitoid larvae on the foliage to lethal concentrations of 2-tridecanone and the physical effects of high trichome densities. Parasitism by *E. bryani* is unaffected by the resistance traits (Farrar *et al.*, 1992; Farrar and Kennedy, 1993).

Effects of plant resistance traits on the third trophic level may be mediated through their effects on the phytophagous hosts or prey of parasitoids, predators, and pathogens. Plant chemicals present in the gut or tissues of host/prey may adversely affect parasitoids and predators. This is illustrated by the effects of the plant produced alkaloids α-tomatine and nicotine.

α-Tomatine is present in the foliage and unripe fruit of tomato and has been implicated in resistance to the tomato fruitworm, *H. zea*. It is a potent growth inhibitor of *H. zea*. For *H. zea* developing in tomato fruits or on artificial diet, α-tomatine levels are positively correlated with larval development time and mortality, and negatively correlated with larval growth rate and adult weight of individuals developing in the fruit (Juvik and Stevens, 1982; Bloem *et al.*, 1989; Elliger *et al.*, 1981). The larval parasitoid *Hyposoter exiguae* is adversely affected when it develops in *H. zea* larvae that feed on diet containing α-tomatine. The effects of α-tomatine are mediated through the host larva and result in prolonged larval development; disruption of pupal eclosion; deformation of antennal, abdominal, and genital structures; and a reduction in adult weight and longevity (Duffey and Bloem, 1986). In addition, there is some evidence that high levels of α-tomatine in foliage of some tomato species may inhibit infection of phytophagous insects by the entomopathogenic fungi *Beauvaria bassiana* and *Nomuraea rileyi* (Costa and Gaugler, 1989; Gallardo *et al.*, 1990; Hare and Andreadis, 1983). There is also evidence that α-tomatine reduces predation on *Manduca sexta* larvae by the predatory stinkbug *Podisus maculiventris*. Predators, which had previously preyed upon α-tomatine-fed *M. sexta* larvae, subsequently rejected *M. sexta* larvae regardless of whether the larvae had fed on a diet containing α-tomatine (Traugott and Stamp, 1996). Additional examples of phytochemically mediated prey rejection are provided by Smiley *et al.* (1985) and Bowers and Larin (1989).

Nicotine is the principal alkaloid in tobacco foliage and occurs in the tissues of tobacco hornworm, *Manduca sexta*, larvae feeding on tobacco. Although parasitism rates of *M. sexta* by *Cotesia congregata* are similar on high and low nicotine tobacco varieties, greater mortality of *C. congregata* larvae in host larvae on high nicotine tobacco results in the production of fewer adult parasitoids per host larva than on low nicotine tobacco (Barbosa *et al.*, 1986; Thorpe and Barbosa, 1986).

Adverse indirect effects of resistance traits on parasitoids and predators may result simply because feeding on hosts or prey that are dead or dying due to plant resistance may provide a poor source of nutrition or a poor milieu for development. These effects can be complex. For example, *H. zea* that ingest 2-undecanone from resistant tomato plants during the last larval instar die during the pupal stage. Under conditions of high relative humidity, the parasitoid *A. marmoratus* is able to complete development within the dead host pupae, but under low relative humidity, the dead host pupae desiccate, killing the parasitoid larvae before they complete development (Farrar *et al.*, 1992; Farrar and Kennedy, 1993).

Individual-level effects of plant resistance traits on predators and parasitoids are not always negative. For example, poor plant quality can also enhance susceptibility to parasitism by impairing the herbivores ability to encapsulate parasitoid eggs (Cheng, 1970; Benrey and Denno, 1997; Vinson and Barbosa, 1987). Additionally, plant resistance traits that reduce growth or prolong development of phytophagous insects can lead to increased parasitism or predation by extending the duration of larval stages or size classes most susceptible to attack by predators and parasitoids, provided that the resistance trait does not otherwise adversely affect the parasitoid or predator (Benray and Denno, 1997; Thaler 1999*a*; *b*). In that these types of effects reflect a preference by the parasitoid or predator for a particular host/prey age or size class, the magnitude of effect may be influenced by the abundance of alternate prey/hosts relative to that of the affected species (Lu *et al.*, 1996).

Increased exposure of phytophagous species to natural enemies may also result if the phytophagous species exhibit increased restlessness on resistant plants. On rice, for example, increased movement on resistant rice plants by brown planthopper, *Nilaparvata lugens*, increases its risk of predation by lycosid spiders (Kartohardjono and Heinrichs, 1984).

In general, individual-level, indirect effects of plant resistance traits are more likely to occur if prey/hosts experience sublethal effects or if they are killed by the resistance after having passed their period of vulnerability to parasitism. Thus, in the case of transgenic plant toxins that kill the target pest quickly but have sublethal effects on non-target phytophagous species, individual-level effects are more likely to be observed on natural enemies associated with the sublethally affected phytophagous species.

9.2.2 *Population-level effects*

In pest management, our focus is on pest populations, not on individuals. We are concerned with the net or combined effects of resistance and biological control on the pest population and the crop damage that it causes. Thus, it is the effects of plant resistance traits on pest populations and on the level of pest mortality caused by the natural enemy populations (i.e. biological control) that are important. These population-level effects involve complex ecological relationships and numerous stochastic elements. These operate within the context of a biotic community that contains a complex of natural enemies and their potential hosts/prey. As a result, population-level effects are difficult to predict based on individual-level effects observed in short-term studies. Hare (1992) has classified the potential combined effects of plant resistance and natural enemies on the equilibrium pest population as additive, synergistic, antagonistic, or complex.

The effects of plant resistance and natural enemies on pest populations are additive if the mortality rate of the pest population attributable to natural enemies is independent of that caused by the resistance trait, such that the contribution of natural enemies to total pest mortality is constant across all levels of resistance. Additive effects of resistance and parasitoids have been reported for lepidoptera resistance in soybean (McCutcheon and Turnipseed, 1981) and for aphid resistance in oats, cantaloupe, alfalfa, and chrysanthemum (Saito *et al.*, 1983; Kennedy *et al.*, 1975; Pimentel and Wheeler, 1973; Wyatt, 1983). Additive effects of resistance and predators have been reported for leafhopper and planthopper resistance in rice (Myint *et al.*, 1986; Salim and Heinrichs, 1986).

Synergistic effects of resistance and natural enemies on pest populations occur when the contribution of natural enemies to total pest mortality increases as the level of pest resistance increases. This type of interaction might occur when the resistance trait increases the pest's vulnerability to attack by predators or parasitoids. Synergistic population-level effects might be observed even in instances where the resistant trait adversely affects the natural enemy, if those effects are less severe than effects on the pest. Studies by Kauffman *et al.* (1985) and Kauffman and Flanders (1985) illustrate this point nicely. When reared on the resistant soybean variety Cutler 71 in the laboratory, Mexican bean beetle, *Epilachna varivestis*, suffered delayed development, increased mortality, and reduced fecundity relative to the susceptible variety Williams. As a consequence, the intrinsic rate of increase *r* was reduced from 0.053 on Williams to 0.012 on Cutler 71. The resistance of Cutler 71 also exerted an indirect, adverse effect on the larval parasitoid

Table 9.1. *Comparison of life table statistics for the Mexican bean beetle,* Epilachna varivestis, *and its parasitoid* Pediobius foveolatus *on Mexican bean beetle resistant and susceptible soybean varieties*

	Resistant soybean		Susceptible soybean	
Parameter	*E. varivestis*	*P. foveolatus*	*E. varivestis*	*P. foveolatus*
R_o (# female progeny/female)	1.8	20.3	12.7	29.1
Generation time	50.5	18.9	47.8	16.5
r	0.012	0.159	0.053	0.205
$r_{P.\ fov.}/r_{E.\ var.}$	13.25		3.87	

Data from Kauffman *et al.* (1985) and Kauffman and Flanders (1985).
Resistant variety = Cutler 71; Susceptible variety = Williams.

Pediobius foveolatus by reducing survival and reproduction of the parasitoid. Successful parasitism of *E. varivestis* larvae fed susceptible foliage was 3.4 times that of larvae fed resistant soybean foliage and the number of adult parasitoids adults produced per parasitized host was 3.5 times greater on susceptible than resistant foliage. Despite the adverse effect of resistance on the parasitoid, a comparison of life table statistics for *E. varivestis* and *P. foveolatus* on resistant and susceptible soybean revealed that the population growth rate of *E. varivestis* was affected more than that of its parasitoid and that the potential for the parasitoid to suppress populations of its host is greater on resistant than susceptible soybean (Table 9.1). Kauffman and Flanders (1985) rightfully cautioned that their results should not be directly extrapolated to the field without confirmation of the observed differences under field conditions. Synergistic effects of resistance and natural enemies have also been reported for Lepidoptera resistance in maize and parasitoids, predators, and entomopathogens (Pair *et al.*, 1986; Isenhour *et al.*, 1989; Kea *et al.*, 1978; Bell, 1978), and for brown planthopper, *Nilaparvata lugens*, resistance in rice and predators (Kartohardjono and Heinrichs, 1984).

Antagonistic interactions between plant resistance and natural enemies affecting pest population density occur when the contribution of natural enemies to total pest mortality decreases with increasing levels of resistance. These types of interactions have been reported for Lepidoptera resistance in maize and parasitoids (Isenhour and Wiseman, 1989), Lepidoptera resistance in tomato and parasitoids and predators (Kennedy, 2003), Lepidoptera and Hemiptera resistance in soybean and parasitoids and predators (Yanes and Boethel, 1983; Orr *et al.*, 1985; Isenhour *et al.*, 1989), and aphid resistance in barley and parasitoids (Starks *et al.*, 1972). Antagonistic interactions can result from either direct or indirect adverse effects of resistance on a natural enemy's survival, reproductive rate, or behavior. In extreme cases, mortality of the resisted pest due to natural enemies might be completely replaced by mortality due to high levels of resistance (e.g. Barbour *et al.*, 1997; Farrar *et al.*, 1994). However, in cases in which the resistance has an antagonistic effect on natural enemies of a second pest species not adversely affected by the resistance trait, the resulting disruption of biological control could lead to higher populations of the second pest species on the resistant variety than on susceptible varieties.

Host density-mediated effects

Population-level effects of resistance traits on both phytophagous insects and their natural enemies will be influenced by the enemy's responses to changes in host density. These responses may be independent of any direct or indirect individual-level effects of the resistance traits on the enemy. For enemies that show a direct density-dependent relationship with their prey/hosts, the percent mortality that they cause varies directly with prey/host density. Thus, reductions in pest density due to resistance will be associated with a decrease in percentage mortality rates due to predation or parasitism (Figure 9.1a). Therefore, the impact of natural enemy species having direct density dependent relationships with their prey/hosts will decline as resistance levels increase. In contrast, for enemies that show an inverse density-dependent relationship with their prey/hosts, percentage mortality due to predation or parasitism varies inversely with prey/host population density. Therefore, reductions in pest density due to plant resistance will be associated with an increase in percentage mortality due to predation or parasitism as resistance levels in the plant increase; consequently, the impact of natural enemies will increase as plant resistance levels increase (Figure 9.1b). In those cases in which the relationship between a pest and its natural enemy is characterized by density-independence, parasitism or predation rates are unaffected by changes in host/prey population density, at least over certain density ranges. In such cases, percentage mortality due to the natural enemy will be unaffected by the resistance level of the plant (Figure 9.1c).

At the population level, density-mediated effects can mimic the antagonistic, synergistic, and additive relationships between resistance and natural enemies described previously (Hare, 1992). However, their occurrence may be completely independent of any direct or indirect individual-level effects of a resistance trait on natural enemies. In addition to their

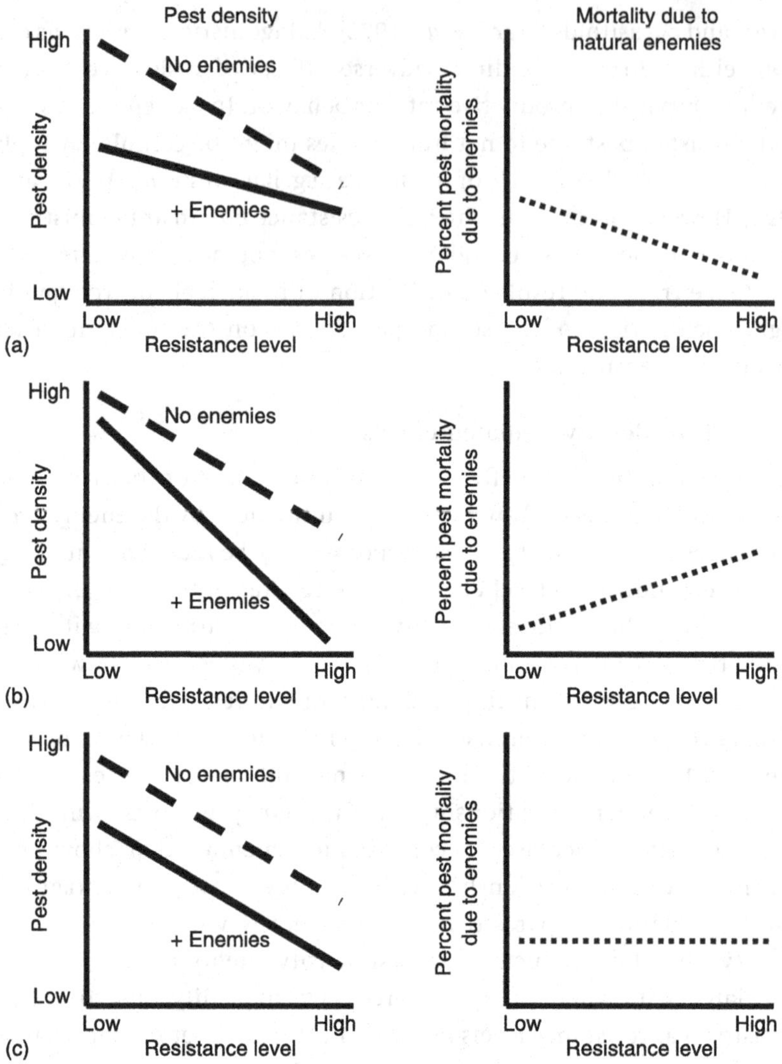

Figure 9.1. Host density-mediated interactions between plant resistance and natural enemies on host mortality. (a) Natural enemy exhibits direct density-dependent response to host/prey density; (b) Natural enemy exhibits inverse density-dependent relationship; (c) Natural enemy exhibits density independent relationship. Relationships shown as linear for simplicity.

importance in determining the population dynamics of natural enemies and pests within a crop, density-mediated effects can influence the distribution and abundance of affected natural enemy species at the landscape level. They can also influence how a natural enemy affects the rate of pest adaptation to resistance traits (Gould *et al.*, 1991; Arpaia *et al.*, 1997).

In general, when a resistance trait affects only one or a few phytophagous species, specialist natural enemies are likely to be affected more by reductions in their host/prey densities than generalists, if alternative prey/host species are available (Riddick *et al.*, 1998). Natural enemies respond to effective prey/host density, which is determined by the total prey/host resource that is available in a particular habitat. The effects of resistance on effective host/prey density depend on the relative proportion of the host/prey complex on the resistant crop that is affected by the resistance trait and by the magnitude of those effects. Reductions in effective prey density below a critical level will cause enemies to disperse or prevent them from ovipositing or aggregating on resistant crops (Riddick *et al.*, 1998).

Although it is difficult to predict population-level consequences of resistance traits based on knowledge of individual-level effects of resistance on parasitoids or predators, there are situations in which population level effects that are significant from a pest management perspective are more likely to occur. These include: when resistance affects the reproductive rate of the natural enemy more than that of the pest; when resistance significantly alters the numerical or functional response of the natural enemy to changes in pest density; when there are few other species of natural enemies present in the resistant crop; or when the principal natural enemy(ies) is a specialist. In general, the net effect of resistance can be expected to benefit the natural enemies if the pest population is suppressed more than the natural enemy population (Gutierrez, 1986). When assessing the impact of resistance traits on natural enemy populations and their impact on pest populations, it is important to distinguish density-mediated effects from those resulting from direct and host/prey mediated effects of the resistance trait on the natural enemies.

9.2.3 *Landscape-level effects*

The agricultural landscape within which pest and natural enemy populations exist can be viewed as an array of patches of habitat and potential habitat that change in relative and absolute suitability over time and that are separated by patches of non-habitat. Resource tracking by natural enemies leads to source/sink relationships among habitats over the course of a season as the quality of some habitats declines while that of others improves. Some crops, which support high populations of prey/hosts or other necessary resources, act as aggregation and reproduction sites for some species of natural enemies. When the resource base provided by these crops is depleted, these enemies disperse to other resource-rich sites (Barbosa, 1998; Letourneau, 1998).

This relationship is illustrated by the coccinellid *Coleomegilla maculata*, a generalist predator that feeds on arthropods as well as pollen (Smith, 1960; Hodek, 1973). *C. maculata* overwinters as aggregations of adults beneath litter in protected areas (Wright and Laing, 1982). During the growing season, *C. maculata* colonizes a sequence of habitats based on the presence of suitable prey and pollen. In Michigan, *C. maculata* populations move in the spring from their overwintering sites to weeds and then to wheat. They then move from wheat to corn and from corn to alfalfa before they finally move to weedy areas along the edges of alfalfa fields prior to overwintering. Reproduction occurs in each of the crop habitats (Maredia *et al.*, 1992). In eastern North Carolina, the sequence of habitats exploited by *C. maculata* is different, reflecting the composition of crops that are grown (Nault and Kennedy, 2000) (Figure 9.2). The populations move from their overwintering sites into wheat, where significant reproduction occurs. In wheat, the adults and larvae feed

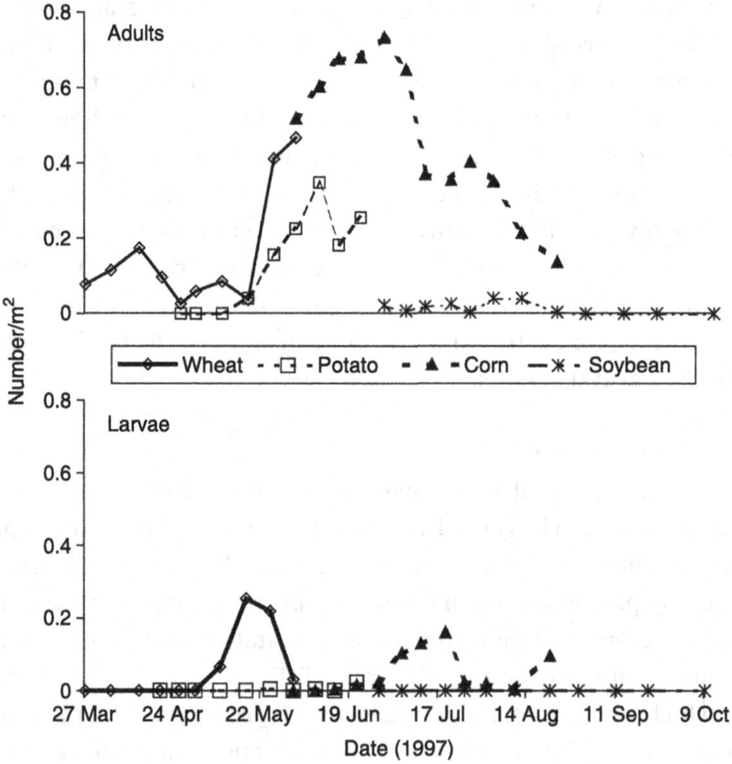

Figure 9.2. Seasonal abundance of *Coleomegilla maculata* adults and larvae in the most prevalent crops in the Tidewater Region of eastern North Carolina in 1995.

primarily on aphids, thrips, and wheat pollen. At the time these first generation larvae are completing their development, the wheat is maturing and the quality of habitat it provides is declining rapidly, forcing the newly emerging adults to disperse in late May and early June. A portion of the first generation adult population colonizes potato where they prey heavily on Colorado potato beetle eggs and small larvae, but there is little or no reproduction by *C. maculata* in potato. (Hilbeck and Kennedy, 1995; Hilbeck *et al.*, 1997.) A large portion of the *C. maculata* adult population dispersing from wheat colonizes maize, where they feed on aphids and a variety of other insects and on pollen (Cottrell and Yeargan, 1998). Second and partial third larval generations are produced in maize before the crop matures in August. Adults begin to disperse from maize after pollen shed but some adults remain in the crop until it begins to senesce. Adults dispersing from corn apparently do not concentrate in any of the predominant crops during the late summer and are not found at high densities again until overwintering aggregations are formed in the fall. In this eastern North Carolina agroecosystem, wheat and corn act as nurseries generating large populations of *C. maculata*, which disperse to other habitats where they effect significant levels of predation on pest populations.

One can readily imagine how in this and similar systems, widespread planting of a highly resistant variety of a crop that serves as a major reproductive habitat for an important natural enemy species could significantly alter the source/sink relationships among crops in the agricultural landscape. If the resistance trait dramatically reduced the abundance or quality of prey/host species in the source crop or if it reduced the quality of a critical resource (e.g. pollen) provided by the crop to the natural enemy, populations of affected parasitoid or predator species produced on the crop could be significantly reduced or delayed, altering the dynamics of natural enemy populations and biological control throughout the landscape. The landscape-level effects of limited, localized plantings of resistant varieties of a crop that has severe adverse effects on natural enemy populations are likely to be of little or no consequence. However, widespread planting across the landscape of crops expressing the same or related resistance mechanisms will increase the potential for landscape-level effects on natural enemy populations and on biological control. It is precisely because transgenic, insect-resistant crop varieties expressing the same or similar resistance mechanisms (Bt δ-endotoxins) are being planted on an unprecedented scale (11.4 million hectares worldwide in 2000: Shelton *et al.*, 2002) that the potential exists for them to cause significant landscape-level effects on natural enemy populations.

9.3 Effects of transgenic crops expressing Bt δ-endotoxins on natural enemies

The transgenic insect resistant crops that have been commercialized to date-derive their resistance to selected pest species from expression of genes coding for Bt δ-endotoxins. As stated previously, there is no reason to expect that the impacts of transgenic crops on natural enemies will differ in kind from those of conventionally bred, insect resistant crops. Indeed the *Bt*-expressing crops that are currently grown commercially act very much like conventional varieties that have toxic effects (antibiosis) on targeted pests. The differences between current transgenic insecticidal crops and conventionally bred crops in impact on targeted pest populations and on natural enemies of pests are quantitative in nature. Here, we examine some of these quantitative differences. Detailed reviews of specific effects of *Bt* crops on natural enemies are provided by Groot and Dicke (2001) and by Hilbeck (2002).

9.3.1 *Direct toxic effects*

At the individual level, specific Bt toxins have toxic effects on a very narrow range of organisms. This specificity is dictated by the chemical properties of each Bt toxin and by the biochemical mode of action of Bt toxins. For example, some Bt toxins are only soluble at very high midgut pH. Because the pH in lepidopteran midguts is in the range of 10–11, these toxins dissolve and interact with the midgut epithelial cells of these insects. However, the lower pH of coleopteran midguts inhibits activity of these toxins. Even among Lepidoptera with similar midgut pH, not all Bt toxins are equally active. The Cry1Ac Bt toxin is very active against the tobacco budworm, *Heliothis virescens*, but is barely active against the silkworm moth, *Bombyx mori*. In contrast, the Cry1Aa toxin, which is chemically very similar to Cry1Ac, is very toxic to *B. mori* but barely toxic to *H. virescens* (Gould, 1998). This differential toxicity is apparently due to the presence of distinct membrane-bound proteins on the external surface of midgut epithelial cells (Ferre and van Rie, 2002).

Although there are over 100 chemically distinct Bt toxins, the high specificity of these toxins has made it difficult to find Bt toxins that are active against a broad spectrum of major pest species. For example, no Bt toxins are known to be highly active against important pests such as whiteflies and aphids. While the specificity of Bt toxins is often seen as a limitation in terms of controlling pests, it is beneficial from the perspective that any given Bt toxin is unlikely to have broad effects on the natural enemy community.

Available data on the effects of Bt on non-target individuals indicates that its direct toxicological impacts should be no greater than that of conventionally bred, insect-resistant crop varieties (Groot and Dicke, 2001; Hilbeck, 2002). Most toxicological studies have found no direct effects of Bt toxins on natural enemies, but this result is not universal. Hilbeck and colleagues (Hilbeck *et al.*, 1998*a*; *b*; 1999) conducted a series of experiments to determine if there were direct toxicological effects of the Bt toxins Cry1Ab and Cry2Aa on the predaceous lacewing *Chrysoperla carnea*. Lacewing larvae were fed artificial diets containing the toxins or Bt toxin-fed larvae of the nontarget pest, *Spodoptera littoralis*, which is highly tolerant of the Bt toxins. In both cases, the Bt toxin-laden food caused higher mortality than unadulterated diet or *S. littoralis* larvae fed Bt toxin-free diet. In a follow-up study, Dutton *et al.* (2002) compared mortality and development effects on *C. carnea* from feeding on *S. littoralis*, a spider mite (*Tetranychus urticae*), and an aphid (*Rhopalosiphum padi*) that had been reared on Bt and non-Bt corn. As in the earlier studies, when *C. carnea* fed on the *S. littoralis* from Bt corn, survival was decreased. However, feeding on Bt corn-reared *T. urticae* or *R. padi* did not negatively impact the lacewings. Biochemical studies indicated that the aphid, *R. padi*, did not have Bt in its body because it fed on phloem, and the phloem of the Bt corn plants did not contain Bt toxin. Therefore, its lack of toxicity to *C. carnea* was reasonable. However, the spider mite, *T. urticae*, contained higher levels of Bt toxin than *S. littoralis*, so its lack of toxicity is not easily explained. The authors of the study suggest that because *S. littoralis* is a non-preferred host for *C. carnea*, even when fed on normal corn, and because *S. littoralis* is slightly impacted by the Bt toxin, there may be a Bt toxin by prey interaction that affects the *C. carnea* larvae.

Ponsard *et al.* (2002) provided four heteropteran predators with Cry1Ac cotton and *Spodoptera exigua* that had fed on Cry1Ac or with control cotton and control-fed *S. exigua* and found that two of the four predators, *Orius tristicolor* and *Geocoris punctipes*, had significantly reduced longevity. It is not clear if the effect seen was due to plant or insect feeding by these insects.

Head *et al.* (2001) examined Bt levels in a number of non-target phytophagous insects. They found that the levels of Cry1Ab toxin in insects fed diet containing the toxin varied with insect species and food source. An aphid, *Rhopalosiphum maidis*, had very low levels of the toxin in its body tissue when fed on toxin-containing artificial diet but had no detectable toxin in its body tissue when it fed on transgenic maize plants expressing the Cry1Ab toxin, potentially due to its phloem-feeding behavior. The levels of Cry1Ab in the tissues of lepidopteran larvae fed transgenic maize leaf tissue were 143, 67, and 59 times less than that of the leaf tissue for European corn borer, corn earworm, and black cutworm larvae, respectively. Because the levels of Bt toxin in corn and cotton are typically hundreds of times higher than the concentration needed to cause insect mortality (Gould, 1998), even these lowered levels could cause toxicity. These results are very important in demonstrating that there may be a general expectation for a prey by toxin interaction when testing for effects of *Bt* crops on natural enemies, so it is important to offer the natural enemies prey species that they are likely to encounter in the field instead of relying on a "model prey species" that is easy to work with in the laboratory.

Most studies of non-target Bt effects have focused on the Cry1 proteins, but recent studies have examined direct toxicity effects of Cry3Bb protein on non-targets. Because Cry1 toxins are specific to Lepidoptera and there are few species of predaceous Lepidoptera, the lack of significant effects in most non-target testing is not surprising. However, Cry3Bb is toxic to a coleopteran pest, the corn rootworm, and there are many predaceous Coleoptera, so from a taxonomic perspective non-target effects are more likely. Lundgren and Wiedenmann (2002), and Duan *et al.* (2002) both tested for a direct impact of Cry3Bb on a predaceous coccinellid, *Coleomegilla maculata*. Interestingly, neither study found any toxic effect on the coccinellid. Clearly, there is a need for tests with a wide variety of natural enemies but these results are intriguing.

Because the only commercialized insecticidal transgenic crops are Bt toxin-based, non-target testing has centered on Bt toxins. Initial experimental work with transgenic plants that produce the insecticidal *Galanthus nivalis* agglutinin (GNA) indicate that such plants may have broader non-target effects than Bt toxin-producing plants. Unlike Bt plants, in which the promotor sequences used to express Bt toxins result in little or no Bt present in the phloem, GNA has been specifically expressed in the phloem (Romeis *et al.*, 2003). To examine potential direct effects of the toxin on parasitoids that might feed on GNA containing aphid honeydew, Romeis *et al.* (2003) fed three parasitoid species a GNA containing sucrose solution and found that each of the parasitoids was negatively affected in some fitness related parameter.

9.3.2 *Indirect effects on natural enemy individuals*

Bt toxins also have the potential to have indirect negative or positive effects on individual natural enemies. For example, Schuler *et al.* (1999) reported that the parasitoid *Cotesia plutellae* was negatively affected by Bt plants because their host *Plutella xylostella* was killed by the Bt endotoxin before the parasitoid larvae completed their development. In contrast, Cloutier and Jean (1998) found that the predaceous stinkbug,

Perillus bioculatus, was better able to attack Colorado potato beetles that had ingested Bt in their diet than those fed on a normal diet. Johnson (1997) found that the parasitoid *Campoletis sonorensis* attacked fewer neonate *H. virescens* larvae on transgenic Bt-tobacco than on conventional tobacco, presumably because the lower feeding rate of the *H. virescens* larvae on Bt tobacco resulted in reduced production of the plant volatiles that attract the parasitoid adults. However, Johnson and Gould (1992) and Johnson (1997) found that after 6 days the parasitism level by *C. sonorensis* was higher on the transgenic plants. They interpreted this result as potentially due to slower larval growth and greater larval movement on the Bt expressing plants. The percent parasitism of *H. virescens* by another parasitoid, *Cardiochiles nigriceps*, did not seem to be affected by presence of the Bt toxin in the tobacco. Differences between the parasitization behavior of these two parasitoids may account for the different results (Johnson and Gould, 1992; Johnson, 1997).

9.3.3 *Population-level effects*

Effects at the individual level do not always translate to effects at the population level and there are conditions under which there can be population level effects that are not detected in short-term laboratory tests with individuals (National Research Council, 2002).

For example, Hilbeck *et al.* (1998*a*; *b*; 1999) demonstrated in laboratory studies that *C. carnea* can be affected by the Bt toxin in corn, but Pilcher *et al.* (1997) found no effect of Bt corn on population density of *C. carnea* in the field. Groot and Dicke (2001) point out that this lack of field effect is not surprising because the density of eggs of the major pest, *O. nubilalis*, did not differ between the Bt and non-Bt plots.

In cases where the Bt crop causes complete mortality of a specialized natural enemy's prey, an effect on the natural enemy's density is more likely to occur. Riddick *et al.* (1998) found that the ground beetle, *Lebia grandis*, which specializes as an adult on Colorado potato beetle eggs and in the larval stage is an ectoparasitoid of Colorado potato beetle pupae larvae, could not be found in plots containing only Bt potato plants, but was found at high densities in plots containing only normal potato plants. Colorado potato beetle densities in the transgenic potato plots were approximately 400 times lower than in the non-Bt plots. In contrast, the generalist predator, *C. maculata* did not differ significantly in abundance between the plots. The authors concluded that the presence of other prey, such as *O. nubilalis* eggs and *Myzus persicae*, could have held the *C. maculata* in the Bt fields. In another study, Arpaia *et al.* (1997) set up potato plots with high and low densities of *L. decemlineata* eggs and found that *C. maculata* aggregated in plots with the high density of potato beetle eggs. In this study alternate prey were not abundant in the plots (Gould, unpublished observation). Reed *et al.* (2001) examined generalist predators in Bt and non-Bt plots where *C. maculata* and other ladybird beetles made up less than 3% of the predators collected. In this system *Nabis* spp., *Geocoris* spp., and *Orius* spp. comprised approximately 95% of the observed predators. None of these three common predators differed in abundance between the Bt and non-Bt plots.

Each of the above studies used a small plot size, so it is important to recognize that different results may be obtained when the same relationships are studied in commercial fields. Furthermore, each of these studies was conducted over a short period of time, so any impact of the Bt crops on rates of natural enemy reproduction could have been missed. Longer-term, large plot studies will be needed in the future.

If a Bt crop causes a reduction in numbers of natural enemies through toxic effects or simply by reduction in prey density, these effects can ripple through the agricultural landscape. As discussed above, *C. maculata* populations move among a sequence of crops during the growing season. If a widely planted variety of one Bt crop removed a major prey item of *C. maculata* early in the season and thereby reduced *C. maculata* reproduction in that crop, fewer of these predators would move to the next crop and their ability to suppress a pest species in the second crop could be diminished. Such landscape level effects of Bt crops have never been noticed or experimentally tested.

Planting transgenic rice varieties that express Bt toxins is under consideration in many tropical countries. Rice insect pests tend to be specialized on rice plants, with some being strictly monophagous (e.g. yellow stemborer), and there are few wild rice hosts available. In this situation, wide-scale planting of Bt-rice could cause substantial alterations in herbivore/natural enemy interactions. If and when Bt rice is commercialized it will be important to monitor such effects.

It seems logical to ask if there would be pest management consequences resulting from the demise of specialized natural enemies of a specialized pest herbivore if Bt-producing varieties were achieving high levels of pest suppression. The potential for such consequences is illustrated by a scenario in which the pest becomes resistant to the Bt toxin in an area where the specialized natural enemy population has been decimated.

9.3.4 *Impacts of natural enemies on efficacy of Bt crops*

While much has been written about the potential of Bt crops to impact natural enemies, less consideration has been given to the potential of natural enemies to impact the efficacy of Bt crops. EPA regulations (US Environmental Protection Agency, 1998) require that each farm set aside some acreage planted to non-Bt varieties as refuges for Bt susceptible pest individuals (this varies from 5% to 50% depending on the crop and location, with 20% being most common), and these areas could also serve to produce natural enemies. In a cropping situation in which 4 out of 5 plants is toxic to the pest (i.e. 20% unsprayed refuge) and the adult female pest does not discriminate between Bt and non-Bt plants during oviposition, only 1 out of 5 of her eggs will survive. If the initial replacement rate of the pest population (Ro) is 5, it would be expected to be reduced to 1 when only 1 out of 5 plants is non-toxic, and there is density independent population regulation. If there are density dependent effects of natural enemies, the Ro would be expected to be greater than 1, but if there was inverse density dependence the Ro would be less than 1. Only empirical studies could determine the type of density dependence caused by the natural enemies.

Riggin-Bucci and Gould (1997) used small plots to examine how varied within-field ratios of collards (non-transgenic) that were and were not protected by Bt toxins would affect the population dynamics of the diamondback moth, *Plutella xylostella*, its generalist predators, and one of its specialized parasitoids, *Diadegma insulare*. Results of field trials in two locations indicated that when there was a refuge of approximately 20%, the diamondback moth populations in the plots declined, the *D. insulare* parasitism rate remained constant, and the predator abundance remained constant. In this system, the effects of *D. insulare* were not density-dependent over a single season. The equal abundance of generalist natural enemies in all plots could have lead to inverse density-dependent mortality of the *P. xylostella* but we did not have data on actual predation rates.

The field plot work of Johnson and Gould (1992) discussed previously indicated that in this system in which Bt-plants have primarily sublethal effects on the pest, at least one specific natural enemy, *C. sonorensis*, and the natural enemy community in general increased the efficacy of the Bt plants in a synergistic manner (i.e. more than additive). Other short-term laboratory studies described previously indicated that similar synergistic effects could be found in the field (e.g. Cloutier and Jean, 1998).

9.4 Natural enemies and management of pest adaptation to transgenic plants

In some of the classical literature on IPM, it is suggested that one advantage to using multiple tools for managing pests would be that a pest confronted with these multiple stresses would not be able to adapt to any single stress or would do so more slowly than if a single control tool was used exclusively. One can envision the pest being pulled in many directions by evolution, and in the end not adapting at all. If in adapting to a parasitoid a pest becomes more susceptible to a pesticide and if adapting to a pesticide is accompanied by increased susceptibility to the parasitoid, then combining the use of both stresses could lead to evolutionary stasis. This perspective may have some validity but ecological genetic experiments have not typically found very clear tradeoffs in adaptation to multiple stresses (Rausher, 1988). We used a simple genetic simulation model to ask whether combining the use of Bt crops with the use of natural enemies would have any impact on the rate that pests adapted to the Bt crop (Gould *et al.*, 1991). These simulations initially focused on Bt crops that did not kill the majority of pest individuals but did slow the growth of survivors that were not resistant to the Bt toxin. Simulations involving Bt crops that did not cause high levels of pest mortality were appropriate in the late 1980s and early 1990s when the first prototypes of Bt crops were not as lethal as the transgenic varieties that have since been commercialized. Indeed, an argument can still be made for the relevance of such simulations in terms of their contribution to general knowledge and because some Bt varieties have high efficacy against some target species and low efficacy against non-target pests (e.g. Cry1Ab corn offers approximately 100% control of European corn borer, but causes low mortality of *Spodoptera frugiperda* [Cabrera, 2001]).

The results of these simulations indicated that natural enemies could increase, decrease, or have no effect on the rate at which a pest adapted to a Bt crop. The type of impact depended on the biology of both the natural enemy and the pest species. In one scenario, the slowed growth (twice as slow) of the susceptible pest individuals compared to resistant individuals on transgenic crops provided parasitoids of larvae with a longer temporal window of opportunity for laying eggs in the susceptible stage individuals. This increased parasitism of Bt toxin-susceptible larvae resulted in a greater difference between the fitness of resistant and susceptible pest genotypes than expected when no parasitoid was present. In this scenario, the pest adapted more rapidly to the Bt crop in the presence of the natural enemy than in its absence (Figure 9.3a).

In a very different scenario where a predator with a type II functional response was modeled, resistance evolved more slowly in the presence of the natural enemy than in its absence (Figure 9.3b). This was due to the fact that the resistant individuals reached the size class fed upon by the predator before the susceptible individuals. Because the resistant individuals were rare

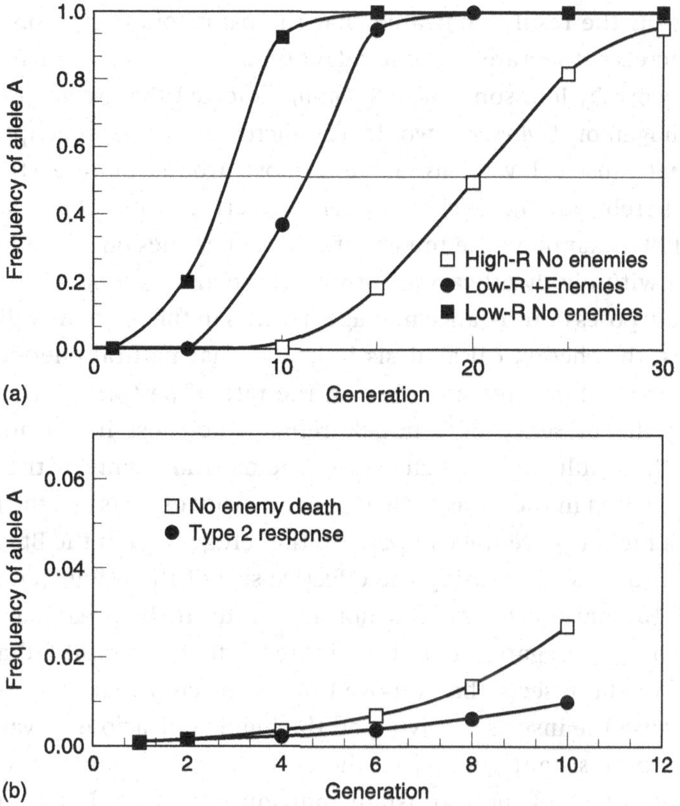

Figure 9.3. Predicted rate of natural enemies on the rate of adaptation to resistant crops with low levels of resistance. (a) The natural enemy, a larval parasitoid, causes more rapid adaptation to a crop with a low level of resistance than is predicted in its absence; the rate of adaptation, however, is slower that the rate of adaptation to a plant with high levels of resistance (and no refuge). (b) The natural enemy, a predator with a type 2 functional response, slows the rate of pest adaptation to the plant with a low level of resistance (from Gould *et al.*, 1991).

when the Bt crop was introduced into the system, a higher proportion of the few fast growing, resistant individuals was attacked by the predator. By the time the mass of susceptible pest individuals reached the predator's preferred-size class, there was such a high density of prey that the predator became satiated and therefore attacked a smaller percentage of these individuals. This dynamic raised the overall fitness of susceptible individuals relative to the fitness of resistant individuals and therefore slowed the rate of adaptation.

The field and laboratory tests by Johnson and Gould (1992) and Johnson (1997) that are described above were used to test these theoretical results.

For *H. virescens* the results indicated that the parasitoid *C. sonorensis* would tend to increase the rate of pest adaptation to the Bt toxins. Further laboratory work by Johnson *et al.* (1997*a*; *b*) indicated that *Nomuraea rileyi*, a fungal pathogen of *H. virescens*, would also increase the rate of adaptation to Bt crops that caused *H. virescens* larvae to move around more on their host plant and thereby have more chance of encountering spores of the pathogen.

Gould (1994) examined the impact of natural enemies on pest adaptation in a system with highly efficacious Bt varieties planted along with a non-Bt refuge for purposes of resistance management. As in the previously discussed simulations, the theoretical analysis indicated that natural enemies could increase, decrease, or have no effect on the rate of pest adaptation to the Bt crop. In the presence of a non-Bt refuge, the most important factor impacting the result was the behavior of the natural enemy. If the natural enemy aggregated in the areas with a higher density of pests (i.e. the refuges) and killed a higher percentage of pests in the refuge than in the Bt crop, the effect is the same as decreasing the effective size of the refuge (Figure 9.4). If, instead, the natural enemy did not aggregate in the area of high pest density, or only aggregated there to a limited extent, it would kill a higher percentage of the insects that survived in the Bt crop than insects in the refuge. Because the insects surviving on the highly efficacious Bt variety are expected to be resistant genotypes, the natural enemy would in effect be lowering the fitness of those resistant individuals and would slow the rate that they increased in the population (Figure 9.4).

Arpaia *et al.* (1997) examined the model results with a field test that used *L. decemlineata* as a pest and *C. maculata* as a natural enemy. Based on the behavior in small plots that mimicked Bt and non-Bt potatoes, they found that *C. maculata* aggregated in the areas of high egg density; however, this aggregation was not very strong and there was a higher percentage mortality of eggs in the non-Bt plots. Under these conditions, predation by *C. maculata* is expected to slow the rate of pest adaptation. Riddick *et al.* (1998) found no aggregation of *C. maculata* in non-Bt plots. If this resulted in higher predation rates in the Bt plots where there would be more *C. maculata* per *L. decemlineata*, it is expected that the predator would slow the rate at which resistance develops. Follow-up work by N. Mallampalli (unpublished data) indicated that when alternate prey were present in both the Bt and non-Bt plots, the predation rates on *L. decemlineata* may sometimes be the same in both types of plots, resulting in the predator having no impact on resistance evolution. The work of Riddick *et al.* (1998) on the specialist natural enemy, *L. grandis*, suggests a different evolutionary outcome. In Bt potato plots with highly reduced densities of *L. decemlineata*, no *L. grandis* were found. Therefore,

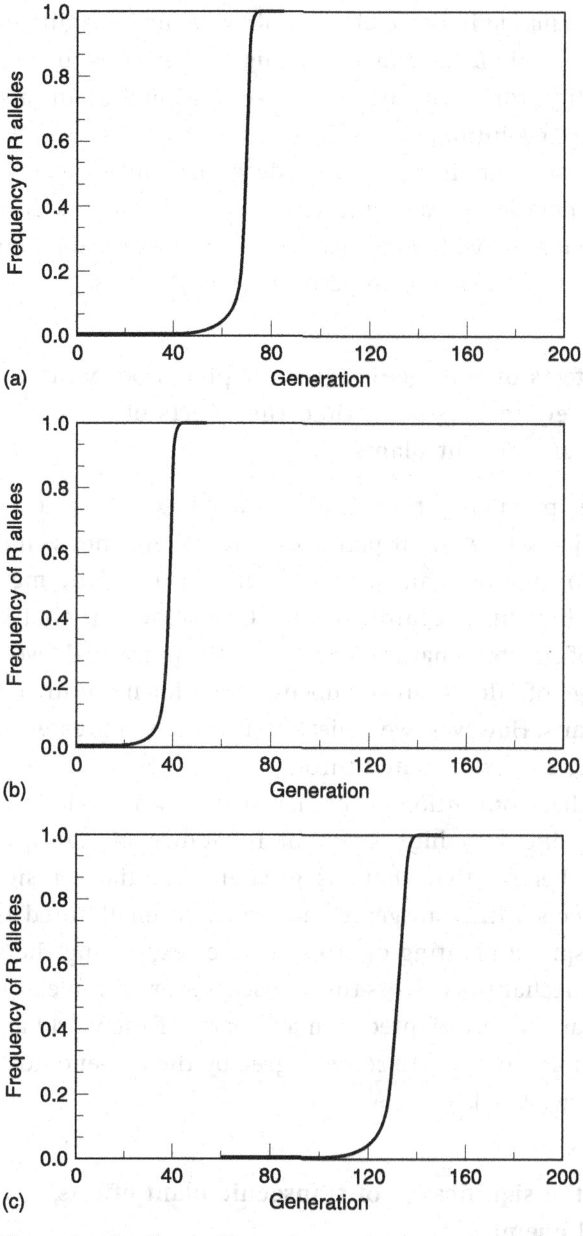

Figure 9.4. Predicted effects of natural enemies on the rate of pest adaptation to resistant crops with high levels of resistance when a refuge is present that contains pests and natural enemies. (a) There are no natural enemies. (b) Natural enemies cause twice as much mortality to pests in the refuge compared to pests that have survived on the resistant crop; this results in more rapid pest adaptation. (c) The natural enemies cause twice as much mortality to pests on the resistant crop compared with pests in the refuge; this type of natural enemy behavior slows rate of pest adaptation (from Gould, 1994).

any resistant *L. decemlineata* in those plots would have no mortality from this specialist, but susceptible *L. decemlineata* at higher densities in the refuges would have mortality from this natural enemy, resulting in an increase in the rate of resistance evolution.

The theoretical and empirical studies described above are still very preliminary. More detailed models and empirical experiments on a larger temporal and spatial scale will be needed to clarify the probable impacts of natural enemies on the rate at which pests adapt to Bt crops.

9.5 Are the effects of transgenic resistant plants on natural enemies likely to be greater than the effects of conventional resistant plants?

As indicated previously, there is no reason to expect a priori that the effects of transgenic, pest-resistant plants on natural enemies will differ in kind from effects of pest-resistant plants developed using the methods of conventional plant breeding. Regardless of the traits used, it is not likely that the type and level of effects on natural enemies at the individual level will fall outside of the range of effects already documented for naturally occurring plant resistance traits. However, we believe that there is a greater potential for population-level effects mediated through prey density with transgenic resistant varieties than conventionally bred, resistant varieties because of the emphasis on achieving very high levels of resistance in transgenic crop varieties. We also believe that there is greater potential for significant landscape-level effects with transgenic than conventionally bred resistant crops due to widespread planting of crop varieties expressing the same or related resistance mechanisms across the landscape. For example, transgenic cotton and maize are widely planted in many parts of the world and share common pests that are affected to some degree by the Bt-δ-endotoxins they express (Shelton *et al.*, 2002).

9.6 Assessing the significance of transgenic plant effects on natural enemies

It is one thing to characterize the effects of transgenic plants on natural enemies. It is quite another to determine their ecological significance and their practical consequences for crop protection unless the population-level effects are large and consistent. Effects can be identified through carefully designed laboratory and field studies. The results of laboratory studies of individual-level effects are valuable in identifying species that warrant in-depth field and modeling studies to characterize any population-level effects (Hansen *et al.*, 1999; Sibly, 1999; Stark and Banks, 2003). The significance of any effects on natural enemies that are identified should be evaluated within the contexts of pest management and ecological risk. In the context of pest management, emphasis is on the consequences of the combined effects of both resistance and natural enemies (i.e. biological control) on crop loss and profitability. In agroecosystems in which transgenic resistant crops sharing common or similar resistance traits are likely to be extensively planted, a full analysis of the effects of transgenic resistant crops must extend beyond the individual fields or crops in which the transgenic crops are planted. A landscape-level focus, which reflects the life systems (sensu Clark *et al.*, 1967) of the pests and their natural enemies is needed because of the potential for altered source/sink relationships among crops for natural enemy populations. In the context of ecological risk, the focus should again be on landscape-level, population effects on the affected natural enemy species and on the prey/host species whose populations they affect significantly. The significance of any observed effects must be evaluated relative to the impact of prevailing agricultural production and pest management practices, including the standard pesticide use practices in place in the affected agroecosystems.

References

Arpaia, S., Gould, F. and Kennedy, G. G. (1997). Potential impact of *Coleomegilla maculata* predation on adaptation of *Leptinotarsa decemlineata* to Bt-transgenic potatoes. *Entomologia Experimentalis et Applicata*, **82**, 91–100.

Barbosa, P. (1998). Agroecosystems and conservation biological control. In P. Barbosa (ed.), *Conservation Biological Control*. New York: Academic Press. pp. 39–54.

Barbosa, P. and Benrey, B. (1998). The influence of plants on insect parasitoids: implications for conservation biological control. In P. Barbosa (ed.), *Conservation Biological Control*. New York: Academic Press. pp. 55–82.

Barbosa, P., Saunders, J.A., Kemper, J. *et al.* (1986). Plant allelochemicals and insect parasitoids: effects of nicotine on *Cotesia congregata* and *Hyposoter annulipes*. *Journal of Chemical Ecology*, **12**, 1319–28.

Barbour, J.D., Farrar, R.R., Jr. and Kennedy, G.G. (1997). Populations of predaceous natural enemies developing on insect-resistant and susceptible tomato in North Carolina. *Biological Control*, **9**, 173–84.

Bell, J.V. (1978). Development and mortality of bollworms fed resistant and susceptible soybean cultivars treated with *Nomuraea rileyi* or *Bacillus thuringiensis*. *Journal of the Georgia Entomological Society*, **13**, 43–50.

Benrey, B. and Denno, R.F. (1997). The slow-growth-high-mortality hypothesis: a test using the cabbage butterfly. *Ecology*, **78**, 987–99.

Bloem, K., Kelley, K. C. and Duffey, S. S. (1989). Differential effect of tomatine and its alleviation by cholesterol on larval growth efficiency of food utilization in *Heliothis zea* and *Spodoptera exigua*. *Journal of Chemical Ecology*, **15**, 387–98.

Bottrell, D. G., Barbosa, P. and Gould, F. (1998). Manipulating natural enemies by plant variety selection and modification: a realistic strategy? *Annual Review of Entomology*, **43**, 347–67.

Bowers, M. D. and Larin, Z. (1989). Acquired chemical defense in the lycaenid butterfly, *Eumaeus atala*. *Journal of Chemical Ecology*, **15**, 133–46.

Cabrera, J. C. (2001). Interactions Between *Spodoptera frugiperda*, (Lepidoptera: Noctuidae) and Bt-Transgenic Sweet Corn. Ph.D. Thesis. Raleigh, NC: North Carolina State University.

Cheng, L. (1970). Timing of attack of *Lyphia dubia* Fall (Diptera: Tachinidae) on the winter moth, *Operophtera brumata* (L.) (Lepidoptera: Geometridae) as a factor affecting parasite success. *Journal of Animal Ecology*, **39**, 313–20.

Clark, L. R., Geier, P. W., Hughes, R. D. and Morris, R. F. (1967). *The Ecology of Insect Populations in Theory and Practice*. London: Methuen.

Cloutier, C. and Jean, C. (1998). Synergism between natural enemies and biopesticides: a test case using the stinkbug *Perillus bioculatus* (Hemiptera: Pentatomidae) and *Bacillus thuringiensis* against Colorado potato beetle (Coleoptera: Chrysomelidae). *Journal of Economic Entomology*, **95**, 1096–108.

Costa, S. D. and Gaugler, R. (1989). Influence of Solanum host plants on Colorado potato beetle (Coleoptera: Chrysomelidae) susceptibility to the entomopathogen *Beauveria bassiana*. *Environmental Entomology*, **18**, 531–6.

Cottrell, T. E. and Yeargan, K. V. (1998). Effect of pollen on *Coleomegilla maculata* (Coleoptera: Coccinellidae) population density, predation, and cannibalism in sweet corn. *Environmental Entomology*, **27**, 1402–10.

Dicke, M. (1999). Are herbivore-induced plant volatiles reliable indicators of herbivore identity to foraging carnivorous arthropods? *Entomologia Experimentalis et Applicata*, **91**, 131–42.

Duan, J. J., Head, G. and McKee, M. J. *et al.* (2002). Evaluation of dietary effects of transgenic corn pollen expressing Cry2Bb1 protein on a non-target ladybird beetle, *Coleomegilla maculata*. *Entomologia Experimentalis et Applicata*, **104**, 271–80.

Duffey, S. S. and Bloem, K. A. (1986). Plant defense–herbivore–parasite interactions and biological control. In M. Kogan (ed.), *Ecological Theory and Integrated Pest Management Practice*. New York: Wiley. pp. 135–83.

Dutton, A., Klein, H., Romeis, J. and Bigler, F. (2002). Uptake of Bt-toxin by herbivores feeding on transgenic maize and consequences for the predator *Chrysoperla carnea*. *Ecological Entomology*, **27**, 441–7.

Elliger, C. A., Wong, Y., Chan, B. G. and Waiss, A. C., Jr. (1981). Growth inhibitors in tomato (*Lycopersicon*) to tomato fruitworm (*Heliothis zea*). *Journal of Chemical Ecology*, **7**, 753–8.

Farrar, R. R., Jr. and Kennedy, G. G. (1993). Field cage performance of two tachinid parasitoids of the tomato fruitworm on insect resistant and susceptible tomato lines. *Entomologia Experimentalis et Applicata*, **67**, 73–8.

Farrar, R.R., Jr., Barbour, J.D. and Kennedy, G.G. (1994). Field evaluation of insect resistance in a wild tomato and its effect on insect parasitoids. *Entomologia Experimentalis et Applicata*, **71**, 211–26.

Farrar, R.R., Jr., Kennedy, G.G. and Kashyap, R.K. (1992). Influence of life history differences of two tachinid parasitoids of *Helicoverpa zea* (Boddie) (Lepidoptera: Noctuidae) on their interactions with glandular trichome/methyl ketone-based insect resistance in tomato. *Journal of Chemical Ecology*, **18**, 499–515.

Ferre, J. and Van Rie, J. (2002). Biochemistry and genetics of insect resistance to *Bacillus thuringiensis*. *Annual Review of Entomology*, **47**, 501–33.

Gallardo, F., Boethel, D.J., Fuxa, J.R. and Richter, A. (1990). Susceptibility of *Heliothis zea* (Boddie) larvae to *Nomuraea rileyi* (Farlow) Samson: effects of alpha-tomatine at the third trophic level. *Journal of Chemical Ecology*, **16**, 1751–9.

Gould, F. (1994). Potential and problems with high-dose strategies for pesticidal engineered crops. *Biocontrol Science and Technology*, **4**, 451–61.

Gould, F. (1998). Sustainability of transgenic insecticidal cultivars: integrating pest genetics and ecology. *Annual Review of Entomology*, **43**, 701–26.

Gould, F., Kennedy, G.G. and Johnson, M.T. (1991). Effects of natural enemies on the rate of herbivore adaptation to resistant host plants. *Entomologia Experimentalis et Applicata*, **58**, 1–14.

Groot, A.T. and Dicke, M. (2001). *Transgenic Crops in an Agro-Ecological Context: Multitrophic Effects of Insect-Resistant Plants*. Wageningen: Ponsen and Looyen BV.

Gutierrez, A.P. (1986). Analysis of the interaction of host plant resistance, phytophagous and entomophagous species. In D.J. Boethel and R.D. Eikenbary (eds.), *Interactions of Plant Resistance and Parasitoids and Predators of Insects*. Chichester, UK: Ellis Horwood. pp. 198–215.

Hagen, K.S. (1987). Nutritional ecology of terrestrial insect predators. In F. Slansky Jr. and J.G. Rodriguez (eds.), *Nutritional Ecology of Insects, Mites, Spiders, and Related Invertebrates*. New York: J. Wiley & Sons. pp. 533–77.

Hansen, F.T., Forbes, V.E. and Forbes, T.L. (1999). Using elasticity analysis of demographic models to link toxicant effects on individuals to the population level: an example. *Functional Ecology*, **13**, 157–62.

Hare, J.D. (1992). Effects of plant variation on herbivore-natural enemy interactions. In R.S. Fritz and E.L. Simms (eds.), *Plant resistance to Herbivores and Pathogens – Ecology, Evolution, and Genetics*. Chicago: University of Chicago Press. pp. 278–98.

Hare, J.D. and Andreadis, T.G. (1983). Variation in the susceptibility of *Leptinotarsa decemlineata* (Coleoptera: Chrysomelidae) when reared on different host plants to the fungal pathogen, *Beauveria bassiana* in the field and laboratory. *Environmental Entomology*, **12**, 1871–96.

Head, G., Brown, C.R., Groth, M.E. and Duan, J.J. (2001). Cry1Ab protein levels in phytophagous insects feeding on transgenic corn: implications for secondary exposure risk analysis. *Entomologia Experimentalis et Applicata*, **99**, 37–45.

Hilbeck, A. (2002). Transgenic host plant resistance and non-target effects. In D.K. Letourneau and B.E. Burrows (eds.), *Genetically Engineered Organisms: Assessing Environmental and Human Health Effects*. Boca Raton, FL: CRC Press. pp. 167–85.

Hilbeck, A. and Kennedy, G.G. (1995). Predators feeding on the Colorado potato beetle in insecticide-free plots and insecticide-treated commercial potato fields in eastern North Carolina. *Biological Control*, **6**, 273–82.

Hilbeck, A., Eckel, C. and Kennedy, G.G. (1997). Predation on Colorado potato beetle eggs by generalist predators in research and commercial plantings. *Biological Control*, **8**, 191–6.

Hilbeck, A., Baumgartner, M., Fried, P.M. and Bigler, F. (1998*a*). Effects of transgenic *Bacillus thuringiensis* corn-fed prey on mortality and development time of immature *Chrysoperla carnea* (Neuroptera: Chrysopidae). *Environmental Entomology*, **27**, 480–7.

Hilbeck, A., Moar, W.J., Pusztai-Carey, J., Filippini, A. and Bigler, F. (1998*b*). Toxicity of the *Bacillus thuringiensis* Cry1Ab toxin and protoxin and Cry 2A protoxin on the predator *Chrysoperla carnea* (Neuroptera: Chrysopidae) using diet incorporated bioassays. *Environmental Entomology*, **27**, 1255–63.

Hilbeck, A., Moar, W.J., Pusztai-Carey, J., Filippini, A. and Bigler, F. (1999). Prey-mediated effects of Cry1Ab toxin and protoxin and Cry2A protoxin on the predator *Chrysoperla carnea*. *Entomologia Experimentalis et Applicata*, **91**, 305–16.

Hodek, I. (1973). *Biology of Coccinellidae*. The Hague, Netherlands: Junk Publ.

Isenhour, D.J. and Wiseman, B.R. (1989). Parasitism of the fall armyworm (Lepidoptera: Noctuidae) by *Campoletis sonorensis* (Hymenoptera: Ichneumonidae) as affected by host feeding on silks of *Zea mays* L. cv. Zapalote Chico. *Environmental Entomology*, **18**, 394–7.

Isenhour, D.J., Wiseman, B.R. and Layton, J.R. (1989). Enhanced predation by *Orius insidiosus* (Homoptera: Anthocoridae) on larvae of *Heliothis zea* and *Spodoptera frugiperda* (Lepidoptera: Noctuidae) caused by prey feeding on resistant corn genotypes. *Environmental Entomology*, **18**, 418–22.

Johnson, M.T. (1997). Interaction of resistant plants and wasp parasitoids of *Heliothis virescens* (Lepidoptera: Noctuidae). *Environmental Entomology*, **26**, 207–14.

Johnson, M.T. and Gould, F. (1992). Interaction of genetically engineered host plant resistance and natural enemies of *Heliothis virescens* (Lepidoptera: Noctuidae) in tobacco. *Environmental Entomology*, **21**, 586–97.

Johnson, M.T., Gould, F. and Kennedy, G.G. (1997*a*). Effects of natural enemies on relative fitness of *Heliothis virescens* genotypes adapted and not adapted to resistant host plants. *Entomologia Experimentalis et Applicata*, **82**, 219–230.

Johnson, M.T., Gould, F. and Kennedy, G.G. (1997*b*). Effect of an entomopathogen on adaptation of *Heliothis virescens* populations selected on resistant host plants. *Entomologia Experimentalis et Applicata*, **83**, 121–35.

Juvik, J.A. and Stevens, M.A. (1982). Physiological mechanisms of host-plant resistance in the genus *Lycopersicon* to *Heliothis zea* and *Spodoptera exigua*, two insect pests of the cultivated tomato. *Journal of the American Society of Horticultural Science*, **107**, 1065–9.

Kartohardjono, A. and Heinrichs, E.A. (1984). Populations of the brown planthopper, *Nilaparvata lugens* (Stal) (Homoptera: Delphacidae), and its predators on rice varieties with different levels of resistance. *Environmental Entomology*, **13**, 359–65.

Kauffman, W. C. and Flanders, R. V. (1985). Effects of variably resistant soybean and lima bean cultivars on *Pediobius foveolatus* (Hymenoptera: Eulophidae) a parasitoid of the Mexican bean beetle, *Epilachna varivestis* (Coleoptera: Coccinellidae). *Environmental Entomology*, **14**, 678–82.

Kauffman, W. C., Flanders, R. V. and Edwards, C. R. (1985). Population growth potentials of the Mexican bean beetle, *Epilachna varivestis* (Coleoptera: Coccinellidae) on soybean and lima bean cultivars. *Environmental Entomology*, **14**, 674–7.

Kea, W. C., Turnipseed, S. G. and Carner, G. R. (1978). Influence of resistant soybeans on the susceptibility of lepidopterous pests to insecticides. *Journal of Economic Entomology*, **71**, 58–60.

Kennedy, G. G. (2003). Tomato, pests, parasitoids, and predators: tritrophic interactions involving the genus *Lycopersicon*. *Annual Review of Entomology*, **48**, 51–72.

Kennedy, G. G., Kishaba, A. N. and Bohn, G. W. (1975). Response of several pest species to *Cucumis melo* L. lines resistance to *Aphis gossypii* Glover. *Environmental Entomology*, **4**, 653–7.

Letourneau, D. K. (1998). Conservation biology: lessons for conserving natural enemies. In P. Barbosa (ed.), *Conservation Biological Control*. New York: Academic Press. pp. 9–38.

Lu, W., Kennedy, G. G. and Gould, F. (1996). Differential predation by *Coleomegilla maculata* on Colorado potato beetle strains that vary in growth on tomato. *Entomologia Experimentalis et Applicata*, **81**, 7–14.

Lundgren, J. G. and Wiedenmann, R. N. (2002). Coleopteran-specific Cry3Bb toxin from transgenic corn pollen does not affect the fitness of a nontarget species, *Coleomegilla maculata* DeGeer (Coleoptera: Coccinellidae). *Environmental Entomology*, **31**, 1213–18.

Maredia, K. M., Gage, S. H., Landis, D. A. and Worth, T. M. (1992). Ecological observations on predatory coccinellidae (Coleoptera) in southwestern Michigan. *Great Lakes Entomologist*, **25**, 265–70.

McCutcheon, G. S. and Turnipseed, S. G. (1981). Parasites of lepidopterous larvaea in insect resistant and susceptible soybeans in South Carolina. *Environmental Entomology*, **10**, 69–74.

Myint, M. M., Rapusas, H. R. and Heinrichs, E. A. (1986). Integration of varietal resistance and predation for the management of *Nephotettix virescens* (Homoptera: Cicadellidae) populations on rice. *Crop Protection*, **5**, 259–65.

National Research Council. (2002). *Environmental Effects of Transgenic Plants: The Scope and Adequacy of Regulation*. Washington, DC: National Academy of Sciences Press.

Nault, B. A. and Kennedy, G. G. (2000). Seasonal changes in habitat preference by *Coleomegilla maculata*: implications for Colorado potato beetle management in potato. *Biological Control*, **17**, 164–73.

Obrycki, J. J. and Tauber, M. J. (1984). Natural enemy activity on glandular pubescent potato plants in the greenhouse: an unreliable predictor of effects in the field. *Environmental Entomology*, **13**, 679–83.

Orr, D.B., Boethel, D.J. and Jones, W.A. (1985). Biology of *Telenomus chloropus* (Hemiptera: Pentatomidae) reared on resistant and susceptible soybean genotypes. *Canadian Entomologist*, **117**, 1137–42.

Pair, S.D., Wiseman, B.R. and Sparks, A.N. (1986). Influence of four corn cultivars on fall armyworm (Lepidoptera: Noctuidae) establishment of parasitization. *Florida Entomologist*, **69**, 566–70.

Pilcher, C.D., Obrycki, J.J., Rice, M.E. and Lewis, L.C. (1997). Preimaginal development, survival, and field abundance of insect predators on transgenic *Bacillus thuringiensis* corn. *Environmental Entomology*, **26**, 446–54.

Pimentel, D. and Wheeler, A.G., Jr. (1973). Influence of alfalfa resistance on a pea aphid population and its associated parasites, predators, and competitors. *Environmental Entomology*, **2**, 1–11.

Ponsard, S., Gutierrez, A.P. and Mills, N.J. (2002). Effect of Bt-toxin (Cry1Ac) in transgenic cotton on the adult longevity of four heteropteran predators. *Environmental Entomology*, **31**, 1197–205.

Price, P.W. (1986). Ecological aspects of host plant resistance and biological control: interactions among three trophic levels. In D.J. Boethel and R.D. Eikenbary (eds.), *Interactions of Plant Resistance and Parasitoids and Predators of Insects*. Chichester, UK: Ellis Horwood. pp. 11–30.

Rabb, R.L. and Bradley, J.R., Jr. (1968). The influence of host plants on parasitism of eggs of the tobacco hornworm. *Journal of Economic Entomology*, **61**, 1249–52.

Rausher, M.D. (1988). Is coevolution dead? *Ecology*, **69**, 898–901.

Reed, G.L., Jensen, A.S., Riebe, J., Head, G. and Duan, J.J. (2001). Transgenic Bt potato and conventional insecticides for Colorado potato beetle management: comparative efficacy and non-target impacts. *Entomologia Experimentalis et Applicata*, **100**, 89–100.

Riddick, E.W., Dively, G. and Barbosa, P. (1998). Effect of a seed-mix deployment of Cry3A-transgenic and nontransgenic potato on the abundance of *Lebia grandis* (Coleoptera: Carabidae) and *Coleomegilla maculata* (Coleoptera: Coccinellidae). *Annals of the Entomological Society of America*, **91**, 647–53.

Riggin-Bucci, T.M. and Gould, F. (1997). Impact of intraplot mixtures of toxic and nontoxic plants on population dynamics of diamondback moth (Lepidoptera: Plutellidae) and its natural enemies. *Economic Entomology*, **90**, 241–51.

Romeis, J., Badendreier, D. and Wackers, F.L. (2003). Consumption of snowdrop lectin (*Galanthus nivalis* agglutinin) causes direct effects on adult parasitic wasps. *Oecologia*, **134**, 528–36.

Saito, C.E., Eikenbary, R.D. and Starks, K.J. (1983). Compatibility of *Lysiphlebus testaceipes* (Homoptera: Aphididae) biotypes "C" and "E" reared on susceptible and resistant oat varieties. *Environmental Entomology*, **12**, 603–4.

Salim, M. and Heinrichs, E.A. (1986). Impact of varietal resistance in rice and predation on the mortality of *Sogatella furcifera* (Horvath) (Homoptera: Delphacidae). *Crop Protection*, **5**, 395–9.

Schuler, T.H., Potting, R.P.J., Denholm, I. and Poppy, G.M. (1999). Parasitoid behavior and *Bacillus thuringiensis* plants. *Nature*, **400**, 825–6.

Shelton, A. M., Zhao, J. Z. and Roush, R. T. (2002). Economic, ecological, food safety, and social consequences of the deployment of Bt transgenic plants. *Annual Review of Entomology*, **47**, 845–81.

Sibly, R. M. (1999). Efficient experimental designs for studying stress and population density in animal populations. *Ecological Applications*, **9**, 496–503.

Smiley, J. T., Horn, J. M. and Rank, N. E. (1985). Ecological effects of salicin at three trophic levels: new problems from old adaptations. *Science*, **229**, 649–51.

Smith, R. C. (1960). A technique for rearing coccinellid beetles on dry foods, and influence of various pollens on the development of *Coleomegilla maculata lengi* Timb. (Coleoptera: Coccinellidae). *Canadian Journal of Zoology*, **38**, 1047–9.

Stark, J. D. and Banks, J. E. (2003). Population-level effects of pesticides and other toxicants on arthropods. *Annual Review of Entomology*, **48**, 505–19.

Starks, K. J., Muniappan, R. and Eikenbary, R. D. (1972). Interaction between plant resistance and parasitism against the greenbug on barley and sorghum. *Annals of the Entomological Society of America*, **65**, 650–5.

Thaler, J. S. (1999*a*). Induced resistance in agricultural crops: effects of jasmonic acid on herbivory and yield in tomato plants. *Environmental Entomology*, **28**, 30–7.

Thaler, J. S. (1999*b*). Jasmonate-inducible plant defenses cause increased parasitism of herbivores. *Nature*, **399**, 686–8.

Thorpe, K. W. and Barbosa, P. (1986). Effects of consumption of high and low nicotine tobacco by *Manduca sexta* on survival of gregarious endoparasitoid *Cotesia congregata*. *Journal of Chemical Ecology*, **12**, 1329–37.

Traugott, M. S. and Stamp, N. E. (1996). Effects of chlorogenic acid and tomatine-fed caterpillars on the behavior of an insect predator. *Journal of Insect Behavior*, **9**, 461–76.

US Environmental Protection Agency. (1998). *The Environmental Protection Agency's White Paper on Bt Plant-Pesticide Resistance Management*. #739-S-98-001. Washington, DC: Author.

van Emden, H. F. (1966). Plant insect relationships and pest control. *World Review of Pest Control*, **5**, 115–23.

van Emden, H. F. (1986). The interaction of plant resistance and natural enemies: effects on populations of sucking insects. In D. J. Boethel and R. D. Eikenbary (eds.), *Interactions of Plant Resistance and Parasitoids and Predators of Insects*. Chichester, UK: Ellis Horwood. pp. 138–50.

Vet, L. E. M. and Dicke, M. (1992). Ecology of infochemical use by natural enemies in a tritrophic context. *Annual Review of Entomology*, **37**, 141–72.

Vinson, S. B. and Barbosa, P. (1987). Interrelationships of nutritional ecology of parasitoids. In F. Slansky and J. G. Rodriguez (eds.), *Nutritional Ecology of Insects, Mites, and Spiders and Related Invertebrates*. New York: J. Wiley & Sons. pp. 673–95.

Wright, E. J. and Laing, J. E. (1982). Stage specific mortality of *Coleomegilla maculata lengi* Timberlake on corn in southern Ontario. *Environmental Entomology*, **11**, 32–7.

Wyatt, I.J. (1983). The distribution of *Myzus persicae* (Sulz.) on year-round chrysanthemums. II. Winter season: the effect of parasitization by *Aphidius matricariae* Hal. *Annals of Applied Biology*, **65**, 31–41.

Yanes, J., Jr. and Boethel, D.J. (1983). Effect of a resistant soybean genotype on the development of the soybean looper (Lepidoptera: Noctuidae) and an introduced parasitoid, *Micropletis demolitor* Wilkinson (Hymenoptera: Braconidae). *Environmental Entomology*, **12**, 1270–4.

10

Modeling the dynamics of tritrophic population interactions

A. P. GUTIERREZ AND J. BAUMGÄRTNER

10.1 Introduction

Increasingly, population modeling and systems analysis are being used to examine the complex issues that are at the heart of CP/IPM (crop production and integrated pest management) and biological control. The design of economically sound and sustainable crop management strategies requires a thorough understanding of the whole production system including arthropod pests, pathogens, and weeds. More than three decades ago, Huffaker and Croft (1976) stressed the need to rely on systems analysis and interdisciplinary collaboration to accomplish this task. Soon the question arose as to how mathematical techniques employed in the analysis of physical systems could be adapted to solve agroecosystem problems that are principally biological in nature and that focus on population management (Gutierrez and Wang, 1977; Getz and Gutierrez, 1982). Simple models of population dynamics often excluded the biological details for mathematical tractability and hence are frequently inadequate instruments for field application. Individual-based models have been used to explore population interactions (e.g. De Angelis and Gross, 1992), but often the rules for the interactions at the individual level are unknown. Simulation approaches stress biological realism and completeness and some show promise for exploring system structure and function, especially physiologically based models (PBM) (Gutierrez and Wang, 1977), sufficient to gain insights into complex quantitative relationships (see Gilbert *et al.*, 1976; Gutierrez and Baumgärtner, 1984*a*; *b*; Graf *et al.*, 1990a; Gutierrez, 1996; Di Cola *et al.*, 1998). In this chapter we will consider only physiologically based multitrophic population dynamics models, or models with the potential to be so extended.

Most of the models we review concern resource-based predator–prey relationships in a general sense and all have three essential components: (a) models for the dynamics of age-mass structured populations (Gutierrez, 1996; Di Cola *et al.*, 1998; Baumgärtner, 1999); (b) functional response models describing the biology of resource acquisition (Hassell *et al.*, 1976); (c) numerical response models dealing with resource allocation including an accounting of predator birth/growth and intrinsic death rates and death rates due to higher trophic levels (Lawton *et al.*, 1975; Beddington *et al.*, 1976; Gutierrez, 1996; Mills and Gutierrez, 1996; Hawkins and Cornell, 1999; Baumgärtner, 1999). Energy is the major currency linking trophic interactions, including the economic one, where the goal is to maximize the capture of the energy fixed as yield and to protect it from pests using environmentally friendly IPM methods. To be useful, systems models must consider both bottom-up and top-down factors affecting each component population (see McQueen and Post, 1988), for, as Mills and Getz (1996) emphasized, "... host parasitoid relationships never take place in isolation ... as variable plant resource may have as great an influence on the pest population as do the action of the parasitoid population" (e.g. Gutierrez *et al.*, 1988*a*; 1994). This idea is older (see Gutierrez and Wang, 1977; Gutierrez and Baumgärtner, 1984*a*; *b*) and has special relevance to IPM. The models used in IPM must be at least tritrophic if we are to develop holistic IPM strategies where natural mortality and/or that due to host plant resistance replace pesticides. This is further complicated because weather and time varying edaphic factors and spatial considerations modify the field dynamics of plant–herbivore–natural enemy interaction, making it difficult to evaluate them using standard field experiments and common static statistical methods that are designed to detect significant differences and trends. Some of the components required to model agroecosystems are depicted in Figure 10.1 and include bionomics, behavior and physiology, population level sampling, and analysis with modeling and geographic information systems (GIS) being the integrating tools. Remote sensing and geopositioning may provide weather data complementary to weather station records and allow geo-referenced positioning of observations. Together with GIS they are known as geographic information technology (GIT) that is increasingly used in ecosystem analysis work.

The question of how to model field tritrophic systems is an ongoing debate and controversy that is beyond the scope of this chapter. The spirit and at times hostility of the debate were captured by Gilbert *et al.* (1976), Gilbert (1984), and Lawton (1977), though considerable middle ground has been gained in recent years. Multitrophic population interactions have been

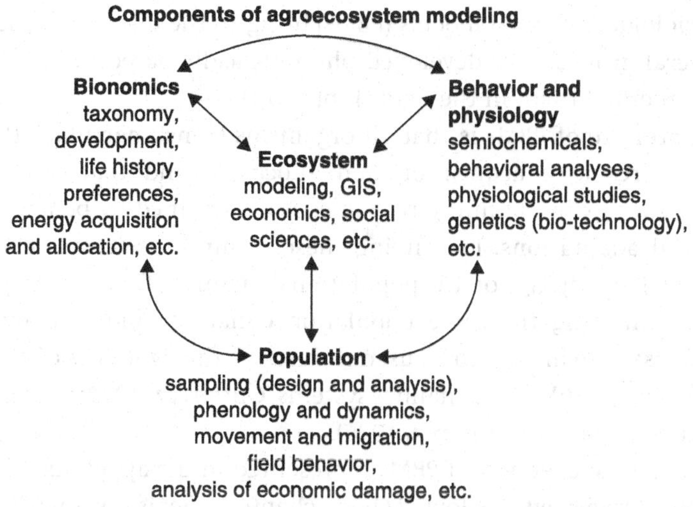

Figure 10.1. Components of agroecosystem research and modeling.

analyzed by Gutierrez and Baumgärtner (1984*a*; *b*) and discussed in the context of bottom-up and top-down population regulation (Gutierrez *et al.*, 1994) and more generally, in the context of food web dynamics (Berryman *et al.*, 1995). The importance of indirect effects of predation on prey assemblages has become a major issue in ecology and experimental work has been undertaken to test the relevant hypotheses (e.g. Spiller and Schoener, 2001). Turchin and Batzli (2001) studied multitrophic interactions involving mammal populations and reviewed, as done previously in Berryman *et al.* (1995), the different methodologies used for modeling population interactions (Baumgärtner, 2001). Of particular interest to us is whether there are ecological principles or constructs that can be considered in model development (see Peters, 1991). For example, are thermodynamics principles relevant (Curry and Feldman, 1987; Severini *et al.*, 1990; 1996) or sufficient for model development? (see Jorgensen, 2001). Most important, the question of how to embed the detail outlined in Figure 10.1 in a population dynamics model is at the heart of the controversy. A major point is whether trophic interactions should be viewed from the prey point of view (i.e. the traditional way) or from the consumer point of view (Berryman and Gutierrez, 1999) as is done here using demand driven models (see below). Most models use population density (numbers) to model the interaction between consumer and resource species ignoring important age, size, and other attributes that change with time (see Gutierrez and Wang, 1977; Gurney *et al.*, 1996; Di Cola *et al.*, 1998). To be really useful, the models must consider such attributes and be independent of time, place, and circumstance, and this is what physiologically based models (PBMs) bring to the discussion (Gutierrez 1996). Several groups have developed physiologically based models but few have implemented them in the field (Table 10.1).

A basic premise of PBMs is that all organisms (consumers) face the same problems of resource acquisition and allocation and that much of an organism's success in obtaining resources is determined by behavioral and physiological adaptations. Identifying these common processes and determining how they impinge on the population dynamics of species simplify the problem by allowing the same population dynamics and functional and numerical response models to be used to describe the dynamics of all species in multitrophic population dynamics systems (Gutierrez, 1992; Gutierrez and Baumgärtner, 1984*a*; *b*; Gutierrez, 1996).

A mathematical overview of PBMs is presented in the appendix with some applications reviewed below. Our chapter deals primarily with poikilothermic organisms whose demands for resources are determined by their age, sex, and other physiological states and the set of abiotic conditions they are experiencing, and temperature-dependent respiration rates in resource allocation processes. Conceptually, the same model has been extended to assess the human or economic trophic level (Regev *et al.*, 1998).

10.2 A conceptual basis for physiologically based tritrophic models (PBMs)

Populations have age, mass, and other kinds of population structure and energy flow links trophic interactions. A plant population (i.e. a crop) may be viewed as consisting of individual plants each containing populations of plant subunits (leaves, stem, root, and fruit (= buds, flowers, growing fruits) having age structure (Harper and White, 1974). Within a plant, the subunit populations are linked by the flow of nutrient fixed during photosynthesis (de Wit and Goudriaan, 1978; Penning De Vries *et al.*, 1989; Gutierrez and Wang, 1977). Similarly, a population of arthropods may have males and females with the latter viewed as individuals each with a developing subpopulation of reproductive structures (ova, embryos) within. The development of the subpopulations in plants and animals is determined by the resource acquisition success of the parent (Gutierrez and Wang, 1977). Simpler or more complex forms of reproduction and growth may be viewed as modifications of this basic scheme. The trophic interactions between plant and herbivore or herbivores and natural enemies reflect all forms of resource acquisition. An herbivore may attack whole plants or plant parts as if a true predator or merely some ages of plant subunit populations (e.g. young leaves,

Table 10.1. *A partial list of physiologically based models used in field IPM and biological control*

System	Trophic levels	Major result
Alfalfa		
Gutierrez *et al.*, 1976	2*	Alfalfa weevil phenology and dynamics and the development of IPM strategies.
Gutierrez and Baumgärtner, 1984*a*; *b*	4*	Interaction of pea and blue alfalfa aphids and their natural enemies. First use of distributed maturation time models in tritrophic systems.
Ruggle and Gutierrez, 1995	3*	Spotted alfalfa aphid and its parasitoid *Tryoxys utilis explaining the lack of SAA resistance to alfalfa* genotypes.
Apple		
Baumgärtner *et al.*, 1986	2*	Model for apple tree developed.
Baumgärtner and Severini, 1987	2*	Apple leafminer – microclimate effects on its population dynamics examined.
Graf *et al.*, 1985	2	Dynamics of apple aphids.
Zahner and Baumgärtner, 1988	2*	Analysis of plant–spider mite interactions
Wermelinger *et al.*, 1990*a*; *b*; 1991*a*; *b*		Linked two-spotted spider mite responses to N and water stress
Cabbage		
Johnsen *et al.*, 1997*b*	2*	Diapause induction and termination in cabbage rootfly.
Cassava		
Gutierrez *et al.*, 1988*a–c*; 1993; Bonato *et al.*, 1994*a*; *b*	4*	Detailed analysis of the factors responsible for the biological control of mealy bug and green mite.

Table 10.1. (*cont.*)

System	Trophic levels	Major result
		Predictions about the role of nutrition on biological control of both pests.
Gutierrez *et al.*, 1999	4*	Metapopulation model of the ten species cassava system.
Gutierrrez *et al.*, 1994; Schreiber and Gutierrez, 1998	3	Analytical models of tritrophic systems.
Coffee		
Gutierrez *et al.*, 1998	3*	Prospective modeling of the effects of biological control agents. Redirected research efforts of large national programs.
Cotton		
Gutierrez and Ponsard, 2005; Gutierrez *et al.*, 2005	2*	Evaluated the effects of temporal refuges in Bt cotton on the development of resistance in pink bollworm.
Gutierrez *et al.*, 1975	2*	First demographic model of plant–herbivore system.
Gutierrez *et al.*, 1977*a*; *b*; 1991*a*; *b*	2*	Studies on *lygus* bug, defoliators, pink bollworm and bollweevil. The effects of pests on plant compensation and yield and the development of IPM strategies.
Gutierrez *et al.*, 1982	2*	Model of verticillium wilt in cotton.
Gutierrez *et al.*, 2007	2*	GIS/model of climate on geographic range of pink bollworm.
Mills and Gutierrez, 1996	3*	Showed the dangers of introducing heteronomous parasitoids for whitefly control.

Stone and Gutierrez, 1986*a*; *b*; Stone *et al.*, 1986	2*	Evaluated the efficacy of pheromones for the control of pink bollworm.
von Ark *et al.*, 1984	3*	Demonstrated the effects of pesticides on outbreaks of whitefly.
Wang *et al.*, 1977	2*	Model of cotton–bollweevil interaction.
Wang and Gutierrez, 1980	2*	Analysis of simple versus complex models.
Cowpea		
Tamo and Baumgärtner, 1993; Tamo *et al.*, 1993	3*	Model of cowpea, thrips, and an egg parasitoid.
Grape		
Cerutti *et al.*, 1989, 1990; 1991*a*; *b*	3*	Interactions between leafhopper and egg parasitoid and *the role of alternate hosts examined*.
Gutierrez *et al.*, 1985	1*	Plant dynamics.
Wermelinger *et al.*, 1990*a*; *b*; 1991*a*; *b*; 1992	2*	(a) Plant dynamics demonstrating, (b) the effects of plant nitrogen on mite dynamics.
Maize		
Bonato *et al.*, 1999	1*	Plant dynamics.
Meikle *et al.*, 1998; 1999	2	Dynamics of larger grain borer in stored grain.
Mosquito		
Fouque *et al.*, 1991, 1992; Fouque and Baumgärtner, 1996	1	Mosquito time-varying life table.
Olive		
Rochat and Gutierrez, 2001	3*	Retrospective analysis showing the effects of weather on regulation of olive scale by two parasitoids.
Gutierrez *et al.*, in press	3*	Retrospective analysis showing the effects of competition between a parasitoid and a coccinelid predator.

Table 10.1. (*cont.*)

System	Trophic levels	Major result
Rice		
Graf *et al.*, 1990*a*; *b*	2*	Plant–weed competition.
Graf *et al.*, 1992	2*	Plant–stem borer interactions.
Wu and Wilson, 1997	2*	Plant–rice weevil interactions.
d'Oltremont and Gutierrez, 2002*a*; *b*	5+*	Linkage of multitrophic fishpond-flooded rice systems – the effects of nutrient recycling on the components of the system–object-oriented approach.
Wheat		
Holst and Ruggle, 1997; Holst *et al.*, 1997	3*	Wheat aphid and natural enemy interactions. Generalized the metabolic pool approach using an object-oriented model.
Yellow starthistle		
Gutierrez, *et al.*, 2005*a*	2*	Prospective analysis of the potential of seven introduced species for biological control of the weed.

*Includes the plant as the base system.

growing fruit, etc). An herbivore may oviposit in a plant structure as if a parasitoid, or it consumes the whole structure (predation). These terms seem more familiar when used to describe relationships at higher trophic levels, but regardless of trophic level, all fall under the ambit of predator–prey interactions. Intra- or interspecific interactions among individuals may facilitate resource acquisition, or individuals may compete for resources. However, the functional response components of the population models must take into account both competition and mutualism.

All organisms have a maximum capacity to assimilate resource(s) under particular abiotic conditions (i.e. a genetic demand), and all seek to meet this demand to satisfy the genetic imperative of maximizing fitness (see Berryman and Gutierrez, 1999 for contrasting views). The demand rate is a major parameter of our model and may be viewed as the sum of the outflow from the metabolic pool model under non-limiting conditions (Figure 10.2, see Petrusewicz and McFayden, 1970; Batzili, 1974; Gutierrez and Wang, 1977; Holst and Ruggle, 1997). The allocation priorities of resources acquired (supply) are analogous in all organisms: first to egestion (gutation by plants or ridding itself of excess sugars or compounds might be analogous processes to egestion in animals or carnivorous plants), then and after correction for

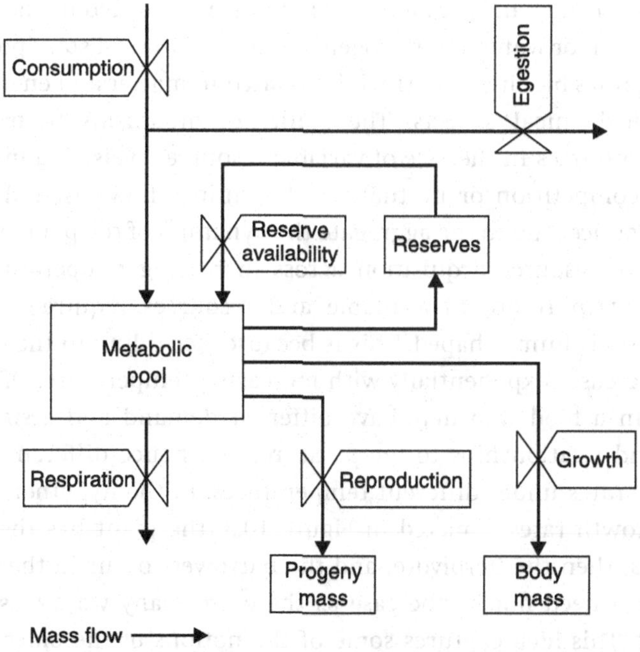

Figure 10.2. The metabolic pool model depicting the rate of resource acquisition and allocation.

conversion costs, allocation to respiration, then to age stage-dependent reproduction, and last to growth across all trophic levels. The level of physiological detail one wishes to incorporate depends on the question to be answered. Examples of the metabolic pool approach in modeling the development of individuals are found in Gutierrez *et al.* (1981) and Baumgärtner *et al.* (1987).

The function describing the resource acquisition (the functional response of predators) has a convex relationship to resource (prey) density. Note that the supply in this sense is what the organism acquires rather than the quantity of resources available, and hence differs from the economic notion of supply. This function applies to photosynthesis per unit of leaf on level of light, to nitrogen and water uptake by roots on water and nitrogen levels in the root zone or more commonly to consumption by herbivores, predators, and parasitoids (Gutierrez *et al.*, 1988*b*; Gilioli *et al.*, 2005). The resource acquisition analogy can be extended to the search for mates, nesting sites, and other resources (Gutierrez *et al.*, 1994). All of these resource acquisition processes have components of resource availability, temperature-dependent and resource-controlled demand for the resource by all consumers, competition, and aspects of search. The cassava system model exemplifies the structure of resource acquisition functions (Gutierrez *et al.*, 1988*a–c*; Bonato *et al.*, 1994*a*; *b*): the demand for biomass or energy is modified by the ratio of supply to demand for water and nitrogen. From a theoretical standpoint, the structure combines biophysical principles related to matter and energy with nutritional biochemical aspects. The ability of organisms to meet the demands for resources in the face of variable resource levels, and inter- and intra-specific competition or mutualism, determines how fast individuals grow and reproduce, and in the aggregate the dynamics of the population. At a given level of resource acquisition across increasing temperatures, the difference between resources available and resources required to cover respiration costs is hump-shaped. This is because, according to the Q_{10} rule, respiration increases exponentially with increasing temperature. Of course, each species in a food web may have different demand and assimilation parameters and relationships to temperature, and hence different growth (reproduction) rates under different temperatures. In the hypothetical relationship of growth rates depicted in Figure 10.3, the plant has the widest thermal limits, then the herbivore, and the narrowest occur in the natural enemy, but this need not be the case as there are many ways to skin the proverbial cat. This idea captures some of the notions of the older physiological ecology (e.g. Shelford, 1931; Fitzpatrick and Nix, 1970; Gutierrez *et al.*, 1974; Suthurst *et al.*, 1991) as well as applications to modern issues concerning

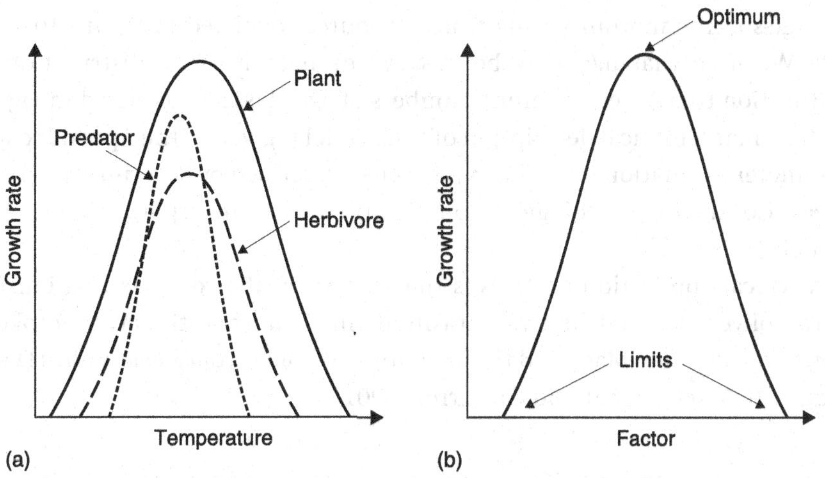

Figure 10.3. Conceptual illustration of poikilotherm growth rates: (a) the different responses of a plant, herbivore, and predator to temperature assuming fixed levels of resource for each, and (b) a general response to the level of any factor.

global warming (Gutierrez, 2001). The ratio of supply (what the organism gets) to demand (what it could assimilate) is a measure of the organism's and in the aggregate the population's success in obtaining resources. Shortfalls reduce maximal birth, growth, and intrinsic survivorship rates (Gutierrez and Baumgärtner, 1984*a*; *b*) but may also affect net emigration rates as populations with large shortfall loose members and those with good resources gain them (a numerical response due to immigration, Gutierrez *et al.*, 1999). These concepts are the basis for much of what follows below.

The dynamics of a system may be modeled at many biological levels (individual, cohort, population or metapopulation). We primarily focus on the population level and seek an explanation based on individual and cohort levels with the significance evaluated at the ecosystem level. The von Foerster (1959) model and its discrete (Leslie, 1945) and stochastic variants (Vansickle, 1977; Severini *et al.*, 1990; 1996) have been used to represent the temporal dynamics of populations with time-varying age structures. However, to include the time-varying dynamics of other population attributes (e.g. energy or biomass, nitrogen content), the discrete (Wang *et al.*, 1977) and stochastic variants (e.g. Gutierrez and Baumgärtner, 1984*b*) of the Sinko and Streifer (1967) model have been used (see Gutierrez, 1996; Di Cola *et al.*, 1998). More recently, the time distributed maturation time model of Vansickle (1977) and its variants have found wide application. In the Vansickle model, births enter the first age class ($x(t)$) and exit as death in the last age class ($y(t)$) while aging between age classes occurs via the transition rates $r_i(t)$ with the net age-specific mortality time from all sources depicted in Figure 10.4a as $\mu_i(t)$. We note that $\mu_i(t)$ may be positive or negative. The distribution of maturation time given different numbers of age classes is depicted in Figure 10.4b. A mathematical description of this model is given in the Appendix, and parameter estimation procedures have been described by Severini *et al.* (1990; 1996). Considerable biological complexity can be incorporated into this model.

A typical application of PBMs is the analysis of the complex relationship of the olive scale and its two parasitoids in California: the arrhenotokous *Aphytis maculicornis* (Masi) and the adephoparasitoid *Coccophagoides utilis* Doutt (Figure 10.5, see Rochat and Gutierrez, 2001).

10.3 Review of applications to crop systems of physiologically based models (PBMs)

It is impossible to review the complete literature on multitrophic models and hence we restrict ourselves to PBMs having demographic properties that have had field applications (Table 10.1). Some of the models are retrospective analyses and others are prospective analyses in biological

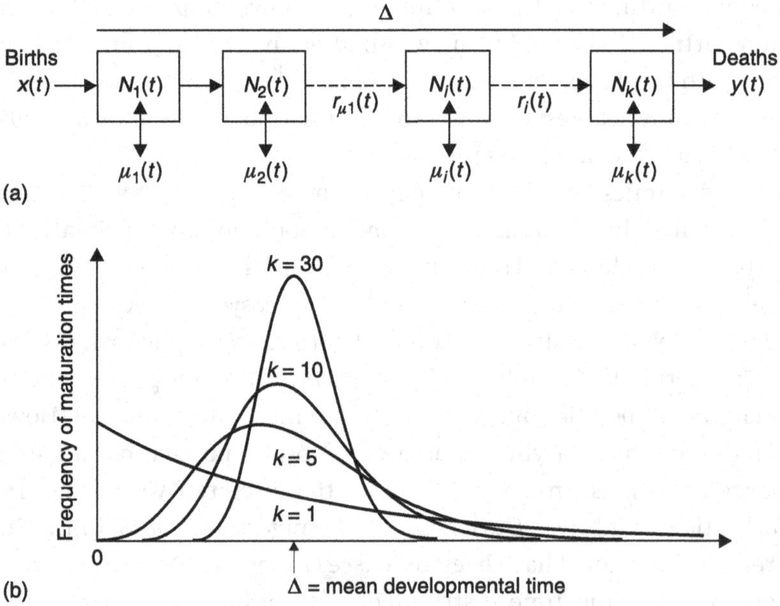

Figure 10.4. The distributed maturation time model: (a) the birth (x(t)), net age (i) specific death ($\mu_i(t)$) rates and (b) the distribution of maturation times with varying numbers of age classes (see Appendix for details).

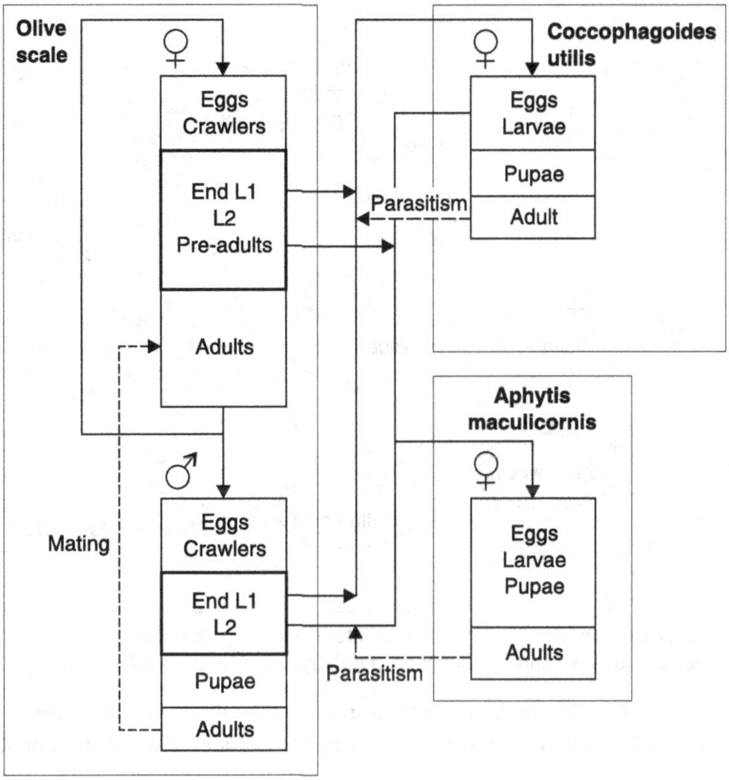

Figure 10.5. The biology of olive scale, and two of its parasitoids (*Aphytis maculicornis* and *Coccophagoides utilis*) (see Rochat and Gutierrez, 2001).

control and IPM. Also included are recent studies dealing with the impact of transgenic crops on pest and natural enemy populations, on resistance management and the efficacy of the technology. Physiologically based models build on the early work of Gilbert *et al.* (1976) and the philosophically allied plant modeling of de Wit and Goudriaan (1978). However, only six crops out of the many cases listed in Table 10.1 will be discussed in this section.

10.3.1 *Alfalfa*

There are more than 1500 species of alfalfa (R. F. Smith, unpublished), but, as proposed by Gilbert *et al.* (1976), most species do not interact except indirectly via the plant which integrates weather and edaphic effects as well as the feeding of herbivores attacking its different subunits. Some of the earliest tritrophic models were developed for aphids and alfalfa weevil in alfalfa, and in all respects follow the methods outlined in the appendix. The interactions of the species are depicted in Figure 10.6 (see Baumgärtner *et al.*,

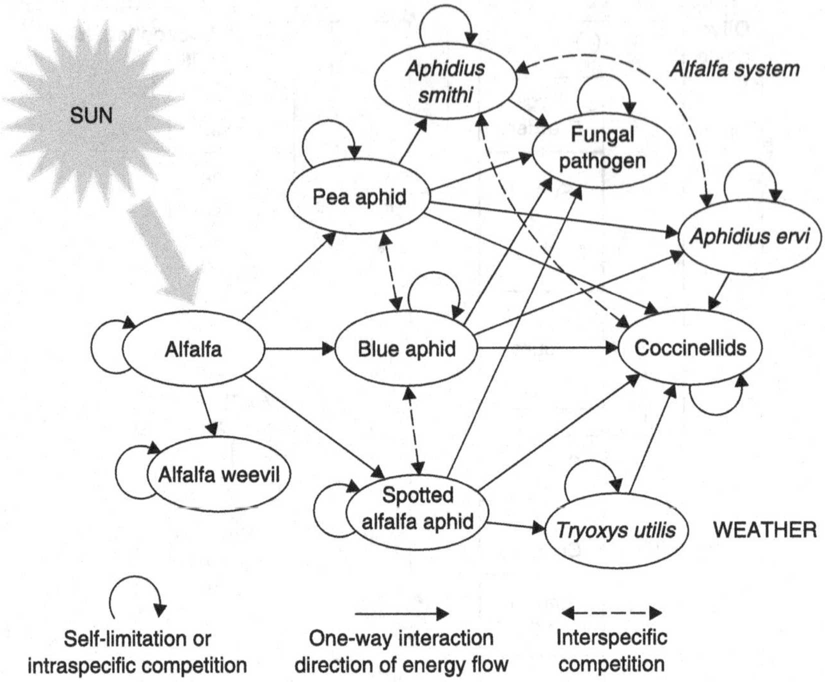

Figure 10.6. Tritrophic interactions in California alfalfa of alfalfa weevil, three aphids (pea aphid, blue aphid, and spotted alfalfa aphid) and their natural enemies.

1981*a*; *b*; Gutierrez *et al.*, 1976, Gutierrez and Baumgärtner, 1984*a*; *b*; Ruggle and Gutierrez, 1995). The relationships are complex but were easily assembled using the methods outlined in the Appendix. Here, we focus on the interactions of pea aphid (*Acyrthosiphon pisum* Harris) and blue alfalfa aphid (*A. kondoii* Shinji) and their natural enemies.

In alfalfa (*A*), pea aphid (*P*) is a superior competitor to blue aphid but is attacked by an effective host-specific parasitoid, *Aphidius smithi* (*S*), that is a superior competitor to a second parasitoid *A. ervi* (*E*) that attacks both pea and blue aphids. This latter parasitoid prefers pea aphid. Coccinellid beetles attack both aphids in a frequency dependent manner depressing their populations. A fourth factor is the aphid pathogen *Pandora neoaphidis* (*F*) that is most active during wet periods and infects pea aphid at 12× greater rate than blue aphid (Pickering and Gutierrez, 1991). All of these species, including alfalfa, are exotic and their order of introduction was *A,P,E,S,B,F*. Pea aphid was a serious pest but has become considerably less common after the introduction of blue aphid as can be explained using the succession diagram in Figure 10.7 starting with alfalfa (*A*) in our diagram (Schreiber and Gutierrez, 1998). Upper case letters indicate dominance and lower case letters

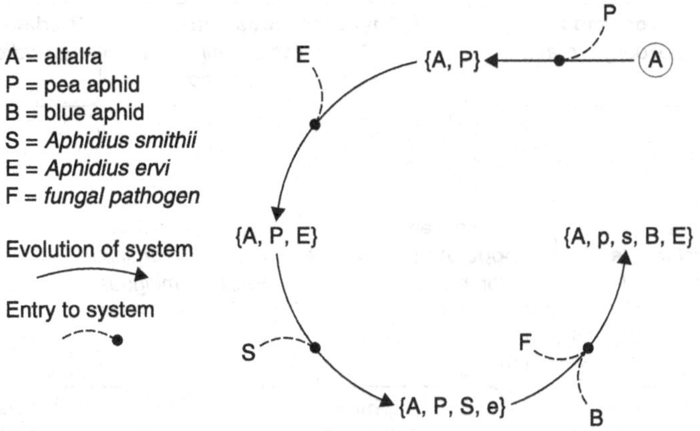

Figure 10.7. An assembly diagram of pea aphid and blue aphid and their natural enemies (after Schreiber and Gutierrez, 1998). Capital letters imply dominance.

indicate persistence of the species but at low densities. The end result is that given the presence of the pathogen (*F*), the less competitive blue aphid and *A. ervi* dominate (upper case), while pea aphid and *A. smithi* (lower case) become less common due to the greater susceptibility of pea aphid to the pathogen and the parasitoids.

10.3.2 *Apple*

Many models have been developed for representing the phenology and growth processes of apple trees as well as for orchard inhabiting diseases and arthropod populations (Winter, 1986; Baumgärtner *et al.*, 1988*b*; Anderson, 1990; Habib and Blaise, 1996; Wagenmakers *et al.*, 1999). The extension phenology modeling system PETE (Welch *et al.*, 1978; Croft and Knight, 1983; Welch, 1984) based on Manetsch's (1976) time distributed delay models has been widely used to model arthropod phenologies for the timing of pest control operations. It was used, for example, to predict the phenologies of economically important tortricids in both Swiss (Baumgärtner *et al.*, 1988*a*) and Italian (Baumgartner and Baronio, 1989; De Berardinis *et al.*, 1993) orchards. To account for mortality, the Vansickle (1977) model was used to study the effect of apple leaf burial by earthworms and of microclimate on the dynamics of leaf miner *Phyllonorycter blancardella* F. populations, factors having a profound influence on the flight patterns of the pest (Baumgärtner and Severini, 1987; Baumgärtner *et al.*, 1990*a*).

The development of apple orchards and the interactions of arthropod populations in them are summarized by Baumgärtner *et al.* (1990*a*; *b*). A physiologically based model for Golden Delicious apple orchards

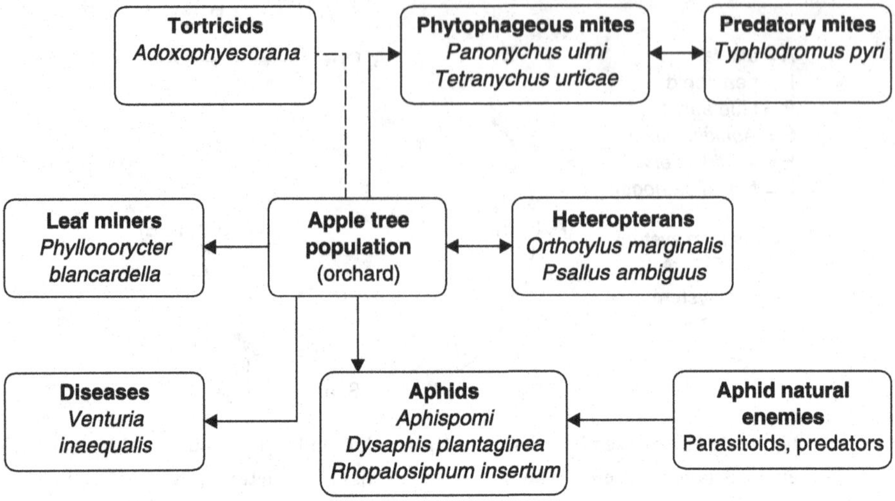

Figure 10.8. Tritrophic interactions in a Swiss apple orchard system.

(Baumgärtner *et al.*, 1986) was combined with population systems composed of aphids (Graf *et al.*, 1985), heteropterans (Schaub and Baumgärtner, 1989) and mites (Zahner *et al.*, 1988; Baumgärtner *et al.*, 1989). The models allowed the separation of the effects of weather, host plant quality, biological control factors and management procedures, including chemical control on population dynamics (Baumgärtner and Zahner, 1984). Among the host plant effects was vegetative growth pattern influence, as controlled by fruit set, on apple scab *Venturia inaequalis* (Cke.) Wint. infestations and on biological control of spider mites (Baumgärtner *et al.*, 1989). Reduced fruit set enhances vegetative growth that increases disease incidence and decreases the activity of spider mite predators. As a consequence, spider mite populations reach higher densities that declined slowly to overwintering densities. This result was considered as one of the possible explanations for the breakdown of natural control of spider mites in Swiss apple orchards during some seasons.

10.3.3 Cassava

Several models for cassava plant growth and development have been developed. Cock *et al.* (1985) used the "spillover hypothesis" for resource allocation wherein cassava root (tuber) filling begins after all other demands are satisfied. Gutierrez *et al.* (1998) used a physiologically based demographic modeling approach whereby allocation to competing demands was satisfied allometrically based on the metabolic pool model. The model of Mathews and Hunt (1994) used empirical relationships to solve the allocation problem while Gray (2000) proposed a mechanistic solution for the allocation problem.

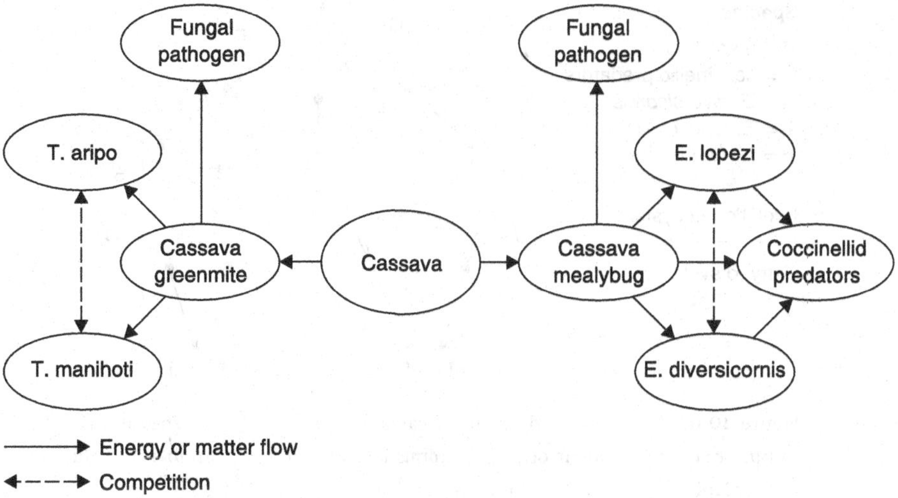

Figure 10.9. The tritrophic sub-Saharan cassava system.

Of these, only the models of Gutierrez *et al.* (1988*a–c*; 1993; 1999) and Bonato *et al.* (1994*a*) have demographic properties that can be easily extended to include the higher trophic level dynamics of the cassava mealybug, *Phenacoccus manihoti* Mat.-Ferr. (CM), and cassava greenmite, *Mononychellus tanajoa* (Bondar) (CGM). Neuenschwander *et al.* (1987) outlined the cassava food web that is larger than depicted in Figure 10.9.

Cassava mealybug (CM)

Gutierrez *et al.* (1988*a*; 1993) developed a unified ratio-dependent supply/demand-driven tritrophic model of the cassava system and used it to explore the basis for the successful control of cassava mealybug (CM) in Africa by the exotic arrhenotokous parasitoid, *Epidinocarsis lopezi* DeSantis, and the failure of a related parasitoid, *E. diversicornis* (Howard), to establish (Herren and Neuenschwander, 1991). The model predicted that the functional and numerical responses of the parasitoids proved insufficient to explain the observed dynamics of the mealybug. However, rainfall and its enhancement of the fungal pathogen *Neozyites fumosa* suppressed CM numbers sufficiently during the wet season for the parasitoid *E. lopezi* to regulate CM density at low levels, when soil factors (e.g. nitrogen and water) slow plant growth rates directly and CM size indirectly. Both parasitoids prefer to lay female eggs in large CM, but *E. lopezi* produces more females in smaller hosts than does *E. diversicornis*, resulting in a greater male-biased sex ratio in *E. diversicornis* relative to *E. lopezi*. Additional factors favoring *A. lopezi* is its high host-finding

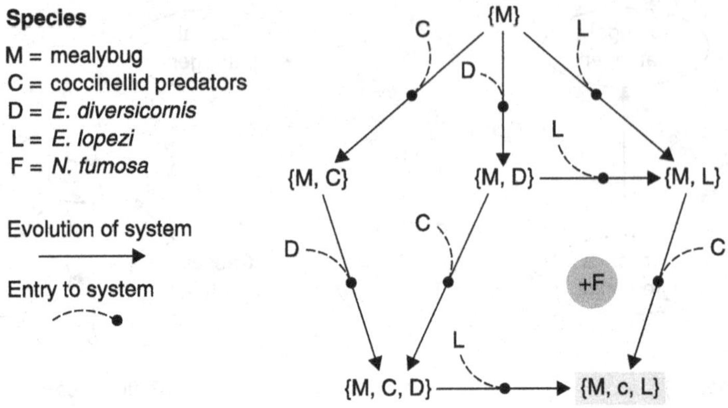

Figure 10.10. An assembly diagram of cassava/mealybug system showing the competitive displacement of *E. diversicornis* by *A. lopezi* (see Schreiber and Gutierrez, 1998). Capital letters imply dominance.

capacity and dominance in cases of multiple parasitism. All of these factors enabled *E. lopezi* to displace *E. diversicornis* during periods of low host density. Coccinellid predators merely reduce population levels but not to regulate them (Gutierrez *et al.*, 1988*a*). The interactions among the species are depicted in the successional diagram in Figure 10.10. We note that a different order of introduction of the parasitoids would have resulted in the same outcome.

Cassava greenmite (CGM)

A program of introducing natural enemies against CGM resulted in suppression of the pest across much of the West African cassava belt (see Hanna *et al.*, 2005). This biological control project was considerably more difficult than the control of the cassava mealybug because of complex taxonomic, biological and predator behavior problems. The development and use of similar simulation models enabled Gutierrez *et al.* (1988*c*) and Bonato *et al.* (1994*b*) to study mite influence on crop growth patterns. Among the factor responsible for control of this pest were low food quality of the host plant, dislodgement and death of mites by rain and wind, and predation. Gutierrez *et al.* (1988*b*; *c*) and Bonato *et al.* (1994*a*; *b*) included the effects of temperature, rainfall, and plant nitrogen and water stress in their model to capture the bottom-up effects of the plant on CGM dynamics and Toko *et al.* (1996) experimentally tested the predictions of the model with regards the effects of nitrogen level. In general, CGM levels fell during the rainy season but increased during the dry season in the absence of effective natural enemies. An important component of suppression during the rainy season was the activity of the fungal pathogen *Neozygitis cf. floridana*. This pathogen was found to reduce the population growth of *M. tanajoa* when the initial ratio of infected to susceptible mites exceeds unity (Oduor *et al.*, 1997). Gutierrez *et al.* (1999) included the biology of two exotic predators (*Tetranychus aripo* DeLeon and *T. manihoti* Moraes) in their model and showed how *T. aripo*, despite having the lowest vital rates, was responsible for the biological control of CGM during the dry season. The biological attributes responsible for this control were its propensity to remain on cassava in the face of low CGM population densities and its ability to complete development and reproduce feeding on cassava exudates and maize pollen (S.J. Yaninek, personal communication). These attributes are especially important at the beginning of the dry season when *T. aripo* is able to attack CGM populations at their lowest most vulnerable stage (Gutierrez *et al.*, 1999; S.J. Yaninek, personal communication; Figure 10.11).

10.3.4 Cotton

Among the earliest physiologically based models are those of cotton (Hesketh *et al.*, 1971; Baker *et al.*, 1972). These were canopy models that lacked

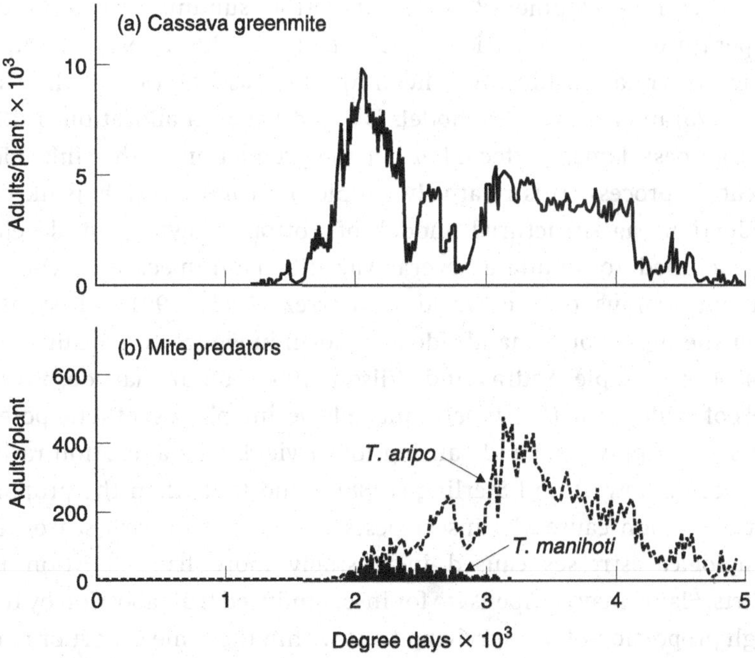

Figure 10.11. Simulated dynamics of cassava greenmite and two of it introduce natural enemies (*Tetranychus aripo and T. manihoti*). Note that *T. aripo* displaces *T. manihoti*.

the demographic properties required for linkage to higher trophic levels. These models owe much to the work of C. T. de Wit, Penning De Vries and colleagues in the Netherlands. Gutierrez *et al.* (1975) used the physiological underpinnings of SIMCOT II (Baker *et al.*, 1972) to develop a demographic model for plant growth and development suitable for linking to higher tropic levels. The initial mathematical underpinnings of the model were based on the Leslie–von Foerster demographic models (Gutierrez *et al.*, 1975; Gutierrez and Wang, 1977; Wang *et al.*, 1977; Gutierrez *et al.*, 1977*a*; *b*). Later, the distributed maturation time model was substituted to provide a more realistic formulation for cotton (Gutierrez *et al.*, 1991b) and most of the other systems models discussed herein (see Table 10.1, Appendix). The metabolic pool paradigm was used to allocate photosynthate to respiration, growth, reproduction and reserves with nitrogen and water acting as scalars for photosynthesis and hence dry matter allocation rates and the initiation of new subunits numbers, and the attrition of susceptible age fruits. What this means is that the plant is the integrating mechanism for weather and pest damage.

Independent studies have confirmed much of the biology used to develop the model. For example, Reddy *et al.* (1997) showed that the potential growth and developmental rates of cotton subunits are a function of temperature and that enriching CO_2 to twice the ambient level did not change the relationships. Another important assumption of the Gutierrez *et al.* (1975) and later related models is that dry matter allocation is allometric and that pest damage affected either the production or the sink side of the allocation process. This relatively simple per capita, supply (sink)-demand-(production) age-structured model of cotton growth and development was modified to simulate several varieties (and species) of cotton from different regions of the world (Gutierrez *et al.*, 1991*b*). Pest attacking either the supply or demand side are shown in Figure 10.12 (Gutierrez *et al.*, 1974). For example, Sadras and Wilson (1997) found that, despite varying levels of spider mite (*T. urticae*) damage lowering photosynthetic potential of leaves (i.e. supply side) and causing lower yields, the allocation ratios were unaffected. Stewart and Sterling (1988) found that when the proportion of fruit abscission caused by insect pests was largest in early-season cohorts, physiological stresses caused increasingly more fruit abortion in later cohorts. Plants may compensate for insect-induced fruit abortion by retaining a high proportion of the surviving fruit within the same cohort or if a cohort is severely depleted, fruit survivorship is increased in subsequent cohorts. These notions have been basic assumptions of physiologically based models

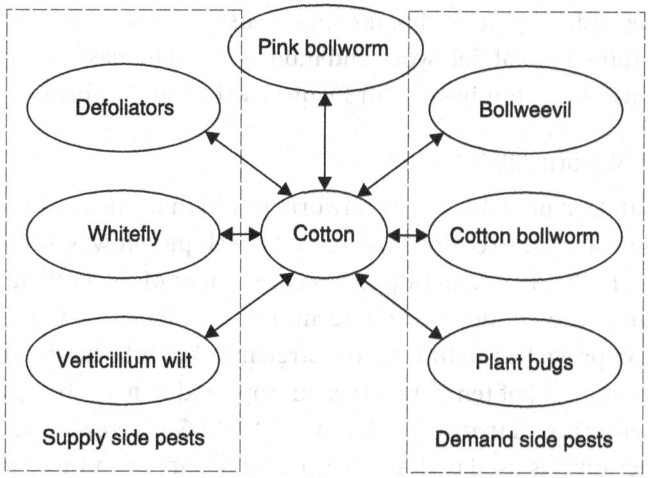

Figure 10.12. California cotton and seven pests. Note that some pests attack the capacity of the crop to produce photsynthate (supply side pests) others attack the sinks (demand side pests), while pink bollworm attacks the standing crop of fruit.

and provided the basis for explaining plant compensation for plant bug damage (Gutierrez *et al.*, 1975). This compensation was also verified experimentally by pruning fruit and from field data on the interaction of cotton and cotton boll weevil and bollworm (Gutierrez *et al.*, 1991*a*; *b*). Simple field IPM decision rules capturing this compensation were formulated based on the ratio of fruit production rates and to damage rate estimates (e.g. Gutierrez *et al.*, 1981c).

Various cotton models have been used to examine the effects of different pests on plant growth and yield, and below some that led to the development of IPM strategies are examined.

Lygus bug

The pest status of *Lygus hesperus* Knight and extant pesticide-based control practices were questioned using extensive fieldwork (Falcon *et al.*, 1971). Simulation modeling explained the field data by enabling the separation of weather and biotic effects that showed that the damage ascribed to *Lygus* was far greater than the actual (Gutierrez *et al.*, 1975; 1977*b*). The analysis showed that cotton could compensate for fruit bud losses due to *Lygus* bug levels common in the area. Furthermore, the model explained that much of the variation in yield was due to weather patterns, variable plant density, planting dates, and pesticide-induced pests (pest resurgence, secondary pest outbreaks and resistance: see van den Bosch, 1978). The combined use of fieldwork and modeling led to massive reductions in insecticide use in cotton in the San Joaquin Valley of California.

Pink bollworm (PBW)

A simulation model of pink bollworm (PBW) that captured its biology and phenology as linked to cotton growth and development was developed by Gutierrez *et al.* (1977*a*; 1986). An important component of the PBW model was a model of diapause induction and termination. Modeling the diapause response of PBW proved straightforward. Large numbers of experiments were conducted across a grid of temperature *x* photoperiod to map the proportion of the larvae entering diapause (Butler *et al.*, 1978; Gutierrez *et al.*, 1981). A Gompertz function was used to describe the pattern of PBW adult emergence from diapause during the period of early spring to mid-summer.[1] Understanding diapause initiation and termination proved to be the key to developing control strategies based on early crop termination (see Chu *et al.*, 1996).

Stone and Gutierrez (1986*a*; *b*) and Stone *et al.* (1986) extended the PBW model to examine the use of pheromones for PBW control. Variants of the PBW model have been implemented in California (Gutierrez *et al.*, 1977*a*; Stone and Gutierrez, 1986*a*; *b*), Brazil (Gutierrez *et al.*, 1986) and Egypt (Russell, 1995).

More recently, the cotton model was extended to simulate genetically modified cotton with genes expressing the toxin of the bacterium (*Bacillus thuringiensis* Berliner). The concentrations of the toxin vary with plant age and with age of individual plant sub-units (see Greenplate, 1999). The model for PBW was modified to include the population genetics for resistance and to reflect its interactions with the time and age varying expression of the toxin in the plant. Non-target effects of the *Bt* toxin on natural enemies on PBW and other pests (see below) were also investigated and included in the model (Ponsard *et al.*, 2002; Gutierrez and Ponsard, 2005; Gutierrez *et al.*, 2005). The model suggests that a small temporal refuge is created in the crop by the time-age varying declining expression of the toxin in the plant (Greenplate, 1999). Spatial refuges consisting of 5% plantings of non-Bt cotton have been mandated by EPA to help preserve susceptibility to *Bt* in the pest. However, in Arizona, there has been a ~45% non-adoption of the technology and yet the frequency of resistance in PBW is twice as high as expected (i.e. 0.16; Tabashnik *et al.*, 2000). A combination of the mandated spatial refuges of non-*Bt* cotton, the large non acceptance and the temporal refuge may have slowed the rate of resistance buildup in PBW in Arizona, but the analysis suggests that the mandated 5% refuge would be insufficient. In the absence of these refuges (a *Bt* cotton monoculture), resistance in PBW would increase rapidly. A major factor increasing the development of resistance in Arizona would appear to be the relatively low rate of long distance movement of PBW adults.

Bollworm

Bollworm is a generic term for heliothine pests that behave as fruit predators usually requiring more than one structure to complete development. Wilson and Gutierrez (1980) studied *Helicoverpa zea* (Hubner) and Hassan and Wilson (1993) studied fruit damage *by H. armigera* (Hubner) and *H. punctigera* (Wallengren). The findings on host dynamics of host (structure) preference, larval and fruit age-specific consumption rates, fruit (age class) availability, and seasonal larvae age class availability were used as a basis for developing a fruit predation model to model the effects on growing cotton and for assessing their economic damage.

Resistance development to *Bt* has not been shown in field bollworm populations. A dynamics model for *H. zea* was modified to include the population genetics of resistance to *Bt*-toxin in bollworm as modified by natural enemies, spatial refuges caused by the polyphagous habits of *H. zea* on non-*Bt* hosts, the increasing tolerance of pest larvae with age, and pesticide use (Gutierrez *et al.*, 2005). Adult *Orius* and *Geocoris* predators and larval lacewings feeding on *Bt* toxin-laden prey larvae have reduced longevity and likely fecundity (Hilbeck *et al.*, 1998, 1999; Ponsard *et al.*, 2002) and the effect was included in the model. Field studies have shown lower average densities of natural enemies in *Bt* cotton (Naranjo, 2005). The simulation analysis shows that the most important factor slowing resistance development is the spatial refuges created in non-*Bt* hosts and the highly migratory habits of bollworm itself. In general, survival of bollworm is extremely low in cotton unless the natural enemies are disrupted by pesticide use. The model shows that the Bt toxin merely substitutes for natural enemy mortality. As in pink bollworm, the model predicts a rapid buildup of resistance in a *Bt* cotton monoculture, but not when the refuges provided by non-*Bt* hosts are included (Gutierrez *et al.*, 2005).

Similar results were obtained for the polyphagous defoliator *Spodoptera exigua* (Hubner, BAW) which has much higher intrinsic tolerance to the *Bt* toxin than bollworm. The model was also used to examine the effects of pyramiding two *Bt* toxins. The results indicate that the development of resistance would be hastened under scenarios such as using insecticides. Work is currently underway to examine the effects of one and two *Bt* toxins systems on multiple target and non-target level species (Figure 10.10).

Whitefly

Whiteflies became a global problem as early as the 1980s (Gerling, 1990). The species, *Bemisia tabaci* Genn., became a serious problem in the Sudan Gezira where it was studied by von Arx *et al.* (1984) using a modified version of the *G. hirsutum* cotton model (Gutierrez *et al.*, 1975) to simulate the phenology and growth of *G. barbadense* varieties. The plant model was used as a basis for understanding the cotton–whitefly interactions as affected by sowing dates, plant spacing, natural enemies and patterns of pesticide use. Whitefly abundance is strongly linked to vegetative growth patterns that can be enhanced by bollworms and inadequate irrigation practices (von Arx *et al.*, 1984; Baumgärtner *et al.*, 1986). The cessation of irrigation at the time of boll opening has been found an efficient strategy to control regrowth and a concomitant build-up of whiteflies (Von Arx *et al.*, 1984).

Later, a tritrophic prospective model of whitefly–aphelinid parasitoids interactions on cotton was developed by Mills and Gutierrez (1996) using the Gutierrez *et al.* (1991*b*) version of the cotton model[2] as the basis. Aphelinid parasitoids of scale insects and mealybugs may be typical primary parasitoid where both males and females develop on whitefly hosts. In other cases, some species are obligate autoparasitoid where males develop only on conspecific females (see Figure 10.5), or a facultative autoparasitoid where males may also develop on other parasitoids as well as conspecific females (heteronomous hyperparasitism). Parasitoids having hyperparasitic habits have been used in classical biological control, but an unresolved question was whether they are compatible with the goals of biological control. The results indicate that arrhenotokous primary parasitoids can substantially suppress whitefly populations, but that the impact of autoparasitoids is constrained by self-parasitism. The combination of a primary parasitoid and an obligate autoparasitoid provide the greatest suppression of cumulative whitefly abundance. However, the addition of a facultative autoparasitoid may be disruptive. The analysis cautioned against the indiscriminate introduction in biological control programs of aphelinid parasitoids with hyperparasitic habits. This result further highlights the need for detailed experimental observations on sex allocation and host discrimination in facultative autoparasitoids prior to their introduction. However, as found with olive scale and its two parasitoids, their differing responses to temperature enabled control of olive scale despite the autoparasitic biology of one of the parasitoids *Coccophagoides utilis* Doutt (Rochat and Gutierrez, 2001). Huffaker and Kennett (1966) predicted this based on field observations.

Bollweevil (*Anthonomus grandis*)

Bollweevil, like pink bollworm, behaves as a parasitoid of cotton fruit. A population dynamics model of the cotton boll weevil under Brazilian conditions was linked to a model for cotton growth and development (Gutierrez *et al.*, 1991*a*; *b*) as modified by the compounding effects of low soil nitrogen. The fruiting dynamics of the plant greatly affected the dynamics of the weevil, and vice versa. Simulation analysis suggested that varieties of cottons that were the mainstay of the Brazilian industry when the bollweevil arrived were poorly suited to compensate for boll weevil damage. The model suggested that short season cottons similar to those developed in North America were more suitable alternatives. A simple IPM decision rule, similar to that derived for field populations in Nicaragua (Gutierrez *et al.*, 1981), was developed for Brazilian cotton that suggested that the use of an economic threshold based on this rule could greatly reduce the number of insecticide applications. Last, area wide pest management using insecticide and insecticide laced adult pheromone traps (*Gossyplure*) has been highly successful in reducing (some think eradicating) this pest in the Southeastern United States (Villavaso *et al.*, 2001; D.D. Hardee, personal communication) and one is vexed why the technology has not spread to Brazil were bollweevil is also reshaping the cotton economy.

Spider mites

Wilson *et al.* (1992) and Trichilo and Wilson (1993) used a physiologically based simulation model allied to that of Gutierrez *et al.* (1991*b*) to assess the factor responsible for observed spider mite damage in cotton following applications of insecticides. Spider mite outbreaks were in part due to physiological stimulation following pyrethroid insecticide use that affected several mite life history parameters, with reduce natural enemy-mediated predation having the greatest impact.

Verticillium wilt

A model of the pathogen Verticillium was added to the cotton model to examine the effects of pathogenicity, time of infection and plant density on cotton yield (Gutierrez *et al.*, 1982). This collaborative effort with pathologists was later used as a basis for developing optimal strategies for the cultural control of the pathogen (Regev *et al.*, 1990).

10.3.5 Grape

A dynamic model for grapevine (*Vitis vinifera* L.) dry matter growth and nitrogen (N) dynamics was developed by Gutierrez *et al.* (1985) based on the Leslie–von Foerster model. Later, Wermelinger *et al.* (1991*a*) developed a systems model based on the distributed developmental times model (see Appendix). In both models, age-structured populations of plant subunits (fruit, leaves, shoots, and roots) were modeled on a perennial frame with the subunit populations having age-structured attributes of numbers, dry matter and N mass.

Mites

The Wermelinger *et al.* model was used as the basis for examining several pests in the vineyard ecosystem in Central Europe. A time invariant distributed delay model with attrition was derived to simulate the dynamics of leaf, shoot, root, fruit and mite populations, a functional response model was used to represent resource acquisition, while the metabolic pool model was used for resource allocation (Wermelinger *et al.*, 1991*a*). Among the first pests linked to the grape model was the spider mite *Tetranychus urticae* Koch (Wermelinger *et al.*, 1992). Detailed life history studies provided the basic data for the development of the mite model. Among the data collected at different temperatures were developmental times, mortality rates, adult oviposition pattern, longevity, and sex ratio. Wemelinger *et al.* (1991*b*) reviewed the literature on nutritional effects of host plants on spider mites and using micro-propagated apple trees determined the effects of temperature and different levels of N, P and K on *T. urticae* developmental time, egg reproduction and longevity. Over a range of 1.5%–3.0% leaf N, r_m increased by a factor of 4, the number of multiplications per generation (R_0) by 11, while the doubling time of the population was prolonged four fold on severely N-deficient leaves (Wermelinger *et al.* 1991*a*; *b*).

Wermelinger *et al.* (1992) also developed a dynamics model for the European red mite *Panonychus ulmi* (Koch) that was validated with independent field data. They found that mite development, survival, and reproduction were mainly driven by temperature and precipitation as modified by the food supply and the nitrogen status of host leaves. The bitrophic relationship between the host and the herbivore has a quantitative and a qualitative component: the plant supplies food (dry matter) and mite-feeding stages reduce the photosynthetically active leaf area. Qualitatively, nitrogen concentration of vine leaves affected oviposition and development of the spider mites. The model successfully stimulated the mite dynamics across different locations and years.

Leafhoppers

The grape leafhoppers, *Erythroneura elegantula* Osborn in North America and *Empoasca vitis* Goethe in Europe are important pests in vineyards as high leafhopper densities cause leaf burn that is thought to affect vine productivity. The use of alternative hosts is a prominent feature of their biology and important components in their pest management (Doutt and Nakata, 1973; Cerutti *et al.*, 1989; 1990; 1991*a*; *b*). In California, the native grape leafhopper *E. elegantula* is attacked by an egg parasitoid *Anagrus epos* that has a winter refuge in the blackberry leafhopper *Dikrella californicus* (Lawson) (Williams, 1981). The variegated leafhopper *E. variabilis* Beamer is a relatively recent invader of California vineyards and has caused a partial displacement of *E. elegantula*. Settle and Wilson (1990*a*; *b*) found differential parasitism by the egg parasitoid *Anagrus epos* on the two leafhoppers. Their model of host preference took into account the details of parasitoid foraging behavior and the spatial heterogeneity of hosts within a patch but the model was not linked to a plant model. They found that host encounter by the parasitoids was directly proportional to the relative density of the two hosts, but *A. epos* did not readily detect *E. variabilis* eggs buried deep within the leaf. The model prospectively compared three different biotypes of *A. epos*, one of which was imported from Mexico, as a biological control agent. The exotic Mexican biotype proved better at encountering and detecting the exotic *E. variabilis* than did the California biotypes, and not unexpectedly was a better biological control agent (Settle and Wilson, 1990*b*). Although the exotic parasitoid biotype readily accepted both leafhoppers it attacked the grape leafhopper at a higher rate, shifting the competitive balance from one of equality to one of strong disadvantage for the native *E. elegantula*.

In Europe, *E. vitis* females overwinter as adults on non-grape hosts and live up to 1.5 months after their return to the vineyards where they lay an average of 10.8 eggs (Cerutti *et al.*, 1990). During summer, non-diapausing females live and reproduce on grape but females reaching the adult stage after mid-August enter a reproductive diapause (Cerutti *et al.*, 1989; 1991*b*). The development times of eggs, larvae and nymphs on grape, as well as the mortality of the larvae and nymphs were studied as a function of temperature. The most important biotic mortality factor was the parasitic wasp, *Anagrus atomus*, which parasitizes up to 90% of leafhopper eggs. A second egg parasitic wasp, *Stethynium triclavatum* Enock, and an unidentified dipterous parasitoid of larvae and adults of *E. vitis* were found to be less important. The Cerutti *et al.* (1991*a*; *b*) model included the above biology of

E. vitis and the egg parasitoid *A. atomus* and explained the underlying factors governing their phenology and dynamics.

10.3.6 *Rice*

Several models of rice have been developed from the perspectives of plant growth physiology (Iwaki, 1975; 1977; Van Keulen, 1976; McMennamy and O'Toole, 1983; Penning De Vries *et al.*, 1989; Graf *et al.*, 1990*c*; Wu and Wilson, 1997; Anastacio *et al.*, 1999*a*), but only the latter two models have the requisite demographic structure. Other models focus on the dynamics of rice pest populations. Among them is the time-varying distributed-delay-based model of the beetle *Dicladispa gestroi* Chap. (Severini *et al.*, 1996). Rather than considering ambient temperatures, development is driven by insect body temperatures calculated by a microclimate simulator. Some models have included rice–insect pest interactions and the effects of the pests on rice production – rice leaf-folders (Graf *et al.*, 1992); rice water weevil (Wu and Wilson, 1997), and the effect of competition from weeds on rice yield (Graf *et al.*, 1990*a*; Caton *et al.*, 1999). Others have used models to explore plant breeding issues in rice (Dingkuhn *et al.*, 1993); the effect of morphological modifications of plants and their response to various stress factors (Mutsaers and Wang, 1999); methane emissions from rice fields (Inubushi *et al.*, 1997; Olszyk *et al.*, 1999); the effects of climate change on rice production (Bachelet and Gay, 1993; Penning De Vries, 1993); crop planning and production strategies (Bakema *et al.*, 1984); and crayfish production in flooded rice (Anastacio *et al.*, 1999*b*).

Rice-fishpond systems

In the tropics, farmers often husband rice and fish in linked rice floodwater–pond systems. The same aquatic food web may occur in both systems, and is composed of species of bacteria that recycle nutrients from organic matter, flagellates–ciliates that fed on the bacteria, autotrophic micro- and macro-phytoplankton that use the nutrients (algae), and higher tropic interactions among zooplankton and fish (the phytophagous Nile tilapia, *Oreochromis niloticus* L., and the carnivorous North African catfish, *Clarias gariepinus* Burchell (Figure 10.13). d'Oultremont and Gutierrez (2002*b*) modeled the mass dynamics of this linked system using the metabolic pool paradigm for most species (Gutierrez and Baumgärtner, 1984*a*; *b*; Gutierrez *et al.*, 1994), models for floodwater rice (Graf *et al.*, 1990*b*), and models from the literature for aspects of the biophysical system (e.g. Chapra, 1997). Nutrient recycling is vital to the trophic interactions as it forms the base

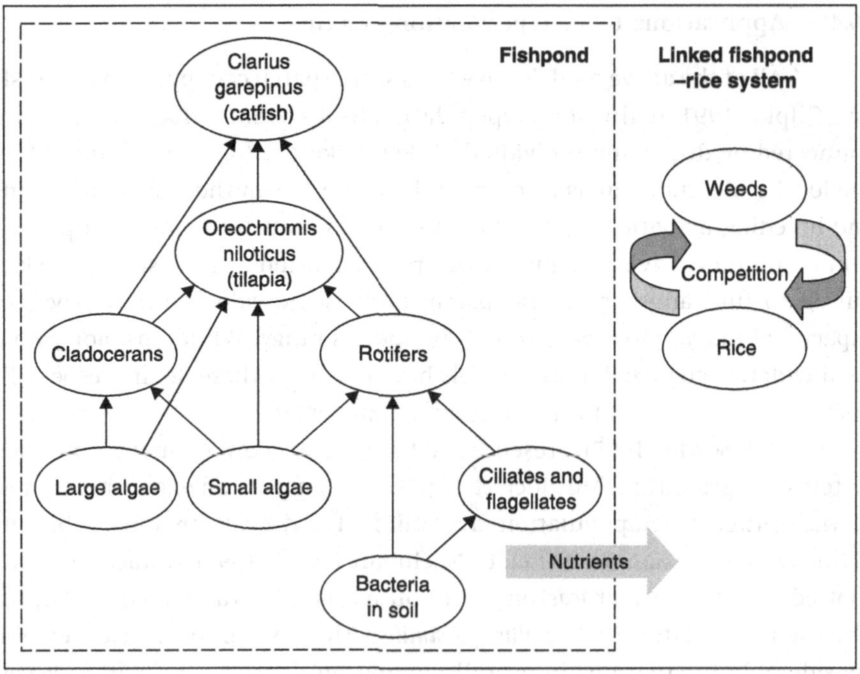

Figure 10.13. The fish pond-flooded rice system (see d'Oultremont and Gutierrez, 2000*a*; *b*).

level regulating the flow of energy and nutrients to higher trophic levels (d'Oultremont and Gutierrez, 2002*a*). Shading of the floodwater by the rice canopy lowers water temperature and decreases the amount of light reaching the water surface and has important effects on aquatic autotrophs. These effects in turn have important bottom-up effects on higher trophic levels.

This model was used to explore the effects of fertilization, fish presence and feeding in ponds on nutrient recycling, phyto- and zooplankton dynamics and rice and tilapia yields (d'Oultremont and Gutierrez, 2002*b*). In the linked rice–fishpond system, a net transfer of nutrients occurs from the pond to rice in the water and in fish feces that increases rice yield. Also emerging was a cautionary note concerning incorporating the genes for *Bt* toxin into rice. In addition to potential direct effects on higher trophic levels, *Bt* toxin leaching from rice roots could adversely affect nutrient recycling due to *Bt* bactericidal effects.

The modular construction of the object-oriented programming models (OOP) readily lends itself for linking models of various pests to the system (e.g. rice brown planthopper, *Nilaparvata lugens* Stal; Kenmore *et al.*, 1986) or the detritivores that serve as prey for predators that build and ultimately control *N. lugens* (see Settle *et al.*, 1996).

10.4 Applications to metapopulation systems

All of the above models have been single patch canopy models. Hanski and Gilpin (1991) define a metapopulation as a system of local populations connected by dispersing individuals. They make the following points: (1) at the local scale, individuals interact with each other in the course of feeding and breeding activities; (2) movement of individuals from one local population to another may occur for various reasons, often at a substantial risk of failing to find another habitat patch; and (3) individuals rarely have the capacity of moving to most parts of the species range. While this biology has been widely recognized, most models have excluded these details, especially knowledge of the rules for between patch movements.

Ives and Settle (1997) presented a theoretical view of metapopulation systems in agriculture. Murdoch *et al.* (1996) attempted an experimental test of the refuges metapopulation dynamics of red scale in citrus, but no definitive conclusion were made. Nachman (1987; 1991) studied the two-spotted spider mite (*T. urticae*) on greenhouse cucumbers and its control by the phytoseiid predator *Phytoseiulus persimilis*. The two mite species exhibit considerable fluctuations in overall population densities but within acceptable limits. The system persisted at the greenhouse scale despite frequent local extinction on individual plants. Experimental evidence indicates that the mites form a metapopulation system characterized by "shifting mosaic" dynamics. A stochastic simulation model was used to analyze the role played by dispersal in the dynamics and persistence of the system. Demographic stochasticity generated sufficient endogenous "noise" to counteract the synchronizing effect of density-dependent dispersal with the proviso that the dispersal rates are not too high and the system is not too small. Low dispersal rates increase the risk of local outbreaks of spider mites that may cause destruction of plants (see also Zemek and Nachman, 1999).

A metapopulation model of the cassava food web was developed that compared well with the field data (Gutierrez *et al.*, 1999). The important innovation beyond that of Gutierrez *et al.* (1993) was the use of patch specific supply/demand relations to regulate the emigration rate of appropriate stage individuals from patches due to resource shortfalls. Migrating individuals of each species were assumed to form a transitory pool of migrants during each time step that were stochastically reallocated among patches at rates corrected for species specific patch finding capacity and the distance between patches. For some species with low patch finding capacities (e.g. mite spiderlings and mealybugs) this resulted in high mortality during migration. In this system, bad patches of resource lost individuals and good patches gained them, resulting in local extinction and re-population. The simulations were in all respects similar to that using a canopy model with stochastic immigration (Gutierrez *et al.*, 1993).

IPM systems have been extended into dimensions of space, time, objects of management and actor organizations (Kogan *et al.*, 1999; Baumgärtner *et al.*, 2003*a*). This extension resulted in the consideration of levels in IPM schemes and applications of hierarchy theory (Allen and Starr, 1982). In fact, the aforementioned models can be seen as operating on patch and field levels and thereby, representing two important levels of the presumed hierarchical organization of nature (Baumgärtner *et al.*, 2002). Presumably, some levels may reflect discontinuities in scale dependencies of ecological variables (Baumgärtner *et al.*, 2002). Hierarchy theory applications and the increasing importance given to scale considerations in ecology may indicate opportunities for further developing multitrophic models. Further development may be possible through more efficient structuring of the actor subsystem operating on ecological systems and combining them into ecosocial systems as defined by Waltner-Toews *et al.* (2003).

10.5 Geographic information system application in tritrophic analysis

PBMs readily lend themselves to examining biological problems across a geographic landscape with different ecosystems. The reason is that PBMs model the population dynamics of organism as modified by weather-driven processes. Hence as weather changes over time and geography, the dynamics of organisms change in response. Gutierrez *et al.* (2005a) developed a windows menu GIS system using ARC/INFO 8 that could embed the executable file for plant/herbivore/natural enemy systems that meet the minimal input-output requirements and run them using weather files. The models of any system can be added to the GIS by simply loading them into a GIS systems folder (i.e. bin) and running them using menu options to select for species using Boolean variables, geographic regions within the USA, and within California ecological regions, evapotranspiration zones, and elevations within zones. In addition, start and stop dates for across year simulation, grid size for kreiging, reporting and mapping intervals, editing the text setup initial conditions for each species, and other factors are included in the set up file. Multitrophic system models for alfalfa, cotton, coffee, cassava, grape, olive, rice, and yellow starthistle are currently in the system.

Among the questions addressed so far, have been the effectiveness of biological control of yellow starthistle (YST) (Figure 10.14), oleander scale (*Aspidiotus nerii* Bouchè) and vine mealybug, *Planococcus ficus* (Signoret)

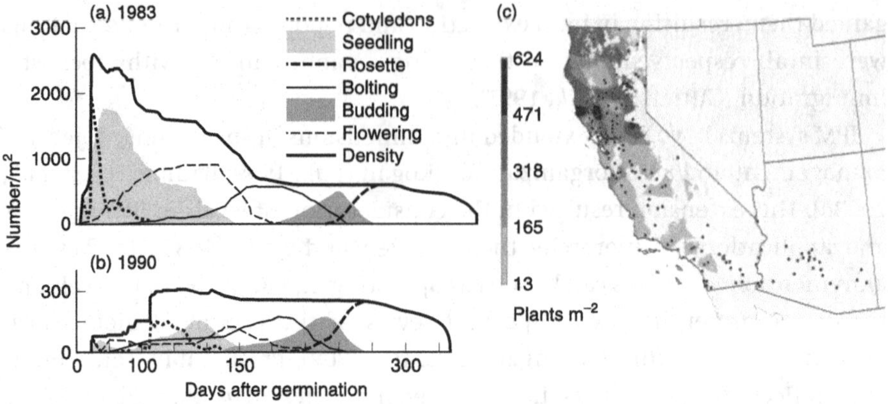

Figure 10.14. The simulated effects of temperature and rainfall on the germination and phenology of yellow starthistle (YST) at Davis, CA during (a) 1983 and (b) the 1990 seasons, CA, and (c) the simulated abundance of mature YST plants m^2 across California during 2003 after eight years of simulation (see Gutierrez *et al.*, 2005*a*).

(Gutierrez *et al.*, 2005; Gutierrez *et al.*, in press). Simulations of yellow starthistle at Davis, CA were run for the period 1985–2003 show that the pattern of germination and maturity of the plants varied radically from year to year due to rainfall and temperature patterns (1982–1983 vs. 1989–1990, Figure 10.14a, b respectively). In addition, the amount of seed produced and remaining in the soil seedbank also varies across years at each location (not shown).

Figure 10.14c shows the abundance of plants across California in 2003 after several years beginning with the same initial seed density at all sites. The model-GIS capture the observed distribution of YST quite nicely (see Gutierrez *et al.*, 2005*a*), but it is multivariate analysis of the output data that provides the most instructive insights (see below). In general, the analysis suggests that the suppressive effects of the different natural enemies introduced for the control of YST plus the added effects of competition from annual grasses are inadequate. The natural enemies attack the seedheads but their attack permits far too many seeds to survive, and the overall effect is inversely density-dependent. Seed survival and a high degree of plant compensation at low density are the dominant factors. The model suggests that for control to occur, natural enemies that kill whole plants and or greatly reduce the capacity of the plant to produce or mature seed are required. Such agents are currently being introduced (e.g. a leaf rust fungus, *Puccinea jaceae var. solstitiali,s* and a root-feeding weevil, *Ceratapion bassicorne*), and their impact could be easily introduced into the model GIS and evaluated.

Another issue examined is the effects of climate warming on the abundance and distribution of the cold intolerant pink bollworm in cotton in California and Arizona (Gutierrez *et al.*, 2006; Figure 10.15) and other species in other systems (e.g. olive and olive fly). The model-GIS analysis suggests that *PBW* would not be a major threat to the Great Central Valley (GCV) of California under current weather conditions as only low populations could persist in the very southern reaches of the GCV (Figure 10.15a, d), but that an average increase in temperatures of 2–2.5°C would greatly increase winter survival and allow expansion of *PBW* into the GCV, where it could become a serious economic pest. In addition, this warming would also increase its pest status in areas already infested; namely the desert valleys of extreme Southern California and Arizona (Figure 10.15 b, c, e, f).

10.6 Multivariate statistical methods in tritrophic analysis

The use of multivariate analysis in the social sciences and its application to the study of complex tritrophic interaction in ecology can also yield significant insights. For example, Neuenschwander *et al.* (1989) used multivariate analyses of field survey data to estimate the impact of cassava

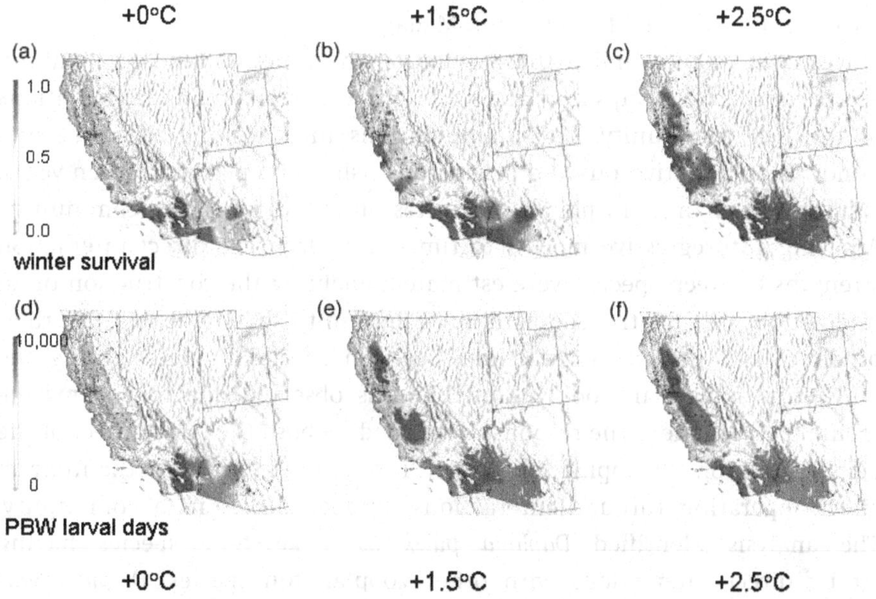

Figure 10.15. Winter survival of *PBW* larvae (range 0–1) and cumulative *PBW* larval days plant^{-1} (range 0–10 000) respectively during 2003 across the ecological zones of southern Arizona and California: (a, d) observed weather, (b, e) 1.5°C increase in average daily temperature and (c, f) 2.5°C increase in average daily temperature.

mealybug populations on the growth and tuber yield of cassava and the benefits of biological control by the exotic parasitoid *E. lopezi* in the savanna and forest ecosystems in Ghana and Ivory Coast. Twenty-nine variables associated with plant growth, agronomic and environmental factors, and insect populations were recorded. Mealybug density was positively correlated with stunting of the cassava shoot tips while average harvest indices declined and populations of *E. lopezi* and of indigenous coccinellids increased in response. The parasitoids were found at lower host levels than were predators even though the predators had been shown not to be effective regulating agents. The length of time *E. lopezi* had been present in an area was the most important factor influencing the decline in mealybug densities. In response to declining mealybug densities, significant tuber yield gains occurred in the savanna zone (2.48 t/ha) but not in the forest region. Similar analyses were made of the output of more than a thousand stochastic runs of a metapopulation model of the cassava system (Gutierrez *et al.*, 1999) with the results being in the same direction as an analysis of the Neuenschwander *et al.* (1989) field data. However, the model enabled evaluation of the role of the various natural enemies in the suppression and regulation of CM and CGM and their interactions. Baumgärtner *et al.* (1989) relied on multiple regression analysis to study the contribution of pests and agronomic measures on rice yield formation in Madagascar.

Ives *et al.* (1999) studied the population dynamics of nine zooplankton species or species groups in a lake subjected to experimental manipulations of the fish community. The manipulations included the additions and removals of planktivorous and piscivorous fish over a period of seven years, resulting in changes in planktivory rates on the zooplankton community. Applying autoregressive models to time-series data, the direct interaction strengths between species were estimated, enabling the construction of an interaction web for the zooplankton community that indicated the roles of direct and indirect interactions between species in determining the long-term changes in zooplankton biomass observed due to an environmental perturbation. The response depended on both the direct effect of the perturbation on the population growth rate of the species and the indirect effects operating through interactions among species in a community. The analysis identified *Daphnia pulex* as a keystone species having strong direct interactions with other zooplankton species. *D. pulex* was strongly affected by changes in planktivory. Changes in the biomasses of other zooplankton species during the planktivory manipulations were influenced strongly by indirect interactions acting through changes in *D. pulex* biomass. Recent dynamic modeling of aquatic systems offers the possibility of examining these interactions on a finer scale (d'Oultremont and Gutierrez, 2002*a*).

10.7 Summary

A wide range of modeling approaches may be used to model field tritrophic dynamics, but of these, physiologically based approaches have yielded the most consistent results. An advantage of the physiologically based approach is its applicability to modeling a wide range of ecological systems using essentially the same model. The use of this modeling paradigm is aided by physiological and behavioral analogies that exist among species and trophic levels including the economic one (Regev *et al.*, 1998; Gutierrez and Regev, 2005). The most basic one is that all organisms have the same imperative: maximize fitness by maximizing resource acquisition and minimize costs and losses to predation. This biology is subsumed under the ambit of the metabolic pool paradigm. Organisms seek to satisfy a genetic demand for resources given current states and environmental conditions. Using a per capita approach, organisms search for resources (biomass, energy) that are allocated to egestion, respiration and growth and reproduction. The various components of this demand are easily identified, aiding in gathering the appropriate data. In population dynamic models based on this paradigm, the vital rates of all species are scaled from the maximum by the ratio of actual rate of consumption of resources to the genetic maximal demand. Estimating growth rates in this manner may be used to predict geographic limits of favorability and phenologies. Good examples of this are in the area of conservation of buckwheat *Fagopyrum esculentum* (Baumgärtner *et al.*, 1998) and wild daffodil *Narcissus radiiflorus* in Switzerland (Baumgärtner and Hartmann, 2000; 2001). Such applications are extensions of older ideas using growth rates in response to abiotic factors to predict areas of favorability (Shelford, 1931; Fitzpatrick and Nix, 1970; Gutierrez *et al.*, 1974).

The plant is the base trophic level and integrates many of the abiotic factors and provides the resource base for higher trophic levels (bottom-up effects), while higher trophic levels provide the top-down regulation on pests. The systems models are modular, making inclusion of different species a simple matter, because it is simply another model with known structure and its linkages to other species is via its role as a consumer and or as a resource. Approaching modeling from this physiological perspective simplifies the problem and makes the model independent of the data one wishes to simulate. Of course, weather and abiotic factors help determine the demand and hence are forcing variables for the system. The use of object oriented programming further facilitates the programming of the system (see Sequeira *et al.* 1991, Holt *et al.*, 1997; d'Oultremont and Gutierrez, 2002*a*; *b*).

The models may be used in prospective and retrospective analyses of complex systems (see Table 10.1). The models may be used to evaluate natural and biological control programs (e.g. cassava, coffee, grape) and to develop IPM strategies (alfalfa, cotton, cowpea). The model may also be used to evaluate the utility of genetically modified crops (GMOs, e.g. cotton, maize, rice) and to assess the potential for resistance development in pests and the collateral effects of the GMOs on natural enemies. The models may furthermore become important elements of adaptive management systems proposed by Holling (1978) and in an IPM context, further discussed and implemented by Baumgärtner *et al.* (2003*b*) and Sciarretta *et al.* (2005). Comiskey *et al.* (1999) regard adaptive management as a systematic, cyclic process for continually improving management policies and practices (tactics, strategies) based on lessons learnt from operational activities. Within the cyclic process, models can continuously be modified to obtain improved understanding of systems behavior and efficiently assist in decision-making (Baumgärtner *et al.*, 2003*b*).

A systems model of crop systems may be viewed as libraries of what is known about the system and like any good library may be augmented as new information becomes available. The models should be independent of time and place because they describe physiological and population processes that are driven by observed weather and edaphic factors. This property allows them to be embedded in geographical information system (GIS) that can access spatially defined weather. Such models enable us to analyze on a regional basis crop-pest phenologies, crop yields, pest damage, biological control efficacy, assess new pest problems quickly as well as various other scenarios. A perceived weakness of the approach is that the models lack analytical solution and formulating the complexities of a biological system requires a minimum of data. This unifying approach greatly simplifies the ecological problems, and in some cases may permit analysis of complex systems where the barest of observational data are available (Rochat and Gutierrez, 2001)

Appendix: formulating physiologically based models

There are numerous ways to formulate demographic models, but a complete review of the subject is beyond the scope of this chapter.

Here we focus on physiologically based models that have had field application. Mass-based models lacking age structure are examined first to illustrate the nature of the multitrophic problem (Gutierrez, 1992) and in later sections age-mass structure are added to the model (see DiCola *et al.*, 1999 for a recent review of related models). Modeling the population dynamics may be viewed from the consumer point of view and asks how much of the resource sought was acquired, or from the prey point of view of the resources and asks how much was attacked (Berryman and Gutierrez, 1999). Here we approach the problem from the consumer per capita point of view.

Mass-based population dynamics

For simplicity, we assume a simple physiologically based mass dynamics model, lacking age structure, to illustrate the conceptual bases of the model. The two essential components of the model for each species are the functional response and population dynamics models. Only the units and interpretation of the parameters of the model vary among species and trophic levels.

The functional response model

The functional response model is the resource acquisition model that must include time varying intraspecific competition for resources, consumer demand, searching capacity, and prey availability. The effects of weather enter the model via the processes of resource acquisition demand and allocation. The functional response model used allows inclusion of metaphysiology and biology (see text, Figure 10.2). The model is the Gutierrez–Baumgärtner variant of Watt's (1959) model [equation (10.1)] modified to include the physiology of resource demand (see Gutierrez, 1992; 1996 for its derivation). We note that this model has the same form as the de Wit and Gouriaan (1976) photosynthesis model that was derived as an empirical fit to data. To illustrate this functional response model, we use it to estimate the photosynthetic rate of an average plant ($S(T(t))$, i.e. the supply) in a population of ρ plants m^{-2} at temperature T at time t of average mass $M(t)$.

$$S(T(t)) = f(u)D(T(t)) = \left(1 - e^{\frac{-\alpha(t)\psi(t)}{D(T(t))M(t)\rho(t)}}\right)D(T(t))M(t). \qquad (10.1)$$

Rearranging terms, $0 \leq f(u) = S(T(t))/\{D(T(t))M(t)\} < 1$ is the proportion of per plant demand ($D(T(t))M(t)$ in g nCH_2O) acquired given the light energy (converted to g biomass equivalents ($\psi(t)$) incident m^{-2}) and competition from other plants (i.e. $\rho(t)$ in the exponent). The efficiency of light interception is a function with constant c of average plant mass ($M(t)$) [equation (10.2)].

$$\alpha(M_L(t)) = (1 - e^{-cM(t)}), \text{ (i.e. Beer's Law, Nicholson–Bailey model)} \quad (10.2)$$

This makes type II model [equation (10.1)] a type III functional response model. The constant c includes search efficiency and converts the ratio of plant mass to area for growth to leaf area index. Note that the demand is the sum of the demand by all plant subunits as affected by temperature (i.e. the sum of maximum outflows in text Figure 10.2 at time t, see Gutierrez, 1996).

Plants (and other organisms) also compete for other essential resources (e.g. nitrogen, water and other factors in the root zone), and a shortfall in acquisition of any of them slows the photosynthetic rate and indirectly decreases growth, reproduction and survival rates. Ignoring T and t, the success of a plant acquiring each resource is also modeled using variants of [equation (10.1)] (e.g. f_η for nitrogen and f_ω for water). The analogous search function α_r in response models f_η and f_ω is a function of root mass and root zone volume. The demand for nitrogen (D_η) is proportional to carbon demand (i.e. D in [equation (10.1)]) and the demand for water (D_ω) is the maximum evapo-transpiration rate computed using a weather-driven biophysical model (e.g. Ritchie, 1972). Simple balance models account for fluxes in soil nitrogen (uptake, applications and decomposition of organic matter) and soil water (uptake and rainfall) (see Gutierrez *et al.*, 1988*b*). The supply/demand ratios for nitrogen and water are:

$$\text{nitrogen:} \quad 0 \leq f_\eta(u_\eta) = S_\eta/D_\eta < 1 \quad (10.3)$$

$$\text{water:} \quad 0 \leq f_\omega(u_\omega) = S_\omega/D_\omega < 1 \quad (10.4)$$

The effects of resources shortfalls on photosynthesis are incorporated in the photosynthesis model as compounding acquisition probabilities (Gutierrez *et al.*, 1994). In general, if there are x resources, their combined effects (ϕ^*) equals

$$0 < \phi^* = f_\eta(u_\eta)f_\omega(u_\omega)\ldots f_x(u_x) < 1, \quad (10.5)$$

and correcting [equation (10.1)] for these shortfall, the overall supply of photosynthate is

$$S^* = \phi^* f(u)D. \quad (10.6)$$

All of the details of this plant model have parallels in the models of higher trophic levels (see Gutierrez, 1996, for a review). We now incorporate the above metaphysiology into our common mass population dynamics model.

Tritrophic population dynamics

Plant level

Assume that $M(t)$ is the mass of a canopy of average plants at time t and temperature T, then their growth rate dM/dt is the photosynthetic rate $f(u)DM$ minus the respiration rate ($r = r(t(T)$, i.e. the Q_{10} rule) corrected for the conversion rate of resource to self (θ and ϕ^* was defined in equation (10.5)).

$$dM/dt = \theta\{\phi^* f(u)DM - rM\} \tag{10.7}$$

This model captures the biology of the metabolic pool (text Figure 10.2). We include the effects of herbivory as follows.

$$dM/dt = \theta\{\phi^* f(u)DM - rM\} - \sum_{k=1}^{K} \phi_k^* g_k(u_k)D_k H_k. \tag{10.8}$$

Feeding by all K herbivores (H_k, $k=1,\ldots,K$) equals

$$\sum_{k=1}^{K} \phi_k^* g_k(u_k)D_k H_k = \sum_{k=1}^{K} \phi_k^* g_k(M, \{H_1, \ldots, H_k\}, \{P_1, \ldots, P_N\})D_k H_k^*. \tag{10.8.1}$$

One can easily add additional complexity such as feeding by parasitized herbivores ($\{P_1, \ldots, P_N\}$) in $g_k(u_k)$. The parameters of the per unit mass functional response $\phi_k^* g_k(u_k)D_k$ of the kth herbivore population (H_k) are analogous to those in $\phi^* f(u)D$ (equations [(10.1)–(10.6)]). For herbivores, ϕ_k^* may include the effects of other essential nutrient (e.g. Wermelinger *et al.*, 1991*a*; Toko *et al.* 1996) as well as mating success, availability of oviposition sites and other factors that may slow animal population growth rates.

Herbivores

The dynamics model for the kth herbivore population (equation (10.9)) has the same form as (10.7) except for a net immigration term I_k, though a similar term for immigration could be introduced for the biology of some plant.

$$dH_k/dt = \theta_k\{\phi_k^* g_k(u_k)D_k - r_k\}H_k + I_k - \sum_{n=1}^{N} \phi_n^* h_n(u_n)D_n P_n \tag{10.9}$$

The herbivory rate ($\phi_k^* g_k(u_k)D_k H_k$) for the kth herbivore is part of equation (10.7) and contributes to the plant mortality rate (see Hassell *et al.*, 1976). Similarly, the death rate of the kth herbivore from the nth predator is $\phi_n^* h_n(u_n)D_nP_n$ and that due to all N predator species is $\sum_{n=1}^{N} \phi_n^* h_n(u_n)D_nP_n$. A similar model is used to describe the dynamics of still higher trophic levels (see Gutierrez *et al.*, 1999).

For generalist predators, there may be multiple resources, and hence the prey biomass density (H_n) of all resource species and substages (H_i, $j=1,\ldots,J$) that may be attacked by the nth of ns consumer (i.e. predators) is

$$H_n(t) = \sum_{n}^{ns}\sum_{j=1}^{J} \xi_{n,j}H_{n,j}(t). \tag{10.10}$$

The parameter ($0 \leq \xi_{n,j} \leq 1$) is the preference of the nth consumer species for the jth resource. Populations (or sub-stages in age structured model) with preference values equal zero are effectively removed from the calculations by $\xi_{n,j}$ (Rochat and Gutierrez, 2001). Preference is an important factor modifying the interaction between species in a food web.

Metapopulation dynamics

Assume we have many patches of plants each of which may have part or the full complement of higher trophic level species, species that may migrate between the patches. The within-patch dynamics is essentially the same as outlined above for canopy systems, except that in this case we have a coordinate system (i,j) to identify each patch and to be able to calculate distances between them. The plants may be distributed evenly on grid points or randomly. If the patches are close, competition for light, water and soil nutrients may occur between patches, and the ability of animal migrants to find other patches may be greatly increased.

We illustrate the development of the metapopulation dynamics models using an intermediate trophic level, say the herbivore level (see Gutierrez *et al.*, 1999), where ij are the coordinates of the patches.

$$dH_{kij}/dt = \theta_k\{\phi_{kij}^* g_{kij}(u_k)D_{kij} - r_k\}H_{kij} + I_{kij} - \sum_{n=1}^{N} \phi_{nij}^* h_{nij}(u_n)D_{nij}P_{nij} \tag{10.11}$$

Each i,j patch may be viewed as a population of plants infested by populations of herbivores and natural enemies that can migrate between patches. The herbivory rate ($\phi_{kij}^* g_{kij}(u_k)D_{kij}\ H_{kij}$) for the kth herbivore contributes to the mortality rate of plants in the patch ij. The net immigration rate $I_{kij}(t)$ of the kth herbivore in the ijth patch (equation (10.12)) is the difference between the immigration ($\phi_k v_{kij}(t) X_k(t)$) and emigration

$$\left(H_{kij}(t)\left(1-\frac{S^*_{kij}(t)}{D_{kij}(t)}\right)\right)$$

rates where the total of all migrants (X_k) of species k across all patches is $X_k = \sum_i^I \sum_j^J H_{kij}(1 - S^*/D)_{kij}$.

$$I_{kij} = \varphi_k v_{kij}(t) X_k(t) - H_{kij}(t)\left(1 - S^*_{kij}(t)/D_{kij}(t)\right) \tag{10.12}$$

The emigration from a patch is linearly related to local resource shortfall ($0 < (1-S^*/D)_{kij} < 1$), while the immigration rate ($\varphi_k v_{kij} X_k$) is influenced by the intrinsic capacity of the kth species to find new patches ($0 < \varphi_k < 1$) and the average proportion (v_{kij}) of the migrant poolX_k expected to find patch ij. Hence, if the maximum number of patches is $\rho_{\max}$, then the maximum patch finding rate is $v_{kij} = 1/\rho_{\max}$. However, if $\rho < \rho_{max}$, then the realized patch finding rate is $v_{kij}(t,\rho) = \rho/\rho^2_{\max}$. In a stochastic setting, v_{kij} would be allowed to vary about the expected value for a given ρ, and in such cases the product sum $\sum_i \sum_j \varphi_{kij} v_{kij}$ may be greater than unity suggesting immigration occurs from outside the patches, but if it is less than unity there is a net loss. Because search is imperfect, it is usually less than unity and may be near zero for very poor searchers and for large distances between patches. The parameter values for the different species may vary considerably with a low value for φ_k implying a high mortality rate during migration.

The key point of the immigration submodel is that patches with higher levels of resource gain migrants and those with poor levels lose them. In age-structured models, only some ages of a species migrate.

Adding age-mass structure

Adding number dynamics and adding age structure (or other attributes) is not difficult (see Sinko and Streifer, 1967; Wang *et al.*, 1977; Gutierrez, 1996). Such models can accommodate the widely varying conditions organisms face as well as the biological adaptations used to solve resource acquisition and allocation problems (Gutierrez and Wang, 1977). Adding age and mass structure is merely a complication of the biology described above as the same model now applies to each age class. Non-feeding stages (e.g. seed, egg, pupal stages) may not grow, but they do respire, die and in some cases migrate. Resources acquired by feeding stages may go to growth or to reproduction. We avoid the complication of metapopulations, and illustrate the components of the number and mass biology for a single population; be they plants or animals.

Births (number or mass) from adults enter the first age class of the model (text Figure 10.4a). Net changes in age class density (number or mass) may occur as inflows due to recruitment, growth, and deaths. Aging occurs as flows rate of individuals between age classes. Individuals may age at different rates, and to capture this biology we use distributed maturation time models (Manetsch, 1976; Vansickle, 1977).

Distributed maturation time models

Assume we are modeling the number or mass of the nth of ns functional populations (N_n) that comprise a tritrophic system, then the dynamics of the ith age class ($N_{n,i}$ (t)) ($i = 1, \ldots, k_n$) for the nth functional populations (subscript n ignored), may be modeled using equation (10.13).

$$\frac{dN_i(t)}{dt} = r_{i-1}(t) - r_i(t) - \mu_i(t)N_i(t) \qquad (10.13)$$

In this model, gains or losses accrue as rates to the ith age class (N_i) are due to aging into ($r_{i-1}(t)$) or out (($r_i(t)$)) of the age class, or as the total proportional loss rates due to mortality, growth and net immigration ($-\infty < \mu_i(t) < +\infty$) (see text, Figure 10.4a). Time-varying age-specific mortality may occur in any age class. Birth and death rates of all functional populations are affected by the interplay of supply-demand ratios (see below).

The distribution of developmental times of cohort members has an Erlang distribution with mean developmental time Δ (subscript n ignored) and variance s^2 (see Vansickle 1977) (text Figure 10.4b). In the absence of mortality, the number of age classes that yields the observed distribution may be computed as $k = \Delta^2/s^2$. If k is small, the variability of developmental times is large and vice versa. The width of each age class is Δ/k. Deaths may occur in any age class and deaths at maximum age exit from the population via the last or kth age class.

Developmental time varies with temperature T at time t and each functional population may have a different response (i.e. $\Delta_n(t) = \Delta_n(T(t))$) (Janisch, 1925; Stinner *et al.*, 1974; Logan *et al.*, 1976; Shape *et al.*, 1977; Lactin *et al.*, 1995; Hilbert, 1995). Developmental time may also vary with nutrition (e.g. age of fruit for herbivore) (Sharpe and Hu, 1980) and other environmental variables.

Flow rates $r_i(t)$ may be converted to numbers

$$N_i(t) = r_i(t)\Delta(t)/k \qquad (10.14.1)$$

with the total number in the population being

$$N(t) = \sum_{i=1}^{k} N_i(t) = \frac{\Delta(t)}{k} \sum_{i=1}^{k} r_i(t). \qquad (10.14.2)$$

The instantaneous solution for the flow rates from the ith age class ($r_i(t)$) is equation (10.15) (see Gutierrez (1996) for the discrete form, and DiCola *et al.* (1999) for the mathematical derivations of this and other potential models).

$$\frac{dr_i(t)}{dt} = \frac{k}{\Delta(t)} \left\{ r_{i-1}(t) - \left[1 + \mu_i(t) \frac{\Delta(t)}{k} + \frac{d\Delta(t)}{kdt} \right] r_i(t) \right\} \qquad (10.15)$$

Age-mass structured metapopulations

The development of metapopulation models using the distributed maturation time approach is essentially the same as that outlined in equations (10.13) and (10.15). The only complication involves the coordinate system and net immigration rates (i.e. equations (10.11)–(10.13)).

Rate of development

Temperature and other factors affect poikilotherm population vital rates. For example, the rate of development in response to temperature varies widely among poikilotherms (plant and most animals). Linear (degree-day, DeCandolle 1855) and non-linear (Janisch, 1925; Stinner *et al.*, 1974; Logan *et al.*, 1976; Sharpe *et al.*, 1977; Lactin *et al.*, 1995; Hilbert, 1995) models have been used to capture this biology.

$$R_n(t(T)) = (\Delta(t(T)))^{-1} \qquad (10.16)$$

The function $R(t(T))$ is the developmental rate per day at time t and temperature T, where $\Delta(t(T))$ is the mean developmental time in days at temperature T. Of course, $R\,(t(T))$ may also depend on nutrition (e.g. soil fertility in plants or host nutritional quality in higher trophic levels). For example, the developmental rate of bollweevil and pink bollworm depends on the nutritional status of fruit of different ages ($\phi_\Delta(a)$ (Isley, 1932; Stone and Gutierrez, 1986*a*).

$$R_n(t(T), \phi_\Delta(a)) = (\Delta(t(T), \phi_\Delta(a)))^{-1} \qquad (10.17)$$

Growth and reproduction

Resource acquisition may be characterized as predation, parasitoidism or parasitism. Plant may be viewed as predators searching for light (water, nutrients) that they use to fix carbon (photosynthesis). Predator attack and consume whole individuals and follow age-specific allocation rules (see Figure 10.2). Parasitoids seek hosts for their progeny to complete the immature part of the lifecycle. In predation and parasitoid attack, death of the prey/host is inevitable. Parasites normally require only one host, death is not inevitable and many individuals usually develop on the host. In some relationships these distinctions blur but are easily handled. Factors that may scale age-specific growth and birth rates include:

$$0 \le \phi_{S/D}(t) < 1 \quad \text{(supply/demand considerations)}$$
$$0 \le \phi_{nutri}(t) \le 1 \quad \text{(food quality)}$$
$$0 \le \phi_{n}(T(t)) \le 1 \quad \text{(the effect of temperature).}$$

and other limiting factors.

In the age-structure model, these scalars enter as age specific parameter, and in this sense are similar to the lumped effects of $0 \le \phi^* < 1$ in our simple mass dynamics model (equation (10.7)).

Mortality

The consumer growth birth rate is related to the resource death rate via the process of predation (Beddington *et al.*, 1976). Within a population, death may also accrue due to adverse temperature, to intrinsic causes, and to shortfalls in resource acquisition that may cause deaths directly or via emigration.

Supply/demand effects

Death (and net emigration) rates are assumed proportional to the shortfall in resource acquisition rates.

$$\mu_{(R)n,i}(t) = 1 - \text{supply/demand} \tag{10.18}$$

Temperature-dependent mortality

Data on the effects of temperature on survivorship ($\phi_{(T)n}(T(t))$ is normally computed over the lifespan of the organism.

$$\mu_{(T)n}(t) = 1 - \phi_{(T)\,n}(T(t)) \tag{10.19}$$

The rate for the ith of the k age classes of the population is

$$\mu_{(T)n,i}(t) = 1 - \left[\phi_{(\mu)\,n}(T(t))\right]^{1/k_n} \tag{10.20}$$

Total age specific mortality of age class i in a continuous model is the sum of the rates due to all factors and is incorporated in the distributed maturation time dynamics model (equation ((10.13))) as

$$\mu_{n,i}(t) = \mu_{(Bt)\,n,i}(t) + \mu_{(T)\,n,i}(t) + \mu_{(Pr\,ed)\,n,i}(t) + \mu_{(R)\,n,i}(t). \tag{10.21}$$

We caution that different integration schemes require different methods of incorporating the mortality.

Notes

1. A model for diapause termination in *Delia radicum* L. was developed by Johnsen *et al.*, 1997*b*. Their model assumed that three phases are involved: two in which no morphological development occurs, followed by a third phase or metamorphosis of the pupae to the adult stage (i.e. post-dormancy) culminating in the emergence of the adult. Each phase was assumed to have a different response to temperature with different optima. The model predicted the emergence of six cohorts of pupae that had been reared under different temperature regimes (Johnsen *et al.*, 1997*b*). The model also predicted the field phenology of diapause in cabbage maggot in coastal northern California where cool coastal temperatures induce diapause in some individuals throughout the year (Johnsen and Gutierrez, 1997*a*).
2. The different versions have the same biological underpinning, only the mathematical construct differs.

References

Allen, T.F.H. and Starr, T.B. (1982). Hierarchy. *Perspectives for Ecological Complexity*. Chicago: The University of Chicago Press.

Anastacio, P.M., Frias, A.F. and Marques, J.C. (1999*a*). CRISP (crayfish and rice integrated system of production): 1. Modelling rice (*Oryza sativa*) growth and production. *Ecological Modelling*, **123**, 17–28.

Anastacio, P.M., Nielsen, S.N. and Marques, J.C. (1999*b*). CRISP (crayfish and rice integrated system of production): 2. Modelling crayfish (*Procambarus clarkii*) population dynamics. *Ecological Modelling*, **123**, 5–16.

Anderson, J.L. (ed.) (1990). Second International Symposium on computer modelling in fruit research and orchard management. Utah State University, Logan, 5–8 September, 1989. *Acta Horticulturae*, **276**, 1–372.

Bachelet, D. and Gay, C.A. (1993). The impacts of climate change on rice yield: A comparison of four model performances. *Ecological Modelling*, **65**, 71–93.

Baker, D.N., Hesketh, J.D. and Duncan, W.G. (1972). Simulation of growth and yield in cotton. *Crop Science*, **11**, 431–5.

Bakema, E., Jansen, D.M. and Penning de Vries, F.W.T. (1984). Systems analysis and stimulation for rice production – A research and training project. In *Workshop on Physical Aspects of Soil Management in Rice-Based Cropping Systems*. Manila, Philippines: International Rice Research Institute.

Batzli, G.O. (1974). Production assimilation and accumulation of organic matter in ecosystems. *Journal of Theoretical Biology*, **45**, 205–17.

Baumgärtner, J. (1999). Population models: bases for the design of control, utilization, and conservation strategies. (Atti delle "giornate di studio sui metodi numerici, statistici ed informatici nella difesa delle colture agrarie e delle foreste"), Sassari, May 19–22. *Frustula Entomologica*, **22**, 7–22.

Baumgärtner, J. (2001). Biodiversity that mitigates pests in agroecosystems. In *Biodiversity Planning Support Programs. Global Environment Facility*, Unep Bpsp Thematic Studies on the Integration of Biodiversity into National Agricultural Sectors. Nairobi, Kenya: UNEP BPSP.

Baumgärtner, J. and Baronio, P. (1989). Modello fenologico di volo di Lobesia botrana Den. & Schiff. (Lep., Tortricidae) Relativo alla situazione ambientale della Emiglia–Romagna. *Boll. Istituto Entom. 'Guido Grandi' Università Bologna*, **43**, 157–70.

Baumgärtner, J., Bieri, M. and Delucchi, V. (1987). Growth and development of immature life stages of *Propylaea-quaturdecimpunctata* and *Coccinella septempunctata* L. (Col., Coccinellidae) simulated by the metabolic pool model. *Entomophaga*, **32**, 415–24.

Baumgärtner, J., Bieri, M., Klay, A., Genini, M., and Zahner, Ph. (1989). Fungicide side effects on the dynamics of an acarine predator–prey system. In C. Gessler, D.J. Butt and B. Koller (eds.), *Apple Orchards: An Explorative Study with Simulation Models*. Integrated Control of Pome Fruit Diseases, *Vol. II. Bull. Srop/Wprs*, **12**, pp. 317–35.

Baumgärtner, J., Delucchi, V., von Arx, R. and Rubli, D. (1986). Whitefly (*Bemisia tabaci* Genn., Stern.: Aleyrodidae) infestation patterns as influenced by cotton, weather and Heliothis: Hypotheses testing by using simulation models. *Agriculture, Ecosystems and Environment*, **17**, 49–59.

Baumgärtner, J.U., Frazer, B.D., Gilbert, N. *et al.* (1981*a*). Coccinellids (Coleoptera) and aphids (Homoptera): The overall relationship. *Canadian Entomologist*, **113**, 975–80.

Baumgärtner, J., Favre, J.J. and Schmid, A. (1988*a*). The use of a time-varying distributed delay model for simulating the flight phenology of the summerfruit tortrix *Adoxophyes orana* F.V.R. in the Valais, Switzerland. Bull. *Srop/Wprs*, **11**, 33–7.

Baumgärtner, J., Genini, M., Graf, B., Gutierrez, A.P. and Zahner, P. (1988*b*). Generalizing a population model for simulating golden delicious apple tree growth and development. *Proc. Intern. Symp. Computer Modelling in Fruit Research and Orchard Management*. Stuttgard–Holenheim.

Baumgärtner, J., Gilioli, G., Schneider, D. and Severini, M. (2002). The management of populations in hierarchically organized systems. *Notiziario sulla protezione delle piante*, **15**, 247–63.

Baumgärtner, J., Graf, B., Zahner, Ph., Genini, M. and Gutierrez, A.P. (1986). Generalizing a population model for simulating 'golden delicious' apple tree growth and development. *Acta Horticulturae*, **184**, 111–22.

Baumgärtner, J., Gutierrez, A.P. and A Klay, A. (1988c). Elements for modelling the dynamics of tritrophic population interactions. *Experimental and Applied Acarology*, **5**, 243–64.

Baumgärtner, J.U., Gutierrez, A.P. and Summers, C.S. (1981*b*). The influence of aphid prey consumption on searching behavior, weight increase, developmental time, and mortality of *Chrysopa carnea* Stephens and *Hippodamia convergens* G.-M. *Canadian Entomologist*, **113**, 1015–24.

Baumgärtner, J. and Hartmann, J. (2000). The use of phenology models in plant conservation programs: the establishment of the earliest cutting date for the wild daffodil *Narcissus radiiflorus*. *Biological Conservation*, **93**, 155–61.

Baumgärtner, J. and Hartmann, J. (2001). The design and implementation of sustainable plant diversity conservation program for alpine meadows and pastures. *Journal of Agricultural and Environmental Ethics*, **14**, 67–83.

Baumgärtner, J., Regev, U., Rahalivavololonoa, N. *et al.* (1989). Rice production in Madagascar: regression analysis with particular reference to pest control. *Agriculture, Ecosystems and Environment*, **30**, 37–47.

Baumgärtner, J., Schilperoord, P., Basetti, P., Baiocchi, A. and Jermini, M. (1998). The use of a phenology model and a risk analyses for planning buckwater (*Fagopyrum esculentum*) sowing dates in alpine areas. *Agricultural Systems*, **57**, 557–69.

Baumgärtner, J., Schulthess, F. and Xia, Y.L. (2003*a*). Integrated arthropod pest management systems for human health improvement in Africa. *Insect Science and Its Application*, **23**, 85–98.

Baumgärtner, J., Severini, M. and Ricci, M. (1990*a*). The mortality of overwintering *Phyllonorycter blancardella* (Lep., Gracillariidae) pupae simulated as a loss in a time-varying distributed delay model. *Mitteilungen Der Schweizerischen Entomologischen Gesellschaft*, **63**, 439–50.

Baumgärtner, J. and Severini, M. (1987). Microclimate and arthropod phenologies: the leaf miner *Phyllonorycter blancardella* F. (Lep.) as an example. In F. Prodi, F. Rossi and G. Cristoferi (eds.), *Agrometeorology*, Fondazione Cesena Agricultura Publ. pp. 225–43.

Baumgärtner, J., Getachew Tikubet, Melaku Girma *et al.* (2003*b*). Cases for adaptive ecological systems management. *Redia*, **LXXXVI**, 165–72.

Baumgärtner, J., Wermelinger, B., Hugentobler, U. *et al.* (1990*b*). Use of a dynamic model on dry matter production and allocation in apple orchard ecosystem research. *Acta Horticulturae*, **276**, 123–39.

Baumgärtner, J. and Yano, E. (1990). Whitefly population dynamics and modelling. In D. Gerling (ed.), *Whiteflies: Their Bionomics, Pest Status and Management*. Andover, UK: Intercept Ltd.

Baumgärtner, J. and Zahner, Ph. (1984). Simulation experiments with stochastic population models as a tool to explore pest management strategies in an apple tree–mite system (*Panonychus ulmi* Koch, *Tetranychus urticae* Koch). In R. Cavalloro (ed.), *Statistical and Mathematical Methods in Population Dynamics and Pest Control*. Proceedings of the Meeting of the EC Expert Group, Parma, October 26–28, 1983. Rotterdam: Balkema. pp. 166–78.

Beddington, J.R., Hassell, M.P. and Lawton, J.H. (1976). The components of arthropod predation. II. The predator rate of increase. *Journal of Animal Ecology*, **45**, 165–85.

Berryman, A.A. and Gutierrez, A.P. (1999). Dynamics of Insect Predator–Prey Interactions. In C.B. Huffaker and A.P. Gutierrez (eds.), *Ecological Entomology*. John Wiley and Sons.

Berryman, A.A., Gutierrez, A.P. and Arditi, R. (1995). Credible parsimonious and useful predator–prey models – a reply to Abrams, Gleeson and Sarnelle. *Ecology*, **76**, 1980–5.

Bonato, O., Baumgärtner, J. and Gutierrez, J. (1994*a*). *A simulation model to analyse the cassava–mites system in Central Africa*, 10th Symposium of the International Society for Tropical Root Crops (ISTRC). Salvador, Bahia, November 13–19.

Bonato, O., Baumgartner, J. and Gutierrez, J. (1994*b*). Impact of *Mononychellus progresivus* and *Oligonychus gossypii* (Acari: Tetranychidae) on cassava growth and yield in Central Africa. *Journal of Horticultural Science*, **69**, 1089–94.

Bonato, O., Schulthess, F.F. and Baumgärtner, J. (1999). Simulation model for maize crop growth based on acquisition and allocation processes for carbohydrate and nitrogen. *Ecological Modelling*, **124**, 11–28.

Butler, G.D., Jr., Hamilton, A.G. and Gutierrez, A.P. (1978). Pink bollworm: diapause in relation to temperature and photoperiod. *American Entomological Society*, **71**, 201–2.

Carrière, Y., Ellers-Kirk, C., Liu, Y.B. *et al.* (2001*a*). Fitness costs and maternal effects associated with resistance to transgenic cotton in the pink bollworm (Lepidoptera: Gelechiidae). *Journal of Economic Entomology*, **94**, 1571–6.

Carrière, Y., Ellers-Kirk, C., Patin, A.L. *et al.* (2001*b*). Overwintering cost associated with resistance to transgenic cotton in the pink bollworm (Lepidoptera: Gelechiidae). *Journal of Economic Entomology*, **94**, 935–41.

Caton, B.P., Foin, T.C. and Hill, J.E. (1999). A plant growth model for integrated weed management in direct-seeded rice. II. Validation testing of water-depth effects and monoculture growth. *Field Crops Research*, **62**, 145–55.

Cerutti, F., Baumgärtner, J. and Delucchi, V. (1990). Research on the grapevine ecosystem in the Tessin, Switzerland : III. Biology and mortality factors affecting Empoasca vitis Goethe (Homoptera, Cicadellidae, Typhlocybinae). *Mitteilungen Der Schweizerischen Entomologischen Gesellschaft*, **63**, 43–54.

Cerutti, F., Baumgärtner, J. and Delucchi, V. (1991*a*). Richerche sull'ecosistema vigneto nel ticino. iv. modellizzazione della dinamica di popolazioni di Empoasca Vitis Göthe (Homoptera, Cicadellidae, Typhlocybinae). *Boll. Ist. Ent. G. Grandi Univ. Bologna*, **46**, 179–200.

Cerutti, F., Baumgärtner, J. and Delucchi, V. (1991*b*). The dynamics of grape leafhopper *Empoasca vitis* Gothe populations in Southern Switzerland and the implications for habitat management. *Biocontrol Science and Technology*, **1**, 177–94.

Cerutti, F., Delucchi, V., Baumgartner, J. and Rubli, D. (1989). Research on the "vineyard" ecosystem in Ticino, Switzerland : I. Colonization of vineyards by the leafhopper *Empoasca vitis* Goethe (Homoptera, Cicadellidae, Typhlocybinae) and its parasitoid *Anagrus atomus* Haliday (Hymenoptera, Myaridae), and the importance of neighboring plants. *Mitteilungen Der Schweizerischen Entomologischen Gesellschaft*, **62**, 253–67.

Chapra, S. C. (1997). *Surface Water-Quality Modeling*. New York: McGraw–Hill.

Chu Chang-Chi, T. J., Henneberry, R. C., Weddle, E. T. *et al.* (1996). Reduction of pink bollworm (Lepidotera: Gelechiidae) populations in Imperial Valley, California, following mandatory short-season management systems. *Journal of Economic Entomology*, **89**, 175–82.

Cock, J. H. (1985). Cassava: physiological basis. In J. H. Cock and J. A. Reyes (eds.), *Cassava, Research, Production and Utilization, UNDP/CIAT*. pp. 33–62.

Comiskey, J. A., Dallmeier, F., and Alonso, A. (1999). Framework for assessment and monitoring of biodiversity. In S. Levin (ed.), *Encyclopedia of Biodiversity*, Vol. 3. New York: Academic Press. pp. 63–73.

Croft, B. A. and Knight, A. L. (1983). Evaluation of the PETE phenology modeling system for integrated pest management of deciduous tree fruit species. *Bulletin of the Entomological Society of America*, **29**, 37–42.

Curry, G. L. and Feldman, R. M. (1987). *Mathematical Foundations of Population Dynamics*, Texas: A&M University Press.

d'Oultremont, T. and Gutierrez, A. P. (2002*a*). An object-oriented multitrophic model of a rice–fish agroecosystem: I. Fishpond foodweb. I. A tropical fishpond food web. *Ecological Modelling*, **156**, 123–42.

d'Oultremont, T. and Gutierrez, A. P. (2002*b*). An object-oriented multitrophic model of a rice–fish agroecosystem: II. Linking the flooded rice–fishpond systems. *Ecological Modelling*, **155**, 159–76.

De Angelis, D. L. and L. J. Gross (eds.) (1992). *Individual Based Models in Ecology, Populations, Communities and Ecosystems*, New York: Chapman and Hall.

De Berardinis, E., Tiso, R., Butturini, A. and Briolini, G. (1993). A phenological forecasting model for the apple and pear leaf-roller *Argyrotaenia pulchellana* (Hw.) (Lepidoptera: Tortricidae). *Bollettino dell'Istituto di Entomologia "Guido Grandi" della Universita degli Studi di Bologna*, **47**, 111–22.

De Wit, C. T. and Gouriaan, J. (1978). *Simulation of Ecological Processes*, 2nd edn. Wageningen; Netherlands: Pudoc Publishers.

Di Cola, G., Gilioli, G., and Baumgärtner, J. (1998). Mathematical models for age-structured population dynamics: an overview. In J. Baumgärtner, P. Brandmayr and B. F. J. Manly (eds.), *Population and Community Ecology for Insect Management and Conservation*. Rotterdam: Balkema. pp. 45–62.

Di Cola, G., Gilioli, G. and Baumgärtner, J. (1999). Mathematical models for age–structured population dynamics. In C. B. Huffaker and A. P. Gutierrez (eds.), *Ecological Entomology*, second edn. New York: John Wiley and Sons. pp. 503–34.

Dingkuhn, M., Penning de Vries, F. W. T., & Miezan, K. M. (1993). Improvement of rice plant type concepts: Systems research enables interaction of physiology and breeding. In F. Penning de Vries, V. Teng and K. Metselaar (eds.), *Systems Approaches for Sustainable Agricultural Development*. Dordrecht: Kluwer Academic Publishers. pp. 19–35.

Doutt, R.L. and Nakata, J. (1973). The Rubus leafhopper and its egg parasitoid: an endemic biotic system useful in grape-pest management. *Environmental Entomology*, **2**, 381–406.

Falcon, L.A., Van Den Bosch, R., Gallagher, J. and Davidson, A. (1971). Investigation of the pest status of *Lygus hesperus* in cotton in Central California. *Journal of Economic Entomology*, **64**, 56–61.

Fitzpatrick, E.A. and Nix, H.A. (1970). The climatic factor in Australian grasslands ecology. In R.M. Moore (ed.), *Australian Grasslands*. Australian National University Press. pp. 3–26.

Foerster, von H. (1959). Some remarks on changing populations. In F. Stahlman, Jr. (ed.), *The Kinetics of Cellular Proliferation*. New York: Grune and Stratton. pp. 382–407.

Fouque, F. and Baumgärtner, J. (1996). Simulating development and survival of *Aedes vexans* (Diptera, Culicidae) preimaginal stages under field conditions. *Journal of Medical Entomology*, **33**, 32–8.

Fouque, F., Baumgartner, J. and Delucchi, V. (1992). Analysis of temperature-dependent stage-frequency data of *Aedes vexans* Meigen populations originated from the Magadino Plain Southern Switzerland. *Bulletin of the Society for Vector Ecology*, **17**, 28–37.

Fouque, F., Delucchi, V. and Baumgartner, J. (1991). Mosquito control in the Magadino Plain: faunistic survey of the culicids and identification of the species harmful to a man. *Mitteilungen Der Schweizerischen Entomologischen Gesellschaft*, **64**, 231–42.

Gerling, D. (1990). *Whiteflies: Their Bionomics, Pest Status and Management*, Andover, UK: Intercept Ltd.

Getz, W. and Gutierrez, A.P. (1982). A perspective on crop systems analysis and insect pest management. *Annual Review of Entomology*, **27**, 447–66.

Gilbert, N. (1984). What they didn't tell you about limit cycles. *Oecologica*, **65**, 112–13.

Gilbert, N.E., Gutierrez, A.P., Frazer, B.D. and Jones, R. (1976). *Ecological Relationships*. Reading, UK and San Francisco: W. H. Freeman and Co.

Gilioli, G., Baumgärtner, J. and Vacante, V. (2005). Temperature influences on the functional response of *Coenosia attenuata* Stein (Diptera, Muscidae) individuals. *Journal of Economic Entomology*, **98**, 1524–30.

Graf, B., Baumgärtner, J. and Delucchi, V. (1985). Simulation models for the dynamics of three apple aphids, *Dysaphis plantaginae*, *Rhopalosiphum insertum*, and *Aphis pomi* (Homoptera, Aphididae) in a Swiss apple orchard. *Zeitschrift fur Angewandte Entomologie*, **99**, 453–65.

Graf, B., Baumgartner, J. and Gutierrez, A.P. (1990*a*). Modeling agroecosystem dynamics with the metabolic pool approach. *Mitteilungen Der Schweizerischen Entomologischen Gesellschaft*, **63**, 465–76.

Graf, B., Gutierrez, A.P., Rakotobe, O., Zahner, P. and Delucchi, V. (1990*b*). A simulation model for the dynamics of rice growth and development. 2. The competition with weeds for nitrogen and light. *Agricultural Systems*, **32**, 367–92.

Graf, B., Lamb, R., Heong, K.L. and Fabellar, L. (1992). A simulation model for the population dynamics of rice leaf-folders Lepidoptera: Pyralidae and their interactions with rice. *Journal of Applied Ecology*, **29**, 558–70.

Graf, B., Rakotobe, O., Zahner, P., Delucchi, V. and Gutierrez, A. P. (1990*c*). A simulation model for the dynamics of rice growth and development. 1. The carbon balance. *Agricultural Systems*, **32**, 341–65.

Gray, V.M. (2000). A comparison of two approaches for modelling cassava (*Manihot esculenta* Crantz.) crop growth. *Annals of Botany London*, **85**, 77–90.

Greenplate, J.T. (1999). Quantification of *Bacillus thurigiensis* insect control protein Cry1Ac over time in bollgard cotton fruits and terminals. *Journal of Economic Entomology*, **92**, 1379–83.

Gurney, W.S.C., Middleton, D.A.J., Nisbet, R.M. *et al.* (1996). Individual energetics and the equilibrium demography of structured populations. *Theoretical Population Biology*, **49**, 344–68.

Gutierrez, A.P. (1992). The physiological basis of ratio dependent theory. *Ecology*, **73**, 1529–53.

Gutierrez, A.P. (1996). *Applied Population Ecology: A Supply–Demand Approach*. New York: John Wiley and Sons.

Gutierrez, A.P. (2001). Climate change: effects on pest dynamics. In K.R. Reddy and H.F. Hodges (eds.), *Climate Change and Global Crop Productivity*. Wallingford, UK: CAB International. pp. 353–74.

Gutierrez, A.P. and Baumgärtner, J.U. (1984*a*). Multitrophic level models of predator–prey energetics: I. Age specific energetics models – pea aphid *Acyrthosiphon pisum* (Harris) (Homoptera: Aphididae) as an example. *Canadian Entomologist*, **116**, 923–32.

Gutierrez, A.P. and Baumgärtner, J.U. (1984*b*). Multitrophic level models of predator–prey energetics: II. A realistic model of plant–herbivore–parasitoid–predator interactions. *Canadian Entomologist*, **116**, 933–49.

Gutierrez, A.P., Baumgärtner, J.U. and Hagen, K.S. (1981*a*). A conceptual model for growth, development, and reproduction in the ladybird beetle, *Hippodamia convergens* (Coleoptera: Coccinellidae). *Canadian Entomologist*, **113**, 21–33.

Gutierrez, A.P., Butler, G.D., Jr. and Ellis, C.K. (1981*b*). Pink bollworm: diapause induction and termination in relation to fluctuating temperatures and decreasing photophases. *Environmental Entomology*, **10**, 936–42.

Gutierrez, A.P., Butler, G., Wang, Y. and Westphal, D. (1977*a*). The interaction of the pink bollworm, cotton and weather. *Canadian Entomologist*, **109**, 1457–68.

Gutierrez, A.P., Christensen, J.B., Merritt, C.M. *et al.* (1976). Alfalfa and the Egyptian alfalfa weevil. *Canadian Entomologist*, **108**, 635–48.

Gutierrez, A.P., Daxl, R., Leon Quant, G. and Falcon, L.A. (1981*c*). Estimating economic thresholds for bollworm, *Heliothis zea* Boddie, and boll weevil, *Anthonomus grandis* Boh., damage in Nicaraguan cotton, *Gossypium hirsutum* L. *Environmental Entomology*, **10**, 872–9.

Gutierrez, A.P., Devay, J.E., Pullman, G.S. and Friebertshauser, G.E. (1982). A model of verticillium wilt in relation to cotton growth and development. *Phytopathology*, **73**, 89–95.

Gutierrez, A.P., Dos Santos, W.J., Pizzamiglio, M.A. *et al.* (1991*a*). Modelling the interaction of cotton and the cotton boll weevil: II. Boll weevil (*Anthonomus grandis*) in Brazil. *Journal of Applied Ecology*, **28**, 398–418.

Gutierrez, A.P., Dos Santos, W.J., Villacorta, A. *et al.* (1991*b*). Modelling the interaction of cotton and the cotton boll weevil: I. A comparison of growth and development of cotton varieties. *Journal of Applied Ecology*, **28**, 371–97.

Gutierrez, A.P., Ellis, C.K., d'Oultremont, T. and Pontic, L. (2007). Climatic limits of pink bollworm in Arizona and California: effects of climate warming. *Acta Oecologica*, **30**, 353–64.

Gutierrez, A.P., Falcon, L.A., Loew, W.B., Leipzig, P. and van den Bosch, R. (1975). An analysis of cotton production in California: a model for Acala cotton and the efficiency of defoliators on its yields. *Environmental Entomology*, **4**, 125–36.

Gutierrez, A.P., Havenstein, D.E., Nix, H.A. and Moore, P.A. (1974). The ecology of *Aphis craccivora* Koch and subterranean clover stunt virus. III. A regional perspective of the phenology and migration of the cowpea aphid. *Journal of Applied Ecology*, **11**, 21–35.

Gutierrez, A.P., Leigh, T.F., Wang, Y. and Cave, R.D. (1977*b*). An analysis of cotton production in California: *Lygus hesperus* injury – an evaluation. *Canadian Entomologist*, **109**, 1375–86.

Gutierrez, A.P., Mills, N.J., Schreiber, S.J. and Ellis, C.K. (1994). A physiologically based tritrophic perspective on bottom up–top down regulation of populations. *Ecology*, **75**, 2227–42.

Gutierrez, A.P., Neuenschwander, P., Schulthess, F. *et al.* (1988*a*). Analysis of biological control of cassava pests in Africa: II. Cassava mealybug, *Phenacoccus manihoti*. *Journal of Applied Ecology*, **25**, 921–40.

Gutierrez, A.P., Neuenschwander, P. and Van Alphen, J.J.M. (1993). Factors affecting biological control of cassava mealybug by exotic parasitoids: a ratio-dependent supply-demand driven model. *Journal of Applied Ecology*, **30**, 706–21.

Gutierrez, A.P., Pitcairn, M.J., Ellis, C.K., Carruthers, N. and Ghezelbash, R. (2005*a*). Evaluating biological control of yellow starthistle (*Centaurea solstitialis*) in California: a GIS based supply–demand demographic model. *Biological Control*, **34**, 115–31.

Gutierrez, A.P. and Pizzamiglio, M.A. (in press). Physiologically based GIS model of weather mediated model of competition between a parasitoid and a coccinellid predator of oleander scale. *Neotropical Entomology*.

Gutierrez, A.P. and Ponsard, S. (2005*b*). Physiologically based model of Bt cotton–pest interactions: I. Pink bollworm: resistance, refuges and risk. *Ecological Modelling*, **191**, 346–59.

Gutierrez, A.P., Ponsard, S. and Adamczyk, J.J., Jr. (2005*c*). A physiologically based model of Bt cotton–pest interactions: II. bollworm–defoliator–natural enemy interactions. *Ecological Modelling*, **191**, 360–82.

Gutierrez, A.P., Pizzamiglio, M.A., Dos Santos, W.J., Villacorta, A. and Gallagher, K.D. (1986). Analysis of diapause induction and termination in field pink bollworm,

Pectinophora gossypilla (Saunders, 1843). *Brazilian Environmental Entomology*, **15**, 494–500.

Gutierrez, A.P. and Regev, U. (2005). The bioeconomics of tritrophic systems: applications to invasive species. *Ecological Economics*, **52**, 382–96.

Gutierrez, A.P., Schulthess, F., Wilson, L.T. *et al.* (1987). Energy acquisition and allocation in plants and insects: a hypothesis for the possible role of hormones in insect feeding patterns. *Canadian Entomologist*, **119**, 109–29.

Gutierrez, A.P., Wermelinger, B., Schulthess, F., Baumgärtner, J.U., Herren, H.R., Ellis, C.K. and Yaninek, J.S. (1988*b*). Analysis of biological control of cassava pests in Africa: I. Simulation of carbon nitrogen and water dynamics in cassava. *Journal of Applied Ecology*, **25**, 901–20.

Gutierrez, A.P., Villacorta, A., Cure, J.R. and Ellis, C.K. (1998). tritophic analysis of the coffee (*Coffee arabica*)–coffee berry borer (*Hypothenemus hampei* Ferrari)–parasitoid system. *Annals of the Entomological Society of Brazil*, **27**, 357–85.

Gutierrez, A.P. and Wang, Y. (1977). Applied population ecology: models for crop production and pest management. In C.S. Holling and G.A. Norton (eds.), *Proceedings of the IIASA Workshop on Pest Management Modelling*, October 25–28 1977, Laxenburg, Austria. pp. 255–80.

Gutierrez, A.P., Williams, D.W. and Kido, H. (1985). A model of grape growth and development: the mathematical structure and biological considerations. *Crop Science*, **25**, 721–8.

Gutierrez, A.P., Yaninek, J.S., Neuenschwander, P. and Ellis, C.K. (1999). A physiologically-based tritrophic metapopulation model of the African cassava foodweb. *Ecological Modelling*, **123**, 225–42.

Gutierrez, A.P., Yaninek, J.S., Wermelinger, B., Herren, H.R. and Ellis, C.K. (1988*c*). Analysis of biological control of cassava pests in Africa: III. Cassava greenmite *Mononychellus tanajoa*. *Journal of Applied Ecology*, **25**, 941–50.

Habib, R. and Blaise, Ph. (eds.). (1996). Third International Symposium on Computer Modelling in Fruit Research and Orchard Management. Fourth International Symposium on Computer Modelling in Fruit Research and Orchard Management. Avignon, 4–8 September, (1995). *Acta Horticulturae*, **416**.

Hanna, R., Onzo, A., Lingeman, R., Yaninek, J.S. and Sabelis, M. (2005). Seasonal cycles and persistence in an acarine predator–prey system on cassava in Africa. *Population Ecology*, **47**, 107–17.

Hanski, I.A. (ed.) (2004). *Ecology, Genetics, and Evolution of Metapopulations*, Amsterdam: Elsevier.

Hanski, I. and Gilpin, M. (1991). Metapopulation dynamics: brief history and conceptual domain. *Biological Journal of the Linnean Society*, **42**, 3–16.

Harper, J.L. and White, J. (1974). The demography of plants. *Annual Review of Ecology and Systematics*, **5**, 419–63.

Hassan, S.T.S. and Wilson, L.T. (1993). Simulated larval feeding damage patterns of *Heliothis armigera* (Huebner) and *H. punctigera* (Wallengren) (Lepidoptera: Noctuidae) on cotton in Australia. *International Journal of Pest Management*, **39**, 239–45.

Hassell, M.P., Lawton, J.H. and Beddington, J.R. (1976). The components of arthropod predation. I. The prey death rate. *Journal of Animal Ecology*, **45**, 135–64.

Hastings, A. (1994). Metapopulation dynamics and genetics. *Annual Review of Ecology and Systematics*, **25**, 167–88.

Hawkins, B.A. and Cornell, V.H. (1999). *Theoretical Approaches to Biological Control*, Cambridge, UK: Cambridge University Press.

Herren, H.R. and Neuenschwander, P. (1991). Biological control of cassava pests in Africa. *Annual Review of Entomology*, **36**, 257–83.

Hesketh, J.D., Baker, N.D. and Duncan, W.G. (1971). Simulation of growth and yield of cotton: respiration and the carbon balance. *Crop Science*, **11**, 39–398.

Hilbeck, A., Baumgartner, M., Martinnad, P.M. and Bigler, F. (1998). Effects of transgenic *Bacillus thuringiensis* corn-fed prey on mortality and development time of immature *Chrysoperla carnea* (Neuroptera: Chrysopidae). *Environmental Entomology*, **27**, 480–7.

Hilbeck, A., Moar, W.J., Pusztai-Carey, M., Filippini, A. and Bigler, F. (1999). Prey-mediated effects of Cry1Ab toxin and protoxin and Cry2A protoxin on the predator Chrysoperla carnea. *Entomologia Experimentalis et Applicata*, **91**, 305–16.

Hilbert, D.W. (1995). Growth-based approach to modeling the developmental rate of arthropods. *Environmental Entomology*, **24**, 771–8.

Holling, C.S. (1978). *Adaptive Environmental Assessment and Management*. Chichester, UK: Wiley.

Holst, N., Axelsen, J.A., Olesen, J.E. and Ruggle, P. (1997). Object-oriented implementation of the metabolic pool model. *Ecological Modelling*, **104**, 175–87.

Holst, N. and Ruggle, P. (1997). A physiologically based model of pest–natural enemy interactions. *Experimental and Applied Acarology*, **21**, 325–41.

Holt, J., Jeger, M.J., Thresh, J.M. and Otim-Nape, G.W. (1997). An epidemiological model incorporating vector population dynamics applied to African cassava mosaic virus disease. *Journal of Applied Ecology*, **34**, 793–806.

Huffaker, C.B. and Croft, B.A. (1976). Integrated pest management in the U.S.: progress and promise. *Environmental Health Perspectives*, **14**, 167–83.

Huffaker, C.B. and Kennett, C.E. (1966). Studies of two parasites in control of the olive scale, *Parltoria oleae* (Colvée). IV. Biological control of *Parlatoria oleae* (Colvée) through the compensatory action of two introduced parasites. *Hilgardia*, **37**, 283–334.

Inubushi, K., Hori, K., Matsumoto, S. and Wada, H. (1997). Anaerobic decomposition of organic carbon in paddy soil in relation to methane emission to the atmosphere. *Water Science and Technology*, **36**, 523–30.

Isley, D. (1932). Abundance of bollweevil in relation to summer weather and food. *Arkansas Experimental Station Bulletin*.

Iwaki, H. (1975). Computer simulation of vegetative growth of rice plants. In Y. Murata (ed.), *Crop Productivity and Solar Energy Utilization in Various Climates in Japan*, Tokyo: University of Tokyo Press.

Iwaki, H. (1977). Computer simulation of growth process of paddy rice. *Japan Agricultural Research Quarterly*, **11**, 6–11.

Ives, A.R., Carpenter, S.R. and Dennis, B. (1999). Community interaction webs and zooplankton responses to planktivory manipulations. *Ecology*, **80**, 1405–21.

Ives, A.R. and Jansen, V.A.A. (1998). Complex dynamics in stochastic tritrophic models. *Ecology*, **79**, 1039–52.

Ives, A.R. and Settle, W.H. (1997). Metapopulation dynamics and pest control in agricultural systems. *Amer. Nat*, **149**, 220–46.

Janisch, E. (1925). Uber die Temperaturabhangigkeit Biologischer Vorgange und ihre Kurvenmabige Analyse. *Pfluger Archiv fur die Gesamte Physiologie des Menschen und dir tiere*, **209**, 414–36.

Johnsen, S. and Gutierrez, A.P. (1997*a*). Induction and termination of winter diapause in a Californian strain of the cabbage maggot (Diptera: Anthomyiidae). *Environmental Entomology*, **26**, 84–90.

Johnsen, S., Gutierrez, A.P. and Jorgensen, J. (1997*b*). Overwintering in the cabbage root fly *Delia radicum*: a dynamic model of temperature-dependent dormancy and post-dormancy development. *Journal of Applied Ecology*, **34**, 21–8.

Jørgensen, S.E. (2001). (Ed.). *Thermodynamics and Ecological Modelling*, Boca Raton, FL: Lewis Publishers.

Jørgensen, S.E. and Svirezhev, Y.M. (2004). *Towards a Thermodynamics Theory for Ecological Systems*. Amsterdam: Elsevier.

Kenmore, P.E., Carino, F., Perez, C.A., Dyck, V.A. and Gutierrez, A.P. (1986). Population regulation of the rice brown planthopper (*Nilaparvata lugens* Stal) within rice fields in the Philippines. *Journal of Plant Protection in the Tropics*, **1**, 19–37.

Kogan, M., Croft, B.A., and Sutherst, R.F. (1999). Applications of ecology for integrated pest management. In C.B. Huffaker and A.P. Gutierrez (eds.), *Ecological Entomology*. John Wiley & Sons. pp. 681–727.

Lactin, D.J., Holliday, N.J., Johnson, D.L. and Craigen, R. (1995). Improved rate model of temperature-dependent development by arthropods. *Environmental Entomology*, **24**, 68–75.

Lawton, J.H. (1977). Spokes missing in ecological wheel. *Nature* (London), **265**, 768.

Lawton, J.H., Hassell, M.P. and Beddington, J.R. (1975). Prey death rates and rate of increase of anthropod predator populations. *Nature*, **225**, 60–2.

Leslie, P.H. (1945). On the use of matrices in certain population mathematics. *Biometrika*, **33**, 183–212.

Logan, J.A., Woollkind, D.J., Hoyt, S.C. and Tanagoshi, L.K. (1976). An analytical model of temperature dependent rate phenomena in arthropods. *Environmental Entomology*, **5**, 1133–40.

Manetsch, T.J. (1976). Time-varying distributed delays and their use in aggregate models of large systems. *IEEE Transactions on Systems Man and Cybernetics*, **6**, 547–53.

Matthews, R.B. and Hunt, L.A. (1994). GUMCAS: a model describing the growth of cassava (*Manihot esculenta* L. Crantz). *Field Crops Research*, **36**, 69–84.

McMennamy, J.A. and O'Toole, J.C. (1983). Ricemod: a physiologically based rice growth and yield model. *IRPS*, **87**, 1–33.

McQueen, D.J. and Post, J.R. (1988). Cascading trophic interactions uncoupling at the zooplankton-phytoplankton link. *Hydrobiologia*, **159**, 277–96.

Meikle, W.G., Holst, N. and Markham, R.H. (1999). Population simulation model of *Sitophilus zeamais* (Coleoptera: Curculionidae) in grain stores in West Africa. *Environmental Entomology*, **28**, 836–44.

Meikle, W.G., Holst, N., Scholz Dagmar, D. and Markham, R. (1998). Simulation model of *Prostephanus truncatus* (Coleoptera: Bostrichidae) in rural maize stores in the Republic of Benin. *Environmental Entomology*, **27**, 59–69.

Mills, N.J. and Getz, W.M. (1996). Modelling the biological control of insect pests: A review of host-parasitoid models. *Ecological Modelling*, **92**, 121–43.

Mills, N.J. and Gutierrez, A.P. (1996). Prospective modelling in biological control: an analysis of the dynamics of heteronomous hyperparasitism in a cotton-whitefly-parasitoid system. *Journal of Applied Ecology*, **33**, 1379–94.

Murdoch, W.W., Swarbrick, S.L., Luck, R.F., Walde, S. and Yu, D.S. (1996). Refuge dynamics and metapopulation dynamics: An experimental test. *American Naturalist*, **147**, 424–44.

Mutsaers, H.J.W. and Wang, Z. (1999). Are simulation models ready for agricultural research in developing countries? *Agronomy Journal*, **91**, 1–4.

Nachman, G. (1987). Systems analysis of acarine predator-prey interactions: II. The role of spatial processes in system stability. *Journal of Animal Ecology*, **56**, 267–82.

Nachman, G. (1991). An acarine predator-prey metapopulation system inhabiting greenhouse cucumbers. *Biological Journal of the Linnean Society*, **42**, 285–304.

Naranjo, S. (2005). Long-term assessment of the effects of transgenic Bt cotton on the abundance of nontarget arthropods natural enemies. *Ecological Entomology*, **34**, 1193–210.

Neuenschwander, P., Hammond, W.N.O., Gutierrez, A.P. *et al.* (1989). Impact assessment of the biological control of the cassava mealybug, *Phenacoccus manihoti* Matile-Ferrero (Hemiptera: Pseudococcidae), by the introduced parasitoid Epidinocarsis lopezi (De Santis) (Hymenoptera: Encyrtidae). *Bulletin of Entomological Research*, **79**, 579–94.

Neuenschwander, P., Hennessey, R.D. and Herren, H.R. (1987). Food web of insects associated with the cassava mealybug, *Phenacoccus manihoti* Matile-Ferrero (Hemiptera: Pseudococcidae), and its introduced parasitoid, *Epidinocarsis lopezi* (DeSantis) (Hymenoptera: Encyrtidae), in Africa. *Bulletin of Entomological Research*, **79**, 177–89.

Oduor, G.I., Sabelis, M.W., Lingeman, R., De Moraes, G. and Yaninek, J.S. (1997). Modelling fungal (*Neozygites* cf. *floridana*) epizootics in local populations of cassava green mites (*Mononychellus tanajoa*). *Experimental and Applied Acarology*, **21**, 485–506.

Olszyk, D.M., Centeno, H.G.S., Ziska, L.H., Kern, J.S. and Matthews, R.B. (1999). Global climate change, rice productivity and methane emissions: comparison of simulated and experimental results. *Agriculture and Forest Meteorology*, **97**, 87–101.

Penning De Vries, F.W.T. (1993). Rice production and climate change. In F. Penning de Vries, P. Teng and K. Metselaar (eds.), *Systems Approaches for Sustainable Agricultural Development*. Boston: Kluwer Academic Publishers. pp. 175–89.

Penning De Vries, F.W.T., Jansen, D.M., Ten Berge, H.F.M. and Bakema, A.H. (1989). *Simulation of Ecophysiological Processes of Growth of Several Annual Crops*. Wageningen, Netherlands: Pudoc.

Peters, R. (1991). *A Critique for Ecology*. Cambridge, UK: Cambridge University Press.

Petrusewicz, K. and MacFayden, A. (1970). *Productivity of Terrestrial Animals: Principles and Methods*, IBP Handbook **13**. Oxford, UK: Blackwell.

Pickering, J. and Gutierrez, A.P. (1991). Differential impact of the Pathogen *Pandora neoaphidis* (R. & H.) Hunber (Zygomycetes: Entomophthorales) on the species composition of Acyrthosiphon aphids in alfalfa. *Canadian Entomologist*, **123**, 315–20.

Ponsard, S., Gutierrez, A.P. and Mills, N.J. (2002). Effects of Bt-toxin (Cry1Ac) in transgenic cotton on the adult longevity of four heteropteran predators. *Environmental Entomology*, **31**, 1197–205.

Reddy, K.R., Hodges, H.J. and Mckinion, J.M. (1997). Modeling temperature effects on cotton internode and leaf growth. *Crop Science*, **37**, 503–9.

Regev, U., Gutierrez, A.P., Devay, J.E. and Ellis, C.K. (1990). Optimal strategies for management of verticillium wilt. *Agricultural Systems*, **33**, 139–52.

Regev, U., Gutierrez, A.P., Schreiber, S.J. and Zilberman, D. (1998). Biological and economic foundations of renewable resource exploitation. *Ecological Economics*, **26**, 227–42.

Ritchie, J.T. (1972). Model for predicting evaporation from a row crop with incomplete cover. *Water Resources Research*, **8**, 1204–13.

Rochat, J. and Gutierrez, A.P. (2001). Weather mediated regulation of olive scale by two parasitoids. *Journal of Animal Ecology*, **70**, 476–90.

Ruggle, P. and Gutierrez, A.P. (1995). Use of life tables to assess host plant resistance in alfalfa to *Therioaphis trifolii* F. *maculata* (Homoptera: Aphidae): hypothesis for maintenance of resistance. *Environmental Entomology*, **24**, 313–25.

Russell, D. (1995). *Report to the Natural Resources Institute on the use of the Moalr/Nri Uk Cotton Model for Pink Bollworm Control in Fayoum and Beni-Suef*. Natural Resources Institute. Portsmouth, UK.

Sabelis, M.W., Diekmann, O. and Jansen, V.A.A. (1991). Metapopulation persistence despite local extinction: Predator–prey patch models of the Lotka–Volterra type. *Biological Journal of the Linnean Society*, **42**, 267–84.

Sadras, V.O. and Wilson, L.J. (1997). Growth analysis of cotton crops infested with spider mites: II. Partitioning of dry matter. *Crop Science*, **37**, 492–7.

Schaub, L. and Baumgärtner, J. (1989). The significance of mortalities and temperatures on the phenology of *Orthotylus marginalis* Reut. (Heteroptera, Miridae). *Mitteilungen der Schweizerischen Entomologischen*, **62**, 235–45.

Schreiber, S. and Gutierrez, A.P. (1998). A supply–demand perspective of species invasions and coexistence: applications to biological control. *Ecological Modelling*, **106**, 27–45.

Sciarretta, A., Girma, M., Belayun, L., Tikubet, G. and Baumgärtner, J. (2005). Development of an adaptive Tsetse fly population management scheme for the Luke community, Ethiopia. *Journal of Medical Entomology*, **42**, 1006–10.

Sequeira, R. A., Sharpe, P. J. H., Stone, N. D., El-Zik, K. M. and Makela, M. E. (1991). Object-oriented simulation: plant growth and discrete organ to organ interactions. *Ecological Modelling*, **58**, 55–90.

Settle, W.H. and Wilson, L.T. (1990*a*). Behavioral factors affecting differential parasitism by *Anagrus epos* (Hymenoptera: Mymaridae) of two species of *erythroneuran* leafhoppers (Homoptera: Cicadellidae). *Journal of Animal Ecology*, **59**, 877–92.

Settle, W. H. and Wilson, L. T. (1990*b*). Invasion by the variegated leafhopper and biotic interaction: parasitism, competition, and apparent competition. *Ecology*, **71**, 1461–70.

Settle, W. H., Ariawan, H., Hartjahyo, E. T. *et al*. (1996). *Managing Tropical Rice Pests Through Conservation of Generalist Natural Enemies and Alternative Prey*, 77, 1975–88.

Severini, M., Baumgärtner, J. and Ricci, M. (1990). Theory and practice of parameter estimation of distributed delay models for insect and plant phenologies. In R. Guzzi, R. A. Navarra and J. Shukla (eds.), *Meteorology and Environmental Sciences*. Singapore: World Scientific and International Publishers. pp. 674–719.

Severini, M., Baumgärtner, J., Seifert, M. and Ricci, M. (1996). The analysis of poikilothermic population development by means of time distributed delay models. In G. Di Cola and G. Gilioli (eds.), Computer Science and Mathematical Methods in Plant Protection, *Quaderni del Dipartimento di Matematica*, 135, 159–76.

Shelford, V. E. (1931). Some concepts of bioecology. *Ecology*, **12**, 455–67.

Sharpe, P. J. H., Curry, G. L., DeMichele, D. W. and Cole, C. L. (1977). Distribution model of organism developmental times. *Journal of Theoretical Biology*, **66**, 21–38.

Sharpe, P.J.H. and Hu, L.C. (1980). Reaction kinetics of nutrition dependent poikilotherm development. *Journal of Theoretical Biology*, **82**, 317–33.

Sinko, J. W. and Streifer, W. (1967). A new model for age structure of a population. *Ecology*, **48**, 910–18.

Spiller, D. A. and Schoener, T. W. (2001). An experimental test for predator-mediated interactions among spider species. *Ecology*, **82**, 1560–70.

Stephens, A. P., Buskirk, S. W., Hayward, G. D. and Del Rio, C. M. (2005). Information theory and hypotheses testing: a call for pluralism. *Journal of Applied Ecology*, **42**, 4–12.

Stewart, S. D. and Sterling, W. L. (1988). Dynamics and impact of cotton fruit abscission and survival. *Environmental Entomology*, **17**, 629–35.

Stinner, R. E., Gutierrez, A. P. and Butler, G., Jr. (1974). An algorithm for temperature-dependent growth rate simulation. *Canadian Entomologist*, **106**, 519–24.

Stone, N. D. and Gutierrez, A. P. (1986*a*). I. A field oriented simulation of pink bollworm in southwestern desert cotton. *Hilgardia*, **54**, 1–24.

Stone, N. D. and Gutierrez, A. P. (1986*b*). II. A management model for pink bollworm in southwestern cotton. *Hilgardia*, **54**, 25–41.

Stone, N. D. and Gutierrez, A. P., Getz, W. M. and Norgaard, R. (1986). III. Strategies for pink bollworm control in southwestern desert cotton: An economic simulation study. *Hilgardia*, **54**, 42–56.

Sutherst, R. W., Maywald, G. F. and Bottomly, W. (1991). From CLIMEX to PESKY, a generic expert system for risk assessment. *EPPO Bulletin*, **21**, 595–608.

Tabashnik, B. E., Patin, A. L., Dennehy, T. J. *et al.* (2000). Frequency of resistance to *Bacillus thuringiensis* in field populations of pink bollworm. *Proceedings of the National Academy of Science USA*, **97**, 12980–4.

Tamò, M. and Bäumgartner, J. (1993). Analysis of the cowpea agro-ecosystem in West Africa. I. A demographic model for carbon acquisition and allocation in cowpea, *Vigna unguiculata* (L.) Walp. *Ecological Modelling*, **65**, 95–121.

Tamò, M., Baumgärtner, J. and Gutierrez, A. P. (1993). Modelling the interaction between cowpea and the bean flower thrips *Megalurothrips sjostedti* (Trybom) (Thysanoptera, Thripidae). *Ecological Modelling*, **70**, 89–113.

Toko, M., O'Neil, R. T. J. and Yaninek, J. S. (1996). Development, reproduction and survival of *Mononychellus tanajoa* (Bondar) (Acari: Tetranychidae) on cassava grown under soils of different levels of nitrogen. *Experimental and Applied Acarology*, **20**, 405–19.

Trichilo, P. J. and Wilson, L. T. (1993). An ecosystem analysis of spider mite outbreaks: physiological stimulation or natural enemy suppression. *Experimental and Applied Acarology*, **17**, 291–314.

Turchin, P. and Batzli, G. O. (2001). Availability of food and the population dynamics of arvicoline rodents. *Ecology*, **82**, 1521–34.

Vansickle, J. (1977). Attrition in distributed delay models. *IEEE Transactions on Systems Man and Cybernetics*, **7**, 635–8.

van den Bosch, R. (1978). *The Pesticide Conspiracy*. New York: Doubleday.

Van Keulen, H. (1976). A calculation method for potential rice production. Contributions from the Central Research Institute for Agriculture. *Bogor*, **21**, 1–26.

Villavaso, E. J., Mulrooney, J. E., McGovern, W. L. and Howard, K. D. (2001). Low dosages of malathion for boll weevil eradication. *Southwestern Entomologist Supplement*.

von Arx, R., Baumgartner, J. and Delucchi, V. (1984). A model to simulate the population dynamics of *Bemisia tabaci* (Aleyrodidae) on cotton in the Sudan Gezira. *Zeitschrift Fuer Angewandte Entomologie*, **96**, 341–63.

Wagenmakers, P. S., van der Werf, W. and Blaise, Ph. (1999). Proceedings of the fifth international symposium on computer modelling in fruit research and orchard management. *Acta Horticulturae*, **499**, 303.

Waltner-Toews, D., Kay, J. K., Neudoerffer, C. and Gitau, T. (2003). Perspective changes everything: managing ecosystem from inside out. *Frontiers in Ecology and Environment*, **1**, 23–30.

Wang, Y. and Gutierrez, A. P. (1980). An assessment of the use of stability analysis in population ecology. *Journal of Animal Ecology*, **49**, 435–52.

Wang, Y., Gutierrez, A. P., Oster, G. and Daxl, R. (1977). General population model for plant growth and development: coupling cotton–herbivore interactions. *Canadian Entomologist*, **109**, 1359–74.

Watt, K.F.E. (1959). A mathematical model for the effect of densities of attacked and attacking species on the number attacked. *Canadian Entomologist*, **91**, 129–44.

Welch, S.M. (1984). Developments in computer-based IPM extension delivery systems. *Annual Review of Entomology*, **29**, 359–81.

Welch, S.M., Croft, B.A., Brunner, J.F. and Michels, M. (1978). PETE: An extension phenology modeling system for multi-species pest complex. *Environmental Entomology*, **7**, 482–94.

Wermelinger, B., Baumgartner, J. and Gutierrez, A.P. (1991*a*). A demographic model of assimilation and allocation of carbon and nitrogen in grapevines. *Ecological Modelling*, **53**, 1–26.

Wermelinger, Baumgartner, B.J., Zahner, P. and Delucchi, V. (1990*a*). Environmental factors affecting the life tables of *Tetranychus urticae* Koch (Acarina): I. Temperature. *Mitteilungen Der Schweizerischen Entomologischen Gesellschaft*, **63**, 55–62.

Wermelinger, B., Candolfi, M.P. and Baumgartner, J. (1992). A model of the European red mite (Acari: Tetranychidae) population dynamics and its linkage to grapevine growth and development. *Journal of Applied Entomology*, **114**, 155–66.

Wermelinger, B., Oertli, J.J. and Baumgartner, J. (1991*b*). Environmental factors affecting the life-tables of *Tetranychus urticae* (Acari: Tetranychidae): III. Host-plant nutrition. *Experimental and Applied Acarology*, **12**, 259–74.

Wermelinger, B., Schnider, F., Oertli, J.J. and Baumgartner, J. (1990*b*). Environmental factors affecting the life tables of *Tetranychus urticae* Koch (Acarina): II. Host plant water stress. *Mitteilungen der Schweizerischen Entomologischen Gesellschaft*, **63**, 347–58.

Williams, D.W. (1981). Ecology of the Blackberry Leafhopper, *Dikrella Californica* (Lawson), and its Role California Grape Agro-Ecosystems. Ph.D. thesis. Berkeley, CA: University of California.

Wilson, L.T., Corbett, A., Trichilo, P.J. *et al.* (1992). Strategic and tactical modelling cotton–spider mite agroecosystem management. *Experimental and Applied Acarology*, **14**, 357–70.

Wilson, L.T. and Gutierrez, A.P. (1980). Fruit predation submodel: Heliothis larvae feeding upon cotton fruiting structures. *Hilgardia*, **48**, 24–36.

Winter, F. (Ed.) (1986). First international symposium on computer modelling in fruit research and orchard management. Hohenheim, Federal Republic of Germany, 10–13 September (1985). *Acta Horticulturae*, **184**, 205.

Wit, de C.T. and Goudriaan, J. (1978). *Simulation of Ecological Processes*, 2nd edn. Wageningen, The Netherlands: Pudoc Publishers.

Wu, G.W. and Wilson, L.T. (1997). Growth and yield response of rice to rice water weevil injury. *Environmental Entomology*, **26**, 1191–201.

Zahner, P., Band, J. and Baumgärtner, J. (1988). Analysis of plant–spider mite interactions: I. Population models for the dynamics of *Panonychus ulmi* and *Tetranychus urticae* in apple orchard. *Acta Oecologica Applicata*, **9**, 311–32.

Zemek, R. and Nachman, G. (1999). Interactions in a tritrophic acarine predator–prey metapopulation system: prey location and distance moved by *Phytoseiulus persimilis* (Acari: Phytoseiidae). *Experimental and Applied Acarology*, **23**, 21–40.

11

Weed ecology, habitat management and IPM

R.F. NORRIS

11.1 Introduction

Pests are organisms that interfere with the goals of humans. Plants that interfere with human activities are typically referred to as weeds. Such plants are unique as a category of pest as they are ecologically producers, in contrast with organisms in all other pest categories, which are consumers. Weeds are thus in a different ecological position from all other categories of pests, resulting in differences in control tactics, and making weeds and their management likely to interact with all other categories of pests.

Weeds, like all plants, are not mobile. Therefore weeds cannot "search", in the sense of moving to a new location, for resources that they need. Lack of mobility also means that weeds cannot move to escape from organisms that would feed on them. These two differences between weeds and organisms in all animal pest categories mean that the approaches to weed management, and their response to such management, are often quite different from those for animal pests.

Most animal pests and aerial pathogens are not equally devastating every year. Their population dynamics lead to outbreak years when losses can be very high, and other years when populations remain low and losses are small. Weeds, on the other hand, like most other soil-borne pests, typically occur in damaging numbers every year once a site is infested, necessitating control action every year in order to achieve the crop yields desired by humans.

Since the late 1960s, there has been increasing utilization of the integrated pest management (IPM) paradigm. An underlying principle of IPM is that the biology and ecology of the target pest(s) must be understood before an economically and ecologically suitable management strategy can be devised. The importance of weed ecology to weed management is presented by Radosevich *et al.* (1997) and by Liebman *et al.* (2001). Weed science has developed the concept of integrated weed management (IWM) as a subset of the broader IPM paradigm (Buhler, 2002; Cardina *et al.*, 1999; Mortensen *et al.*, 2000; Swanton and Murphy, 1996; Thill *et al.*, 1991; Vangessel, 1996). This chapter reviews some of the more important aspects of weed biology and ecology needed to utilize integrated weed management programs within the IPM framework.

11.2 Ecosystems and weeds

11.2.1 Natural succession

Crop yield that is adequate to meet human needs is eliminated for many crops when weeds are not controlled. Many modern crops cannot survive in the absence of human input. Weeds are, in fact, the early manifestation of succession; they are the first stages in the progression from open soil to climax vegetation. In the absence of human input, weeds represent establishment of the early successional stages that take over as the crop plants die out due to lack of human intervention. When Gilbert Lawes withdrew all intervention from the wheat crop on part of the long-term experiment on Broadbalk at Rothamsted Experimental Station in 1883, the wheat survived three years (Kerr *et al.*, 2000; Lawes, 1884). The fundamental reason that late successional stages cannot meet human needs is due to the decline in net productivity per unit area that occurs as succession progresses towards the climax vegetation. There is no net productivity once the climax stage has been reached, as carbon gain from photosynthesis is balanced by carbon loss due to respiration. From the anthropocentric human viewpoint, weed management is therefore ecologically mandated.

11.2.2 Trophic relationships

Trophic dynamics attempts to understand the flow ecological resources, such as energy and mineral nutrients, through the various organisms that comprise an ecosystem. The concept of food webs has been developed to explain the complex interactions between organisms in an ecosystem based on their food source(s) and to whom they, in turn, serve as food (e.g. Pimm, 2002; Pimm *et al.*, 1991; Polis and Winemiller, 1996). The food web concept results in a hierarchy of organisms based on their position in the food chain.

Plants are the only organisms that can directly utilize mineral nutrients, energy from the sun, carbon dioxide in the atmosphere, and water to synthesize the biochemicals needed for life. Ecologically plants are referred to as producers (Figure 11.1). All other organisms are consumers; those that eat plants are herbivores or are plant pathogens, and those that eat other consumers are carnivores. For a more complete presentation of these concepts, see pages 71–80 in Norris *et al.* 2003.

Weeds are plants. Therefore, ecologically, they are producers (Figure 11.1). All other organisms that are considered to be pests are primary consumers. The unique ecological position of weeds among pests means that their biology and ecology is often different from that of animal pests. Techniques to monitor and manage weeds are also often different than those used for animal pests. Due to the trophic position of weeds in ecosystems, controlling them can remove resources for consumer organisms. This means that management of weeds has the potential to interact with organisms in all other pest and beneficial categories.

Because weeds are producers, higher trophic order organisms play different roles in the IPM context in comparison with other pest organisms (Figure 11.1). Primary consumers are beneficial, not pests, when they use weeds as their food source. They provide biological control of the weed. Secondary consumers, instead of being beneficial, are a problem for biological weed control as they are organisms that feed on the primary consumer biocontrol agent.

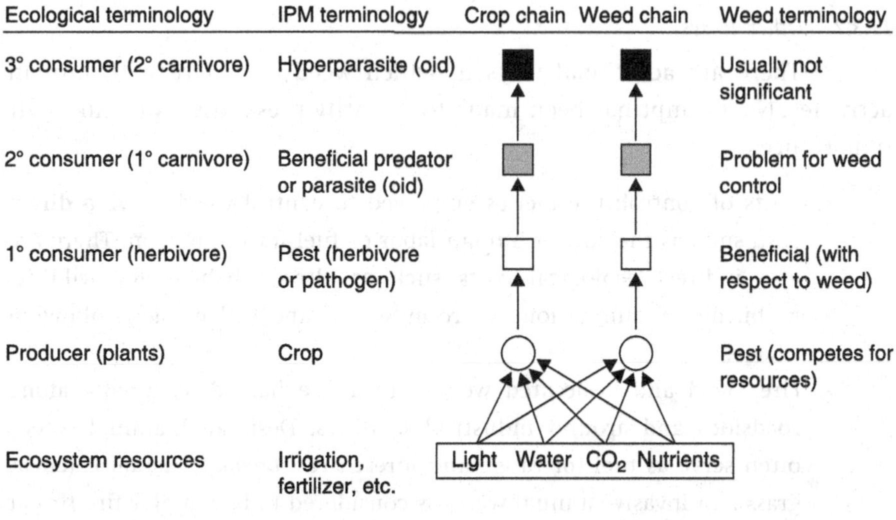

Figure 11.1. Simple crop and weed food chains with associated terminology.

11.3 Impacts of weeds

The first component of IPM decision-making is to ascertain that the pest is actually causing damage and economic loss. Due to the trophic position of weeds the type of damage inflicted on crops is different from that caused by all other categories of pests. Pests in all categories other than weeds directly damage crops, such as by consuming plant parts, injecting toxins, or causing wounds. Weeds do not directly damage crop plants, but rather compete for ecosystem resources, or serve as a resource for other organisms.

11.3.1 Crop losses to competition

Competition with crops is the most serious impact of weeds (Bridges, 1992). The crop experiences a reduction in resources in response to competition by weeds, leading to less growth and to loss of yield. Even with current best management practices it was estimated that monetary losses due to weeds in crops exceeded $4 billion in the USA and Canada in 1992. This figure does not include losses attributed to other impacts of weeds that are noted in the following sections.

Most annual crop yields are typically reduced between 50% and 90% in the absence of weed control. As noted earlier, in the absence of weed control wheat did not survive beyond three years. The weed science literature is replete with examples of crop losses due to competition (Zimdahl, 1980), and there are many more reports in the intervening years.

11.3.2 Other losses

There are additional ways in which weeds interfere with human activities. No attempt has been made to quantify these losses or rank their importance.

- Costs of control: the tactics employed to control weeds have a direct cost, such as the cost of human labor or fuel for cultivation. There are also indirect ecological costs such as altered habitat for wildlife, herbicide contamination of groundwater, and soil erosion following tillage.
- Fire: dead and desiccated weeds are a fire hazard, especially along roadsides and around industrial facilities. Desiccated annual weeds often serve as fuel for range and forest fires; *Bromus tectorum* L. (cheat grass), an invasive annual weed, is considered to be a major fire threat to western rangelands in the USA.

- Allergic reaction: pollen from many weed species causes allergic reaction in sensitive persons; pollen of *Ambrosia trifida* L. (giant ragweed) is a notorious example. Many plants contain chemicals that cause reaction when sensitive individuals touch them; *Toxicodendron diversilobum* (T. & G.) Greene (poison oak) and T. *radicans* (L.) Ktze. (poison ivy) are examples.
- Poisoning: many weeds contain chemicals that are toxic to herbivores. This is a manifestation of the chemical protection developed in response to lack of mobility to escape attack. Large numbers of livestock are poisoned by weeds every year; *Halogeton glomeratus* (Stephen ex Bieb.) C. A. Mey (halogeton) is an example.
- Aquatic: many plants grow in aquatic systems. When they interfere with human goals they are considered to be weeds. For example, weeds such as *Eichornia crassipes* (Mart.) Solms (water hyacinth, a flowering plant), *Salvinia auriculata* Aubl. (giant salvinia, a fern), and *Caulerpa* sp. (an alga) create enormous losses to navigation, water supplies, and recreation.
- Rights-of-way safety: weeds can obscure safety signs and lights, and can physically damage paved surfaces.
- Alteration of natural ecosystems: degradation of natural ecosystems in response to invasion of non-indigenous species is now recognized as a major impact of weeds (see next section).

11.3.3 Plant invasion/habitat alteration

Spread of plants from their native habitat into areas in which they had not previously occurred is recognized as a serious threat to agriculture, and to natural ecosystems. It has been estimated that over 1200 plants, that are now classified as weeds, have been introduced into the USA since the mid-nineteenth century (United States Congress and Office of Technology Assessment, 1993).

Agriculture, on a worldwide basis, continually faces invasion by non-indigenous weed species, which can lead to disruption of current weed management practices. For example, most of the weeds in regions where irrigated agriculture is practiced are not native; they were either introduced as crops, as crop contaminants, or were accidentally imported by human activity. It has been estimated that non-indigenous weeds annually cause over $34 billion in losses (lost production and cost of control) in the USA (Pimentel *et al.*, 2000).

The invasion of native communities by non-indigenous species is much more insidious than invasion in agricultural ecosystems. Such invasion often goes unrecognized, and can ultimately lead to substantial modification of the ecology of the system being invaded due to vegetation displacement (competition) by the invading species. For example, in the USA, it is estimated that 50% of the species on the endangered plant species lists are so endangered by invasive plants as a result of habitat modification. Invasive plant species threaten habitats from the Galapagos Islands to marinas in southern California in addition to the problems that they cause commercial agriculture.

Attempts are now being made to establish biological characteristics of plants that may lead to prediction of invasive ability (e.g. Daehler and Carino, 2000; Rejmanek, 2000). These characteristics are discussed in greater detail in the following section on weediness life traits, as they are the biology of plants from the viewpoint of what makes a plant a weed.

11.3.4 *Loss prediction*

Prior to initiating an IPM action it is often desirable to predict the losses that can be anticipated to occur. Predictive models are being developed both for invasion traits and for crop losses. Most data are still site-specific, which limits the utility in attempting to predict yield losses in different ecosystems.

Interference is a complicated process that integrates several different aspects of plant growth. Mechanistic models that simulate competition have been developed that predict losses due to weeds (Kropff and van Laar, 1993; Lotz *et al.*, 1995), but use of such models as an aid to weed management decision-making remains minimal.

As crop yield loss occurs at low weed densities (see below) a second approach has been to estimate the impacts of a single weed plant on the surrounding crop plants. This is referred to as an area of influence, which is similar to neighborhood models used in plant ecology. Once the impact of single weed is known, it is possible to predict yield loss from any known density, provided the weeds are not sufficiently dense to experience intraspecific competition. The area of influence method for estimating yield loss is used in several weed control decision aids (e.g. Kwon *et al.*, 1995; Lindquist, 2001; Renner and Black, 1991; White and Coble, 1997; Wilkerson *et al.*, 1991).

A third approach to predicting yield loss caused by weeds involves estimation of early canopy development. In this technique the crop yield is correlated with estimated weed and crop leaf area relatively early in crop development. This technique remains at the experimental level.

11.4 Weed biology: weediness life traits

Weeds are plants. What makes some plants weedy, while others typically are not, is a fundamental ecological question. Some have argued that there are no general weedy characteristics, claiming that each ecosystem will have different ability of weeds to invade. However, most books on weed biology and control (Aldrich and Kremer, 1997; Anderson, 1996; Radosevich *et al.*, 1997; Ross and Lembi, 1999) provide lists of weediness traits based on the "ideal weed" list developed by Baker (1965).

More recently, lists of plant characteristics have been developed as part of attempts to predict if a plant will be an invasive weed (Pheloung *et al.*, 1999). The following discussion is based on an annotated list developed from the two sources noted above (Table 11.1). The table is divided into overall categories of climatic traits, undesirable traits, reproductive characteristics, dispersal mechanisms, persistence traits, and genetic structure. These form a useful framework for considering weediness of plants.

To be a weed, a plant must have broad climatic requirements, so that it can flourish under varied conditions. Probably the single most important characteristic is the ability to germinate and grow over a relatively wide range of conditions. The other requirements listed in Table 11.1 are probably less important.

Plants must possess one to several undesirable traits in order to be classified as weeds by humans. Morphological and physiological traits that confer on the weed the ability to outcompete other plants are probably the most important. These are the characteristics that lead to the ability of weeds to cause yield loss in crops. From the aspect of invasion of non-agricultural habitats, the ability to form monospecific stands that displace other vegetation is a critical manifestation of the competitive ability of weeds. All the other undesirable characteristics listed in Table 11.1 become important in specific habitats, but are probably not universal indicators of weediness.

Certain reproductive characteristics are necessary for a plant to become a weed (Table 11.1). Minimum generative time before achieving reproductive status is universally considered an important trait in weediness. Self-fertilization, coupled with some outcrossing is common, as this removes the necessity for pollination by other plants, yet maintains the ability to hybridize naturally. When weeds do need pollination it should not require specialist pollinators, or the plant should be wind pollinated. The presence of a vegetative reproduction mechanism provides a very important

Table 11.1. *Characteristics of plants that are associated with weediness*[a]

Climatic requirements
Capable of germination and growth under wide variation in conditions.
Can withstand periods of adverse conditions, especially drought.
Shade tolerant at some stage in its development.
Undesirable traits
Possess spines, thorns, burrs, etc. capable of causing injury to animals.
Allelopathic against other plants, but does not impact its own growth.
Unpalatable and/or toxic to grazing animals.
Causes allergic or toxic reaction in humans.
Creates fire hazard in natural ecosystems.
Rapid seedling establishment and growth.
Morphological and physiological traits providing ability to outcompete other plants, can form monospecific stands that displace other vegetation.
Climbing, vining, scrambling, or smothering habit of growth.
*Alternative host for known arthropod pests, nematodes, and pathogens.
Reproductive characteristics
Minimum generative time before achieving reproductive status.
Self-fertilized, but has some outcrossing and can also hybridize naturally.
Does not require specialist pollinators, or wind-pollinated.
Has vegetative reproductive mechanism; brittle, with capability of fragmentation and regeneration from fragments.
Dispersal mechanisms
Capable of wind dispersal, floating on water, being dispersed by birds or other animals.
*Likely to be dispersed unintentionally or intentionally by people.
*Likely to be dispersed as a product contaminant.
Persistance attributes
Prolific seed production under favorable conditions; can still produce propagules under unfavorable conditions.
Innate discontinuous seed germination.
Capable of forming a propagule bank (seeds, tubers, rhizomes, etc.) that persist beyond one year.
Resistant to damage, e.g. by cultivation, by herbivory, by fire, by pathogens, or by herbicides.
Genetic characteristics
Many genotypes that permit development of local strains.
Plastic phenotypic expression in relation to climatic and edaphic factors.

[a]Based on lists prepared by Baker (1965) and Pheloung *et al.* (1999), with additions and modifications.

weediness trait. Vegetative reproductive parts that are brittle and possess the capability of fragmentation, coupled with regeneration from fragments, will enhance the plant's ability as a weed.

Although dispersal mechanisms are listed as a weediness trait in Table 11.1, the characteristics listed are probably of greater importance for predicting an invasive species than they are of universal weediness traits. It is apparent from the lists of the world's worst weeds (Holm *et al.*, 1977) that the majority do not have any special adaptation for dispersal (e.g. *Cyperus* spp., *Echinochloa* spp., *Amaranthus* spp.). Thus the likelihood to be dispersed unintentionally or intentionally by people is an important consideration, but this is not a plant trait. The likelihood of being dispersed as a product contaminant is also important in attempts to predict an invasive species. Propagules that are adapted for wind dispersal are capable of floating on water, or can be dispersed by birds or other animals, are traits that can confer weediness.

The persistence of reproductive propagules is a major contributor to weediness. Prolific seed production under favorable conditions occurs in most weed species, but the ability to produce propagules even under unfavorable conditions is essential. Discontinuous seed germination due to innate dormancy is probably an essential attribute of weedy species; there are no common weeds in which all seeds germinate at the first favorable germination event. Discontinuous germination, coupled with long propagule longevity, leads to the formation of a propagule bank (seeds, tubers, rhizomes, etc.); even for weeds with short-lived seeds the seedbank persists at least two or more years. In order to persist weeds must also be able to resist damage, such as by cultivation, by herbivory, by fire, by pathogens, or by herbicides.

For a plant to be a weed it must possess a wide genotypic base. This permits development of locally adapted biotypes, which can tolerate either specific environmental conditions (climatic and edaphic) or control tactics devised by humans. It is unlikely that plants with a narrow genetic base would achieve the status of being a weed; such a plant would succumb too easily to natural controls or those devised by humans.

The life form of the plant does not appear to be an attribute related to weediness. Many weeds occur within plants having either annual or perennial life forms.

It is unlikely that any single plant species will show all the characteristics discussed above. However, when several of the traits occur together the plant is often one that humans consider to be a weed, or possesses strong invasive abilities.

Invasion into natural ecosystems may also be related to lack of biological control. The numbers of parasites and predators present on naturalized alien plants in their new location was lower than present on the same species in their native habitat (Clay, 2003; Mitchell and Power, 2003; Torchin *et al.*, 2003), suggesting that lack of biological control may be the reason that the plant is invasive in its new location. Work carried out using North American invasive species present in France concluded that whether the plant was a serious weed in North America was the single characteristic that was most likely to predict invasion in France (Maillet and Lopez-Garcia, 2000). This suggests that understanding the biology and ecology of weeds in their native habitat may be the key to predicting their invasive ability, and that attempting to predict invasion based on botanical traits may not be completely successful.

11.5 Interference

11.5.1 Concepts and terms

Interference implies that one plant can have effects on another plant, but does not define how the effects are brought about. Interactions between plants are a complex of factors, and it is often difficult to determine which are the more important. The following list of factors is modified from that provided by Harper (1977):

1. interception of light, thus modifying quantity available
2. alteration of light quality
3. utilization of limited water
4. absorption of limited nutrients
5. secretion of selective toxins (allelopathy)
6. changing the humidity profile
7. altering the soil reaction (long-term changes in pH and organic matter)
8. providing limited nitrogen
9. factors impacting consumers that alter interactions between plants:
    9.1. shelter or exclusion of pests and beneficials
    9.2. favoring or reducing pathogenic activity
    9.3. encouraging defecation or urination in neighborhood
    9.4. providing rubbing posts or "play" objects which result in local trampling.

Interference can occur between individuals in the same species (within or intraspecific) versus between individuals in different species (between or interspecific). Intraspecific competition within a weed species is probably important to weed management only at densities that are higher than those at which crop yield impacts have already become unacceptable. Thus, interspecific competition in a weed species is primarily involved in regulation of weed populations at agriculturally unacceptable densities. Interspecific competition is important to crop management, as it is what occurs between weeds and crops. The following is a non-exhaustive discussion of the main components of competition.

11.5.2 *Competition*

The concept of competition is different for a stationary organism than for one that is mobile. Mobile organisms can overcome depletion of local resources by moving to resources in a new location; competition for these organisms thus includes a searching component. Plants are not mobile and thus cannot move to new resources; once resources have been locally depleted plants experience competition until new resources become available.

Competition is probably the most important type of interference between weeds and crops. Plants, crops or weeds, require adequate resources in order to grow. If the number of plants present in an ecosystem is such that the resources available fall below the combined requirements of all the individuals then competition for one or more resources is experienced. If weeds utilize resources that otherwise would have been available for crop plants then the latter cannot grow to their full potential. The result, from the human anthropocentric view, is yield loss.

Attempting to ascertain which resources are limiting can be difficult. Much of the literature involving competition between weeds and crops does not establish which resources were limiting ones.

11.5.3 *Light*

Light is essential for plant growth. It provides the energy for photosynthesis. Duration and quality of light modify plant growth responses. Lack of adequate light leads to reduction in plant growth and yield.

Plants experience competition for light when the canopy of one plant intercepts light that would have otherwise struck the leaf of another plant lower in the canopy. Interception of light by weeds is a universal form of competition. Unlike other resources, light is an environmental resource that is not depleted; there will be as much tomorrow as there was today. The importance to competition for light between plants therefore is differential interception.

Plants exhibit one of two photosynthetic pathways. One type of plant species utilize 3-carbon sugar precursors in photosynthesis, and are referred to as C_3 plants. Other species use a 4-carbon sugar precursor in photosynthesis, and are referred to as C_4 plants. Plants with the C_4 pathway utilize energy and water more efficiently under warm sunny conditions. In warm weather C_4 plants are typically more competitive than C_3 plants.

Light quality can also impact plant growth and competition. Plants are capable of changing growth response to the wavelength of light. Changes in the ratio of red:far-red light has been shown to have a role in how plants respond to the presence of neighbors (Ballare *et al.*, 1988; 1990). The response allows the plants to alter how they will interact with their neighbors.

11.5.4 *Water*

The impact of competition for water is twofold. First, some weeds can better tolerate drought stress than the crop plants. Under conditions of limited water availability such weeds can outcompete crop plants, such as a C_4 weed competing with a C_3 crop. Second, regardless of competitive ability, weeds utilize large amounts of water (Norris, 1996*a*); C_3 weeds such as mustard require about 650 g of water for every gram of tissue, and C_4 weeds such as pigweed require about 300 g of water for every gram of tissue. Weed biomass in fields often exceeds 1000 kg/ha, meaning that a great deal of water has been utilized. Water that has been used by weeds is no longer available for crop plants. The overall supply for the crop has been decreased until more water is added to the system by precipitation or irrigation.

Competition for water can be the most limiting factor in rain-fed agricultural systems. Competition for water may not be as important in irrigated agricultural systems as water is typically added to the system before competition occurs.

11.5.5 *Nutrients*

Weeds take up, and utilize, nutrients in order to grow. When weed densities are high such nutrient utilization results in substantial removal of nutrients from the soil. Nutrients incorporated into weed biomass are not available to crop plants until the weed dies and the nutrients have been released into the soil as the weed tissues are decomposed. This means that such nutrients are typically not available to the current crop.

Competition is probably primarily for nitrogen. Lack of mobility of phosphorus and potassium means that competition for these nutrients only occurs at a microscale. Addition of the latter can, however, change competitive relationships between weeds and crop plants. *Amaranthus retroflexus* L. redroot pigweed., for instance, becomes much more competitive with crops when phosphorus is readily available.

Timing of application, and placement, of fertilizers can be used to alter the competitive relationship between the crop and the weed. The concept is to apply the fertilizer in such a way that, or at a time when, the crop is more likely to have access to the nutrients than the weeds. Placement of nitrogen fertilizer in the drill line of wheat, has, for example, been shown to reduce competition from wild oats in comparison with broadcast application.

11.5.6 Allelopathy

Chemicals released from one plant can cause reduction in growth of neighboring plants (Inderjit and Foy, 1999; Kohli *et al.*, 2001; Rice, 1995; Weston, 1996), which is a form of interference. Allelopathy is an example of how semiochemicals play a role in mediating interference between plants. When a crop plant releases a chemical that causes reductions in weed growth this can be used as a form of weed management. Decomposing rye mulch inhibits germination of weeds, an effect attributed to allelopathy. Utilization of allelopathy to manipulate weeds has not been widely adopted.

11.6 Density-related phenomena

The degree to which competition is experienced by plants is dependent on plant size and plant density. Within a species, intraspecific competition, increasing density leads to the concept of self-thinning. As the plants increase in size the weaker plants die, which allows the stronger plants to continue growth; see pages 171 to 194 in Harper (1977) and pages 93 to 105 in Silvertown and Charlesworth (2001) for an in depth discussion of the topic. Self-thinning of weeds occurs when weed densities are sufficiently high that they experience strong intraspecific competition. The significance to weed management is in attempting to predict population dynamics at high weed densities. Such densities are much higher than those that are acceptable to commercial agriculture, and are thus primarily of theoretical rather than practical interest.

In interspecific competition, the magnitude of competition between crops and weeds is dependent on both crop and weed density. Increasing weed density leads to increasing crop loss. Figure 11.2 provides examples of such density-related losses.

In an IPM framework, the more significant point to losses caused by weeds involves impacts at low weed density. It is possible, using information derived

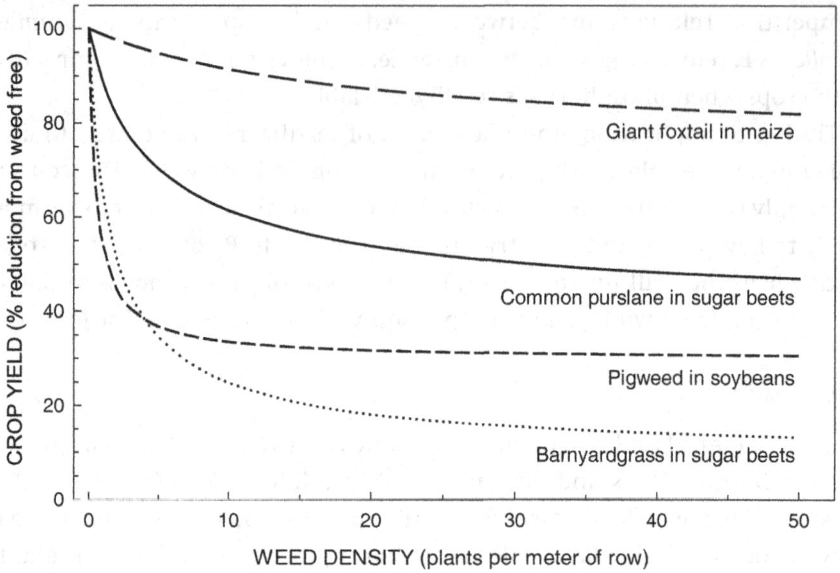

Figure 11.2. Effect of increasing weed density on crop yield.

from equations like those used to develop the graphs in Figure 11.2, to predict the density at which unacceptable yield loss starts to occur. The concept of weed density in relation to crop yield loss forms the basis for using thresholds for weed control decision-making. For many crops the economic threshold weed density is less than one plant m^{-2} for more competitive species.

11.6.1 *Temporal aspects of competition*

The longer that weeds interfere with the crop the greater will be the losses experienced by the crop. This phenomenon led to the development of the concept of the critical period (Berti *et al.*, 1996; Knezevic *et al.*, 2002; Singh *et al.*, 1996; Weaver *et al.*, 1992; Zimdahl, 1988). This period is defined as the time in the crop development in which weeds must be controlled if yield losses are to be avoided, and is depicted as the gray area in Figure 11.3. The concept is based on the fact that weeds which emerge with the crop are small and typically do not utilize enough resources to compete with the crop. There is thus a time during which no crop loss occurs between crop emergence and the time when weeds have grown larger enough to compete. Once the weeds have achieved sufficient size to compete they must be removed to avoid yield loss (T_1 on Figure 11.3). Due to the nature of the weed seedbank seeds continue to germinate during the crop growth cycle; weeds can thus invade a crop even if it was

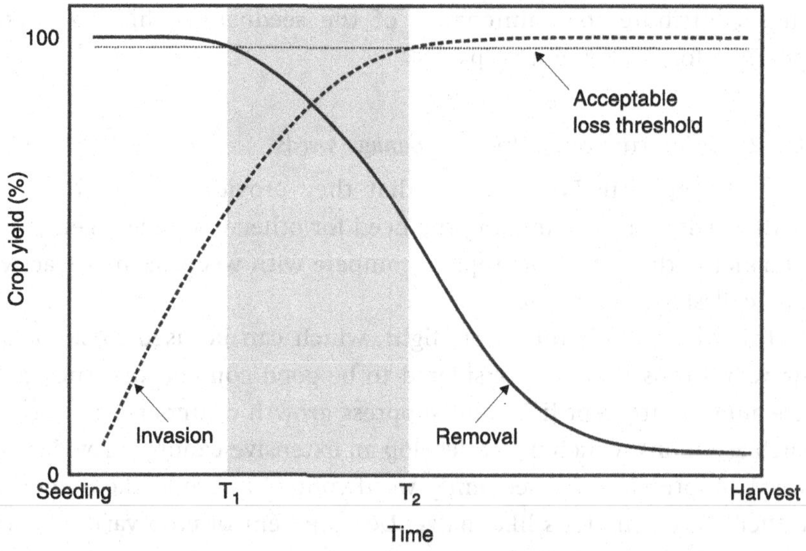

Figure 11.3. Graphic presentation of the critical period for weed control.

maintained weed free until T_1. Weeds that emerge soon after crop emergence compete strongly with the crop if not controlled, but the later a weed emerges the less its impact becomes on the crop. There is thus a time in the crop development (T_2 in Figure 11.3) after which later emerging weeds do not cause unacceptable yield loss. The critical period concept implies that weeds only need to be controlled between the times T_1 and T_2, and that those growing at other times in the crop cycle do not need to be controlled.

Although widely discussed in the literature, see above, there has been little implementation of the critical period concept to field-level weed management. There are several reasons for this lack of implementation. The onset and duration of the critical period is density dependent; for most weeds this relationship has not been elucidated. The onset and length of the critical period is also dependent on the acceptable yield loss; this is crop and market dependent and thus the critical period for a crop in one year may not be the same in the following year. A major difficulty in utilizing the critical period results from innate differences between competitive abilities of weed species. Fields typically have a mixture of weed species present; differing competitive abilities of the different species make defining the critical period difficult. Finally, weeds that emerge after the critical period (T_2 on Figure 11.3) may still be capable of producing seeds although they do not impact crop yield. Such late emerging weeds can thus contribute to maintenance of the seedbank, which is a potential problem for subsequent crops.

11.6.2 *Using crop competition to manage weeds*

Crops can be grown so that they provide increased competition with weeds, thus minimizing the need for other weed management tactics. Enhancing the ability of crops to compete with weeds is an attractive IWM tactic (Pester *et al.*, 1999).

The crop canopy intercepts light, which can be used to suppress weed growth. Crops that are considered to be good competitors, such as maize, are able to intercept light and suppress growth of understory weeds. Crops such as onions, which never develop an extensive canopy, provide very little weed suppression. Consequently weed control in crops like onions is more difficult than in crops like maize. Development of crop varieties that have more rapid canopy development is considered to be a weed management tactic. Increasing crop planting density can achieve partial suppression of weeds. Increased interception of light by the crop is probably the underlying mechanism in both cases. Characteristics such as increased rate of canopy development can be utilized.

Increasing crop density is another technique that can be used to suppress weeds by preempting resources. Increasing crop density can, in more competitive crops such as cereals and maize, lead to greater competition with weeds, resulting in suppression of the latter (see Table 6.1, pages 271–3 in Liebman *et al.*, 2001).

11.6.3 *Modification of competitive interactions by organisms in other pest categories*

There is an old axiom that states "the best weed control is a healthy, vigorous crop". This statement is based on the ability of the crop to suppress the growth of weeds through its competitive ability. Due to the trophic position of weeds any reduction in resource use by the crop has the potential to release such resources for use by weeds.

Damage to the crop canopy by insect feeding or attack by pathogens can result in reduction in the ability of the crop to compete with weeds, leading to greater weed growth when the pests are not adequately controlled. An example of this type of interaction is the increased weed growth in alfalfa when the crop was subjected to cutworm attack (Buntin and Pedigo, 1986). Cutworms and birds feeding in seedling crops, such as corn, can result in stand loss; weeds then grow in the resulting gaps in the crop.

Underground pests damage crop roots, leading to reduced growth, which in turn leads to reduction in resource use by the crop. A dramatic example of such damage would be death of alfalfa plants when gophers sever the taproot.

Weed management is thus more difficult when the competitive ability of crops is compromised by uncontrolled attack by primary consumer pests. Part of the decision regarding necessity to treat such pests should include the decrease in weed suppression that will result from lack of control. Most current IPM programs have not achieved such a level of integration.

11.7 Weed populations: dynamics and regulation

Monitoring population development is a major component of the IPM strategy. The concept requires that the size, and rate of development, of the pest population must be known before the correct timing for a management tactic can be chosen. Although monitoring of plant populations is carried out at the ecological research level, the techniques available do not lend themselves well to the rapid, cheap monitoring required for IPM.

Due to lack of mobility it is not possible to trap weeds as a monitoring technique. As the majority of the population is in the soil it is very difficult to acquire accurate estimates of the total weed population in the field. Unlike most arthropod pests, the presence of the weed seedbank means that, in order to devise the best weed management strategy, it is often necessary to know the population prior to sowing the crop. Accurately determining the number of seeds in the soil seedbank is time consuming and requires considerable expertise to identify the seeds. Counting emerged weeds is relatively easy, but also requires expertise to identify small seedlings. For all these reasons, and others, accurate determination of weed populations for IPM purposes has not been utilized to any great extent.

The use of life-tables is often applied to population dynamics of insect populations. The same concept can be applied to weeds. The basic components of the weed life table are fecundity (seed production; buds on vegetative structures), seedbank, seedling recruitment (germination), and reproductive adults (mature plants). The factors that regulate these states are most easily shown in a diagram (Figure 11.4). Discussion of plant and weed population dynamics can be found in Harper (1977), in Cousens and Mortimer (1995) and in Silvertown and Charlesworth (2001). Each state is briefly reviewed here.

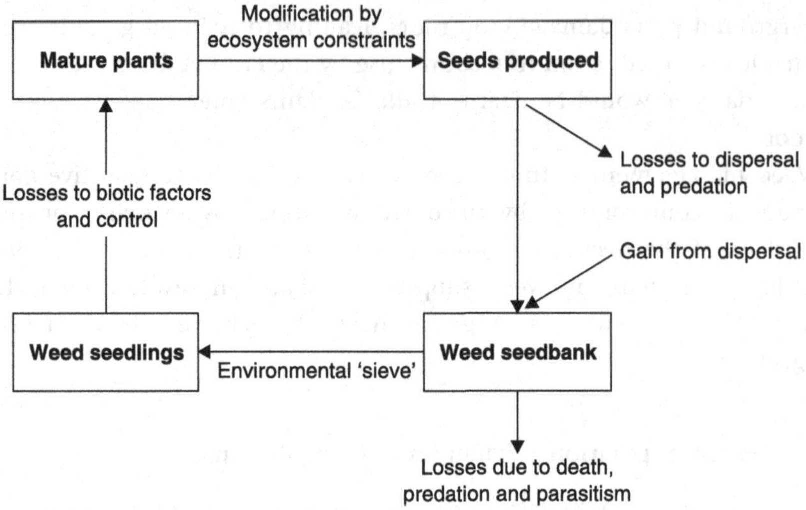

Figure 11.4. Diagram of the main components of the life cycle of an annual plant.

11.7.1 *Fecundity and seed production*

Accurate knowledge of fecundity is essential if population dynamics are to be predicted with any degree of accuracy. Earlier work (Stevens, 1932; 1957) has been widely cited as representative of weed seed production. Research reported for individual weed species in recent years shows that seed production is generally much higher than was reported in the earlier work (Cavers and Benoit, 1989; Norris, 1996*b*). Documentation of seed rain is limited for many species of weeds, and accurate data are often not available for many field-grown weeds. Examples of estimated seed production by some common weeds are listed in Table 11.2. Control strategies must be based on accurate knowledge of seed production; incorrect information may result in poor weed management programs.

11.7.2 *Seed dispersal*

Many plants produce fruit and seed structures that aid in dispersal of the propagules away from the parent plant. Botany texts describe these adaptations, and textbooks on weed ecology use the same descriptions. It is implied that these mechanisms lead to the spread of weeds from one place to another. This has led to the misconception that weeds spread by themselves. If the mechanisms of spread are reviewed it is clear that virtually all major weeds are primarily spread by human activity. The nutsedge species *Cyperus esculentus* L. and *C. rotundus* L. are arguably the most important weeds

Table 11.2. *Examples of estimated seed production by some common weeds*

Latin name	WSSA common name	Estimated seed production per plant	Conditions	Reference
Portulaca oleracea	Common purslane	Up to 2 500 000	Isolated plants	Norris, unpublished
Echinochloa crusgalli	Barnyardgrass	Up to 2 250 000	Isolated plants	(Norris, 1992*b*)
		100 000	In sugarbeets	(Norris, 1992*a*)
			In tomatoes	(Norris *et al.*, 2001*b*)
		2000	In corn	(Norris, 1996*b*)
Solanum ptycanthum	Eastern black nightshade	800 000	In soybeans	(Quakenbush and Anderson, 1984)
Tribulus terrestris	Puncturevine	100 000	Single plants	(Boydston, 1990)
Abutilon theophrasti	Velvetleaf	44 000	Isolated plant	(Steinamus and Norris, 2002)
		349	In corn	
Amaranthus	Pigweed	800 000	Intraspecific	(Hauptli and Jain, 1978)
Chenopodium album	Lambsquarters	Up to 175 000	In soybeans	(Harrison, 1990)
Polygonum persicaria	Ladysthumb	Up to 47 000	In cotton	(Askew and Wilcut, 2002)
Capsella bursapastoris	Shepherd's purse	Up to 61 200	Isolated plants	(Norris, 1996*b*)

on a worldwide basis, yet they have no inherent mechanism of long distance spread. Many of the more important weeds are in the Poaceae (e.g. *Avena fatua* L., *Echinochloa crus-galli* (L.) Beauv., *Elytrigia repens* (L.) Nevski, *Imperata cylindrical* (L.) Beauv.) but do not possess special adaptations for long distance dispersal. Important weeds in the Amaranthaceae, Chenopodiaceae, Portulacaceae, Malvaceae and many other families likewise have no special seed dispersal mechanisms. In the absence of human intervention these weeds spread reproductive propagules no more than a few meters from the parent plant.

The fact that most weeds are spread through human activities has led to the concept that stopping spread should be a major component of agricultural weed management programs (Ghersa and Roush, 1993). In the absence of human-induced spread of weeds it is argued that their control would be simplified.

11.7.3 *Seedbanks*

Seed dormancy, leading to the development of the weed seedbank, is one of the most important attributes of weeds. Plants that do not have dormant seeds, such as most crops, cannot develop a soil seedbank. No attempt will be made here to review the physiology and ecology of seed dormancy, as the topic is adequately addressed elsewhere (Baskin and Baskin, 1998; Cavers and Benoit, 1989; Champion *et al.*, 1998; Dekker, 1999; Fenner, 2000). The weed seedbank in arable fields is enormous. One thousand seeds/m^2 (1 million seeds/ha) is typical of well-managed fields with a "low" weed problem; poorly managed fields often have greater than 50 000 seeds/m^2 (50 million seeds/ha).

Longevity of seeds in the soil is also a major necessity for the development of the weed seedbank. Longevity is an interaction between the dormancy characteristics, morphology, physiological resistance to death, and inherent ability to resist attack by predators, parasites and pathogens. On an overall basis seed size would appear to be an indicator of seed longevity; large seeds persist for short times and small seeds persist much longer (Thompson *et al.*, 1993). Also, hardseed persist longer than seed with less physical protection.

Seeds of most weeds persist for several years, and many species persist for decades. Several long-term studies have been set up to investigate seed longevity; the utility of these to understanding longevity of weed seeds is questionable, as the conditions under which the seeds in such experiments were buried do not duplicate conditions found in agricultural fields. Most botany texts refer to the longevity of seeds based on these long-term experiments, or utilize data from artificial conditions, such as seeds removed from herbarium specimens or extracted from clay or mud in adobe or wattle walls. Work conducted by weed scientists under field conditions now indicates that the longevity of many weed seeds is shorter than predicted by the data from the more artificial conditions. Weeds with short lived seedbanks, such as *Echinochloa crus-galli*, persist for two to five years, but weeds with long-lived seedbanks, such as *Rumex* spp. and *Abutilon theophrasti*, persist for decades.

Provision of a "genetic memory" is a characteristic of the seedbank that has implications for IPM. Seeds that persist in the soil for many years preserve the genetic structure of the population at the time of seed production. Characteristics such as susceptibility or resistance to pesticides are thus preserved; only that part of the seedbank that had germinated is exposed to the current weed management tactic; dormant seeds maintain the pre-existing genetic structure.

11.7.4 *Biological control*

Biological control is a major component of arthropod management, and is considered to be a cornerstone of many IPM programs. The use of plants as a food source by higher trophic level organisms can regulate the size of the plant population. However, except for a few frequently quoted examples there is little human-managed biological control for most weed species. The exceptions are for perennial weeds in extensive ecosystems with low annual economic return, such as *Opuntia* spp. (prickly pear), *Hypericum perforatum* L. (Klamath weed), and *Lantana camara* L. (lantana). Managed biological control for annual weeds in annual cropping systems is limited. Removal of a single weeds species, which is typical of managed biological control, usually does not resolve the total weed control needs for a crop. In the mixed weed populations typically present in most agricultural fields, other weed species can readily replace the one that is subjected to the biocontrol.

All plants are probably subjected to some level of biological control, as there are organisms that utilize them as food. Seed predation before and after shatter results in a major loss of seeds (Louda, 1989), and constitutes an endemic biological control. Such endemic biological control occurs for weed species, but its extent is not well documented. Although many organisms use weed seeds as a food source, it is primarily rodents, birds, and some ground beetles that cause the greatest losses. Seed predation cannot easily be manipulated, and is thus of limited value as a management tool. As noted earlier in this chapter, weeds produce high numbers of seeds (Table 11.2). Most of these seeds never persist to germinate. They are either eaten by seed predators, or are attacked and killed by pathogens. This is a form of natural biological control that goes essentially unrecognized. However, such endemic biological control, although typically high, is not adequate in annual cropping systems to negate the need for other forms of weed management.

There is an inherent ecological problem with biological control of weeds. The organisms involved must be herbivores or plant pathogens. Tactics aimed at managing pests in these categories have the potential to disrupt the biological weed control provided by such organisms.

11.7.5 *Spatial aspects of weed population dynamics*

Weeds do not usually occur uniformly distributed over fields, but rather are present in discontinuous clumps of relatively high density interspersed with areas of low density of where weeds are absent (Cardina *et al.*, 1997; Hughes, 1990; Van Groenendael, 1988). Lack of spatial uniformity causes several problems for integrated management of weeds. Population assessment may be in error unless patchiness of the population is considered; patchiness is particularly important for weed seedbank determination. Most studies of competition by weeds in crops are conducted using uniform spacing of weeds and do not address impacts of variation in weed spatial arrangement. It is now apparent that uniform spacing of weeds results in the maximum level of competition; weeds in a clumped spatial pattern cause a lower overall level of competition (Norris *et al.*, 2001*a*). Changes in competition in relation to weed spatial arrangement are of importance to decision-making based on crop loss thresholds, as single season economic thresholds are higher than those predicted by studies using uniform spatial arrangement of weeds.

Weed management tactics are usually applied on a uniform basis to the entire field, especially herbicide application. Lack of uniformity in weed populations means that herbicide may be applied to areas where it is not required, or insufficient herbicide may be applied in areas of high weed density to control all the plants present in the clump.

11.8 Thresholds and weed management

The concept of the economic threshold (ET) has become central to the management of several arthropod pests under an IPM program (Higley and Pedigo, 1996). It has been widely suggested that the ET concept be applied to weed management (Auld and Tisdell, 1986; Berti *et al.*, 1996; Coble and Mortensen, 1992; Cussans *et al.*, 1986; Heitefuss *et al.*, 1987; Mortimer *et al.*, 1978; Swanton *et al.*, 1999). The main reason for adopting an ET-based approach to weed control is to provide an improved basis of selection of herbicides. Application of the economic threshold concept implies that control of weeds below a specified numerical density in a crop will not result in any net economic gain in that crop. It is a short-term decision based on the economics of a single crop. An extension of the economic threshold concept for weed management has been suggested to accommodate the multi-year aspect of weed management resulting from the presence of the persistent seed bank. It is called the economic optimum threshold (EOT) (Cousens, 1987), but has seen little adoption at the farming level.

The concept of the critical period has been developed for weed management (see earlier discussion). It is an extension of the economic threshold to include a time-factor into the decision. Although this concept is applicable to other categories of pests, it has not been developed. As noted previously the critical period threshold has not been widely adopted for weed management.

The ecological implications of using economic thresholds for weed management must be considered, and probably explain why the use of economic thresholds has not been widely adopted (Buhler, 2002). It is ecologically unlikely that a decision-making process for herbivores (primary consumers) would be the same as that for plants (producers).

Adoption of economic thresholds for management of major agricultural weeds should be approached with caution (Jones and Medd, 2000; Norris, 1999; O'Donovan, 1996; Wallinga and van Oijen, 1997). A major reason that thresholds are questioned for management of major agricultural weeds is the high seed production potential per weed plant (Table 11.2). It is argued that seed production from weeds at or below the single season economic threshold, and thus not controlled when economic thresholds are used for decision-making, is adequate to maintain the seedbank in the soil. The use of single-season economic thresholds therefore mandates high levels of weed control activity in every crop. Because herbicides are the only currently economically feasible method for controlling large weed populations, the use of economic thresholds, in fact, relies on the extensive use of herbicides.

The extreme plasticity of plant growth is also a complicating factor. The use of an economic threshold utilizing the numerical density of a pest is based on the assumption that each pest individual has essentially the same impact on the crop. Variation in the size of individual weed plants, resulting from plastic growth responses, means that numerical density is not a reliable predictor of anticipated loss.

Fluctuation in crop value is another difficulty in implementing any pest control economic thresholds, as the economic threshold varies when the value of the crop changes. Such fluctuation in crop value becomes less significant when some form of long-term population management using multiple control tactics drives the decision making process.

Finally, economic threshold-based management decisions require that the size of the pest population be determined prior to initiating control action. As noted earlier, accurate determination of the size and spatial variability of weed populations is time consuming and therefore expensive. Unless the seedbank is assessed any population determination based on growing plant counts is only partial.

I have suggested that the management decision for weeds should be designed around stopping seed production (Norris, 1999). This is similar to the proposal by Jones and Medd (2000). Due to the high reproductive capacity of many of the more serious weed species (Table 11.2), a better management strategy may be to not permit seed production. This can be called a no-seed threshold (NST); utilizing such a threshold does not mean all weeds must be controlled, but rather only those that are capable of producing seed. This has some similarity with the critical period concept, as it suggests that weeds that invade late in the crop need not be controlled if they will not reach reproductive status or cause economic loss. Adoption of an NST approach to weed management has resulted in decreased weed control costs for some farms (Norris, 1999).

The economic threshold concept leads to disaster when applied to invading species. However, some form of economic threshold seems to represent the approach taken by many farmers and land managers. An invading species is usually not managed until it has impacted crops or ecosystems to the point where it is causing economic loss or altering ecosystem function. At this time the invading species has established a propagule bank that will make it difficult to control.

All the preceding discussions are based on short term experiments or modeling. There is urgent need within the weed research community to establish long-term experiments to elucidate the long-term economic viability of different weed management decision-making strategies.

11.9 Impacts of weeds in IPM systems

Due to the trophic position of weeds in food webs the presence or absence of weeds can have impacts on all other organisms present in the ecosystem. Topics explored in this section are: the importance of weeds on other pest organisms; their role in supporting beneficial organisms; and how changes in weed diversity can alter provision of food and cover for wild birds and animals.

11.9.1 Interactions between weeds and other pest and beneficial organisms in IPM systems

Diversity is frequently presented as a means to stabilize pest populations (e.g. Barbosa, 1998; Collins and Qualset, 1999; Tilman *et al.*, 1999). Weeds are often listed as a component of increased diversity. Just two examples are presented here to indicate the ecological dilemmas presented by interactions between weeds and all categories of pests, and demonstrate that diversity is a proverbial "two-edged sword" from a true IPM perspective. The best recommendation for management of most ground-living vertebrate pests is to remove the cover and the food source; leaving weeds therefore has the potential to increase vertebrate pests. Rotations to non-host crops are designed to decrease populations of several soil-borne pathogens and nematodes; allowing alternate host weeds to grow for diversity purposes could negate the rotation for such pathogen and nematode reduction. When diversity is considered for management of pests it is essential that it be done at level II IPM, not at level I (see pp. 478–80, Norris *et al.*, 2003; see also Chapter 1 in this book), otherwise the potential for unintended side-effects is high.

11.9.2 *Weed interactions with organisms in other pest categories*

Due to their trophic position, weeds can serve as alternative hosts for all categories of pests. Many aspects of such interactions have been reviewed elsewhere (Burdon and Leather, 1990; Cousens and Croft, 2000; Kranz, 1990; Norris and Kogan, 2000; Thresh, 1981). Consideration of interactions at the within-field level and external to the managed field makes a useful framework.

Within-field interactions

Weeds growing within the field can support populations of pathogens, nematodes, mollusks, and pest arthropods. Weeds that support pest populations within the field are particularly important because they can maintain or increase populations of non-mobile organisms, such as *Verticillium* and *Fusarium* wilt pathogens, and rootknot and cyst nematodes. A standard recommendation for control of nematodes is to rotate to non-host crops; allowing alternative host weeds to grow in the non-host crop can maintain the nematode population and negate the use of the rotation as a management tool.

Recent tactics for arthropod management include leaving weedy strips, coined "beetle banks", within fields to maintain beneficial insect populations. Such changes must be assessed at level II IPM.

Interactions with weeds external to fields

Weeds growing external to fields can serve as food and shelter for many animal pests and plant pathogens. This is important for pests that are mobile or are capable of being moved. Such pests can then spread from the weeds back into the adjacent crops. Overwintering of viruses in weeds, with subsequent vectoring back to crops, is common; removal of the alternative host weeds is a strategy used for managing such viruses. Slugs and snails can overwinter in weed vegetation surrounding fields. Likewise, many insects use weeds as a bridge between crops. Lygus bugs build up on weeds and move to crops as the weeds die. The first generation *Heliothis zea* (cotton bollworm) develop on weeds external to fields before moving to the cotton; control of the insect while it is on the weeds is a regional approach to managing the pest.

Vertebrate pests, such as ground squirrels, voles (meadow mice), and gophers often live in weedy field margins from which they either re-infest fields or forage in them. Removal of the weedy vegetation to destroy their habitat is the typical management tactic that is recommended.

11.9.3 Weed interactions with beneficial organisms

There is considerable evidence supporting that increased diversity in agricultural ecosystems leads to decreased arthropod pest outbreaks (e.g. Altieri and Nicholls, 1999; Andow, 1990; Barbosa, 1998; Pickett and Bugg, 1998). Beneficial insects can be supported on prey living on weeds, such as syrphid flies and ladybeetles on aphids on various weeds. Weeds can also provide nectar and pollen that is used by adults of beneficial insects, such as many parasitic wasps. Provision of resources used by beneficial insects by non-crop vegetation leads to the conclusion that high diversity is beneficial to arthropod pest management.

However, utilization of high diversity may be valuable when considering level I IPM for arthropod pests, but it may be of limited value, or even detrimental when level II or higher IPM is considered. Level II IPM implies that all pest organisms in the ecosystem be considered, not just management of those in one category (see Chapter 1 in this book). If impacts on all organisms are not considered there is potential for a management strategy for one category of pest to compromise management of pests in another category.

Weed interactions with wild game

Weed species growing both within and around agricultural fields can serve as a food source for wild birds and animals. Removal of weeds as a component of IPM programs thus has the potential to decrease the feed available for such wild game (Chiverton, 1999). Reductions in wild game populations have been attributed to lower levels of food availability following areawide weed management, especially in Europe. Proposals have been made that weeds not be controlled along fence lines and hedges around fields in order to increase food for wild game. This poses both ecological and ethical dilemmas for IPM. Any practice that enhances wild game is ecologically likely to increase pest vertebrate species that can also use the increased vegetation. Leaving weeds uncontrolled also has the potential to elevate reservoir alternate hosts for pest arthropods and especially for pathogens such as viruses. Finally, uncontrolled weeds may be a source of weed re-invasion into adjacent fields. These types of considerations emphasize the ecological complexity of attempting to put weed management into an IPM context.

11.10 Conclusion

Improvements in knowledge of weed biology and ecology are leading to increased adoption of integrated weed management programs (Buhler, 2002). Cardina *et al.* (1999) argue that IWM needs to move to higher levels of integration. Many current IPM programs are designed only for management of arthropod pests, which has been called level I IPM. In order to move IPM to higher levels, it is essential that weed management become part of the integrated program; level II IPM requires that all pests be managed in an integrated fashion. The interactions between weeds and other pests emphasize the need to move to higher levels of IPM, so that the right arm knows what the left arm is doing.

References

Aldrich, R.J. and Kremer, R.J. (1997). *Principles in Weed Science*. Ames, Iowa: Iowa State University Press.

Altieri, M.A. and Nicholls, C.I. (1999). Biodiversity, ecosystem function, and insect pest management in agricultural systems. In W.W. Collins, and C.O. Qualset (eds.), *Biodiversity in Agroecosystems*. Boca Raton, FL: CRC Press. pp. 69–84.

Anderson, W.P. (1996). *Weed Science: Principles and Applications*, 3rd edn. Minneapolis/St. Paul, Minn: West Pub. Co.

Andow, D.A. (1990). Control of arthropods using crop diversity. In D. Pimentel (ed.), *CRC Handbook of Pest Management in Agriculture*. Boca Raton, FL: CRC Press. pp. 257–85.

Askew, S.D. and Wilcut, J.W. (2002). Ladys' thumb interference and seed production in cotton. *Weed Science*, **50**, 326–32.

Auld, B.A. and Tisdell, C.A. (1986). Economic threshold/critical density models in weed control. Proc. EWRS Symp. 1986 *Economic Weed Control*, 261–8.

Baker, H.G. (1965). Characteristics and modes of origin of weeds. In H.G. Baker and G.L. Stebbins (eds.), *The Genetics of Colonizing Species*. New York: Academic Press Inc. pp. 147–72.

Ballare, C.L., Sanchez, R.A., Scopel, A.L. and Ghersa, C.M. (1988). Morphological responses of *Datura ferox* L. seedlings to the presence of neighbours. Their relationships with canopy microclimate. *Oecologia*, **76**, 288–93.

Ballare, C.L., Scopel, A.L. and Sanchez, R.A. (1990). Far-red radiation reflected from adjacent leaves: an early signal of competition in plant canopies. *Science*, **247**, 329–32.

Barbosa, P. (ed.). (1998). *Conservation Biological Control*. San Diego, CA: Academic Press.

Baskin, C.C. and Baskin, J.M. (1998). *Seeds: Ecology, Biogeography, and Evolution of Dormancy and Germination*. San Diego, CA: Academic Press.

Berti, A., Dunan, C., Sattin, M., Zanin, G. and Westra, P. (1996). A new approach to determine when to control weeds. *Weed Science*, **44**, 496–503.

Boydston, R. A. (1990). Time of emergence and seed production of longspine sandbur (*Cenchurus longispinus*) and puncturevine (*Tribulus terrestris*). *Weed Science*, **38**, 16–21.

Bridges, D. C. (1992). *Crop losses due to weeds in Canada and the United States*. Champaign, IL: Weed Science Society of America.

Buhler, D. D. (2002). Challenges and opportunities for integrated weed management. *Weed Science*, **50**, 273–80.

Buntin, G. D. and Pedigo, L. P. (1986). Enhancement of annual weed populations in alfalfa after stubble defoliation by variegated cutworm Lepidoptera: Noctuidae. *Journal of Economic Entomology*, **79**, 1507–12.

Burdon, J. J. and Leather, S. R. (eds.). (1990). *Pests, Pathogens, and Plant Communities*. Cambridge, Mass: Blackwell Scientific Publications, Inc.

Cardina, J., Johnson, G. A. and Sparrow, D. H. (1997). The nature and consequence of weed spatial distribution. *Weed Science*, **45**, 364–73.

Cardina, J., Webster, T. M., Herms, C. P. and Regnier, E. E. (1999). Development of weed IPM: levels of integration for weed management. In D. D. Buhler (ed.), *Expanding the Context of Weed Management*. New York: Food Products Press, The Haworth Press Inc. pp. 239–67.

Cavers, P. B. and Benoit, D. L. (1989). Seed banks in arable land. In M. A. Leck, V. T. Parker and R. L. Simpson (ed.), *Ecology of Soil Seed Banks*. San Diego, CA: Academic Press Inc. pp. 309–28.

Champion, G. T., Grundy, A. C., Jones, N. E., Marshall, E. J. P. and Froud-Williams, R. J. (eds.). (1998). *Weed Seedbanks: Determination, Dynamics and Manipulation*. Wellesbourne, Warwick, UK: Association of Applied Biologists.

Chiverton, P. A. (1999). The benefits of unsprayed cereal crop margins to grey partridges *Perdix perdix* and pheasants *Phasianus colchicus* in Sweden. *Wildlife Biology*, **5**, 83–92.

Clay, K. (2003). Parasites lost. *Nature*, **421**, 585–6.

Coble, H. D. and Mortensen, D. A. (1992). The threshold concept and its application to weed science. *Weed Technology*, **6**, 191–5.

Collins, W. W. and Qualset, C. O. (ed.). (1999). *Biodiversity in Agroecosystems*. Boca Raton, FL: CRC Press.

Cousens, R. (1987). Theory and reality of weed control thresholds. *Plant Protection Quarterly*, **2**, 13–20.

Cousens, R. and Croft, A. M. (2000). Weed populations and pathogens. *Weed Research*, **40**, 63–82.

Cousens, R. and Mortimer, M. (1995). *Dynamics of Weed Populations*. Cambridge, UK: Cambridge University Press.

Cussans, G. W., Cousens, R. D. and Wilson, B. J. (1986). Thresholds for weed control – the concepts and their interpretation. In *Economic Weed Control*. European Weed Research Society. pp. 253–60.

Daehler, C. C. and Carino, D. A. (2000). Predicting invasive plants: prospects for a general screening system based on current regional models. *Biological Invasions*, **2**, 93–102.

Dekker, J. (1999). Soil weed seed banks and weed management. In D. D. Buhler (ed.), *Expanding the Context of Weed Management*. New York: Food Products Press, The Haworth Press Inc. pp. 139–66.

Fenner, M. (2000). *Seeds: the Ecology of Regeneration in Plant Communities*, 2nd edn. Wallingford, UK; New York: CABI Publishing.

Ghersa, C.M. and Roush, M.L. (1993). Searching for solutions to weed problems. *Bioscience*, **43**, 104–9.

Harper, J.L. (1977). *Population Biology of Plants*. London: Academic Press.

Harrison, S.K. (1990). Interference and seed production by common lambsquarters (*Chenopodium album*) in soybeans (*Glycine max*). *Weed Science*, **38**, 113–18.

Hauptli, H. and Jain, S.K. (1978). Biosystematics and agronomic potential of some weedy and cultivated Amaranths. *Theoretical and Applied Genetics*, **52**, 177–85.

Heitefuss, R., Gerowitt, B. and Wahmoff, W. (1987). Development and implementation of weed economic thresholds in the F. R. Germany. In *British Crop Protection Conference – Weeds*. pp. 1025–33.

Higley, L.G. and Pedigo, L.P. (ed.). (1996). *Economic Thresholds for Integrated Pest Management*. Lincoln, NE: The University of Nebraska Press.

Holm, L.G., Plucknett, D.L., Pancho, J.V. and Herberger, J.P. (1977). *The Worlds Worst Weeds: Distribution and Biology*. Honolulu, HA: University of Hawaii Press.

Hughes, G. (1990). The problem of weed patchiness. *Weed Research*, **30**, 223–4.

Inderjit, D.K.M.M. and Foy, C.L. (1999). *Principles and Practices in Plant Ecology. Allelochemical Interactions*. Boca Raton, FL: CRC Press.

Jones, R.E. and Medd, R.W. (2000). Economic thresholds and the case for longer term approaches to population management of weeds. *Weed Technology*, **14**, 337–50.

Kerr, G., Hermer, R. and Moss, S.R. (2000). A century of vegetation change at Broadbalk wilderness. *English Nature*, **34**, 41–7.

Knezevic, S.Z., Evans, S.P., Blankenship, E.E., Van Acker, R.C. and Lindquist, J.L. (2002). Critical period for weed control: the concept and data analysis. *Weed Science*, **50**, 773–86.

Kohli, R.K., Singh, H.P. and Batish, D. (2001). *Allelopathy in Agroecosystems*. New York: Food Products Press.

Kranz, J. (1990). Tansley review no. 28. Fungal diseases in multispecies plant communities. *New Phytologist*, **116**, 383–405.

Kropff, M.J. and van Laar, H.H. (eds.). (1993). *Modelling Crop–Weed Interactions*. Wallingford, UK: CAB International.

Kwon, T.J., Young, D.L., Young, F.L. and Boerboom, C.M. (1995). Palweed-wheat: a bioeconomic decision model for postemergence weed management in winter wheat *Triticum aestivum*. *Weed Science*, **43**, 595–603.

Lawes, J.G. (1884). In the sweat of thy face shalt thou eat bread. *The Agricultural Gazette*, **23**, 427–8.

Liebman, M., Mohler, C.L. and Staver, C.P. (2001). *Ecological Management of Agricultural Weeds*. Cambridge, UK: Cambridge University Press.

Lindquist, J.L. (2001). Performance of INTERCOM for predicting corn–velvetleaf interference across north-central United States. *Weed Science*, **49**, 195–201.

Lotz, L.A.P., Wallinga, J. and Kropff, M.J. (1995). Crop–weed interactions: quantification and prediction. In D.M. Glen, M.P. Greaves and H.M. Anderson (eds.), *Ecology and Integrated Farming Systems*. London: John Wiley and Sons Ltd. pp. 31–47.

Louda, S.M. (1989). Predation in the dynamics of seed regeneration. In M.A. Leck, V.T. Parker, R.L. Simpson (eds.), *Ecology of Soil Seed Banks*. San Diego, CA: Academic Press Inc. pp. 25–51.

Maillet, J. and Lopez-Garcia, C. (2000). What criteria are relevant for predicting the invasive capacity of a new agricultural weed? The case of invasive American species in France. *Weed Research*, **40**, 11–26.

Mitchell, C.E. and Power, A.G. (2003). Release of invasive plants from fungal and viral pathogens. *Nature*, **421**, 625–7.

Mortensen, D.A., Bastiaans, L. and Sattin, M. (2000). The role of ecology in the development of weed management systems: an outlook. *Weed Research*, **40**, 49–62.

Mortimer, A.M., Putwain, P.D. and McMahon, D.J. (1978). A theoretical approach to the prediction of weed population sizes. In *Proc. 14th Brit. Crop Prot. Conf. – Weeds.*, Vol. **14**. pp. 467–74.

Norris, R.F. (1992*a*). Case history for weed competition/population ecology: Barnyardgrass (*Echinochloa crus-galli*) in sugarbeets (*Beta vulgaris*). *Weed Technology*, **6**, 220–7.

Norris, R.F. (1992*b*). Predicting seed rain in barnyardgrass *Echinochloa crus-galli*. In *IX Colloquium on Weed Biology and Ecology*. Dijon, France: European Weed Research Society. pp. 377–86.

Norris, R.F. (1996*a*). Water use efficiency as a method for predicting water use by weeds. *Weed Technology*, **10**, 153–5.

Norris, R.F. (1996*b*). Weed population dynamics: seed production. In *Proceedings, 2nd International Weed Control Congress*. pp. 15–20. Copenhagen, Denmark.

Norris, R.F. (1999). Ecological implications of using thresholds for weed management. In D.D. Buhler (ed.), *Expanding the Context of Weed Management*. New York: Food Products Press, The Haworth Press Inc. pp. 31–58.

Norris, R.F., Caswell-Chen, E.P. and Kogan, M. (2003). *Concepts in Integrated Pest Management*. Upper Saddle River, NJ: Prentice-Hall.

Norris, R.F., Elmore, C.L., Rejmanek, M. and Akey, W.C. (2001*a*). Spatial arrangement, density, and competition between barnyardgrass and tomato: I. Crop growth and yield. *Weed Science*, **49**, 61–8.

Norris, R.F., Elmore, C.L., Rejmanek, M. and Akey, W.C. (2001*b*). Spatial arrangement, density, and competition between barnyardgrass and tomato: II. Barnyardgrass growth and seed production. *Weed Science*, **49**, 69–76.

Norris, R.F. and Kogan, M. (2000). Interactions between weeds, arthropod pests, and their natural enemies in managed ecosystems. *Weed Science*, **48**, 94–158.

O'Donovan, J.T. (1996). Weed economic thresholds: useful agronomic tool or pipe dream? *Phytoprotection*, **77**, 13–28.

Pester, T.A., Burnside, O.C. and Orf, J.H. (1999). Increasing crop competitiveness to weeds through crop breeding. *Journal of Crop Production*, **2**, 31–58.

Pheloung, P.C., Williams, P.A. and Halloy, S.R. (1999). A weed risk assessment model for use as a biosecurity tool evaluating plant introductions. *Journal of Environmental Management*, **57**, 239–51.

Pickett, C.H. and Bugg, R.L. (1998). *Enhancing Biological Control: Habitat Management to Promote Natural Enemies of Agricultural Pests*. Berkeley, CA: University of California Press.

Pimentel, D., Lach, L., Zuniga, R. and Morrison, D. (2000). Environmental and economic costs of nonindigenous species in the United States. *Bioscience*, **50**, 53–65.

Pimm, S.L. (2002). *Food webs*. Chicago: University of Chicago Press.

Pimm, S.L., Lawton, J.H. and Cohen, J.E. (1991). Food web patterns and their consequences. *Nature*, **350**, 669–74.

Polis, G.A. and Winemiller, K.O. (1996). *Food Webs: Integration of Patterns and Dynamics*. New York: Chapman and Hall.

Quakenbush, L.S. and Anderson, R.L. (1984). Effect of soybean (*Glycine max*). interference on eastern black nightshade (*Solanum ptycanthum*). *Weed Science*, **32**, 638–45.

Radosevich, S., Holt, J. and Ghersa, C. (1997). *Weed Ecology: Implications for Management*, 2nd edn. New York: John Wiley & Sons, Inc.

Rejmanek, M. (2000). Invasive plants: approaches and predictions. *Austral Ecology*, **25**, 497–506.

Renner, K.A. and Black, J.R. (1991). SOYHERB – A computer program for soybean herbicide decision making. *Agronomy Journal*, **83**, 921–5.

Rice, E.L. (1995). *Biological Control of Weeds and Plant Diseases: Advances in Applied Allelopathy*. Norman, OK: University of Oklahoma Press.

Ross, M.A. and Lembi, C.A. (1999). *Applied Weed Science*, 2nd edn. Upper Saddle River, NJ: Prentice-Hall.

Silvertown, J. and Charlesworth, D. (2001). *Introduction to Plant Population Biology*, 4th edn. Oxford, UK; Malden, MA: Blackwell Science.

Singh, M., Saxena, M.C., Abuirmaileh, B.E., Althahabi, S.A. and Haddad, N.I. (1996). Estimation of critical period of weed control. *Weed Science*, **44**, 273–83.

Steinmaus, S.J. and Norris, R.F. (2002). Growth analysis and canopy architecture of velvetleaf grown under light conditions representative of irrigated Mediterranean-type agroecosystems. *Weed Science*, **50**, 42–53.

Stevens, O.A. (1932). The number and weight of seeds produced by weeds. *American Journal of Botany*, **19**, 784–94.

Stevens, O.A. (1957). Weights of seeds and numbers per plant. *Weeds*, **5**, 46–55.

Swanton, C.J. and Murphy, S.D. (1996). Weed science beyond the weeds – the role of integrated weed management, IWM, in agroecosystem health. *Weed Science*, **44**, 437–45.

Swanton, C.J., Weaver, S.E., Cowan, P. *et al.* (1999). Weed thresholds: theory and applicability. In D.D. Buhler (ed.), *Expanding the Context of Weed Management*. New York: Food Products Press, The Haworth Press Inc. pp. 9–29.

Thill, D.C., Lish, J.M., Callihan, R.H. and Bechinski, E.J. (1991). Integrated weed management – a component of pest management. *Weed Technology*, **5**, 648–56.

Thompson, K., Band, S.R. and Hodgson, J.G. (1993). Seed size and shape predict persistence in the soil. *Functional Ecology*, **7**, 236–41.

Thresh, J.M. (ed.). (1981). *Pests, Pathogens and Vegetation*. London, UK: Pitman Books.

Tilman, D.C., Duvick, D.N., Brush, S.B. *et al.* (1999). *'Benefits of Biodiversity,'* Task Force Report no. 133. Ames, Iowa: Council for Agricultural Science and Technology.

Torchin, M.E., Lafferty, K.D., Dobson, A.P., McKenzie, V.J. and Kuris, A.M. (2003). Introduced species and their missing parasites. *Nature*, **421**, 628–30.

United States Congress and Office of Technology Assessment. (1993). *Harmful Non-indigenous Species in the United States*. Washington, DC: US Printing Office.

Van Groenendael, J.M. (1988). Patchy distribution of weeds and some implications for modeling population dynamics: a short literature review. *Weed Research*, **28**, 437–41.

Vangessel, M.J. (1996). Successes of integrated weed management – a symposium. *Weed Science*, **44**, 408–408.

Wallinga, J. and van Oijen, M. (1997). Level of threshold weed density does not affect the long-term frequency of weed control. *Crop Protection*, **16**, 273–8.

Weaver, S.E., Kropff, M.J. and Groeneveld, R.M.W. (1992). Use of ecophysiological models for crop–weed interference: the critical period of weed interference. *Weed Science*, **40**, 302–7.

Weston, L.A. (1996). Utilization of allelopathy for weed management in agro-ecosystems. *Agronomy Journal*, **88**, 860–6.

White, A.D. and Coble, H.D. (1997). Validation of HERB for use in peanut (*Arachis hypogaea*). *Weed Technology*, **11**, 573–9.

Wilkerson, G.G., Modena, S.A. and Coble, H.D. (1991). HERB: decision model for postemergence weed control in soybean. *Agronomy Journal*, **83**, 413–17.

Zimdahl, R.L. (1980). *Weed–crop Competition: A Review, International Plant Protection Center*. Corvallis, OR: Orgeon State University.

Zimdahl, R.L. (1988). The concept and application of the critical weed-free period. In M.A. Altieri and M. Liebman (eds.), *Weed Management in Agroecosystems: Ecological Approaches*. Boca Raton, FL: CRC Press, Inc. pp. 145–55.

12

The ecology of vertebrate pests and integrated pest management (IPM)

G. WITMER

12.1 Introduction

Across the world, vertebrates cause considerable annual damage to agriculture, property, human health and safety, and natural resources. Although some species of all vertebrate groups have been implicated in damage, the species most often involved in serious amounts of damage are birds and mammals. Agroecosystems have provided many new opportunities for vertebrates to exploit, resulting in their becoming serious "pests" with humans taking various steps to protect their agricultural resources. This conflict has intensified as the human population has increased, efforts to get more production out of traditional croplands have intensified, and marginal lands have been placed into crop production. Additionally, as the human population has increased, people have moved into lands occupied by wildlife, resulting in more human–wildlife encounters and conflicts.

Worldwide, the kind of damage caused by wildlife is most often related to the life history strategy of the species, although the actual species and crop involved varies greatly from region to region. In most cases, the conflict arises when wildlife are trying to acquire adequate food resources (i.e. meet nutritional needs) and forage on resources important to, and "reserved" by, humans. Examples can be identified from almost any region of the world for (1) carnivore predation on livestock, highly valued game animals, and endangered wildlife; (2) grain losses to flocking, seed-feeding avian species; (3) grassland rodents and lagomorphs consuming seeds and green foliage that would otherwise be available to livestock; (4) herding ungulates trampling and consuming crops and seedlings planted for reforestation; (5) aquaculture losses to fish-eating birds; and (6) disease transmission from wildlife to humans or their livestock (Conover, 2002; Dolbeer *et al.*, 1994). Another major problem area around the world is the consumption and contamination of stored food stuffs by rodents. In this latter case, the species most often involved are introduced, commensal rodents.

Rodent damage and its management will be emphasized in this chapter because rodents have, historically, been the major, worldwide, vertebrate pest group and there has been, and continues to be, major effort expended to reduce their numbers and damage (Witmer *et al.*, 1995). Rodents are implicated in all major types of damage, including significant predation on native species of animals and plants on islands to which rodents have been accidentally introduced (Witmer *et al.*, 1998). Numerous books have appeared in the last decade from all continents or regions of the world, addressing rodent damage and its management (notably, Singleton *et al.*, 1999). Two large tomes have been written, one from the United States and the other from Russia, dealing exclusively with the family Microtinae (voles). On the other hand, dealing with the problems caused by birds, ungulates, and carnivores pose additional "challenges" to pest management because, for example, those species are more highly visible and "important" to the general public and are usually regulated under the authority of state wildlife agencies (Conover, 2002).

While vertebrate integrated pest management (IPM) has perhaps not been as fully explored and implemented as has IPM for invertebrate, weed, and plant disease pests (e.g. Way and van Emden, 2000), there has been considerable progress in recent decades. Rodenticide application continues to be an important tool in rodent damage management by rapid and large-scale population reduction. These reductions, however, are short-term and there is a growing concern with the environmental hazards and safety issues associated with rodenticide use (Jackson, 2001). Great strides have been made to better understand the nature of rodent populations, why damage occurs, how damage can be predicted and lessened by non-lethal approaches (physical, chemical, behavioral, and cultural), and how to apply ecologically based rodent management strategies (e.g. Singleton *et al.*, 1999). The general equipment, methods, and strategies used to manage rodents, including rodenticides, have been presented in detail by Buckle and Smith (1994) and Hygnstrom *et al.* (1994). Many new approaches (use of disease agents and fertility control) are only in the preliminary development or testing phases for vertebrates. Many technical, regulatory, and sociopolitical hurdles need to be overcome. Additionally, much less investment is being made in solving the problems of vertebrate pests than for other agricultural pests because vertebrate

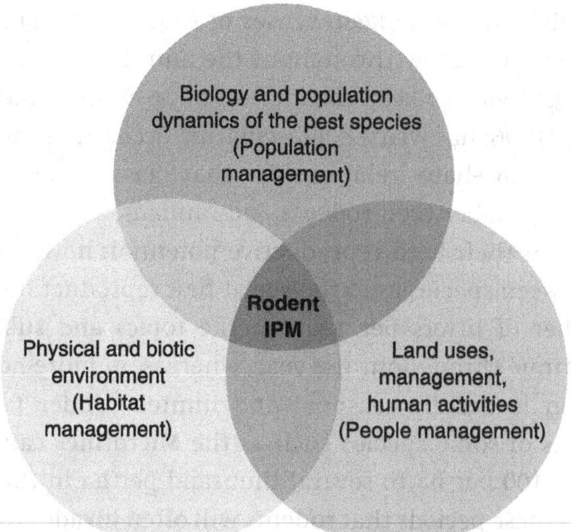

Figure 12.1. An understanding of (1) the biology and population dynamics of the pest species, (2) the ecology of the species within its physical and biotic environment, and (3) an understanding of the relationships of the pest species to the activities of humans, including land uses, management practices, and other human activities (Conover, 2002) are important so that we can develop effective (IPM) strategies for rodent population and damage management.

pest products are considered a "minor" part of the pest product industry (Jackson, 2001).

Solving vertebrate pest problems requires a careful consideration of (1) the biology and population dynamics of the pest species; (2) the ecology of the species within its physical and biotic environment; and (3) an understanding of the relationships of the pest species to the activities of humans, including land uses, management practices, and other human activities (Conover, 2002). It is only when we have an adequate background in those three areas that we can develop effective (IPM) strategies for rodent population and damage management (Figure 12.1). This is true because our main areas of focus are rodent population management, habitat management, and people management (e.g. Witmer *et al.*, 2000).

12.2 Biology and dynamics of the rodent populations

Over a third of all mammalian species in the world are rodents. They occur on all continents with the possible exception of Antarctica; although, I venture to say, commensal rodents may have been accidentally introduced even there! Species have adapted to all lifestyles: terrestrial, aquatic, arboreal, and fossorial. Most rodent species are small, secretive, nocturnal, and adaptable, and have keen senses of touch, taste, and smell. For most species, the incisors grow throughout the animal's life, requiring them to be constantly gnawing to keep the incisors at an appropriate length and position. Alderton (1996) has written a fascinating account of this group of mammals and the love-hate relationship that has always existed, and presumably always will, between rodents and humans.

Rodents are known for their high reproductive potential; however, there is much variability between species as to the age at first reproduction, size of litters, and the number of litters per year. In the topics and subtropics, reproduction can continue throughout the year, whereas, in more northerly latitudes, reproduction is usually seasonal and limited. Under favorable conditions, populations of some species such as the Microtines can irrupt, going from fewer than 100 per ha to several thousand per ha in the period of a few months. It is in these periods that rodents will often invade crop fields and cause severe damage. It is important to note for management purposes that most rodent populations will exhibit a compensatory response to a severe population reduction with earlier age to sexual maturity, higher pregnancy rates, larger litter sizes, more litters per year, and a higher survival rate of young. Currently, there are no commercial products available to reduce the fertility of rodents although research in this area continues.

As part of this life strategy, individuals of most rodent species have short lifespans and the annual mortality rate in a population is high, often about 70%. Although rodents, generally, have good dispersal capabilities, unless conditions are very favorable, mortality rates during dispersal are quite high. Rodents succumb to starvation, predation, disease, drowning and other accidents, and various other mortality factors. Hence, it can be seen that most rodent species exhibit a classic r-selected life strategy (i.e., high innate reproductive potential). An important management message is that while we can quickly reduce a rodent population with lethal means (usually with rodenticides as discussed later), it will quickly rebound if no other actions are taken.

There are many interesting dynamics to various rodent populations that should be understood to better facilitate their management and to reduce damage (Batzli, 1992). The population goes through an annual cycle that may include high and low densities, active and inactive periods, reproductive and non-reproductive periods, and dispersal periods. To avoid inclement periods, some species exhibit a winter dormancy (hibernation), and some species have a summer dormancy (estivation) during hot, dry periods. Some species exhibit multi-year cycles; for example, the microtines often reach population peaks (irruptions) every 3–5 years. While these cycles have been studied for decades, the driving factor(s) has not been definitively identified, but may involve long-term weather patterns, long-term nutrient cycles, predation, and intra-specific social interactions (Krebs, 1996). For effective management of rodent populations to occur, one must take into consideration these cycles and periods of inactivity.

12.2.1 *Monitoring rodent populations*

An important principle of IPM is pest scouting. This holds true for vertebrate pests as well, although the monitoring of vertebrate populations (especially small, nocturnal, secretive species) is problematic (Engeman and Witmer, 2000*a*). Monitoring allows one, first of all, to determine just what species of rodents occur in the area. Several to numerous rodent species may occur in any given area, but in many situations only one species is causing the damage. Knowing what species are in the area is important in designing a control strategy, to allow for the complications of baiting and trapping that other rodents may cause, and to plan for minimizing non-target losses. Monitoring rodent populations is also very important because densities can fluctuate dramatically within a year and between years.

Obtaining accurate estimates of population density is even more difficult, as well as costly, in terms of labor, time, and resource requirements. There is considerable discussion within the wildlife profession as to the importance or need for highly accurate population density estimates. Often, an index that efficiently tracks the pest population is used. The index allows one to document changes in the population through time and space, helps define the potential magnitude and geographical extent of damage that might result from population increases, and sets the stage for the implementation of an IPM strategy. Pest population monitoring is also an important component of the assessment of the efficacy of control methods. There are a number of desirable properties to consider in the selection of wildlife population indexing methods, including some associated with the planning stage, the in-field application of the index, and the analytical phase (Engeman and Witmer, 2000*a*).

A wide array of methods exist for monitoring rodent populations, including trap grids or transects, plot occupancy, open and closed hole indices for burrowing species, bait station or chew card activity and food removal, and runway or burrow opening counts (Engeman and Witmer, 2000*a*; Witmer and VerCauteren, 2001). Unfortunately, we rarely have a very good understanding of the relationship between the index and the actual population density or with the amount of crop/resource damage to expect.

There are advantages and disadvantages to each index that one must carefully consider before using one. For example, the result of many indices will vary with the soil and habitat type, weather conditions, and the time of year. In some cases, it is best to use two or more indices. Additionally, some regulatory agencies may require that two indices are used, for example, when data sets for efficacy are submitted for registration purposes.

12.2.2 *Population management of rodents*

Many species of wildlife may occur in a given area and this is especially true with regard to rodents. Rodents play important ecological roles, however, and most species are not major pests. Some of the roles include soil mixing and aeration, seed and spore dispersal, influences on plant species composition and abundance, and a prey base for many predatory vertebrates. Consequently, it is important to not indiscriminately decimate rodent communities even when an effective, efficient method, such as a rodenticide bait, is available. An exception would be the control of commensal rodents in structures where the operator has much control and non-target hazards can be minimized (Corrigan, 2001). Another example is the use of rodenticides on islands where introduced rodents are causing severe damage to natural resources, there are no or few native terrestrial mammalian species present, and measures can be taken to reduce other potential hazards to very low levels (Witmer *et al.*, 1998).

Populations of rodents can be reduced by a variety of means. While methods such as trapping, burning, flooding, and drives have been (and are still being) used in developing countries, many parts of the world have come to rely on rodenticide baits for rodent control (Singleton *et al.*, 1999; Witmer *et al.*, 1995). Most rodenticides were initially derived from naturally occurring plant materials; however, most are now produced synthetically. There are two general classes of oral rodenticides. Acute rodenticides (including zinc phosphide and strychnine) usually kill with a single feeding. In contrast, chronic or multiple-feeding rodenticides (including warfarin, chlorophacinone, and diphacinone) usually require a period (days) of feeding before killing. The distinction has become somewhat blurred because the second group includes first (examples given) and second (bromadiolone, brodifacoum, difethialone) generation anticoagulants. Second generation anticoagulants are very toxic and usually kill with a single feeding. An additional group of rodent toxicants includes the fumigants (e.g. gas cartridges, aluminum phosphide, methyl bromide) which are used in building fumigation or in burrow systems that are closed after application.

Considerable development has gone into making rodenticides effective, efficient, and relatively safe for use in buildings or the environment. This includes the development of ecologically based baiting strategies to assure safer and more effective use of rodenticides in cropland settings (Jackson, 2001; Ramsey and Wilson, 2000). In many countries, the use of rodenticides is carefully regulated by federal and/or state and provincial governments. Authorities decide who can use rodenticides and what training and record-keeping is required, along with which rodenticides and concentrations can be used and the where, when, and how of their use. Research is underway (although arguably not enough) to find new rodenticides as well as ways to make existing rodenticides more effective and less hazardous to non-target species and the environment. This is especially important in light of the fact that some rodenticides are being removed from the market and there are increasing restrictions on the use of many of these materials (Jackson, 2001).

There are many aspects of the biology and ecology of a rodent species that must be understood in order to effectively use rodenticides (or even traps or bait stations). I will present only a few examples. Most rodents are neophobic, exhibiting a fear of new objects, odors, or tastes in their surroundings. As such, materials may have to be placed out for a few days to allow rodents to adjust to them. Traps may have to be placed in a locked-open position and baited for a few days before they are effective in catching rodents. Some traps are more effective in catching rodents than others and this varies widely by species. Some rodents become trap-shy after an initial capture and are difficult to re-capture, while others become "trap-happy" and can be readily re-captured. These become important considerations for rodent researchers using mark-recapture techniques.

Most rodents have a good sense of taste and smell and a relatively long memory. Consequently, baits must be fresh and not moldy or rancid. Additionally, some acute rodenticides are rather unpalatable (e.g. strychnine is bitter) and others (zinc phosphide) cause sickness so quickly that the animals may become bait-shy after an initial, non-lethal exposure. To avoid this, it is sometimes necessary to pre-bait with the grain or other base material before applying the toxic bait to help assure that the rodents will consume a lethal dose in a single feeding. This is not a problem with the anticoagulants whereby the animal slowly becomes ill over time (i.e. as internal hemorrhaging begins), but continues to feed on the toxic bait which the rodent does not associate with the gradual illness. On the other hand, some populations of rodents that have been repeatedly exposed to an anticoagulant rodenticide, such as warfarin, have become resistant to the toxicant, requiring the use of a different rodenticide or a different control strategy.

The feeding habits and food preferences of rodents may shift during the course of the year, so that a bait used to deliver a toxicant or to lure rodents into traps may be much less effective during some parts of the year. For example, some rodents switch from a diet of green, succulent plant material early in the growing season to a diet primarily consisting of seeds once plants become senescent.

Another way to reduce rodent populations, theoretically, is through the introduction of disease agents or parasites. This approach has not found the success, however, that has been achieved in the control of some insect and plant populations. A major concern of using vertebrate biocides is that the agent may affect non-target species, including humans and livestock. This has been the case with the use of *Salmonella* spp. to control rats. Research continues, however, and a blood protozoan parasite, *Trypanosoma evansi*, and a liver nematode, *Capillaria hepatica*, are being evaluated for their ability to safely control rats and mice. Another difficulty has been the maintenance of the disease agent or parasite in the environment after the target species of vertebrate pest has been greatly reduced. There have been some successes with rabbit population control in Australia with the use of a myxoma virus and a rabbit calicivirus (Pech, 2000).

Fertility control is often considered an attractive alternative to the lethal control of rodents. There have been small-scale trials with various chemical compounds and some of these materials have shown promise. There are, however, many difficulties to overcome before any of these materials become available on the commercial market, including the need for a remote delivery system and the need to get a federal, state or provincial registration that would allow the use of compounds in the field, especially given that the effects of such compounds would probably not be species-specific (Fagerstone *et al.*, 2002).

12.3 Ecological relationships: physical and biotic environment

Effective management of rodent pests also requires a thorough knowledge of the species' ecological relationships, not only in natural or seminatural settings, but especially in human-altered settings. For rodents, the physical environment is comprised of various structural features (e.g. soil, water bodies, rocks, plants, buildings, roads) and weather parameters. The biotic environment consists of animals of other species which can serve as competitors (e.g. other wildlife species, livestock, or humans), especially for food, or as predators (e.g. carnivores, raptors, snakes, humans), and the many endo- and ectoparasites and disease organisms that can debilitate or kill a rodent.

All rodents require food, shelter, and water. The shelter provides protection from predators and inclement weather as well as a favorable place to bear and rear their young. While rodents require water, those water requirements vary greatly by species. Some require no free-standing water at all and can meet their water needs through the metabolism of solid foods or the moisture on vegetation or other surfaces.

Some rodents can significantly alter their physical and biotic environment. Examples include beaver that flood areas by building dams across streams or by plugging culverts, creating sizable water bodies (Naiman *et al.*, 1988). Also, pocket gophers can successfully prevent forest regeneration (after harvest, windstorm, or fire) by clipping and feeding on large numbers of tree seedlings (Engeman and Witmer, 2000*b*). On a smaller scale, rodents are very adept at creating burrow systems or sheltered nests (in trees, fallen logs, rock piles) to provide for their most basic cover needs. However, for the most part, rodents are at the mercy of the vagaries of their physical and biotic environment (Batzli, 1992).

Availability and palatability of foods and quantity and quality of vegetative cover vary greatly between habitats and seasons, and sometimes between years. Consequently, rodents may switch their foraging preference and strategy one or more times during the year as well as between years. The success of many management activities directed towards rodents depends upon whether or not alternative foods are available. Additionally, rodents will often retreat to certain habitats or more sheltered areas when cover or food becomes sparse (e.g. after crop harvest) or weather conditions more severe. These areas serve as refugia and can be important as source populations for future increases, dispersal, or irruptions.

Of course, the amount and quality of food and vegetative cover are greatly influenced by precipitation, temperatures, photo-period, and other climatic parameters. There has been some progress in predicting and modeling rodent population responses to long-term weather patterns (e.g. house mouse irruptions in Australia; Pech *et al.*, 2000). As a general rule, however, there are so many factors involved and we have such a poor understanding of those factors and rodent responses that we rarely know how many rodents will show up, where or when they will appear, how high their population densities will get, and if or when the population will crash. This is why so much of our rodent management strategies have been reactive rather than proactive. Only with a better understanding of these underlying relationships will we begin to more successfully predict rodent populations and damage and be able to design and implement effective proactive strategies (Singleton *et al.*, 1999).

On the other hand, because of complex, and often poorly understood, ecological interactions between species, a focused attack on one rodent species will often result in the unexpected (e.g. Sullivan *et al.*, 1998). This also applies to efforts to alter or influence predator-prey relationships.

12.3.1 *Influencing food and shelter to reduce rodent populations or damage*

Because rodent food and cover (i.e. vegetation) can be greatly influenced by human activities, there has been considerable development of strategies to reduce populations and damage by manipulating vegetation. Many of these manipulations are not done just to reduce rodent habitat (which may be an incidental benefit) but for other reasons such as to reduce vegetative competition with crops or trees, to reduce soil pathogens, or to prepare the site for planting. Burning, plowing, disking, or herbicide application all reduce vegetative cover, at least for the short term, and hence, usually greatly reduce rodent populations. Plowing or disking has the additional advantage of disrupting rodent burrows. All of these methods have been used extensively in reforestation, orchards, and traditional agriculture. It is interesting to note that, with a trend towards no-till agriculture to reduce erosion and water loss and improve soil fertility, the benefits of reduced rodent populations are not occurring to the same extent because the soil is not disturbed to an adequate depth and plant stubble (residues) are left on the surface (Witmer and VerCauteren, 2001). Rodent problems are actually compounded when grassy refugia or fallow areas occur around the periphery of crop fields that rodents can make use of when crop fields are rather bare. Additionally, a nutritious winter food supply for rodents is created by the fall (autumn) planting of crops such as wheat, barley, and legumes. These young plants, even under snow cover, are readily available to rodents, such as voles, that are active throughout the winter.

There has been some success in the use of lure crops or supplemental feeding to reduce damage by rodents or other vertebrates. Cracked corn or soybeans have been broadcast after drill-seeding on no-till cropland so that voles and other rodents will feed on those plants rather than feeding on newly emerged crop seedlings or digging up and feeding on planted seed (Witmer and VerCauteren, 2001). Sunflower seeds were broadcast on forest stands subject to tree squirrel damage with a subsequent reduction in tree damage (Sullivan and Klenner, 1993). A trap-barrier-system has been developed that uses some early planted crop fields to lure rodents into them (Singleton *et al.*, 1998). The lure fields are surrounded by a rodent barrier, but there are regularly spaced openings into multiple-capture rodent traps. Periodically, the rodents in the traps are collected and killed. This method has reduced rodent invasion into the surrounding crop fields that are planted at a somewhat later date.

Another approach to vegetation manipulation still under investigation is the use of endophytic grasses. These are grass varieties that contain an alkaloid-producing fungus that can improve the hardiness of the grass and reduce herbivory. Some preliminary studies suggest that endophytic grass fields support lower rodent densities (Pelton *et al.*, 1991). These grasses could potentially be used in a variety of settings, but might be very valuable around cropfields and orchards where grassy areas have served as a traditional refugia for rodents and, hence, a source of dispersing individuals.

Rodents compete for food with a variety of herbivores, including other rodent species, other wildlife, and livestock. There is some evidence that rodent populations can be reduced by intensive cattle or sheep grazing (Hunter, 1991). In some cases, the intensive grazing is done to reduce vegetative competition with tree saplings. In addition to reducing the food available to rodents, the livestock grazing may also compact the soil and disrupt burrow systems.

As a side issue, several rodent species usually occur in an area and these may be in strong competition with each other. Hence, when one species is controlled or removed, another species which only occurred in low numbers may become much more numerous and begin to cause damage. This has been noted with control or eradication of introduced rats, whereby house mice populations suddenly irrupt once their competition is gone (Corrigan, 2001).

12.3.2 *Influencing natural predation rates of rodents*

The cover needs for most rodents are quite high because of the constant threat of predation, both day and night. Because predation rates on rodents can be high, people have tried to increase predator densities as a way to reduce populations and damage. Unfortunately, prey populations usually drive predator populations, not the other way around. Artificial perches and nest boxes have been constructed to attract hawks and owls near croplands, orchards, and grasslands. These structures, especially where natural perches were limited, were used by raptors and those raptors did prey on rodents, among other animals, in the area. There was no evidence, however, that the rodent population or rodent damage was substantially reduced as a result (e.g. Howard *et al.*, 1985).

Extirpated medium- to large-sized carnivores are being re-introduced to many formerly occupied ranges in the USA. As these populations become established and expand, they may begin to help reduce the rapid increases in some vertebrate populations that have occurred in recent decades.

It should be noted, however, that the introduction of a non-native predator can have unexpected and adverse ecological impacts as has occurred on many islands with the introduction of the mongoose. It was hoped that the mongoose would help control introduced rats. Unfortunately, rats are primarily nocturnal, while the mongoose is primarily diurnal. In a number of situations, the mongoose has decimated the populations of native ground-nesting birds on islands (Witmer *et al.*, 1996).

12.3.3 Exclusion of rodents from areas

I have discussed the modification of habitats to make them less supportive of rodent populations. An alternative approach to reduce or eliminate rodent damage is to exclude them from areas. This sounds attractive because it is a non-lethal approach and could, potentially, solve the problem once and for all. Exclusion devices might be physical barriers such as fencing, sheet metal, or electric wires, but could also be frightening devices, ultrasonic or vibrating devices, or chemical repellents (Buckle and Smith, 1994; Hygnstrom *et al.*, 1994). Unfortunately, it is very difficult to keep rodents out of an area that they want to access. They can usually get over, around, under, or through any kind of barrier put in their way. Their small size, flexibility and agility, and gnawing capability, along with their climbing and digging abilities, make them a formidable adversary. They also habituate rather quickly to noxious odors, sounds, or lights. There are detailed guides available on how to rodent-proof buildings, but success is achieved only with much effort, expense, diligence, and maintenance (Corrigan, 2001; Hygnstrom *et al.*, 1994). In open settings such as croplands or orchards, the task is much more difficult and the chance of success is much smaller. Although research in this area continues, there are few successes to report at this time.

Short, electric fences have been used with some success to exclude rodents from areas, but there are a number of concerns such as non-target hazards and excessive maintenance to keep the fences operating properly (Buckle and Smith, 1994). Physical barriers around individual seedlings have shown some success in protecting seedlings, but, again, there are concerns about cost and maintenance as well as adverse effects on seedling growth (Pipas and Witmer, 1999). Predator odors have shown some effectiveness in some trials for repelling rodents and other herbivores from areas or individual plants (Mason, 1998; Sullivan *et al.*, 1988), but little effectiveness in other trials. It has been speculated that the sulfurous odors in predator urine, feces, glandular excretions, blood/bone meal, and putrescent eggs derive from the breakdown of animal protein and serve as a cue to herbivores that a predator may actually be in the area and pose a threat to the herbivore (i.e. the potential prey; Mason, 1998). Another repellent that has shown some promise is capsaicin (the active ingredient in hot chilli peppers), but a fairly high concentration (~ 2%) of this expensive material is usually needed for a reasonable level of effectiveness (Mason, 1998).

12.4 Influence of land uses, management practices, and human activities on rodent populations

There are many things that landowners or managers can do to help reduce the risk of damage by rodents. An important first step is to familiarize themselves with the biology and ecology of the rodents (and other vertebrates that may cause damage) in the area, along with their "sign" (burrow openings, mounds, runways, nests, tracks, droppings) and how to identify damage by those species (e.g. Dolbeer *et al.*, 1994; Hygnstrom *et al.*, 1994). In North America, often information of this kind can be obtained at local or county extension offices or from other state, provincial, or federal agencies. University wildlife damage specialists are also important sources of information. Unfortunately, in developing countries, wildlife damage management expertise is much less readily available.

Proper sanitation around one's property can significantly reduce food and cover available to rodents (Corrigan, 2001; Singleton *et al.*, 1999). Rubbish piles, uncovered garbage receptacles, wood and metal debris piles, rock piles, piles or bales of hay, heavy mown grass, silage and other exposed livestock feed, grain spills, and mature tree fruit on the ground are all very supportive of rodent populations. A reduction in the availability of water (e.g. standing water or wet areas) can help, but is often difficult to achieve in an outdoor setting. Within buildings, food sanitation and removal of water sources are very important in the management of commensal rodents (Corrigan, 2001).

In some cases, agricultural producers have some discretion in the crops or crop varieties that they grow, the timing of planting, and the location and size of specific crop fields (Singleton *et al.*, 1999). Cereal grains are more likely to be damaged by rodents than some crops such as soybeans or sunflowers.

In many cases, large monoculture crop fields will receive less rodent damage overall with most damage only occurring at the periphery of the crop fields. Valuable crops should not be grown near fallow areas, grasslands, or brushy areas that support rodents year around and serve as refugia from which rodents can rapidly disperse into crops.

In a region that is prone to periodic and substantial rodent damage, it is beneficial to have adjoining landowners cooperate in an overall strategy of reducing activities that support rodents and in rodent control activities (Jackson, 2001; Singleton *et al.*, 1999). Otherwise, a landowner may suffer continuous rodent damage despite rodent control efforts because the surrounding landowners' properties are rodent infested with no or few control activities taking place. Landowner cooperation can also help spread the costs of rodent management activities and materials over more people, thus reducing the cost to each individual landowner. In some situations and in some places, local, state or federal government support is available where vertebrate damage to agricultural production is severe.

12.5 Rodent IPM: bringing it all together

As the above discussion implies, developing a rodent IPM strategy requires the careful consideration of many factors. Once the rodent species is correctly identified, it is important to monitor its population status and associated damage, using one or more of the many methods that exist. Is the rodent abundance related to the amount of damage that occurs and can a threshold be identified for when action should be initiated? Next, one should consider the nature of the rodent species, its biology and ecology, in the setting in which the damage is occurring. How is the animal using its habitat? How is it interacting with other species? What are our actions doing to support the rodent population and to increase the amount of damage that occurs? What are our management options in terms of manipulating the rodent population, its habitat, and our activities and land use practices so that damage can be avoided or greatly reduced? What are the advantages and disadvantages of each of those management options? In general, there is a trend to start with the least invasive techniques before moving to more invasive techniques. Finally, how do we mold all those considerations into a comprehensive rodent IPM strategy that we can apply to the landscape? The strategy under consideration should be evaluated for its potential ability to achieve the objective of rodent damage reduction within the set of real world constraints, including method effectiveness and duration, the associated cost and benefits, the legality, the sociopolitical acceptability, and the environmental benignness of the proposed actions (Engeman and Witmer, 2000*b*). Of course, once we apply the strategy, we should monitor the results to see if we have achieved the desired goal of damage avoidance or reduction and, importantly, not just rodent population reduction and whether or not there were unexpected results. Because relatively little is known about dealing with rodent damage situations in complex landscapes (i.e. agroecosystems), we are, in essence, conducting large-scale experimental field trials. It is only with adequate monitoring and documentation that we can interpret and learn from those trials and, ultimately, improve the comprehensive rodent IPM strategy.

In some cases, decision support systems have been developed to help the landowner or manager formulate and implement a rodent damage control strategy once the rodent population or damage threshold levels are approached or exceeded. Unfortunately, there are relatively few such systems available and most are simple dichotomous keys or simple computer programs. There is a large variability in the goals, complexity, and input and output requirements and capabilities of existing rodent decision support systems.

Important components (or modules) of a comprehensive rodent decision support system include an overview of the species biology and ecology, population and damage identification and monitoring methods, a description of damage potential and associated factors, a mechanism to evaluate alternative management techniques and the integration of techniques, a cost–benefit analysis component, computer user "friendliness" (for computer-assisted programs), and sources of additional information. An interactive training and resource package called "Mouser" (provided as a CD-ROM; Brown *et al.*, 2001), developed for mouse irruptions in Australia, is the most complete rodent decision support system that I have encountered, containing most, if not all, of the desirable components. There is a great need, however, to improve most existing decision support systems and to develop many more for other rodent species, crops, and situations.

It should be evident that effective rodent IPM strategies and decision support systems require substantial information that only long-term research of the given pest species and situation can provide. Furthermore, that research should be an integration of basic and applied studies. Adequate information not only can result in more effective strategies, but also better predictive power, greater support and acceptance by the parties providing the funding, and credibility of the end-users (e.g. farmers) all of which are important to assure the application and sustainability of new strategies (Singleton *et al.*, 1999). Unfortunately, there is relatively little support for long-term rodent research, and, in fact, there are relatively few rodent research scientists. This situation is especially evident when one considers food losses to rodents in developing countries.

While some new tools are being developed, many traditional tools for the control of vertebrate pests and their damage are being lost as the general public and legislators take an increasingly active role in land and resource management (Conover, 2002; Jackson, 2001). Examples include rodenticide baits, traps, and field burning. As suggested in the examples of this chapter, much more research is needed in both lethal and non-lethal means of resolving vertebrate damage situations. The research should include, but not be limited to, rodenticide, repellent, and barrier development and improvement; biological control; fertility control; and habitat manipulation. Another important research need is greater evaluation of the effectiveness of combinations of techniques, given that combinations could potentially be much more effective in the reduction of damage and may be more acceptable to the public.

An additional concern, receiving more attention in recent years, is who should pay for the cost of vertebrate pest population and damage management activities that benefit the general public or the agriculturalists of a region? Unfortunately, vertebrate damage, the cost of population and damage management, and management benefits are not evenly distributed across segments of the public and private sectors. Additional research, increased public education, and increased sensitivity by public and private sector persons involved in vertebrate pest management may help resolve some of these problems.

Rodents, the damage they cause, and the diseases they transmit have plagued human populations since the beginning of civilization. There is no reason to believe that adverse interactions will not continue for the foreseeable future as these two groups vie for resources and co-evolve in natural and human-altered landscapes, especially in agroecosystems. Therein lies the challenge for practitioners of vertebrate IPM.

References

Alderton, D. (1996). *Rodents of the World*. London: Blandford Books.

Batzli, G. O. (1992). Dynamics of small mammal populations: a review. In D. McCullough and R. Barrett (eds.), *Wildlife 2001: Populations*. New York: Elsevier Applied Science. pp. 831–50.

Brown, P. R., Singleton, G. R., Norton, G. A. and Thompson, D. (2001). Mouser (Version 1.0): a decision tool for management. *Proceedings of the Australasian Vertebrate Pest Conference*, **12**, 199–201.

Buckle, A.P. and Smith, R.H. (1994). *Rodent Pests and Their Control*. Wallingford, UK: CAB International.

Conover, M. (2002). *Resolving Human–Wildlife Conflicts*. Boca Raton, FL: Lewis Publishers.

Corrigan, R.M. (2001). *Rodent Control*. Cleveland, OH: GIE Media.

Dolbeer, R.A., Holler, N.R. and Hawthorne, D.W. (1994). Identification and control of wildlife damage. In T.A. Bookout (ed.), *Research and Management Techniques for Wildlife and Habitats*. Bethesda: The Wildlife Society. pp. 474–506.

Engeman, R.M. and Witmer, G.W. (2000*a*). IPM strategies: indexing difficult to monitor populations of pest species. *Proceedings of the Vertebrate Pest Conference*, **19**, 183–9.

Engeman, R.M. and Witmer, G.W. (2000*b*). Integrated management tactics for predicting and alleviating pocket gopher damage to conifer reforestation plantings. *Integrated Pest Management Reviews*, **5**, 41–55.

Fagerstone, K.A., Coffey, M.A., Curtis, P.D. *et al.* (2002). *Wildlife Fertility Control. Technical Review 02-2*. Bethesda: The Wildlife Society.

Howard, W.E., Marsh, R.E. and Corbett, C.W. (1985). Raptor perches: their influence on crop protection. *Acta Zoologica Fennica*, **173**, 191–2.

Hunter, J.E. (1991). Grazing and pocket gopher abundance in a California annual grassland. *Southwestern Naturalist*, **36**, 117–18.

Hygnstrom, S.E., Timm, R.M. and Larson, G.E. (1994). *Prevention and Control of Wildlife Damage*. Lincoln, NE: University of Nebraska Cooperative Extension.

Jackson, W.B. (2001). Current rodenticide strategies. *International Biodeterioration and Biodegradation*, **48**, 127–6.

Krebs, C.J. (1996). Population cycles revisited. *Journal of Mammalogy*, **77**, 8–24.

Mason, J.R. (1998). Mammal repellents: options and considerations for development. *Proceedings of the Vertebrate Pest Conference*, **18**, 325–9.

Naiman, R.J., Johnston, C.A. and Kelley, J.C. (1988). Alteration of North American streams by beaver. *BioScience*, **38**, 753–62.

Pech, R.P. (2000). Biological control of vertebrate pests. *Proceedings of the Vertebrate Pest Conference*, **19**, 206–11.

Pelton, M.R., Fribourg, H.A., Laundre, J.W. and Reynolds, T.W. (1991). Preliminary assessment of small wild mammal populations in tall fescue habitats. *Tennessee Farm and Home Science*, **160**, 68–71.

Pipas, M.J. and Witmer, G.W. (1999). Evaluation of physical barriers to protect ponderosa pine seedlings from pocket gophers. *Western Journal of Applied Forestry*, **14**, 164–8.

Ramsey, D.S.L. and Wilson, J.C. (2000). Towards ecologically based baiting strategies for rodents in agricultural systems. *International Biodeterioration and Biodegradation*, **45**, 183–97.

Singleton, G.R., Hinds, L.A., Herwig, H. and Zhang, Z. (eds.) (1999). *Ecologically-Based Management of Rodent Pests*. Canberra, Australia: Australian Centre for International Agricultural Research.

Singleton, G.R., Sudarmaji and Suriapermana, S. (1998). An experimental field study to evaluate a trap-barrier system and fumigation for controlling the rice rat in rice crops in West Java. *Crop Protection*, **17**, 55–64.

Sullivan, T. P., Crump, D. R. and Sullivan, D. S. (1988). Use of predator odors as repellents to reduce feeding damage by herbivores IV: northern pocket gophers. *Journal of Chemical Ecology*, **14**, 379–89.

Sullivan, T. P. and Klenner, W. (1993). Influences of diversionary food on red squirrel populations and damage to crop trees in young lodgepole pine forest. *Ecological Applications*, **3**, 708–18.

Sullivan, T. P., Sullivan, D. S., Hogue, E. J., Lautenschlager, R. A. and Wagner, R. G. (1998). Population dynamics of small mammals in relation to vegetation management in orchard agroecosystems: compensatory responses in abundance and biomass. *Crop Protection*, **17**, 1–11.

Way, M. J. and van Emden, H. F. (2000). Integrated pest management in practice: pathways towards successful application. *Crop Protection*, **19**, 81–103.

Witmer, G. W., Bucknall, J. L., Fritts, T. H. and Moreno, D. G. (1996). Predator management to protect endangered avian species. *Trans. North American Wildlife and Natural Resources Conference*, **61**, 102–8.

Witmer, G. W., Campbell, III, E. W. and Boyd, F. (1998). Rat management for endangered species protection in the U.S. Virgin Islands. *Proceedings of the Vertebrate Pest Conference*, **18**, 281–6.

Witmer, G. W., Fall, M. W. and Fiedler, L. A. (1995). Rodent control, research, and technology transfer. In J. Bissonette and P. Krausman (eds.), *Integrating People and Wildlife for a Sustainable Future*. Proceedings of the First International Wildlife Management Congress. Bethesda: The Wildlife Society. pp. 693–7.

Witmer, G. W. and VerCauteren, K. C. (2001). Understanding vole problems in direct seeding – strategies for management. In R. Veseth (ed.), *Proceedings of the Northwest Direct Seed Cropping Systems Conference*. Northwest Direct Seed Conference, Spokane. pp. 104–10.

Witmer, G. W., VerCauteren, K. C., Manci, K. M. and Dees, D. M. (2000). Urban-suburban prairie dog management: opportunities and challenges. *Proceedings of the Vertebrate Pest Conference*, **19**, 439–44.

13

Ecosystems: concepts, analyses and practical implications in IPM

T. D. SCHOWALTER

13.1 Introduction

A major principle of integrated pest management (IPM) is that strategies and tactics be consistent with ecological processes. IPM advanced traditional pest control by recognizing economic thresholds (below which population size does not warrant suppression), and addressing aspects of herbivore–plant and predator–prey interactions amenable to manipulation for pest control purposes. However, the premise that herbivorous insects and pathogens in general are detrimental to plant growth and reproduction has persisted.

In recent years, our perspective of insect herbivores has begun to change from this traditional view to a view that recognizes the role of native insect herbivores in maintaining plant productivity, vegetative diversity, and other ecosystem properties at ecosystem carrying capacity. Recent advances in studies of integrated ecosystems has led to an emerging view of herbivores as a negative feedback (regulatory) mechanism, triggered by environmental changes, that may stabilize ecosystem structure and function in natural ecosystems.

Commodity systems are maintained intentionally in an artificial condition to maximize commodity production, generally on an annual basis. This triggers herbivore responses to resource quality or abundance that may be undesirable for some management goals. Nevertheless, understanding the factors that trigger herbivore outbreaks and their consequences for ecosystems conditions can lead to improved management decisions, particularly strategies for preventing outbreaks and methods for evaluating the need for pest suppression, consistent with the objective that IPM supports ecological principles. IPM tactics will still be necessary, especially in the artificial environment of crop systems, or for managing exotic pests.

The ecosystem approach emphasizes evaluation of pests within the context of integrated ecosystems over landscape and long time scales. This chapter addresses two aspects of ecosystems fundamental to IPM approach and practice. First, herbivorous insect populations change within constraints established by variable ecosystem conditions, as affected by resource management practices. Second, herbivorous insects influence many environmental parameters in natural and agricultural ecosystems. Insect effects on some parameters may compensate for effects on others in ways that should be considered in deciding whether, and how, to suppress insect populations.

13.2 Ecosystem conditions that affect insect populations

Ecosystem conditions establish the environmental template with which insects, as well as other organisms, interact, through physiological and behavioral responses in the short term and through evolutionary changes in the long term. A number of environmental factors, including climate, predation, and resource conditions, are particularly important in determining biotic responses. These factors can be manipulated to some extent for IPM purposes.

13.2.1 Climate

All organisms have particular tolerance ranges for the various physical parameters of their environment. The natural range of occurrence of any species reflects its adaptation to conditions characteristic of the region within which it evolved and the natural barriers (e.g. mountains or water bodies) that prevent its spread. The combination of conditions to which organisms are adapted can occur in areas beyond these natural barriers, hence the ease with which many organisms become established in new locations when aided by human transport. Clearly, one objective of IPM is assessment of potential colonists from similar habitats in other areas and prevention of such introductions.

Climate change often may trigger insect outbreaks (e.g. White, 1969; 1976; Waring and Cobb, 1992), although the effect on insect populations is mediated by changes in host plant conditions (White, 1984; Mattson and Haack, 1987; Schowalter *et al.*, 1999). Drought, in particular, often increases populations of defoliating insects, especially Orthoptera and Lepidoptera (White, 1969; 1976; Mattson and Haack, 1987; Waring and Cobb, 1992).

Climate variability is largely beyond manipulation for IPM purposes. However, cold water mists, sprinklers, heaters, and other techniques for preventing frost or disrupting the synchrony of plant growth and insect development can increase insect mortality and reduce losses of plant resources (Miller, 1983).

13.2.2 *Predation and parasitism*

Predation and parasitism are a primary factor regulating prey populations and often are exploited for biological control of pest species (Holling, 1959; van den Bosch *et al.*, 1982; Tanada and Kaya, 1993; van Driesche and Bellows, 1996; see also Chapters 8, 9, 11, and 15 in this volume). Population sizes or ranges of many herbivorous insects are limited by predators and parasites, hence their explosive population growth when introduced into new habitats where natural predators and parasites are absent.

The efficiency with which predators and parasites regulate prey populations depends, in part, on their ability to disperse across the landscape and locate prey populations in habitat patches. Once prey are discovered, predation efficiency also depends on predator satiation and the time it takes predator populations to increase in response to prey availability (Holling, 1959). A number of studies have suggested that predator populations may be more sensitive to habitat fragmentation than are herbivore populations (Kruess and Tscharntke, 1994; 2000; Roland, 1993; Roland and Taylor, 1997; Zabel and Tscharntke, 1998; Thies and Tscharntke, 1999). Fragmentation of natural ecosystems may hinder host discovery by predators and parasitoids and thereby provide more predator-free space for herbivores.

A major tactic of IPM is reducing the time required for predators and parasitoids to find prey and to reproduce, by introducing more predators or more specialized predators into the predator–prey system. This requires identification of the most efficient predator(s) and artificial rearing of sufficient numbers for release into the target ecosystem. Augmentation of predation and parasitism through various biocontrol strategies is addressed elsewhere in this volume.

13.2.3 *Resource conditions*

Although the relative importance of top-down (predation) vs. bottom-up (resource conditions) controls has been a topic of considerable debate (see Hunter and Price, 1992), population growth clearly is limited by resource availability, as well as by predation. Agricultural and silvicultural practices typically promote insect populations associated with crop species by breeding for varieties without distasteful or toxic defenses and by planting crop species at high density over large areas. However, in at least some commodity production systems, host plant condition, density, and apparency to host-seeking insects could be manipulated to reduce pest problems.

Host condition

Healthy plants produce a variety of physical and chemical defenses (e.g. spines, hairs, phenolics, terpenoids, alkaloids) against herbivores (Bernays and Chapman, 1994; Harborne, 1994; Chapter 6 in this volume). Plants are adapted to balance risk of herbivory against the cost of defense and often respond to injury by herbivores with new (induced) or increased concentrations of chemical defenses (e.g. Coley *et al.*, 1985). Plants also can communicate injury to neighboring plants through emission of volatile signals that induce production of more targeted defenses, even among unrelated plant species (Farmer and Ryan, 1990; Dolch and Tscharntke, 2000).

Disturbances and other environmental changes injure organisms and alter water and nutrient fluxes and availability. These changes stress intolerant plants, limiting their ability to produce defenses that often depend on adequate water and nutrient resources (Coley *et al.*, 1985). Wound repair and replacement of lost foliage or root tissues to meet metabolic demands require redirection of carbohydrates and nitrogen from energetically expensive defensive compounds, such as phenols, terpenes, and alkaloids (e.g. Coley *et al.*, 1985). Early successional plants, such as most crop species, generally are sensitive to competitive stress, and insects or pathogens may accelerate replacement of stressed plants by plant species more tolerant of resource limitation (Davidson, 1993; Schowalter, 1985).

Herbivorous insects selectively feed on plants for which they have adapted mechanisms to detoxify or avoid defenses. These adaptations also place demands on energy resources. Therefore, herbivorous insects face an evolutionary tradeoff between specific adaptations that circumvent defenses of one or a few specific host species and more general adaptations that permit exploitation of a wider range of host species, but with low efficiency under conditions favorable for plant defense. Reduced defensive capacity makes plants more vulnerable to both specialist and generalist herbivores and pathogens.

Physical and chemical defenses often have been bred out of crop plants used for human consumption, making them more vulnerable to a variety of native and exotic pests. Two remedies have been (a) exploration for original genotypes to reintroduce defenses and (b) genetic modification to introduce novel defenses, such as the *Bacillus thuringiensis* (Bt) toxin genes into several crop species. However, plants must have sufficient resources to allocate to defenses, so that reduced concentrations do not facilitate insect population growth and development of resistance genes in the insect population. Furthermore, if genetically modified cultivars are planted widely, insect population exposure to the introduced defenses will facilitate adaptations for resistance (Tabashnik, 1994; Alstad and Andow, 1995; Chapter 10 in this volume; see also below).

Host abundance and apparency

Herbivorous insects are highly sensitive to host spacing and to disruption of host detection by association with non-hosts (e.g. Risch, 1980; 1981; Kareiva, 1983; Stanton, 1983; Visser, 1986; Schowalter and Turchin, 1993). Each plant species produces a characteristic blend of volatile chemicals that many insect herbivores and their predators use as cues to orient toward suitable hosts (Visser, 1986; Turlings *et al.*, 1990; Cardé, 1996). Volatiles from non-host plant species are non-attractive, or even repellent, to a given insect species. Host spacing affects insect ability to perceive host cues (Figure 13.1), to reach a potential host with limited time and energy reserves, and to avoid the attention of predators while searching for hosts (Kareiva, 1983; Schultz, 1983; Stanton, 1983). Hence, closely spaced hosts in crop monocultures require low costs in energy and mortality for insect colonization, compared to sparse hosts or hosts mixed with non-hosts (Risch, 1980; 1981). Dense hosts also are more likely to be stressed (and susceptible to insects) as a result of competition for limited resources (as described above). By contrast, relatively small increases in plant spacing or in interspersion of non-host plants can greatly reduce insect and pathogen ability to detect and colonize their hosts (Risch, 1980; 1981; Schowalter and Turchin, 1993; Garrett and Mundt, 2000*a*; *b*). However, insect species adapted to scattered or unpredictable resources may show high efficiency in finding and colonizing these resources (Dubbert *et al.*, 1998).

Landscape effects

Landscape patches dominated by healthy and diverse vegetation (typical of natural ecosystems) provide scattered, isolated suitable host plants that must be discovered by host-seeking insects, with considerable expense in energy and mortality. These conditions keep insect populations low on the landscape and permit many susceptible plants to escape detection and

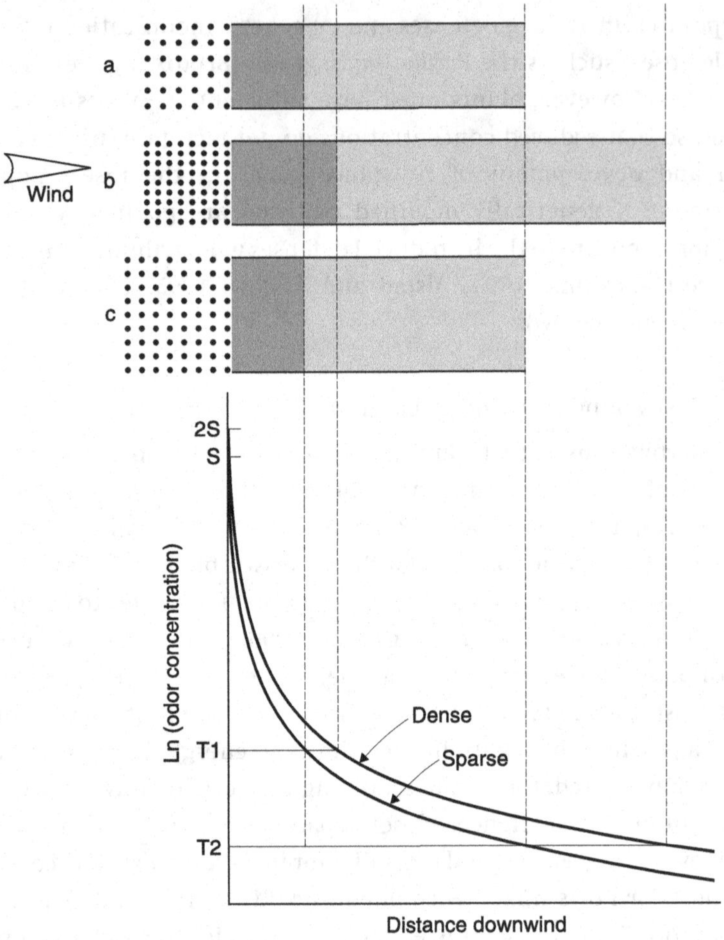

Figure 13.1. Contrasting odor concentration gradients downwind from (a) a small patch of low host density, (b) a small patch of high density, and (c) a large patch of low host density. The curves reflect an ideal situation in which convection due to wind prevails over simple diffusion. In still air, odor concentration is independent of host density. Attractive areas (stippled for low-sensitivity herbivores, threshold T1; shaded for high-sensitivity herbivores, threshold T2) are shown as rectangles for simplicity, but are actually irregular in shape. Reprinted from Stanton (1983) with permission from Elsevier.

colonization. Similarly, specialist predators must search for those plants that are colonized by their prey, permitting some isolated insect populations to escape detection and predation.

Landscape pattern affects species differentially. Kruess and Tscharntke (1994), Dubbert *et al.* (1998) and Thies and Tscharntke (1999) reported that predator species diversity and abundances were greatly reduced in agroecosystems (largely due to loss of protected habitats and food resources not present in crop fields), releasing herbivore populations from control by predation. Zabel and Tscharntke (1998) and Kruess and Tscharntke (2000) found that herbivore diversity and abundance and percent parasitism were positively related to habitat area, whereas predator diversity and abundance were negatively related to habitat isolation, potentially allowing some prey species to increase in abundance in isolated habitats. Steffan-Dewenter and Tscharntke (1999) reported that pollinator activity and seed production declined substantially with increasing plant isolation (Figure 13.2); the abundance of wild bees was a better predictor of number of seeds per plant than was abundance of honey bees. Roland (1993) reported increased duration of a defoliator outbreak in fragmented forests, perhaps reflecting reduced abundances of parasitoids in remnant forest patches (Roland and Taylor, 1997).

Fragments of otherwise resistant vegetation can be subjected to constant inundation by herbivores dispersing from adjacent patches conducive to herbivore population growth. For example, Schowalter and Stein (1987) reported that populations of *Lygus hesperus* dispersing from agricultural crops, such as alfalfa and buckwheat, were responsible for considerable damage to conifer seedlings in adjacent nurseries. Therefore, the particular matrix of crop species or cultivars on the landscape affects insect population distribution and crop damage.

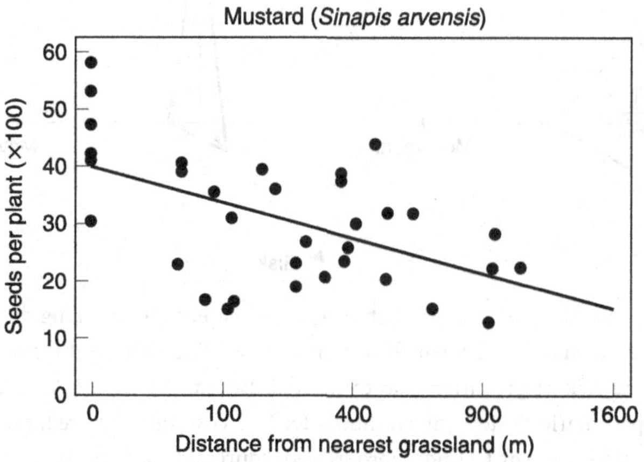

Figure 13.2. Relationship between distance of potted mustard plants from the nearest seminatural grassland in Germany and seed production per plant and flower visitation by wild bees and seed production per plant. Regression line is significant at P = 0.001. Note the square root scale for distance. Reprinted with permission from Springer-Verlag GmbH & Co. from Steffan-Dewenter and Tscharntke (1999).

13.2.4 *Tradeoffs*

In natural ecosystems, the diversity of species and environmental conditions requires that insects adjust life history strategies to maximize fitness under suboptimal conditions. For example, Schultz (1983) described the tradeoff space representing resource quality in one dimension, resource distribution in a second dimension, and vulnerability to predation in the third dimension (Figure 13.3). An herbivore can seek the highest quality resources to maximize growth and reproduction, but this requires movement

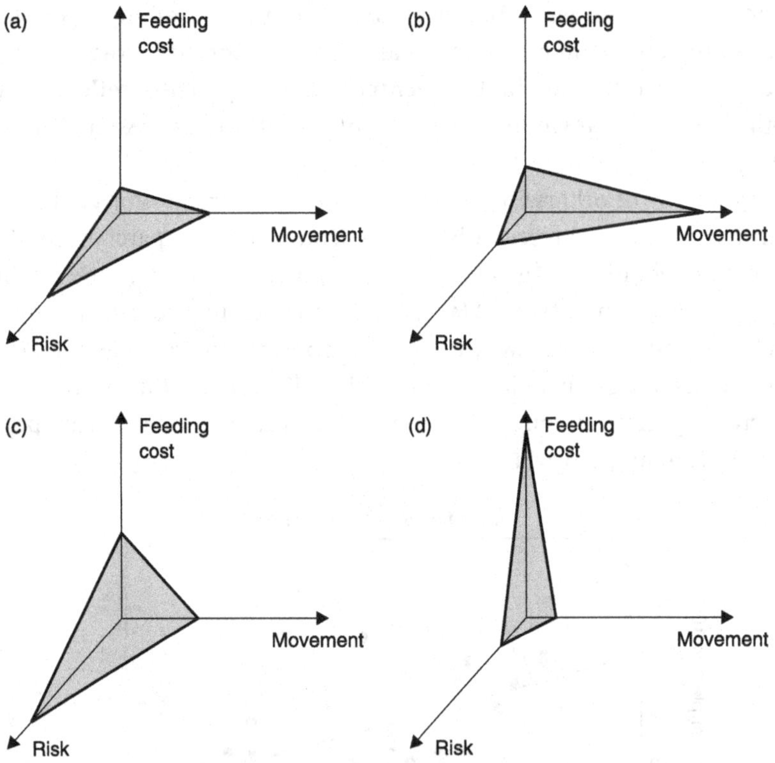

Figure 13.3. Tradeoff planes representing selected caterpillar foraging strategies: (a) a selective diurnal feeder adapted to minimize feeding costs and movement, but with moderate risk of predation due to limited diurnal activity, (b) a selective nocturnal aposomatic feeder, with minimal feeding cost and low predation risk but greater cost of movement to find appropriate resources, (c) a diurnal cryptic feeder, with moderate cost of feeding on resources that permit minimal movement but reduced growth due to time spent hiding, and (d) a food mimic, with minimal movement and exposure, but high cost of feeding due to restricted diet. Reprinted from variable plants and herbivores in natural and managed systems, J. C. Schultz, "Habitat selection and foraging tactics of caterpillars in heterogenous trees" pp. 61–90, copyright 1983, with permission from Elsevier.

(among resources of varying quality) that can attract predators. Alternatively, an herbivore can accept lower quality resources, minimizing the need for movement and allowing the insect to hide from predators, but limiting growth and reproduction. Kogan (1975) and Courtney (1985) reported that some species exhibit tradeoffs between life stages. Short-lived adults preferentially oviposit on acceptable, but low quality, plant species that are abundant and/or apparent, whereas larval survival is higher on less abundant and/or apparent host species. Fitness may be increased by oviposition on less favorable plants that can be found easily, compared to oviposition on more favorable plants that might not be discovered during the adult lifetime.

13.2.5 *Application to agricultural systems*

In contrast to most natural ecosystems, commodity production systems are characterized by chronically high density, apparency, and habitat area of crop plants, conditions most conducive to rapid colonization and exploitation of susceptible plants by their associated herbivores (Kogan, 1975; Risch, 1980; 1981; Kareiva, 1983; Schultz, 1983; Courtney, 1985) and pathogens and least conducive to colonization by entomophagous predators and parasitoids (Roland, 1993; Kruess and Tscharntke, 1994; 2000; Roland and Taylor, 1997; Zabel and Tscharntke, 1999). Closely spaced plants become even more susceptible when genetic heterogeneity is reduced through widespread planting of genetically modified cultivars, or when plant resistance is compromised by cultivation techniques or environmental changes that damage roots or affect water and nutrient availability. Pest management can be enhanced by maximizing plant health and the diversity of cultivars and crop species on the landscape (Risch, 1980; 1981; Garrett and Mundt, 2000*a*; *b*). Predation also can be enhanced by intercropping and no-tillage practices and by retaining or introducing other elements of landscape diversity, such as windrows or hedgerows (House and Stinner, 1987; Kruess and Tscharntke, 1994).

13.3 Herbivore effects on ecosystems

Herbivorous insects affect a variety of ecosystem parameters, including primary productivity, vegetation dynamics, decomposition, soil development and fertility, and hydric and nutrient fluxes. Herbivorous insects traditionally have been viewed as destabilizing factors in ecosystems. However, several effects, including increased vegetation diversity and/or increased growth of plant species more tolerant of current conditions, and increased water flux and nutrient turnover resulting from herbivory, may contribute to long-term stability/sustainability of ecosystems (Schowalter, 2000*a*; *b*). Herbivore effects on soil fertility (for example) may contribute to long-term sustainability of some agricultural systems, especially perennial crops or agroforestry systems. Suppression under such circumstances might be counterproductive, even if commodity production is reduced in the short term. While regulatory effects of insects may not always contribute to the goals of commodity production (e.g. increasing vegetation diversity at the expense of crop plant production), IPM strategies and tactics should be designed to enhance these regulatory functions, where possible. The following section describes ways in which herbivorous insects affect ecosystem parameters.

13.3.1 *Insect effects on ecosystem parameters*

Ecosystems can be described in terms of their structure and function. Structure for terrestrial ecosystems includes vegetation height, diversity of canopy layers, and leaf area index which, in turn, determine biomass; carbon storage; capacity to intercept light, rainfall, and wind; interception of material from the airstream; vertical and horizontal gradients of temperature and relative humidity; rainfall impact and erosion severity; litter depth and quality; and soil temperature and organic matter profiles. Function includes photosynthetic efficiency; primary productivity; and biotic modification of carbon, water, and nutrient fluxes, and climate. Insect herbivores are key contributors to processes of primary productivity, pollination, and seed dispersal, and also affect soil processes (Schowalter, 2000*a*; Schowalter and Withgott, 2001).

Agroecosystems, like natural ecosystems, require inputs of energy and matter that fuel biological functions that determine energy flux and nutrient cycling patterns, and subsequent export of energy and nutrients. Unlike natural ecosystems, maintenance of agroecosystems for human commodity production goals currently involves extensive fossil fuel combustion (for plowing, harvest, and transport), mechanical and chemical disturbances (soil disruption, fertilizer and pesticide application), and removal (harvest) of organic matter. Agricultural and silvicultural activities alter natural rates and seasonal and spatial patterns of primary production, energy flux, and water and nutrient cycling. These alterations affect plant energy and nutrient uptake and allocation to growth vs. defense, thereby affecting plant productivity and vulnerability to herbivores and pathogens. Furthermore, exotic crops, livestock, and pests influence the structure and function of surrounding remnants of natural ecosystems.

Herbivorous insects affect ecosystem parameters largely through their consumption of plant tissues and resulting changes in plant allocation of water, energy and nutrients, plant competitive or mutualistic interactions, and fluxes of energy, water, and nutrients. These changes affect agroecosystems in the following ways.

Herbivorous insects function in natural ecosystems to control host plant density and vegetation diversity. In the same way that predators are recognized as regulators of prey populations, herbivores regulate their host (prey) populations. As described above, outbreaks of native species typically are triggered by stress and/or high density of host plants. Under these circumstances, insect outbreaks are self-limiting, with insects targeting the most stressed (therefore, vulnerable) host plants, reducing host density, mitigating competition for resources, and ultimately reducing their own resource base, leading (along with predator responses) to population decline. The proportion of the host population targeted by the insect population increases as host density or the severity of host stress increases (Figure 13.4). Therefore, the ecosystems most vulnerable to insect population outbreaks are commodity production systems with high plant density and potential stress from competition or cultivation practices.

Nevertheless, some natural pruning and thinning by insects may be tolerable or even beneficial, particularly in agroforestry or rangeland applications in which these functions are necessary for long-term

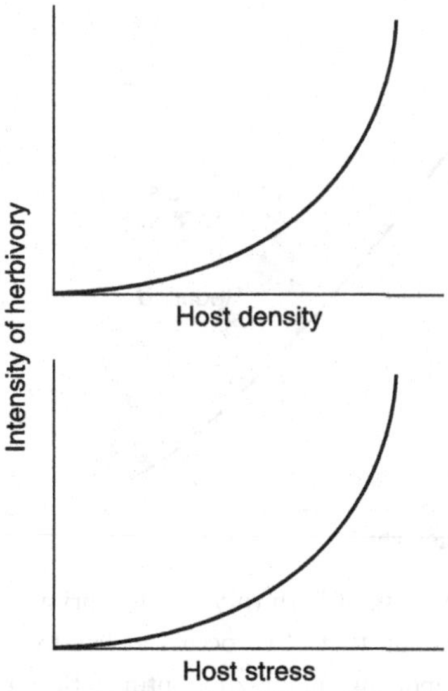

Figure 13.4. Generalized relationship between the intensity of herbivory and (a) host plant density and (b) host plant stress.

productivity of perennial crops. Low-to-moderate grazing typically stimulates productivity of rangelands (Figure 13.5) (Seastedt, 1985; Georgiadis and McNaughton, 1988; Williamson *et al.*, 1989), through stimulation by salivary compounds (Dyer *et al.*, 1993), and/or removal of dead standing material (Knapp and Seastedt, 1986), and forests over longer time periods (Romme *et al.*, 1986; Alfaro and Shepherd, 1991).

Insect herbivores increase the fluxes of light, water, and nutrients from host plants to the soil. In many cases, light, water, or nutrient limitation is the factor that stresses plants and triggers insect outbreaks (Waring and Cobb, 1992). Accordingly, canopy opening and transfer of water and nutrients to the soil may alleviate plant stress (Knapp and Seastedt, 1986; Schowalter *et al.*, 1991) and may be one mechanism underlying compensatory growth of plants subjected to light-to-moderate levels of grazing (Williamson *et al.*, 1989; Pedigo *et al.*, 1986; Trumble *et al.*, 1993). Dyer *et al.* (1995) also found that herbivore salivary and digestive fluids may stimulate plant growth. Obviously, the resources released from host plants via herbivores can be exploited by non-host plants that increase vegetation diversity and interfere with host discovery by insects. In natural ecosystems, this diversity of plant species uses available resources more efficiently, stores more carbon, and

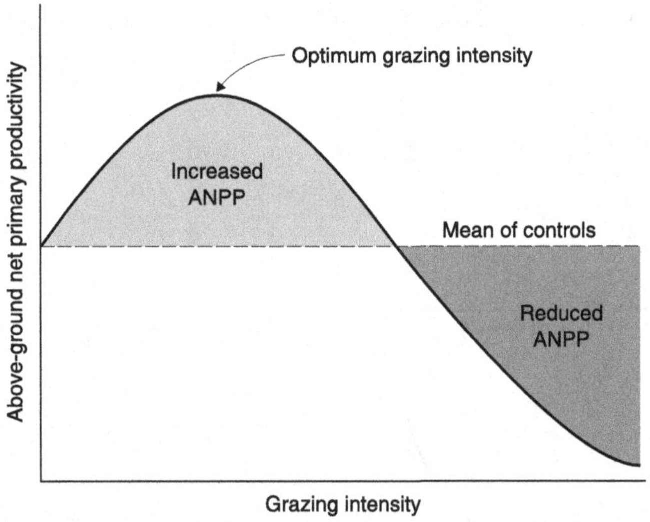

Figure 13.5. Relationship between intensity of herbivory and net primary productivity, showing the peak production that often occurs at low to moderate intensities of herbivory, supporting the grazing optimization hypothesis. From Williamson *et al.* (1989) by permission from the Society for Range Management.

provides necessary habitats for predators and detritivores, thereby maintaining soil structure and fertility, as well as other ecosystem processes. Although replacement of crop plants by non-crop plants via herbivores generally will be undesirable in commodity systems, the cost and benefits of pest control in crop monocultures should be compared to costs and benefits of mixing crop species or varieties, or crop rotation, that mimic natural processes and reduce insect population growth (e.g. Risch, 1981).

Most research on arthropods in agroecosystems has focused on IPM for herbivorous species or on augmentation or conservation of predaceous arthropods (see Chapters 8, 9, 11, and 15 this volume). However, detritivorous species also can affect crop plant production and are sensitive to IPM practices. Litter and soil fauna are instrumental in the breakdown of organic matter, often retained as mulch following harvest to provide resources for the next crop. Although the direct benefit of detritivores to crop production has rarely been measured (Ingham *et al.*, 1985), Setälä and Huhta (1991) demonstrated that soil fauna significantly increased both the growth and nitrogen content of birch seedlings (*Betula pendula*) in pots (Figure 13.6). IPM practices that interfere with the soil/litter community are potentially counterproductive.

13.3.2 Stabilization or destabilization?

Natural ecosystems are composed of co-adapted species that mutually regulate one another, from below via factors that limit resource quality or availability, from above via negative feedback from predation, and laterally via negative feedback from competition, cannibalism, etc. Insects represent ideal potential regulators because their small biomass is highly responsive to cues indicating host condition and abundance and can be rapidly amplified to provide negative feedback (Schowalter, 2000*a*; *b*).

Some studies have indicated that herbivory may function to stabilize primary production (Mattson and Addy, 1975; Romme *et al.*, 1986), an ecosystem variable that determines others such as ecosystem structure, soil conditions, nutrient cycling, and climate modification. Stimulation of (compensatory) plant growth at low-to-moderate intensities of feeding and rapid depression of host growth at high intensities of feeding (Figures 13.4 and 13.5) should maintain primary productivity and vegetation structure within a narrower range than occurs when plants exceed their carrying capacity (Figure 13.7).

The potential regulatory role of insects is a topic of heated debate between those who argue that natural selection at the individual level should not produce attributes that stabilize ecosystems, and those who argue that

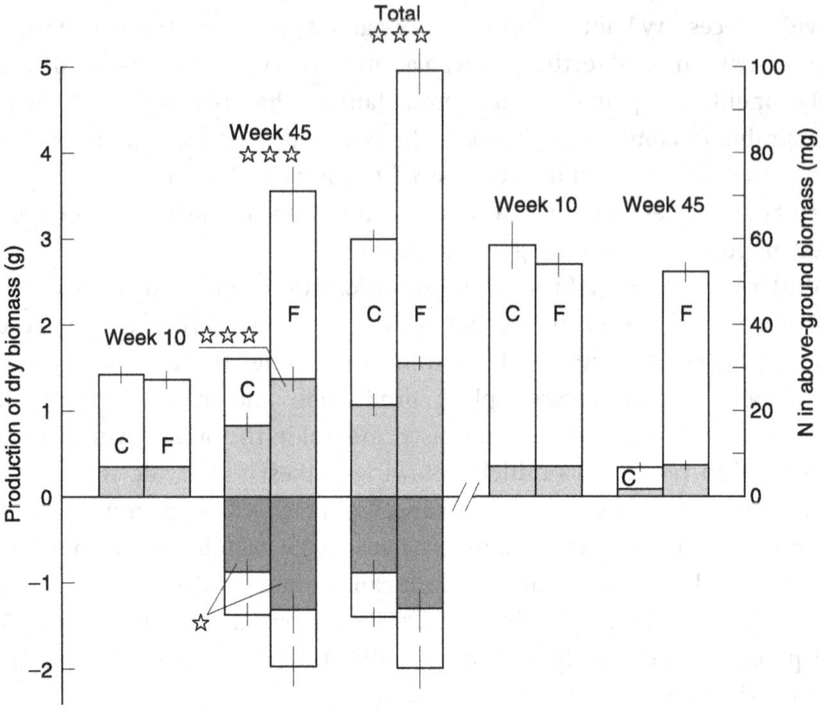

Figure 13.6. Effect of soil fauna on biomass production (left of break in horizontal axis) and nitrogen accumulation (right of break) of birch, *Betula pendula*, seedlings. Bars above the horizontal axis are stems (stippled) and leaves (clear); bars below the horizontal axis are roots in humus (hatched) and in mineral soil (clear). C, fauna removed; F refaunated. Vertical lines represent one standard deviation for all data (except nitrogen at week 45, where vertical lines represent minimum and maximum values). For C versus F, * = $P < 0.05$; *** = $P < 0.001$. Stem nitrogen was not measured at week 10. Reproduced from Setälä and Huhta (1991) by permission from the Ecological Society of America.

an organism's effects on its environment (ecosystem) are subject to selection that should favor individuals whose activities contribute to ecosystem stability (Schowalter, 2000*a*). However, outbreaks of native insect species appear to be relatively rare in ecosystems subject to minimal disturbance or human interference, and in those ecosystems herbivores appear to maintain a diversity of healthy plant populations, facilitate recovery from disturbances, and maintain more consistent fluxes of energy, water, and nutrients. By contrast, where disturbance or human interference has altered vegetation structure, insect outbreaks appear to function to restore more historic conditions (Schowalter, 2000*a*).

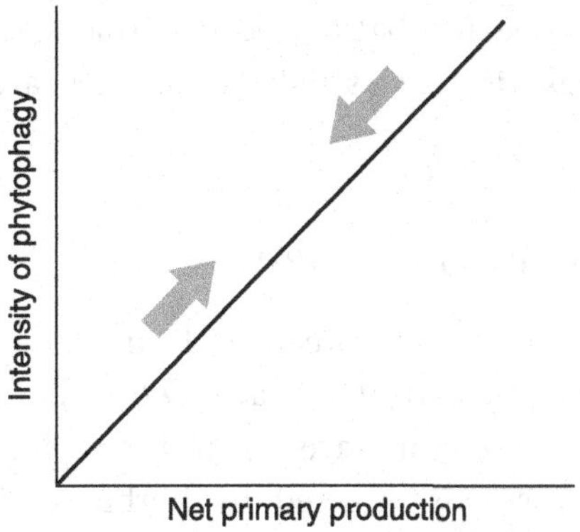

Figure 13.7. Potential stabilization of net primary production by herbivores as a result of stimulation of productivity at low levels of productivity and depression of production at high levels of productivity (see Figure 13.5). From Schowalter (2000*b*) in D. C. Coleman and P. F. Hendrix (2000) Invertebrates as webmasters in ecosystems, by permission from CABI Publishing, Wallingford, UK.

In large part, this argument may be based on our perspective of stability, which has not been clearly defined. The stability of a natural ecosystem may be based on a landscape scale diversity of community types that provide colonists for rapid recovery of disturbed patches. Commodity production systems focused on maintaining a particular species valued for rapid growth are not stable, because natural processes, including feeding by herbivorous insects, function to diversify the ecosystem. Artificially maintaining commodity production systems represents a destabilization of the ecosystem and requires considerable energy and nutrient inputs. Production of annual crops also requires more diligent management than does production of perennial crops that are more tolerant of competition and stresses likely to occur over longer life cycles.

13.3.3 *Exotic species*

Exotic species do not function as regulators (although they may be important regulators in their native habitat). Exotic species have not adapted to the conditions (especially to host chemistry and abundance) of their new ecosystem in the same way as have native species, and their presence alters conditions to which native species have adapted. Furthermore, they are released from predation by their adapted predators, and the predators in their new habitat are not adapted to the attributes of exotic prey. Whereas native species are regulated by negative feedbacks from higher and lower trophic levels, exotic species can exploit suitable conditions without these restrictions and may initiate positive feedback processes that destabilize crop–insect interactions. Hence, exotic species will be pests regardless of management goals.

13.4 Conclusions and recommendations for IPM

Ecosystem structure and function affect and are affected by insect herbivores. Host plant density and stress are two major factors affecting insect populations. These factors are subject to changes in water and nutrient cycling processes as affected by herbivory, as well as by disturbances and environmental changes. Agroecosystems are characterized by high crop (species or cultivar) density and altered defensive capability.

Two questions largely address the ways in which IPM can incorporate ecosystem principles. What agroecosystem conditions lead to unacceptable abundances of insects, and how can these be mitigated? How do insects affect agroecosystem values, and do some effects compensate for others?

Clearly, crop plant diversity can be built into commodity production systems to enhance agroecosystem regulation of crop-feeding insects and pathogens. Many agricultural situations are amendable to reducing the dominance of landscapes by particular crop species and varieties. Intercropping and crop rotation are two established methods of maintaining soil conditions and reducing growth of pest populations. Farming cooperatives could coordinate landscape patterns of commodity production that minimize vulnerability to spread of particular pests. Landscape diversity further enhances IPM by providing refuges for various predators that augment ecosystem regulation of herbivore populations.

Insects affect virtually all ecosystem processes through their effect on plant growth and vegetation structure. Thresholds for suppression will remain low on many, especially annual, crops, but could be raised in some, especially perennial crop, rangeland or forest, systems where herbivore-induced turnover of plant matter, energy, and nutrients may contribute to sustainability of soil conditions and agricultural productivity over longer time periods. However, non-adapted insect species alter rates and seasonality of ecosystem processes and represent serious threats to the stability/sustainability of all ecosystems. Hence, IPM will continue to be necessary to combat invasive species, in both agricultural and "natural" ecosystems.

References

Alfaro, R. I. and Shepherd, R. F. (1991). Tree-ring growth of interior Douglas-fir after one year's defoliation by Douglas-fir tussock moth. *Forest Science*, **37**, 959–64.

Alstad, D. N. and Andow, D. A. (1995). Managing the evolution of insect resistance to transgenic plants. *Science*, **268**, 1894–6.

Bernays, E. A. and Chapman, R. F. (1994). *Host-Plant Selection by Phytophagous Insects*. New York: Chapman & Hall.

Cardé, R. T. (1996). Odour plumes and odour-mediated flight in insects. In *Olfaction in Mosquito–Host Interactions (Ciba Foundation Symposium 200)*. Chichester, UK: John Wiley and Sons. pp. 54–70.

Coley, P. D., Bryant, J. P. and Chapin III, F. S. (1985). Resource availability and plant antiherbivore defense. *Science*, **230**, 895–9.

Courtney, S. P. (1985). Apparency in coevolving relationships. *Oikos*, **44**, 91–8.

Davidson, D. W. (1993). The effects of herbivory and granivory on terrestrial plant succession. *Oikos*, **68**, 23–35.

Dolch, R. and Tscharntke, T. (2000). Defoliation of alders (*Alnus glutinosa*) affects herbivory by leaf beetles on undamaged neighbors. *Oecologia*, **125**, 504–11.

Dubbert, M., Tscharntke, T. and Vidal, S. (1998). Stem-boring insects of fragmented *Calamagrostis* habitats: herbivore–parasitoid community structure and the unpredictability of grass shoot abundance. *Ecological Entomology*, **23**, 271–80.

Dyer, M. I., Turner, C. L. and Seastedt, T. R. (1993). Herbivory and its consequences. *Ecological Applications*, **3**, 10–16.

Dyer, M. I., Moon, A. M., Brown, M. R. and Crossley, Jr., D. A. (1995). Grasshopper crop and midgut extract effects on plants: an example of reward feedback. *Proceedings of the National Academy of Sciences USA*, **92**, 5475–8.

Farmer, E. E. and Ryan, C. A. (1990). Interplant communication: airborne methyl jasmonate induces synthesis of proteinase inhibitors in plant leaves. *Proceedings of the National Academy of Sciences USA*, **87**, 7713–16.

Garrett, K. A. and Mundt, C. C. (2000*a*). Host diversity can reduce potato late blight severity for focal and general patterns of primary inoculum. *Phytopathology*, **90**, 1307–12.

Garrett, K. A. and Mundt, C. C. (2000*b*). Effects of planting density and the composition of wheat cultivar mixtures on stripe rust: an analysis taking into account limits to the replication of controls. *Phytopathology*, **90**, 1313–21.

Georgiadis, N. J. and McNaughton, S. J. (1988). Interactions between grazers and a cyanogenic grass, *Cynodon plectostachyus*. *Oikos*, **51**, 343–50.

Harborne, J. B. (1994). *Introduction to Ecological Biochemistry*, 4th edn. London: Academic Press.

Holling, C. S. (1959). Some characteristics of simple types of predation and parasitism. *Canadian Entomologist*, **91**, 385–98.

House, G. J. and Stinner, B. R. (1987). *Arthropods in Conservation Tillage Systems* (Miscellaneous Publications, No. 65). Lanham, MD: Entomological Society of America.

Hunter, M.D. and Price, P.W. (1992). Playing chutes and ladders: heterogeneity and the relative roles of bottom-up and top-down forces in natural communities. *Ecology*, **73**, 724–32.

Ingham, R.E., Trofymow, J.A., Ingham, E.R. and Coleman, D.C. (1985). Interactions of bacteria, fungi, and their nematode grazers: effects on nutrient cycling and plant growth. *Ecological Monographs*, **55**, 119–40.

Kareiva, P. (1983). Influence of vegetation texture on herbivore populations: resource concentration and herbivore movement. In R.F. Denno and M.S. McClure (eds.), *Variable Plants and Herbivores in Natural and Managed Systems*. New York: Academic Press. pp. 259–89.

Knapp, A.K. and Seastedt, T.R. (1986). Detritus accumulation limits productivity of tallgrass prairie. *BioScience*, **36**, 662–8.

Kogan, M. (1975). Plant resistance in pest management. In R.L. Metcalf and W.H. Luckmann (eds)., *Introduction to Insect Pest Management*. New York: Wiley. pp. 103–46.

Kruess, A. and Tscharntke, T. (1994). Habitat fragmentation, species loss, and biological control. *Science*, **264**, 1581–4.

Kruess, A. and Tscharntke, T. (2000). Species richness and parasitism in a fragmented landscape: experiments and field studies with insects on *Vicia sepium*. *Oecologia*, **122**, 129–37.

Mattson, W.J. and Addy, N.D. (1975). Phytophagous insects as regulators of forest primary production. *Science*, **190**, 515–22.

Mattson, W.J. and Haack, R.A. (1987). The role of drought in outbreaks of plant-eating insects. *BioScience*, **37**, 110–18.

Miller, G.E. (1983). Evaluation of the effectiveness of cold-water misting of trees in seed orchards for control of Douglas-fir cone gall midge (Diptera: Cecidomyiidae). *Journal of Economic Entomology*, **76**, 916–19.

Pedigo, L.P., Hutchins, S.H. and Higley, L.G. (1986). Economic injury levels in theory and practice. *Annual Review of Entomology*, **31**, 341–68.

Risch, S. (1980). The population dynamics of several herbivorous beetles in a tropical agroecosystem: the effect of intercropping corn, beans and squash in Costa Rica. *Journal of Applied Ecology*, **17**, 593–612.

Risch, S.J. (1981). Insect herbivore abundance in tropical monocultures and polycultures: an experimental test of two hypotheses. *Ecology*, **62**, 1325–40.

Roland, J. (1993). Large-scale forest fragmentation increases the duration of tent caterpillar outbreak. *Oecologia*, **93**, 25–30.

Roland, J. and Taylor, P.D. (1997). Insect parasitoid species respond to forest structure at different spatial scales. *Nature*, **386**, 710–13.

Romme, W.H., Knight, D.H. and Yavitt, J.B. (1986). Mountain pine beetle outbreaks in the Rocky Mountains: regulators of primary productivity? *American Naturalist*, **127**, 484–94.

Schowalter, T.D. (1985). Adaptations of insects to disturbance. In S.T.A. Pickett and P.S. White (eds.), *The Ecology of Natural Disturbance and Patch Dynamics*. Orlando, FL: Academic Press. pp. 235–52.

Schowalter, T. D. (2000*a*). *Insect Ecology: An Ecosystem Approach*. San Diego, CA: Academic Press.

Schowalter, T. D. (2000*b*). Insects as regulators of ecosystem development. In D. C. Coleman and P. F. Hendrix (eds.), *Invertebrates as Webmasters in Ecosystems*. Wallingford, UK: CAB International. pp. 99–114.

Schowalter, T. D. and Stein, J. D. (1987). Influence of Douglas-fir seedling provenance and proximity to insect population sources on susceptibility to *Lygus hesperus* (Heteroptera: Miridae) in a forest nursery in western Oregon. *Environmental Entomology*, **16**, 984–6.

Schowalter, T. D. and Turchin, P. (1993). Southern pine beetle infestation development: interaction between pine and hardwood basal areas. *Forest Science*, **39**, 201–10.

Schowalter, T. D. and Withgott, J. (2001). Rethinking insects: what would an ecosystem approach look like? *Conservation Biology in Practice*, **2**, 10–16.

Schowalter, T. D., Sabin, T. E., Stafford, S. G. and Sexton, J. M. (1991). Phytophage effects on primary production, nutrient turnover, and litter decomposition of young Douglas-fir in western Oregon. *Forest Ecology and Management*, **42**, 229–43.

Schowalter, T. D., Lightfoot, D. C. and Whitford, W. G. (1999). Diversity of Arthropod Responses to Host-Plant Water Stress in a Desert Ecosystem in Southern New Mexico. *American Midland Naturalist*, **142**, 281–90.

Schultz, J. C. (1983). Habitat selection and foraging tactics of caterpillars in heterogeneous trees. In R. F. Denno and M. S. McClure (eds.), *Variable Plants and Herbivores in Natural and Managed Systems*. New York: Academic Press. pp. 61–90.

Seastedt, T. R. (1985). Maximization of primary and secondary productivity by grazers. *American Naturalist*, **126**, 559–64.

Setälä, H. and Huhta, V. (1991). Soil fauna increase *Betula pendula* growth: laboratory experiments with coniferous forest floor. *Ecology*, **72**, 665–71.

Stanton, M. L. (1983). Spatial patterns in the plant community and their effects upon insect search. In S. Ahmad (ed.), *Herbivorous Insects: Host-Seeking Behavior and Mechanisms*. New York: Academic Press. pp. 125–57.

Steffan-Dewenter, I. and Tscharntke, T. (1999). Effects of habitat isolation on pollinator communities and seed set. *Oecologia*, **121**, 432–40.

Tabashnik, B. E. (1994). Evolution of resistance to *Bacillus thuringiensis*. *Annual Review of Entomology*, **39**, 47–79.

Tanada, Y. and Kaya, H. (1993). *Insect Pathology*. San Diego, CA: Academic Press.

Thies, C. and Tscharntke, T. (1999). Landscape structure and biological control in agroecosystems. *Science*, **285**, 893–5.

Trumble, J. T., Kolodny-Hirsch, D. M. and Ting, I. P. (1993). Plant compensation for arthropod herbivory. *Annual Review of Entomology*, **38**, 93–119.

Turlings, T. C. J., Tumlinson, J. H. and Lewis, W. J. (1990). Exploitation of herbivore-induced plant odors by host-seeking parasitic wasps. *Science*, **250**, 1251–3.

van den Bosch, R., Messinger, P. S. and Gutierrez, A. P. (1982). *An Introduction to Biological Control*. New York: Plenum Press.

van Driesche, R. G. and Bellows, T. (1996). *Biological Control*. New York: Chapman and Hall.

Visser, J. H. (1986). Host odor perception in phytophagous insects. *Annual Review of Entomology*, **31**, 121–44.

Waring, G. L. and Cobb, N. S. (1992). The impact of plant stress on herbivore population dynamics. In E. A. Bernays (ed.), *Plant–Insect Interactions* (Vol. 4), Boca Raton, FL: CRC Press. pp. 167–226.

White, T. C. R. (1969). An index to measure weather-induced stress of trees associated with outbreaks of psyllids in Australia. *Ecology*, **50**, 905–9.

White, T. C. R. (1976). Weather, food and plagues of locusts. *Oecologia*, **22**, 119–34.

White, T. C. R. (1984). The abundance of invertebrate herbivores in relation to the availability of nitrogen in stressed food plants. *Oecologia*, **63**, 90–105.

Williamson, S. C., Detling, J. K., Dodd, J. L. and Dyer, M. I. (1989). Experimental evaluation of the grazing optimization hypothesis. *Journal of Range Management*, **42**, 149–52.

Zabel, J. and Tscharntke, T. (1998). Does fragmentation of *Urtica* habitats affect phytophagous and predatory insects differentially? *Oecologia*, **116**, 419–25.

14

Agroecology: contributions towards a renewed ecological foundation for pest management

C. I. NICHOLLS AND M. A. ALTIERI

14.1 Introduction

The integrated pest management concept (IPM) arose in the early 1970s in response to concerns about impacts of pesticides on the environment. By providing an alternative to the strategy of unilateral intervention with chemicals, it was hoped that IPM would change the philosophy of crop protection to one that entailed a deeper understanding of insect and crop ecology, thus resulting in a strategy which relied on the use of several complementary tactics. It was envisioned that ecological theory should provide a basis for predicting how specific changes in production practices and inputs might affect pest problems. It was also thought that ecology could aid in the design of agricultural systems less vulnerable to pest outbreaks. In such systems pesticides would be used as occasional supplements to natural regulatory mechanisms. In fact, many authors wrote papers and reviews depicting the ecological basis of pest management (Southwood and Way, 1970; Price and Waldbauer, 1975; Pimentel and Goodman, 1978; Levins and Wilson, 1979). But despite all this early work, which provided much of the needed ecological foundations, most IPM programs deviated to become schemes of "intelligent pesticide management" and failed in putting ecologically based theory into practice.

Lewis *et al.* (1997) argue that the main reason why IPM science has been slow to provide the productive understanding that will assist farmers to move beyond the current production methods is that IPM strategies have long been dominated by quests for "magic bullet" products to control pest outbreaks. IPM approaches have not addressed the ecological root causes of pest problems in modern agriculture. There still prevails a narrow view that specific pests affect productivity, and overcoming such limiting factors via new technologies continues to be the main goal. Emphasis is now placed on purchased biological inputs such as microbial pesticides now widely applied in place of chemical insecticides. This type of technology pertains to a dominant technical approach called *input substitution*. The thrust is highly technological, characterized by a limiting-factor mentality that has driven conventional agricultural research in the past. Agronomists and other agricultural scientists have for generations been taught the "law of minimum" as a central dogma. According to this dogma, at any given moment there is a single factor limiting yield and the only way to overcome that factor is to use an appropriate external input. Once the hurdle of the first limiting factor has been surpassed (aphids, for example) with a specific insecticide as the correct input, then yields may rise until another factor (e.g. mites) becomes limiting, owing to the elimination of predaceous mites. The factor then requires another input, miticide in this case, and so on, perpetuating a process of treating symptoms rather than dealing with the real causes that evoked ecological imbalance (Altieri and Rosset, 1996). Thus, while understanding the insects' abilities and needs, which explain why pests quickly adapt and succeed in agroecosystems is important, it is more crucial is to pinpoint what makes agroecosystems susceptible to pests. By designing agroecosystems that on the one side work against the pests' performance and on the other are less vulnerable to pest invasion, farmers can substantially reduce pest numbers.

It is herein argued that long-term solutions to pest problems can be achieved only by restructuring and managing agricultural systems in ways that maximize the array of "built-in" preventive strengths, with therapeutic tactics serving strictly as back-ups of natural regulator processes. Lewis *et al.* (1997) suggested three approaches to bringing pest populations within acceptable bounds by harnessing the inherent strengths within ecosystems: (1) ecosystem management; (2) crop attributes and multitrophic level interactions; and (3) therapeutic treatments with minimal disruptions. These approaches suppose a deep knowledge of the underlying processes of the managed ecosystem, including the natural factors that suppress pest populations, with the final goal of designing agricultural practices that augment these pest natural regulatory processes. Borrowing concepts from landscape ecology, Thies and Tscharntke (1999) argue that approaches to insect management should involve working at a regional scale, recognizing the spatial heterogeneity of the landscape. This strategy requires a cooperative areawide approach, because the sources of pests extend beyond field boundaries and include a variety of landscape elements.

14.2 Agroecology and pest management

One way of further advancing the ecosystem management approach in IPM is through the understanding that crop health and sustainable yields in the agroecosystem derives from the proper balance of crops, soils, nutrients, sunlight, moisture, and co-existing organisms. The agroecosystem is productive and healthy when this balance of rich growing conditions prevail, and when crop plants remain resilient to tolerate stress and adversity. Occasional disturbances can be overcome by vigorous agroecosystems, which are adaptable, and diverse enough to recover once the stress has passed (Altieri and Rosset, 1996). If the cause of disease, pest, soil degradation, etc. is understood as imbalance, then the goal of agroecological treatment is to recover the balance, setting in motion the agroecosystem's natural tendency toward repairing itself. This tendency is known in ecology as homeostasis, the maintenance of the system's internal functions and defense mechanisms to compensate for external stress factors. But achieving and maintaining homeostasis requires a deep understanding of the nature of the agroecosystems and the principles by which they function. Fortunately, there is a new integrative scientific approach which allows for such understanding. Agroecology provides basic ecological principles on how to study, design and manage agroecosystems that are productive, enduring, and natural resource conserving (Altieri, 1995). Agroecology goes beyond a one-dimensional view of agroecosystems, their genetics, agronomy, edaphology, etc. to embrace an understanding of ecological and social levels of co-evolution, structure, and function. Instead of focusing on one particular component of the agroecosystem, agroecology emphasizes the interrelatedness of all agroecosystem components and the complex dynamics of ecological processes such as nutrient cycling and pest regulation (Gliessman, 1999).

From a management perspective, the agroecological objective is to provide a balanced environment, sustainable yields, biologically mediated soil fertility, and natural pest regulation through the design of diversified agroecosystems and the use of low-input technologies (Altieri, 1994). The strategy is based on ecological principles that lead management to optimal recycling of nutrients and organic matter turnover, close energy flows, water and soil conservation, and balanced pest–natural enemy populations. The strategy exploits the complementation that results from the various combinations of crops, trees, and animals in spatial and temporal arrangements (Altieri and Nicholls, 1999). These combinations determine the establishment of a planned and associated functional biodiversity which, when correctly assembled, delivers key ecological services which subsidize processes that underlie agroecosystem health.

In other words, ecological concepts are utilized to favor natural processes and biological interactions that optimize synergies, so that diversified farms are able to sponsor their own soil fertility, crop protection, and productivity through the activation of soil biology, the recycling of nutrients, the enhancement of beneficial arthropods and antagonists. Based on these principles, agroecologists involved in pest management have developed a framework to achieve crop health through agroecosystem diversification and soil quality enhancement, key pillars of agroecosystem health. The main goal is to enhance the immunity of the agroecosystem (i.e. natural pest control mechanisms) and regulatory processes (i.e. nutrient cycling and population regulation) through management practices and agroecological designs that enhance plant species and genetic diversity in time and space, and the enhancement of organic matter accumulation and biological activity of the soil (Altieri, 1999).

Agroecosystems can be manipulated to improve production and produce more sustainability, with fewer negative environmental and social impacts and fewer external inputs (Altieri, 1995). The design of such systems is based on the application of the following ecological principles (Reinjntjes *et al.*, 1992).

1. Enhancing recycling of biomass and optimizing nutrient availability and balancing nutrient flow.
2. Securing favorable soil conditions for plant growth, particularly by managing organic matter and enhancing soil biotic activity.
3. Minimizing losses due to flows of solar radiation, air, and water by way of microclimate management, water harvesting, and soil management through increased soil cover.
4. Promoting species and genetic diversification of the agroecosystem in time and space.
5. Enhancing beneficial biological interactions and synergisms among agrobiodiversity components, thus resulting in the promotion of key ecological processes and services.

These principles can be applied by way of various techniques and strategies. Each of these will have different effects on productivity, stability, and resiliency within the farm system, depending on the local opportunities, resource constraints, and, in most cases, on the market. The ultimate goal of agroecological design is to integrate components so that overall biological efficiency is improved, biodiversity is preserved, and agroecosystem productivity and its self-sustaining capacity are maintained.

14.2.1 *The ecology of modern mechanized agroecosystems: understanding pest vulnerability*

Contemporary agriculture is highly mechanized and has implied the simplification of the structure of the environment over vast areas, replacing nature's diversity with a small number of cultivated plants. In such systems, genetic manipulation replaces natural processes of plant evolution and selection. Throughout the crop domestication process, humans tended to select plants with fewer morphological and chemical defenses. Such intense human selection for fast growth and high reproductive output resulted in a general lowering of the plants' allocation to defense. Of course, significant amounts of toxic secondary compounds remain in many edible crops, but the general trend has been the gradual reduction of those chemicals and morphological features that protected plants from arthropod herbivores. This often left crop plants more vulnerable than their wild relatives, and it largely explains the widespread belief that there are more outbreaks of insects in agroecosystems than in natural ecosystems (Feeney, 1976). Even decomposition is altered since plant growth is harvested and soil fertility maintained, not through nutrient recycling but with fertilizers. It is well known that cultivated plants grown in genetically homogeneous monocultures do not possess the necessary ecological defense mechanisms to tolerate outbreaks of pest populations. Modern agriculturalists have selected crops for high yields and high profitability, making them more susceptible to pests by sacrificing natural resistance for productivity (Robinson, 1996). Due to the simplification of the environment and a reduction in trophic interactions, populations of crop plants in agroecosystems are rarely self-reproducing or self-regulating. Biological diversity is reduced, trophic structures tend to become simplified, and many niches are left unoccupied. The danger of increased invasions and catastrophic pest or disease outbreak is high, despite the intensive human input in the form of agrochemicals. On the other hand, modern agricultural practices (principally pesticides) negatively affect natural enemies (predators and parasites), which do not thrive well in toxic environments, or do not find the necessary environmental resources and opportunities in monocultures to effectively suppress pests (Altieri, 1994). As long as monocultures are maintained as the structural base of modern agricultural systems, pest problems will continue to be the result of a negative treadmill that reinforces

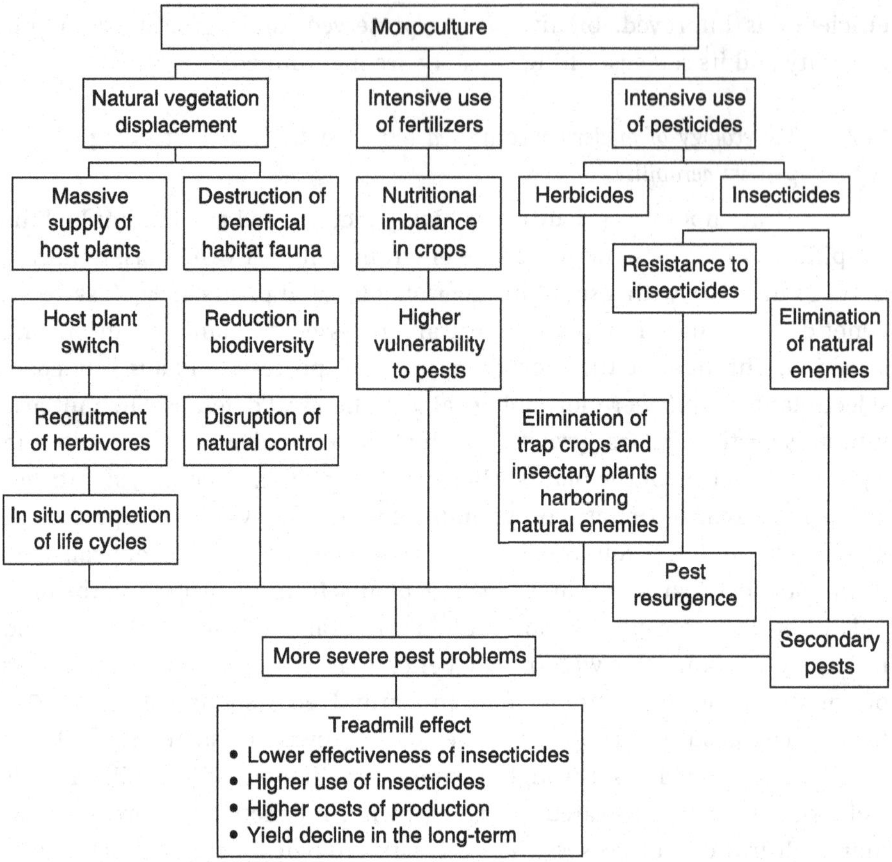

Figure 14.1. The ecological consequences of monoculture with special reference to pest problems and the agrochemical treadmill.

itself, as vegetational simplification, pesticides residues, and nutrition imbalances caused by excess fertilizers compound pest invasions (Figure 14.1). Thus the major challenge for those advocating ecologically based pest management (EBPM) is to find strategies to overcome the ecological limits imposed by monocultures.

Human manipulation and alteration of ecosystems for the purpose of establishing agricultural production makes agroecosystems structurally and functionally very different from natural ecosystems (Table 14.1). Understanding such differences is instructive, according to several researchers who have posited that agroecosystems should mimic the structure and function of natural communities (a practice followed by thousands of indigenous farmers for centuries), as these systems exhibit tight nutrient cycling, resistance to pest invasion, and vertical structure, and preserve

Table 14.1. *Structural and functional differences between natural ecosystems and agroecosystems*[a]

Characteristics	Agroecosystem	Natural ecosystem
Net productivity	High	Medium
Trophic chains	Simple, linear	Complex
Species diversity	Low	High
Genetic diversity	Low	High
Mineral cycles	Open	Closed
Stability (resilience)	Low	High
Entropy	High	Low
Human control	Definite	Not needed
Temporal permanence	Short	Long
Habitat heterogeneity	Simple	Complex
Phenology	Synchronized	Seasonal
Maturity	Immature, early successional	Mature, climax

[a]Modified after Gliessman, 1999.

biodiversity (Ewel, 1986; Soule and Piper, 1992). Browning and Frey (1969) suggested that the exclusive study of agricultural systems can lead to spurious or at least skewed conclusions about pest–crop interactions. He advised researchers to study a natural ecosystem from which knowledge can be gained that is readily applicable to agroecosystems. Most ecologists agree that any pest management approach should try to develop an agroecosystem that emulates later stages of succession (that is, mature communities) as much as possible, for this is how biological stability can be achieved (Root, 1973; van Emden and Williams, 1974; Price and Waldbauer, 1975). This is particularly true in the tropics, where the ecological futility of promoting mechanized monocultures in areas of overwhelming biotic intricacy and where pests flourish year-round, has been amply demonstrated. A more reasonable approach is to imitate natural cycles rather than struggle to impose horticultural simplicity in ecosystems that are inherently complex. Ewell (1986) argues that successional ecosystems can be particularly appropriate templates for the design of sustainable tropical agroecosystems.

14.2.2 *Ecological features of modern agroecosystems*

Agroecosystems are artificial ecosystems that are solar powered, as are natural ecosystems, but differ in that (1) the auxiliary energy sources that enhance productivity are processed fuels (along with animal and human labor) rather than natural energies; (2) diversity is greatly reduced by human management in order to maximize yield of specific food and other products; (3) the dominant plants and animals are under artificial rather than natural selection; and (4) control is external and goal oriented rather than internal via subsystem feedback as in natural ecosystems. Some processes that occur in natural ecosystems can be observed, albeit in a much altered way in agroecosystems. Main features of modern agroecosystems relevant to pest management include (Cox and Atkins, 1974; Landis *et al.*, 2000):

1. Landscape simplification: of major significance is the fact that, with agriculture, the original flora and fauna are completely replaced over a vast area. Where patches of natural vegetation persist they often occur on sites unsuitable for agriculture and contribute only minimally to the ecological stability of the area. Such agroecosystems are unfavorable environments for natural enemies; therefore their capacity to control invaders is lower than in more diverse agricultural landscapes. Due to their reduced structural and functional diversity in relation to natural ecosystems, agroecosystems have generally lower resiliency than natural ecosystems.
2. Continuity: one of the most obvious features of intensive agricultural systems is their well-defined spatial arrangement. Agroecosystems usually have clearly defined boundaries. The boundary between cultivated land and the natural ecosystems from which it was carved is also typically well delineated and ecotones are abrupt. Within the converted area there are plots clearly separated by fences, ditches, canals, windbreaks, and access roads. Sometimes a single crop extends over many square miles, but at times a large agricultural area is broken up into a mosaic of plots or a quilt of agroecosystems defining diverse landscapes.
3. Disruption of succession: intensive agriculture prevents normal succession from taking place. Agricultural fields usually represent the first stage of succession where an existing community is disrupted by deforestation and/or plowing, developing a new, simple, immature, human-made plant community at the site that is neither persistent nor steady-state. All tendencies towards complexity (i.e. insect and weed colonization) are suppressed with agrochemicals, keeping succession in check via maintenance of monocultures.

The objective of growing a crop is to obtain the greatest possible harvest. The best means for this is to establish a system for which the ratio between primary production and biomass is at its highest level; that is, one in which little primary production is required for maintenance. To maintain a system of this type, it is necessary for humans to assume responsibility for the costs of maintenance and regulation normally taken care of by the natural processes that lead to the establishment of a climax ecosystem.

4. Adaptability: in natural ecosystems the assemblage of organisms is the result of natural selection and coevolution. Agroecosystems consist of unnatural assemblages of human-selected domesticated species and an assortment of native or imported opportunistic species that manage to invade the site. These two groups have not been integrated into a steady-state system by the process of co-evolution, and the opportunistic species frequently constitute weed, insect, and disease pests that must be dealt with by the farmer.

14.3 Factors triggering insect pest outbreaks in agroecosystems

Given the major differences between mechanized agroecosystems and natural ecosystems, especially the prevalence of monocultures and the high levels of disturbance, modern agricultural systems lack a suitable ecological infrastructure to resist pest invasions and outbreaks (Altieri, 1994; Landis *et al.*, 2000). Many factors underlie the vulnerability of monocultures to pest invasions, as outlined below.

14.3.1 Decreased landscape diversity

The spread of modern agriculture has resulted in tremendous changes in landscape diversity. There has been a consistent trend toward simplification that entails (a) the enlargement of fields; (b) the aggregation of fields; (c) an increase in the density of crop plants; (d) an increase in the uniformity of crop population age structure and physical quality; and (e) a decrease in inter- and intraspecific diversity within the planted field.

Although these trends appear to exist worldwide, they are more apparent, and certainly best documented, in industrialized countries. Increasingly, evidence suggests that these changes in landscape diversity have led to more insect outbreaks due to the expansion of monocultures at the expense of natural vegetation, through decreasing habitat diversity. One of the main characteristics of the modern agricultural landscape is the large size and homogeneity of crop monocultures, which fragment the natural landscape. Massive expansion and increases in production of crops such as coffee, cocoa, rice, soybean, and oil palm in various regions of the developing world led to major environmental degradation and loss of biodiversity, especially in bird populations (Donald, 2004). Such agricultural intensification can also directly affect the abundance and diversity of natural enemies, as the larger the area under monoculture the lower the viability of a given population of beneficial fauna. At hand is also the issue of colonization of crop "islands" by insects. In the case of annual crops, insects must colonize from the borders each season, and the larger the field, the greater is the distance that must be covered. Several studies suggest (not surprisingly) that natural enemies tend to colonize after their hosts/prey and that the lag tie between the arrival of pest and natural enemy increases with distance from border (source pool). For instance, Price (1976) found that the first occurrence of an herbivore and that of a predatory mite in a soybean field were separated by one week on the edge versus a three week lag in the center. To the extent that this is a general phenomenon, increased field size should lead to more frequent insect outbreaks.

14.3.2 *Decreased on-farm plant diversity*

Many ecologists have conducted experiments testing the theory that decreased plant diversity in agroecosystems allows greater chance for invasive species to colonize, subsequently leading to enhanced herbivorous insect abundance. Many of these experiments have shown that mixing certain plant species with the primary host of a specialized herbivore gives a fairly consistent result: specialized species usually exhibit higher abundance in monocultures than in diversified crop systems (Andow, 1983).

Several reviews have been published documenting the effects of within-habitat diversity on insects (Altieri and Letourneau, 1984; Risch *et al.*, 1983). Two main ecological hypotheses (the natural enemy hypothesis and the resource concentration hypothesis), have been offered to explain why insect populations tend to explode in monocultures and how in agroecosystems insect populations can be stabilized by constructing vegetational architectures that support natural enemies and/or that directly exert inhibitory effects on pest attacks (Root, 1973). A recent study in Portugal illustrates the effects of decreased field plant diversity on increased pest incidence. As new policy and market forces prompt the conversion of traditional complex agroforest vineyard systems to monocultures, Altieri and Nicholls (2002) found higher prevalence of grape herbivores and *Botrytis* bunch rot. Although monocultures may be productive, such gains occurred at the expense of biodiversity and agricultural sustainability, reflected on higher pest vulnerability.

14.3.3 *Pesticide-induced insect outbreaks*

Many examples are reported in the literature of insect pest outbreaks and/or resurgence following insecticide applications (Pimentel and Perkins, 1980). Pesticides either fail to control the target pests or create new pest problems. Development of resistance in insect pest populations is the main way in which pesticide use can lead to pest control failure. More than 500 species of arthropods have become resistant to one or more insecticide or acaricide (Van Driesche and Bellows, 1996).

Another way in which pesticide use can foster outbreaks of pests is through the elimination of the target pest's natural enemies. Predators and parasitoids often experience higher mortality than herbivores following a given spray (Morse *et al.*, 1987). This is due, in part, to the greater mobility of many natural enemies, which exposes them to more insecticide per unit time following a spray.

In addition, natural enemies appear to evolve resistance to insecticides much more slowly than herbivores. As a consequence there is a lower probability that some individuals in populations of natural enemies will have genes for insecticide resistance. This in turn is due to the much smaller size of the natural enemy population relative to the pest population and the different evolutionary history of natural enemies and herbivores.

Pesticides also create new pest problems when natural enemies of ordinarily non-economic species are destroyed by chemicals. These "secondary pests" then reach higher density than normal and begin to cause economic damage (Pimentel and Lehman, 1993).

14.3.4 *Fertilizer-induced pest outbreaks*

Luna (1988) suggested that the physiological susceptibility of crops to insects might be affected by the form of fertilizer used (organic vs. chemical fertilizer). Studies documenting lower abundance of several insect herbivores in organic farming systems have partly attributed such reduction to low nitrogen content in the organically farmed crops. In comparative studies, conventional crops (treated with chemical fertilizer) tend to develop a larger infestation of insects (especially Homoptera) than organic counterparts.

Interestingly, it has been found that certain pesticides can also alter the nutritional biochemistry of crop plants by changing the concentrations of nitrogen, phosphorus, and potassium, by influencing the production of sugars, free amino acids, and proteins, and by influencing the aging process which affects surface hardness, drying, and wax deposition (Oka and Pimentel, 1976; Rodriguez *et al.*, 1957).

14.3.5 *Weather-induced insect pest outbreaks*

Some authors argue that weather can be the most important factor triggering insect outbreaks (Milne, 1957). For example, Miyashita (1963), in reviewing the dynamics of seven of the most serious insect pests in Japanese crops, concluded that weather was the principal cause of the outbreaks in each case. There are several ways in which weather can trigger insect outbreaks. Perhaps the most straightforward mechanism is direct stimulation of the insect and/or host plant physiology. The development and widespread use of degree-day models to predict outbreaks of particular pests and appropriate control strategies are an indication of the importance of the linkage between temperature and growth and the development of herbivorous insects and their host plants. Gutierrez *et al.* (1974) have shown that weather plays a key role in the development of cowpea-aphid populations in southeast Australia. In this case, a series of climatic events favors complex changes in aphid physiological development, migration, and dispersal in such a way as to cause localized outbreaks.

14.3.6 *Transgenic crops and insect pest outbreaks*

In the last six years, transgenic crops have expanded in area, reaching today about 42 million hectares worldwide. Such areas are dominated by monocultures of few crop varieties, mainly herbicide-resistant soybeans and *Bt* corn, with a clear tendency towards decreased agricultural habitat diversity (Marvier, 2001). Agroecologists have argued that such massive and rapid deployment of transgenic crops will exacerbate the problems of conventional modern agriculture (Rissler and Mellon, 1996; Altieri, 2000). At issue is the genetic homogeneity of fields with bioengineered crops which in turn can make such systems increasingly vulnerable to pest and disease problems (NAS, 1972). Transgenic crops affect natural enemies in several ways: the enemy species may feed directly on corn tissues (e.g. pollen) or on hosts that have fed on Bt corn, or host populations may be reduced. By keeping Lepidoptera pest populations at extremely low levels, Bt crops could potentially starve natural enemies, as predators and parasitic wasps that feed on pests need a small amount of prey to survive in the agroecosystem. Among the natural enemies that live exclusively on insects which the transgenic crops are designed to kill (Lepidoptera), egg and larval parasitoids would be most affected because they are totally dependent on live hosts for development and survival, whereas some predators could theoretically thrive on dead or dying prey (Schuler *et al.*, 1999). In a two-year field study in Iowa, abundance of parasitoid species *Macrocentris cingulum*, which is specific to corn borer larvae, was found to be lower in *Bt* cornfields than in non-Bt fields, as might be expected because of significant reductions in larval hosts in Bt corn (Groot and Dicke, 2002).

Natural enemies could also be affected directly through inter-trophic level effects of the toxin. The potential of Bt toxins moving through arthropod food chains poses serious implications for natural biocontrol in agricultural fields. Recent evidence shows that the Bt toxin can affect beneficial insect predators that feed on insect pests present in Bt crops. According to Groot and Dicke (2002) natural enemies may come in contact more often with Bt toxins via non-target organisms, because the toxins do not bind to receptors on the midgut membrane in the non-target herbivores. Studies in Switzerland showed that mean total mortality of predaceous lacewing larvae (Chrysopidae) raised on Bt-fed prey was 62% compared to 37% when raised on Bt-free prey. These Bt prey fed Chrysopidae also exhibited prolonged development time throughout their immature life stage (Hilbeck *et al.*, 1998). The observed sublethal effect shows scope for the fitness of natural enemies to be indirectly affected by Bt toxins expressed in transgenic crops via feeding on suboptimal food or because of host death and scarcity (Groot and Dicke, 2002). Moreover, the toxins produced in Bt plants may be passed on to predators and parasitoids in pollen or leaf tissue.

These findings are of concern to small and/or organic farmers who rely for insect pest control on the rich complex of predators and parasitoids associated with the mixed cropping systems (Altieri, 1994). Inter-trophic level effects of the Bt toxin raise serious concerns about the potential of the disruption of natural pest control. Polyphagous predators that move within and between mixed crop cultivars will encounter Bt-containing non-target prey throughout the crop season (Altieri, 2000). Disrupted biocontrol mechanisms may result in increased crop losses due to invasive pests or to the increased use of pesticides by farmers, with consequent health and environmental effects.

Despite all the pressures for US farmers to adopt this technology, benefits of using transgenic corn are not assured because population densities of the European corn borer (ECB) are not predictable. Due to this and other factors, the use of transgenic corn has not significantly reduced insecticide use in most of the corn-growing areas of the Midwest. Until 2001, the percentage of field corn treated with insecticides in the United States remained at approximately 30%, despite a significant increase in the hectares of Bt corn planted (Obrycki *et al.*, 2001). Moreover, the potential benefits of Bt crops are now in question given that the high-dose/refuge strategy for delaying pest resistance is threatened from the contamination of refuges by transgenic maize. Variable Bt toxin production in seeds of refuge plants undermines the refuge strategy and could accelerate pest resistance to Bt crops (Chilcutt and Tabashnik, 2004).

14.4 Reinstating ecological rationale in modern agriculture

The instability of agroecosystems, manifesting as the worsening of most insect pest problems (and therefore greater dependence on external inputs), is increasingly linked to the expansion of crop monocultures (Altieri, 1994). Plant communities that are modified to meet the special needs of humans become subject to heavy pest damage and generally the more intensely such communities are modified, the more abundant and serious the pests. The inherent self-regulation characteristics of natural communities are lost when humans modify such communities by promoting monocultures. Some agroecologists maintain that this breakdown can be repaired by the addition or enhancement of plant biodiversity at the field and landscape level (Gliessman, 1999; Altieri, 1999).

Emergent ecological properties develop in diversified agroecosystems allowing biodiversity to thrive and establish complex food webs and interactions. But biodiversification must be accompanied by improvement of soil quality, as the link between healthy soils and healthy plants is fundamental to truly ecologically based IPM. The lower pest levels widely reported in organic-farming systems may, in part, arise from plant–insect resistance mediated by biochemical or mineral–nutrient dynamics typical of crops under such management practices. Results from such studies provide interesting evidence to support the view that the long-term joint management of plant diversity and soil organic matter can lead to better plant resistance against insect pests (Letourneau and Goldstein, 2001).

14.5 Harmonizing soil and plant health in agroecosystems

Although the integrity of the agroecosystem relies on synergies of plant diversity and the continuing function of the soil microbial community, and its relationship with organic matter (Altieri and Nicholls, 1999), the evolution of IPM and integrated soil fertility management (ISFM) have proceeded separately. This has prevented many scientists from realizing that most pest management methods used by farmers can also be considered soil fertility management strategies and vice versa. There are positive interactions between soils and pests that once identified can provide guidelines for optimizing total agroecosystem function (Figure 14.2). Increasingly, new

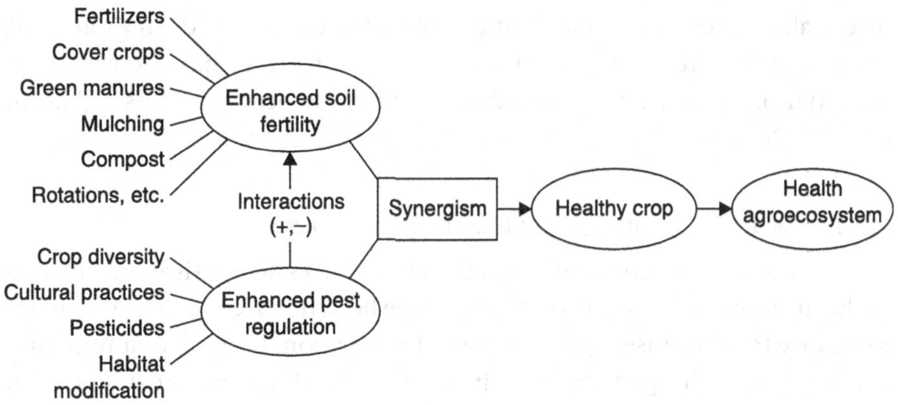

Figure 14.2. Interactions of soil and pest management practices used by farmers, some of which may result in synergism leading to a healthy and productive crop.

research suggests that the ability of a crop plant to resist or tolerate insect pests and diseases is tied to optimal physical, chemical, and mainly biological properties of soils. Soils with high organic matter and active soil biological activity generally exhibit good soil fertility as well as complex food webs and beneficial organisms that prevent infection (Magdoff *et al.*, 2000). Studies in tropical Asian irrigated rice agroecosystems by Settle *et al.* (1996) showed that, by increasing organic matter in test plots, researchers could boost populations of detritivores and plankton-feeders, and in turn significantly boost the abundance of generalist predators. Surprisingly, organic matter management proved to be a key mechanism in the support of high levels of natural biological control. Organically managed soils usually exhibit a richer community of natural enemies that feed on alternative prey found above or within the soil. Collembola have been shown to be an important prey group for generalist predators, such as the ground beetle *Bembidion lapros* (Bilde *et al.*, 2000) and linyphiid spiders in arable fields, helping to sustain and retain these predators as pest control agents within the crop (Bilde *et al.*, 2000). Such relationships and mechanisms have been ignored by scientists as important elements in pest management.

Much of what we know today about the relationship between crop nutrition and pest incidence comes from studies comparing the effects of organic agricultural practices and modern conventional methods on specific pest populations. Soil fertility practices can impact the physiological susceptibility of crop plants to insect pests by either affecting the resistance of individual plants to attack or by altering plant acceptability to certain herbivores. Some studies have also documented how the shift from organic soil management to chemical fertilizers has increased the potential of certain insects and diseases to cause economic losses (Phelan *et al.*, 1995).

14.5.1 *The effects of nitrogen fertilization on insect pests*

The indirect effects of fertilization practices acting through changes in the nutrient composition of the crop have been reported to influence plant resistance to many insect pests. Among the nutritional factors that influence the level of arthropod damage in a crop, total nitrogen (N) has been considered critical for both plants and their consumers (Mattson, 1980; Scriber, 1984; Slansky and Rodriguez, 1987).

In most studies evaluating aphid and mite response to N fertilization, increases in nitrogen rates dramatically increased aphid and mite numbers. According to van Emden (1966), increases in fecundity and developmental rates of the green peach aphid, *Myzus persicae*, were highly correlated to increased levels of soluble nitrogen in leaf tissue. Several other authors have also indicated increased aphid and mite populations from nitrogen fertilization (Tables 14.2 and 14.3). Herbivorous insect populations associated with *Brassica* crop plants have also been reported to increase in response to increased soil nitrogen levels (Table 14.4).

In a two-year study, Brodbeck *et al.* (2001) found that populations of the thrips *Frankliniella occidentalis* were significantly higher on tomatoes that received higher rates of nitrogen fertilization. Seasonal trends in *F. occidentalis*

Table 14.2. *Summary of effects of inorganic fertilizers on mite abundance from selected studies (Luna, 1988)*

Nutrient	Mite species	Crop	Numerical response of insect
N	*Panonychus ulmi*	Apple	+
N	*Tetranychus telarius*	Apple	+
N	*T. telarius*	Beans	+
N, P, K	*Two-spotted spider mite*	Beans/peaches	+
N	*T. telarius*	Tomato	−
N, P	*T. telarius*	Apples	+/−
N, K	*Bryobia praetiosa*	Beans	+/−
N, Ca	*Heliothrips haemorrhoidalis*	Beans	+/−

+ Indicates increase in density with increasing rates of fertilizer element.
− Indicates decrease in density with increasing rates of fertilizer element.

Table 14.3. *Summary of effects of inorganic fertilizers on aphid abundance from selected studies (Luna, 1988)*

Nutrients	Insect	Crop	Numerical response of insect
N, P, K	*Myzus persicae*	Tobacco	+/^/+
N	*Schizaphis graminum* (greenbug)	Oats/rye	−
N, Lime	*S. graminum*	Oats	−
N	*Rhopalosiphum maidis*	Sorghum	+
N, K, Ca	*Myzus persicae*	Brussels sprouts	+/−/−
N, P	*Therioaphis maculate* (spotted alfalfa aphid)	Alfalfa	−/+

+ Indicates increase in density with increasing rates of fertilizer element.
− Indicates decrease in density with increasing rates of fertilizer element.
^ Indicates highest density occurred at intermediate rates of fertilizer element.
v Indicates lowest density occurred at intermediate rates of fertilizer element.
/ Separates the effects of fertilizer elements listed in column 1.

Table 14.4. *Response of herbivores to increased soil nitrogen levels on* Brassica *host plant (Letourneau, 1988)*

Host plant	Herbivore species	Factor	Response
Brussels sprouts	*Myzus persicae*	No. progeny	Increase
Brussels sprouts	*Brevicoryne brassicae*	No. progeny	Small increase, dependent on factors such as K
Rape	*Artogeia rapae*	Oviposition frequency	Increase
Kale and cabbage	*A. rapae*	Oviposition frequency	Increase
Kale	*A. rapae*	Oviposition frequency	Increase
Cabbage	*A. rapae*	Growth rate	Increase
Cabbage	*A. rapae*	Growth rate ultimate size	Increase
Cabbage	*Plutella xylostella*	Feeding preference	Increase

on tomato were found to be correlated to the number of flowers per host plant and changed with the nitrogen status of flowers. Plants subjected to higher fertilization rates produced flowers that had higher nitrogen content as well as variations in several amino-acid profiles that coincided with peak thrip population density. Abundance of *F. occidentalis* (particularly adult females) was most highly correlated to flower concentrations of phenylalanine during population peaks. Other insect populations found to increase following N fertilization included fall armyworm in maize, corn earworm on cotton, pear psylla (*Cacopsylla piricola*) on pear, Comstock mealybug (*Pseudococcus comstocki*) on apple, and European cornborer (*Ostrinia nubilalis*) on field corn (Luna, 1988).

In contrast, because plants are a source of nutrients to herbivorous insects, an increase in the nutrient content of the plant may be argued to increase its acceptability as a food source to pest populations. Variations in herbivore response may be explained by differences in the feeding behavior of the herbivores themselves (Pimentel and Warneke, 1989). For example, with increasing nitrogen concentrations in creosote bush (*Larrea tridentata*) plants, populations of sucking insects were found to increase, but the number of chewing insects declined. With higher nitrogen fertilization, the amount of nutrients in the plant increases, as well as the amount of secondary compounds that may selectively affect herbivores feeding patterns. Thus protein digestion inhibitors that are found to accumulate in plant cell vacuoles are not consumed by sucking herbivores, but will harm chewing herbivores (Mattson, 1980).

In reviewing 50 years of research relating to crop nutrition and insect attack, Scriber (1984) found 135 studies showing increased damage and/or growth of leaf-chewing insects or mites in N-fertilized crops, vs. fewer than 50 studies in which herbivore damage was reduced by normal fertilization regimens. In aggregate, these results suggest a hypothesis with implications for fertilizer use patterns in agriculture, namely that high N inputs can precipitate high levels of herbivore damage in crops. As a corollary, crop plants would be expected to be less prone to insect pests and diseases if organic soil amendments are used, these generally resulting in lower N concentrations in the plant tissue.

Letourneau (1988), however, questioned if the "nitrogen-damage" hypothesis, based on Scriber's review, could be extrapolated to a general warning about fertilizer inputs associated to insect pest attack in agroecosystems. Of 100 studies of insects and mites on plants treated experimentally with high and low N fertilizer levels, Letourneau found two-thirds (67–100) of the insect and mite studies to show an increase in growth, survival, reproductive rate, population densities or plant damage levels in response to increased N fertilizer. The remaining third of the arthropods studied showed either a decrease in damage with fertilizer N or no significant change. The author also noted that experimental design can affect the types of responses observed, suggesting that more reliable data emerged in experiments conducted in field plots, using damage level, population levels, and reproductive rate in individual insect species as best predictors of insect response to increase N.

14.5.2 *The dynamics of insect herbivores in organically managed systems*

Studies documenting lower abundance of several insect herbivores in low-input systems have partly attributed such reductions to the lower nitrogen content in organically farmed crops (Lampkin, 1990). In Japan, density of immigrants of the planthopper species *Sogatella furcifera* was significantly lower and the settling rate of female adults and survival rate of immature stages of ensuing generations were generally lower in organic, compared with conventional, rice fields. Consequently, the density of planthopper nymphs and adults in the ensuing generations was found to decrease in organically farmed fields (Kajimura, 1995). In India, the introduction of high-yielding Green Revolution rice varieties was accompanied by increased and more frequent inputs of fertilizers. In Tamil Nadu, the consumption of NPK (nitrate/phosphate/potassium) fertilizers increased from 296 000 MT in 1970–1 to 791 000 MT in 1996–7. Surprisingly, those changes unexpectedly influenced mosquito breeding and thereby affected the incidence of mosquito-borne disease. Victor and Reuben (2000) found that the application of urea in rice fields significantly increased the population densities of mosquito larvae and pupae (anophelines as well as culicines) in a dose-related manner. In contrast fields treated with organic fertilizers (farmyard manure or green manure from blue-green algae) exhibited significantly lower population densities of mosquito immatures.

In England, conventional winter wheat fields exhibited a larger infestation of the aphid *Metopolophium dirhodum* than their organic counterpart. The conventionally fertilized wheat crop also had higher levels of free protein amino acids in its leaves during June, which were attributed to a nitrogen top dressing applied early in April. However, the difference in the aphid infestations between crops was attributed to the aphid's response to the relative proportions of certain non-protein to protein amino acids present in the leaves at the time of aphid settling on crops (Kowalski and Visser, 1979). The authors concluded that chemically fertilized winter wheat was more palatable than its organically grown counterpart; hence the higher level of infestation.

In greenhouse experiments, when given a choice of maize grown on organic versus chemically fertilized soils collected from three nearby farms, European corn borer (*Ostrinia nubilalis*) females laid significantly more eggs in the chemically fertilized plants (Phelan *et al.*, 1995). Interestingly, there was significant variation in egg laying among chemical fertilizer treatments within the conventionally managed soil, but in plants under the organic soil management, egg laying was uniformly low. Pooling results across all three farms showed that variance in egg laying was approximately 18 times higher among plants in conventionally managed soil than among plants grown under an organic regimen. The authors suggested that this difference is evidence for a form of biological buffering characteristically found more commonly in organically managed soils.

Altieri *et al.* (1998) conducted a series of comparative experiments on various growing seasons between 1989 and 1996 in which broccoli was subjected to varying fertilization regimes (conventional vs. organic). The goal was to test the effects of different nitrogen sources on the abundance of the key insect pests, cabbage aphid (*Brevicoryne brassicae*) and flea beetle (*Phyllotreta cruciferae*). Conventionally fertilized monoculture consistently developed a larger infestation of flea beetles and in some cases of the cabbage aphid, than the organically fertilized broccoli systems. The reduction in aphid and flea beetle infestations in the organically fertilized plots was attributed to lower levels of free nitrogen in the foliage of plants. This further supports the view that insect pest preference can be moderated by alterations to the type and amount of fertilizer used.

By contrast, a study comparing the population responses of *Brassica* pests to organic versus synthetic fertilizers, measured higher *Phyllotreta* flea beetles populations on sludge-amended collard (*Brassica oleracea*) plots early in the season compared with mineral-fertilizer-amended and unfertilized plots (Culliney and Pimentel, 1986). However, later in the season, in these same plots, insect population levels were lowest in organic plots for beetles, aphids and lepidopteran pests. This suggests that the effects of fertilizer type vary with plant growth stage and that organic fertilizers do not necessarily diminish pest populations but, at times, may increase them. For example, in a survey of California tomato producers, despite the pronounced differences in plant quality (N content of leaflets and shoots) both within and among tomato fields, Letourneau *et al.* (1996) found no indication that greater concentrations of tissue N in tomato plants were associated with higher levels of insect damage.

14.6 Conversion

In reality, the implementation of an IPM strategy usually occurs while an agroecosystem is undergoing a process of conversion from a high-input conventional management system to a low-external-input system. This conversion can be conceptualized as a transitional process with three marked phases (MacRae *et al.*, 1990).

1. Increased efficiency of input use as emphasized by traditional integrated pest management.
2. Input substitution or substitution of environmentally benign inputs for agrochemical inputs as practiced by many organic farmers.
3. System redesign: diversification with an optimal crop/animal assemblage, which encourages synergism so that the agroecosystem may sponsor its own soil fertility, natural pest regulation, and crop productivity.

Many of the practices currently being promoted as components of IPM fall in categories 1 and 2. Both of these stages offer clear benefits in terms of lower environmental impacts as they decrease agrochemical input use and often can provide economic advantages compared to conventional systems. Incremental changes are likely to be more acceptable to farmers as drastic modifications that may be viewed as highly risky or requiring complicated management. But does the adoption of practices that increase the efficiency of input use or that substitute biologically based inputs for agrochemicals, which leaves the monoculture structure intact, really have the potential to lead to the productive redesign of agricultural systems?

In general, the fine-tuning of input use through IPM does little to move farmers toward an alternative to high input systems. In most cases IPM translates to "intelligent pesticide management" as it results in selective use of pesticides according to a pre-determined economic threshold, which pests often "surpass" in monoculture situations. On the other hand, input substitution follows the same paradigm of conventional farming; overcoming the limiting factor, but this time with biological or organic inputs. Many of these "alternative inputs" have become commodified, therefore farmers continue to be dependent on input suppliers, many of a corporate nature (Altieri and Rosset, 1996). Clearly, as it stands today, "input substitution" has lost much of its ecological potential.

System redesign, on the contrary, arises from the transformation of agroecosystem function and structure by promoting management guided to ensure fundamental agroecosystem processes. Promotion of biodiversity within agricultural systems is the cornerstone strategy of system redesign, as research has demonstrated that higher diversity (genetic, taxonomic, structural, resource) within the cropping system leads to higher diversity in associated biota, usually leading to more effective pest control and tighter nutrient cycling. As more information about specific relationships between biodiversity, ecosystem processes, and productivity in a variety of agricultural systems is accumulated, design guidelines can be developed further and used to improve agroecosystem sustainability and resource conservation.

14.7 Syndromes of production

One of the frustrations of research in sustainable agriculture has been the inability of low-input practices to outperform conventional practices in side-by-side experimental comparisons, despite the success of many extant commercial organic and low-input production systems (Vandermeer, 1997). A potential explanation for this paradox was offered by Andow and Hidaka (1989) in their description of "syndromes of production". These researchers compared the traditional shizeñ system of rice (*Oryza sativa*) production with the contemporary Japanese high input system. Although rice yields were comparable in the two systems, management practices differed in almost every respect: irrigation practice, transplanting technique, plant density, fertility source and quantity, and management of insects, diseases, and weeds. Andow and Hidaka (1989) argued that systems like shizeñ function in a qualitatively different way than conventional systems. This array of cultural technologies and pest management practices result in functional differences that cannot be accounted for by any single practice.

Thus a production syndrome is a set of management practices that are mutually adaptive and lead to high performance. However, subsets of this collection of practices may be substantially less adaptive; that is, the interaction among practices leads to improved system performance that cannot be explained by the additive effects of individual practices. In other words, each production system represents a distinct group of management techniques and, by implication, ecological relations. This re-emphasizes the fact that agroecological designs are site-specific and what may be applicable elsewhere are not the techniques but rather the ecological principles that underlie sustainability. It is of no use to transfer technologies from one site to another, if the set of ecological interactions associated with such techniques cannot be replicated.

14.8 Diversified agroecosystems and pest management

Diversified cropping systems, such as those based on intercropping and agroforestry or cover cropping of orchards, have been the target of much research recently. This interest is largely based on the emerging evidence that these systems are more stable and more resource conserving (Vandermeer, 1995). Much of these attributes are connected to the higher levels of functional biodiversity associated with complex farming systems. As diversity increases, so do opportunities for coexistence and beneficial interference between species that can enhance agroecosystem sustainability (van Emden and Williams, 1974). Diverse systems encourage complex food webs which entail more potential connections and interactions among members, and many alternative paths of energy and material flow through it. For this and other reasons a more complex community exhibits more stable production and less fluctuations in the numbers of undesirable organisms. Studies further suggest that the more diverse the agroecosystems and the longer this diversity remains undisturbed, the more internal links develop to promote greater insect stability. It is clear, however, that the stability of the insect community depends not only on its trophic diversity, but also on the actual density-dependence nature of the trophic levels (Southwood and Way, 1970). In other words, stability will depend on the precision of the response of any particular trophic link to an increase in the population at a lower level. Recent studies conducted in grassland systems suggest, however, that there are no simple links between species diversity and ecosystem stability, despite empirical evidence that increasing the richness of a particular guild of natural enemies can reduce the diversity of a widespread group of herbivorous pests (Cardinale *et al.*, 2003). What is apparent is that functional characteristics of component species are as important as the total number of species in determining processes and services in ecosystems (Tilman *et al.*, 1996). This latest finding has practical implications for agroecosystem management. If it is easier to mimic specific ecosystem processes rather than duplicate all the complexity of nature, then the focus should be placed on a specific biodiversity component that plays a specific role, such as a plant that fixes nitrogen, provides cover for soil protection or harbors resources for natural enemies. In the case of farmers without major economic and resource limits and who can withstand a certain risk of crop failure, a crop rotation or a simple polyculture may be all it takes to achieve a desired level of stability. But in the case of resource-poor farmers, who can not tolerate crop failure, highly diverse cropping systems would probably be the best choice. The obvious reason is that the benefit of complex agroecosystems is low risk; if a species falls to disease, pest attack or weather, another species is available to fill the void and maintain full use of resources. Thus there are potential ecological benefits to having several species in an agroecosystem: compensatory growth, full use of resources and nutrients, and pest protection (Ewel, 1999).

14.8.1 *Plant diversity and insect pest incidence*

An increasing body of literature documents the effects that plant diversity has on the regulation of insect herbivore populations by favoring the abundance and efficacy of associated natural enemies (Altieri and Letourneau, 1984). Research has shown that mixing certain plant species usually leads to density reductions of specialized herbivore. In a review of 150 published investigations, Risch *et al.* (1983) found evidence to support the notion that specialized insect herbivores were less numerous in diverse systems (53% of 198 cases). In another comprehensive review 209 published studies that deal with the effects of vegetation diversity in agroecosystems on herbivore arthropod species, Andow (1991) found that 52% of the 287 total herbivore species examined in these studies were less abundant in polycultures than in monocultures, while only 15.3% (44 species) exhibited higher densities in polycultures. In a more recent review of 287 cases, Helenius (1998) found that the reduction of monophagous pests was greater in perennial systems, and that the reduction of polyphagous pest numbers was less in perennial than in annual systems. Helenius (1998) concluded that monophagous (specialists) insects are more susceptible to crop diversity than polyphagous insects. He cautioned about the increased risk of pest attack if the dominant herbivore fauna in a given agroecosystem is polyphagous.

The ecological theory relating to the benefits of mixed versus simple cropping systems revolves around two possible explanations of how insect pest populations attain higher levels in monoculture systems compared with diverse ones. The two hypotheses proposed by Root (1973) are as follows.

1. The natural enemy hypothesis, which argues that pest numbers are reduced in more diverse systems because the activity of natural enemies is enhanced by environmental opportunities prevalent in complex systems.
2. The resource concentration hypothesis argues that the presence of a more diverse flora has direct negative effects on the ability of the insect pests to find and utilize its host plant and also to remain in the crop habitat.

The resource concentration hypothesis predicts lower pest abundance in diverse communities because a specialist feeder is less likely to find its host plant due to the presence of confusing masking chemical stimuli, physical barriers to movement or other environmental effects such as shading; it will tend to remain in the intercrop for a shorter period of time simply because the probability of landing on a non-host plant is increased; it may have a lower survivorship and/or fecundity (Bach, 1980). The extent to which these factors operate will depend on the number of host plant species present and the relative preference of the pest for each, the absolute density and spatial arrangement of each host species, and the interference effects from more host plants.

The natural enemy hypothesis attributes lower pest abundance in intercropped or more diverse systems to a higher density of predators and parasitoids (Bach, 1980). The greater density of natural enemies is caused by an improvement in conditions for their survival and reproduction, such as a greater temporal and spatial distribution of nectar and pollen sources, which can increase parasitoid reproductive potential and abundance of alternative host/prey when the pest species are scarce or at inappropriate stages (Risch, 1981). Diversification can increase the effectiveness of specialist parasitoids on a given target herbivore as a result of them spilling over from herbivores on one host plant species to herbivores on the crop (Stiling *et al.*, 2003). These factors can, in theory, combine to provide more favorable conditions for natural enemies and thereby enhance their numbers and effectiveness as control agents.

Some researchers have been busy figuring out which of the two hypotheses is the most important for influencing the relative abundance of pest insects in diverse systems. The question has been approached in two ways: (a) reviews of the literature relating to crop diversity and pest abundance; and (b) by experimentation. Risch *et al.* (1983) concluded that the resource concentration hypothesis was the most likely explanation for reductions in pest abundance in diverse systems. However, 19 studies that tested the natural enemy hypothesis were reviewed by Russell (1989), who found that, of these 19 studies, mortality rates from predators and parasitoids in diverse systems were higher in nine, lower in two, unchanged in three and variable in five. Russell (1989) concluded that the natural enemy hypothesis is an operational mechanism, but he considered the two hypotheses complementary. In studies of crop/weed systems, Baliddawa (1985) found that 56% of pest reductions in weed diversified cropping systems were caused by natural enemies. A recent review by Sunderland and Same (2000) showed that spider abundance was increased by diversification in 63% of studies. The literature supports that spiders tend to remain in diversified patches and that extending diversification throughout the whole crop offers the best prospects for improving pest control.

One of the major problems has been predicting which cropping systems will reduce pest abundance, since not all combinations of crops will produce the desired effect and blind adherence to the principle that a more diversified system will reduce pest infestation is clearly inadequate and often totally wrong (Gurr *et al.*, 1998). To some researchers this indicates the need for caution and a greater understanding of the mechanisms involved to explain how, where, and when such exceptions are likely to occur. It will only be through more detailed ecological studies that such an understanding can be gained and an appropriate predictive theory developed. This means that a greater emphasis has to be placed on ecological experiments rather than on purely descriptive comparative studies.

14.8.2 Recent practical case studies

Despite some of the above mentioned knowledge gaps, many studies have transcended the research phase and have found applicability to regulate specific pests. Examples include:

1. Researchers working with farmers in ten townships in Yunnan, China, covering an area of 5350 hectares, encouraged farmers to switch from rice monocultures to planting variety mixtures of local rice with hybrids. Enhanced genetic diversity reduced blast incidence by 94% and increased total yields by 89%. By the end of two years, it was concluded that fungicides were no longer required (Zhu *et al.*, 2000; Wolfe, 2000).
2. In Africa, scientists at ICIPE (International Center of Insect Physiology and Ecology) developed a habitat management system for stem borer control, which uses two kinds of crops planted together with maize: a plant that repels these borers (the push) and another that attracts (pulls) them (Khan *et al.*, 1998). The push-pull system has been tested on over 450 farms in two districts of Kenya and has now been released for uptake by the national extension systems in East Africa. Participating farmers in the breadbasket of Trans Nzoia are reporting a 15–20% increase in maize yield. In the semi-arid Suba district, plagued by both stemborers and the parasitic weed *Striga*, a substantial increase in milk yield has occurred in the last four years, with farmers now being able to support grade cows on the fodder produced. When farmers plant maize, napier, and *Desmodium* together, a return of US$2.30 for every dollar invested is made, as compared to only US$1.40 obtained by planting maize as a monocrop. Two of the most useful trap crops that pulls in the borers' natural enemies are napier grass (*Pennisetum purpureum*) and Sudan grass

(*Sorghum vulgare sudanese*), both important fodder plants; these are planted in a border around the maize. Two excellent borer-repelling crops which are planted between the rows of maize are molasses grass (*Melinis minutifolia*), which also repels ticks, and the leguminous silverleaf (*Desmodium*). This plant can also suppress *Striga* by a factor of 40 compared to maize monocrops, its N-fixing ability increases soil fertility, and it is an excellent forage. As an added bonus, sale of *Desmodium* seed is proving to be a new income-generating opportunity for women in the project areas.

3. Several researchers have introduced flowering plants in strips within crops as a way to enhance the availability of pollen and nectar, necessary for optimal reproduction, fecundity, and longevity of many natural enemies of pests. *Phacelia tanacetifolia* strips have been used in wheat, sugar beets, and cabbage leading to enhanced abundance of aphidophagous predators, especially syrphid flies, and reduced aphid populations. In England, in an attempt to provide suitable overwintering habitat within fields for aphid predators, researchers created "beetle banks" sown with perennial grasses such as *Dactylis glomerata* and *Holcus lanatus*. When these banks run parallel with the crop rows, great enhancement of predators (up to 1500 beetles per square meter) can be achieved in only two years (Landis *et al.*, 2000).
4. In perennial cropping systems the presence of flowering undergrowth enhances the biological control of a series of insect pests. The beneficial insectary role of *Phacelia* in apple orchards was well demonstrated by Russian and Canadian researchers more than 30 years ago (Altieri, 1994). Maintenance of floral diversity by organic farmers throughout the growing season in California vineyards, in the form of summer cover crops of buckwheat (*Fagopyrum esculentum*) and sunflower (*Helianthus annuus*), had a substantial impact on the abundance of western grape leafhopper, *Erythroneura elegantula* (Homoptera: Cicadellidae), and western flower thrips, *Frankliniella occidentalis* (Thysanoptera: Thripidae), and associated natural enemies. During two consecutive years, vineyard systems with flowering cover crops were characterized by lower densities of leafhoppers and thrips, and larger populations and more species of general predators, including spiders. Although *Anagrus epos* (Hymenoptera: Mymaridae), the most important parasitoid, achieved high numbers and inflicted noticeable mortality of grape leafhopper eggs, no differences in egg parasitism rates were observed between cover cropped and monoculture systems.

Mowing of cover crops forces movement of *Anagrus* and predators to adjacent vines resulting in the lowering of leafhopper densities in such vines. Results indicated that habitat diversification, using summer cover crops that bloom most of the growing season, supports large numbers of predators and parasitoids thereby favoring enhanced biological control of leafhoppers and thrips in vineyards (Nicholls *et al.*, 2000).

5. In Washington state (USA), researchers reported that organic apple orchards that retained some level of plant diversity in the form of weeds mowed as needed, gave apple yields similar to those of conventional and integrated orchards. Their data showed that the low external-input organic system ranked first in environmental and economic sustainability as this system exhibited higher profitability, greater energy efficiency, and lower negative environmental impact (Reganold *et al.*, 2001).
6. In Central America, Staver *et al.* (2001) designed pest-suppressive multistrata shade-grown coffee systems, selecting tree species and associations, density, and spatial arrangement, as well as shade management regimes, with the main goal of creating optimum shade conditions for pest suppression. For example, in low-elevation coffee zones, 35–65% shade promotes leaf retention in the dry seasons and reduces *Cercospora coffeicola*, weeds and *Planococcus citri*; at the same time, it enhances the effectiveness of microbial and parasitic organisms without contributing to increased *Hemileia vastatrix* levels or reducing yields.
7. Several entomologists have concluded that the abundance and diversity of predators and parasitoids within a field are closely related to the nature of the vegetation in the field margins. There is wide acceptance of the importance of field margins as reservoirs of the natural enemies of crop pests. Many studies have demonstrated increased abundance of natural enemies and more effective biological control where crops are bordered by wild vegetation from which natural enemies colonize. Parasitism of the armyworm, *Pseudaletia unipuncta*, was significantly higher in maize fields embedded in a complex landscape than in maize fields surrounded by simpler habitats. In a two-year study researchers found higher parasitism of *Ostrinia nubilalis* larvae by the parasitoid *Eriborus terebrans* in edges of maize fields adjacent to wooded areas than in field interiors (Landis *et al.*, 2000). Similarly, in Germany, parasitism of rape pollen beetle was about 50% at the edge of the fields, while at the center

of the fields parasitism dropped significantly to 20% (Thies and Tscharntke, 1999).

8. One way to introduce the beneficial biodiversity from surrounding landscapes into large-scale monocultures is by establishing vegetationally diverse corridors that allow the movement and distribution of useful arthropod biodiversity into the center of monocultures. Nicholls *et al.* (2001) established a vegetational corridor which connected to a riparian forest and cut across a vineyard monoculture. The corridor allowed natural enemies emerging from the riparian forest to disperse over large areas of otherwise monoculture vineyard systems. The corridor provided a constant supply of alternative food for predators, effectively decoupling predators from a strict dependence on grape herbivores and avoiding a delayed colonization of the vineyard. This complex of predators continuously circulated into the vineyard interstices, establishing a set of trophic interactions leading to natural enemy enrichment, and consequently, lower numbers of leafhoppers and thrips on vines located up to 30–40 m from the corridor.

All of the above examples constitute forms of habitat diversification that provide resources and environmental conditions suitable for natural enemies. The challenge is to identify the type of biodiversity that is desirable to maintain and/or enhance to carry out ecological services of pest control, and then to determine the best practices that will encourage such desired biodiversity components.

14.9 Designing pest-stable agroecosystems

The key challenge for the 21st century pest managers is to translate ecological principles into practical alternative systems to suit the specific needs of farming communities in different agroecological regions of the world. A major strategy emphasized in this paper to design a more sustainable agriculture is to restore agricultural diversity in time and space by following key agroecological guidelines:

- Increase biodiversity in time and space through multiple cropping and agroforestry designs.
- Increase genetic diversity through variety mixtures, multilines, and use of local germplasm and varieties exhibiting horizontal resistance.
- Include and improve fallow through legume-based rotations, use of green manures, cover crops, and livestock integration.

- Enhance landscape diversity with biological corridors, crop-field boundary areas that are vegetationally diverse crops, or by creating a mosaic of agroecosystems and maintaining areas of natural or secondary vegetation as part of the agroecosystem matrix.

We emphasize that diversification schemes should be complemented by soil organic management as both strategies are the pillars of agroecosystem health (Figure 14.3).

Different options to diversify cropping systems are available, depending on whether the current monoculture systems to be modified are based on annual or perennial crops. Diversification can also take place outside the farm, for example, in crop-field boundaries with windbreaks, shelter belts, and living fences, which can improve habitat for wildlife and beneficial insects, provide resources of wood, organic matter, resources for pollinating bees, and, in addition, modify wind speed and microclimate (Altieri and Letourneau, 1982). Plant diversification can be considered a form of conservation biological control, with the goal of creating a suitable ecological infrastructure within the agricultural landscape to provide resources such as pollen and nectar for adult natural enemies, alternative prey or hosts, and shelter from adverse conditions. These resources must be integrated into the landscape in a way that is spatially and temporally favorable to natural enemies and practical for producers to implement.

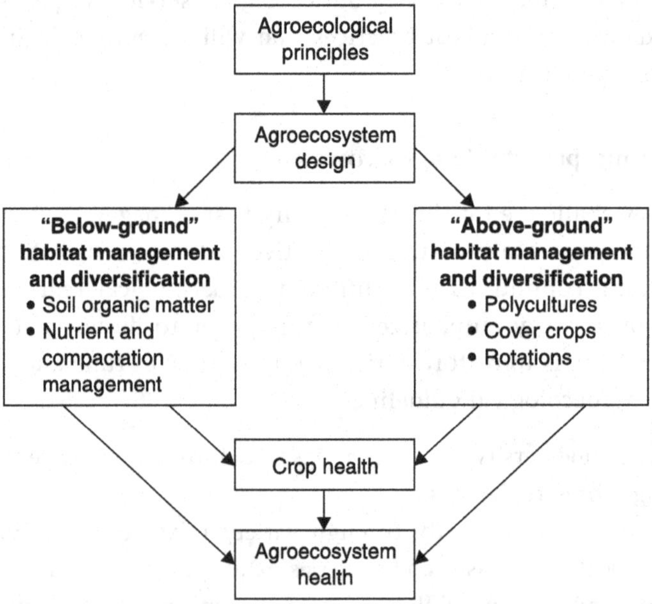

Figure 14.3. The pillars of agroecosystems health.

Landis *et al.* (2000) recommended the following guidelines to be considered when implementing habitat management strategies:

- Selection of the most appropriate plant species.
- The spatial and temporal arrangement of such plants within and/or around the fields.
- The spatial scale over which the habitat enhancement operates, with implications at the field or landscape level.
- The predator/parasitoid behavioral mechanisms which are influenced by the habitat manipulation.
- Potential conflicts that may emerge when adding new plants to the agroecosystem (e.g.) in California, *Rubus* blackberries around vineyards have increased populations of grape leafhopper parasitoids but can also enhance abundance of the sharpshooter which serves as a vector for Pierce's disease).
- Develop ways in which added plants do not upset other agronomic management practices, and select plants that preferentially have multiple effects such as improving pest regulation but at the same time improve soil fertility, weed suppression, etc.

What is crucial is the identification of the type of biodiversity worthwhile maintaining or enhancing to carry out ecological services, and then determining those practices that will best encourage the desired biodiversity components. Figure 14.4 shows that there are many agricultural practices and designs with the potential for enhancing functional biodiversity and those having negative effect. The idea is to apply the best management practices for enhancing or regenerating the kind of biodiversity that not only subsidizes the sustainability of agroecosystems by providing ecological services such as biological control, but also enhances nutrient cycling, water and soil conservation, etc. (Nicholls *et al.*, 2001).

If one or more alternative diversification schemes are used, the possibilities of complementary interactions between agroecosystem components are enhanced, resulting in one or more of the following effects:

- continuous vegetation cover for soil protection
- constant production of food, ensuring a varied diet and several marketing items
- closing nutrient cycles and effective use of local resources
- soil and water conservation through mulching and wind protection
- enhanced biological pest control by providing through diversification resources to beneficial biota

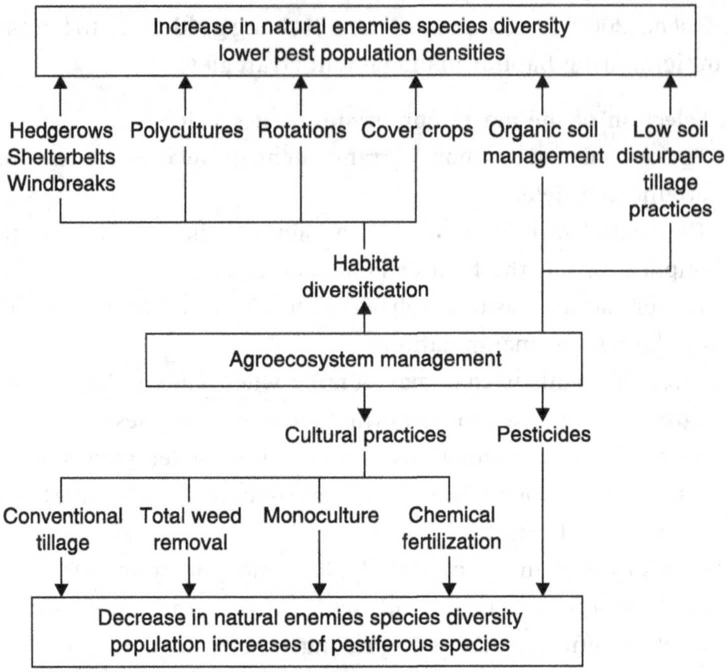

Figure 14.4. The effects of agroecosystem management and associated cultural practices on the biodiversity and natural enemies and the abundance of insect pests.

- increased multiple use capacity of the landscape
- sustained crop production without relying on environmentally degrading chemical inputs.

In summary, key ecological principles for the design of diversified and sustainable agroecosystems include:

- Increasing species diversity as this promotes fuller use of resources (nutrient, radiation, water, etc.), pest protection, and compensatory growth. Researchers have highlighted the importance of various spatial and temporal plant combinations to facilitate complementary resource use or to provide intercrop advantage, such as in the case of legumes facilitating the growth of cereals by supplying extra nitrogen. Compensatory growth is another desirable trait as, if one species succumbs to pests, weather or harvest, another species fills the void maintaining full use of available resources.
- Enhancing longevity through the addition of perennials that contain a thick canopy, thus providing continual cover that can also protect the soil. Constant leaf fall builds organic matter and allows uninterrupted

nutrition circulation. Dense, deep root systems of long-lived woody plants are an effective mechanism for nutrient capture offsetting the negative losses through leaching. Perennial vegetation also provides more habitat permanence and contributing to pest–enemy complexes.

- Imposing a fallow to restore soil fertility through biologically mediated mechanisms, and to reduce agricultural pest populations as life cycles are interrupted with forest regrowth or legume-based rotations.
- Enhancing additions of organic matter by including high biomass-producing plants. Accumulation of both "active" and "slow fraction" organic matter is key for activating soil biology, improving soil structure and macroporosity, and elevating the nutrient status of soils. Moreover, organic matter forms the foundation of complex food webs which influence the abundance and diversity of natural enemies.
- Increasing landscape diversity by having in place a mosaic of agroecosystems representative of various stages of succession. Risk of complete failure is spread among, as well as within, the various cropping systems. Improved pest control is also linked to spatial heterogeneity at the landscape level.

References

Altieri, M.A. (1994). *Biodiversity and Pest Management in Agroecosystems*. New York: Haworth Press.

Altieri, M.A. (1995). *Agroecology: The Science of Sustainable Agriculture*. Boulder, CO: Westview Press.

Altieri, M.A. (1999). The ecological role of biodiversity in agroecosystems. *Agriculture, Ecosystems and Environment*, **74**, 19–31.

Altieri, M.A. (2000). The ecological impacts of transgenic crops on agroecosystem health. *Ecosystem Health*, **6**, 13–23.

Altieri, M.A. and Letourneau, D.K. (1982). Vegetation management and biological control in agroecosystems. *Crop Protection*, **1**, 405–30.

Altieri, M.A. and Letourneau, D.K. (1984). Vegetation diversity and outbreaks of insect pests. *CRC, Critical Reviews in Plant Sciences*, **2**, 131–69.

Altieri, M.A. and Nicholls, C.I. (1999). Biodiversity, ecosystem function and insect pest management in agricultural systems. In W.W. Collins and C.O. Qualset (eds.), *Biodiversity in Agroecosystems*. Boca Raton, FL: CRC Press. pp. 69–84.

Altieri, M.A. and Nicholls, C.I. (2002). The simplification of traditional vineyard base agroforests in northwestern Portugal: some ecological implications. *Agroforestry Systems*, **56**, 185–91.

Altieri, M.A. and Nicholls, C.I. (2004). *Biodiversity and Pest Management in Agroecosystems*. 2nd edn. New York: Haworth Press.

Altieri, M.A. and Rosset, P. (1996). Agroecology and the conversion of large-scale conventional systems to sustainable management. *International Journal of Environmental Studies*, **50**, 165–85.

Altieri, M.A., Rosset, P. and Thrupp, L.A. (1998). *The Potential of Agroecology to Combat Hunger in the Developing World. 2020 Brief.* Washington, DC: IFPRI.

Andow, D.A. (1983). The extent of monoculture and its effects on insect pest populations with particular reference to wheat and cotton. *Agriculture, Ecosystems and Environment*, **9**, 25–35.

Andow, D.A. (1991). Vegetational diversity and arthropod population response. *Annual Review of Entomology*, **36**, 561–86.

Andow, D.A. and Hidaka, K. (1989). Experimental natural history of sustainable agriculture: syndromes of production. *Agriculture, Ecosystems and Environment*, **27**, 447–62.

Bach, C.E. (1980). Effects of plant diversity and time of colonization on an herbivore-plant interaction. *Oecologia*, **44**, 319–26.

Baliddawa, C.W. (1985). Plant species diversity and crop pest control: an analytical review. *Insect Science and Its Application*, **6**, 479–87.

Bilde, T., Axelsen, J.A. and Toft, S. (2000). The value of Collembola from agricultural soils as food for a generalist predator. *Journal of Applied Ecology*, **37**, 672–83.

Brodbeck, B., Stavisky, J., Funderburk, J., Andersen, P. and Olson, S. (2001). Flower nitrogen status and populations of *Frankliniella occidentalis* feeding on *Lycopersicon esculentum*. *Entomologia Experimentalis et Applicata*, **99**, 165–72.

Browning, J.A. and Frey, K.J. (1969). Multiline cultivars as a means of disease control. *Annual Review of Phytopathology*, **7**, 355–82.

Cardinale, B.J., Harvey, C.T., Gross, K. *et al.* (2003). Biodiversity and biocontrol: emergent impacts of a multi-enemy assemblage on pest suppression and crop yield in an agroecosystem. *Ecology Letters*, **6**, 857–65.

Chilcutt, C.F. and Tabashnik, B.E. (2004). Contamination of refuges by *Bacillus thuringiensis* toxin genes from transgenic maize. *Proceedings of the National Academy of Science*, **101**, 7526–9.

Cox, G.W. and Atkins, N.D. (1974). *Agricultural Ecology: An Analysis of World Food Production Systems*. San Francisco: W.H. Freeman and Co.

Culliney, T.W. and Pimentel, D. (1986). Ecological effects of organic agricultural practices on insect populations. *Agriculture, Ecosystems and Environment*, **15**, 253–66.

Donald, P.F. (2004). Biodiversity impacts of some agricultural commodity production systems. *Conservation Biology*, **18**, 17–37.

Ewel, J.J. (1986). Designing agricultural ecosystems for the humid tropics. *Annual Review of Ecology and Systematics*, **17**, 245–71.

Ewel, J.J. (1999). Natural systems as models for the design of sustainable systems of land use. *Agroforestry Systems*, **45**, 1–21.

Feeney, P.P. (1976). Plant apparency and chemical defense. *Recent Advances in Phytochemistry*, **10**, 1–40.

Gliessman, S.R. (1999). *Agroecology: Ecological Processes in Agriculture*. Michigan: Ann Arbor Press.

Groot, A.T. and Dicke, M. (2002). Insect-resistant transgenic plants in a multi-trophic context. *The Plant Journal*, **31**, 387–406.

Gutierrez, A.P., Havenstein, D.E., Nix, H.A. and Moore, P.A. (1974). The ecology of *Aphis craccivora* Koch and subterranean clover stunt virus in Southeast Australia. II. A model of cowpea aphid populations in temperate pastures. *Journal of Applied Ecology*, **11**, 1–20.

Gurr, G.M., Wratten, S.D., Irvin, N.A. *et al.* (1998). Habitat manipulation in Australasia: recent biological control progress and prospects for adoption. In M.P. Zaluki, R.A.I. Drew and G.G. White (eds.), *Pest Management – Future Challenges: Proceedings of the 6th Australian Applied Entomology Research Convention* 29 Sept.–2 Oct. Vol. 2, Brisbane: University of Queensland. pp. 225–35.

Helenius, J. (1998). Enhancement of predation through within-field diversification. In E. Pickett and R.L. Bugg (eds.), *Enhancing Biological Control*. Berkeley, CA: University of California Press. pp. 121–60.

Hilbeck, A., Baumgartner, M., Fried, P.M. and Bigler, F. (1998). Effects of transgenic *Bacillus thuringiensis* corn fed prey on mortality and development time of immature *Chrysoperla carnea* (Neuroptera: Chrysopidae). *Environmental Entomology*, **27**, 460–87.

Kajimura, T. (1995). Effect of organic rice farming on planthoppers: Reproduction of white backed planthopper, *Sogatella furcifera* (Homoptera: Delphacidae). *Researches on Population Ecology*, **37**, 219–24.

Kahn, Z.R., Ampong-Nyarko, K., Hassanali, A. and Kimani, S. (1998). Intercropping increases parasitism of pests. *Nature*, **388**, 631–2.

Khan, Z.R., Pickett, J.A., van der Berg, J. and Woodcock, C.M. (2000). Exploiting chemical ecology and species diversity: stemborer and *Striga* control for maize in Africa. *Pest Management Science*, **56**, 1–6.

Kowalski, R. and Visser, P.E. (1979). Nitrogen in a crop–pest interaction: cereal aphids. In J.A. Lee (ed.), *Nitrogen as an Ecological Parameter*. Oxford, UK: Blackwell Scientific Publications. pp. 67–74.

Lampkin, N. (1990). *Organic Farming*. Ipswich, UK: Farming Press Books.

Landis, D.A., Wratten, S.D. and Gurr, G.A. (2000). Habitat management to conserve natural enemies of arthropod pests in agriculture. *Annual Review of Entomology*, **45**, 175–201.

Letourneau, D.K. (1988). Soil management for pest control: a critical appraisal of the concepts. In *Global Perspectives on Agroecology and Sustainable Agricultural Systems. Sixth International Science Conference of IFOAM*. Santa Cruz, CA. pp. 581–7.

Letourneau, D.K., Drinkwater, L.E. and Shennon, C. (1996). Effects of soil management on crop nitrogen and insect damage in organic versus conventional tomato fields. *Agriculture, Ecosystems, and Environment*, **57**, 174–87.

Letourneau, D.K. and Goldstein, B. (2001). Pest damage and arthropod community structure in organic vs. conventional tomato production in California. *Journal of Applied Ecology*, **38**, 557–70.

Levins, R. and Wilson, M. (1979). Ecological theory and pest management. *Annual Review of Entomology*, **25**, 7–29.

Lewis, W.J., van Lenteren, J.C., Phatak, S.C. and Tumlinson, J.H. (1997). A total system approach to sustainable pest management. *Proceedings of the National Academy of Science, USA*, **94**, 12243–8.

Luna, J.M. (1988). Influence of soil fertility practices on agricultural pests. In *Global Perspectives on Agroecology and Sustainable Agricultural Systems*. Proceedings of the Sixth International Conference of IFOAM, Santa Cruz, CA. pp. 589–600.

Magdoff, F. and van Es, H. (2000). *Building Soils for Better Crops*. Washington, DC: SARE.

Marvier, M. (2001). Ecology of transgenic crops. *American Scientist*, **89**, 160–7.

Mattson, W.J., Jr. (1980). Herbivory in relation to plant nitrogen content. *Annual Review of Ecology and Systematics*, **11**, 119–61.

McRae, R.J., Hill, S.B., Mehuys, F.R. and Henning, J. (1990). Farm scale agronomic and economic conversion from conventional to sustainable agriculture. *Advances in Agronomy*, **43**, 155–98.

Milne, A. (1957). The natural control of on insect populations. *Canadian Entomologist*, **89**, 193–213.

Miyashita, K. (1963). Outbreaks and population fluctuations of insects, with special reference to agricultural insect pests in Japan. *Bulletin of National Agricultural Science Series*, **C15**, 99–170.

Morse, J.G., Bellows, T.S. and Gaston, L.K. (1987). Residual toxicity of acaricides to three beneficial species on California citrus. *Journal of Experimental Entomology*, **80**, 953–60.

National Academy of Sciences. (1972). *Genetic Vulnerability of Major Crops*. Washington, DC: NAS.

Nicholls, C.I., Parrella, M.P. and Altieri, M.A. (2000). Reducing the abundance of leafhoppers and thrips in a northern California organic vineyard through maintenance of full season floral diversity with summer cover crops. *Agricultural and Forest Entomology*, **2**, 107–13.

Nicholls, C.I., Parrella, M.P. and Altieri, M.A. (2001). The effects of a vegetational corridor on the abundance and dispersal of insect biodiversity within a northern California organic vineyard. *Landscape Ecology*, **16**, 133–46.

Obrycki, J.J., Losey, J.E., Taylor, O.R. and Jessie, L.C.H. (2001). Transgenic insecticidal maize: beyond insecticidal toxicity to ecological complexity. *BioScience*, **51**, 353–61.

Oka, I.N. and Pimentel, D. (1976). Herbicide (2,4-D) increases insect and pathogen pests on corn. *Science*, **143**, 239–40.

Phelan, P.L., Mason, J.F. and Stinner, B.R. (1995). Soil fertility management and hostpreference by European corn borer, *Ostrinia nubilalis*, on *Zea mays*: a comparison of organic and conventional chemical farming. *Agricdture, Ecosystems and Environment*, **56**, 1–8.

Pimentel, D. and Goodman, N. (1978). Ecological basis for the management of insect populations. *Oikos*, **30**, 422–37.

Pimentel, D. and Lehman, H. (1993). *The Pesticide Question*. New York: Chapman and Hall.

Pimentel, D. and Perkins, J.H. (1980). *Pest Control: Cultural and Environmental Aspects*. AAAS Selected Symposium 43. Boulder, CO: Westview Press.

Pimentel, D. and Warneke, A. (1989). Ecological effects of manure, sewage sludge and other organic wastes on arthropod populations. *Agricultural Zoology Reviews*, **3**, 1–30.

Price, P. W. (1976). Colonization of crops by arthropods: non-equilibrium communities in soybean fields. *Environmental Entomology*, **5**, 605–12.

Price, P. W. and Waldbauer, G. P. (1975). Ecological aspects of pest management. In R. L. Metcalf and W. H. Luckmann (eds.), *Concepts of Pest Management*. New York: John Wiley & Sons. pp. 37–73.

Reganold, J. P., Glover, J. D., Andrews, P. K. and Hinman, H. R. (2001). Sustainability of three apple production systems. *Nature*, **410**, 926–30.

Reinjtes, C., Haverkort, B. and Waters-Bayer, A. (1992). *Farming for the Future*. London: MacMillan.

Risch, S. J. (1981). Insect herbivore abundance in tropical monocultures and polycultures: an experimental test of two hypotheses. *Ecology*, **62**, 1325–40.

Risch, S. J., Andow, D. and Altieri, M. A. (1983). Agroecosystem diversity and pest control: data, tentative conclusions, and new research directions. *Environmental Entomology*, **12**, 625–9.

Rissler, J. and Mellon, M. (1996). *The Ecological Risks of Engineered Crops*. Cambridge, MA: MIT Press.

Robinson, R. A. (1996). *Return to Resistance: Breeding Crops to Reduce Pesticide Resistance*. Davis, CA: Agriculture Access.

Rodriguez, J. G., Chen, H. H. and Smith, W. T. (1957). Effects of sol insecticides on beans, soybeans, and cotton and resulting effects on mite nutrition. *Journal of Economic Entomology*, **50**, 587–93.

Root, R. B. (1973). Organization of a plant–arthropod association in simple and diverse habitats: the fauna of collards (*Brassica oleraceae*). *Ecological Monographs*, **43**, 94–125.

Russell, E. P. (1989). Enemies hypothesis: a review of the effect of vegetational diversity on predatory insects and parasitoids. *Environmental Entomology*, **18**, 590–9.

Schuler, T. H., Potting, R. P. J., Dunholm, I. and Poppy, G. M. (1999). Parasitoid behavior and Bt plants. *Nature*, **400**, 525.

Scriber, J. M. (1984). Nitrogen nutrition of plants and insect invasion. In R. D. Hauck (ed.), *Nitrogen in Crop Production*. Madison, WI: American Society of Agronomy.

Settle, W., Ariawan, H., Tri Astuti, E. *et al.* (1996). Managing tropical rice pests through conservation of generalist natural enemies and alternative prey. *Ecology*, **77**, 1975–88.

Slansky, F. (1990). Insect nutritional ecology as a basis for studying host plant resistance. *Florida Entomologist*, **73**, 354–78.

Slansky, F. and Rodriguez, J. G. (1987). *Nutritional Ecology of Insects, Mites, Spiders and Related Invertebrates*. New York: Wiley.

Southwood, T. R. E. and Way, M. J. (1970). Ecological background to pest management. In R. L. Rabb and F. E. Guthrie (eds.), *Concepts of Pest Management*. Raleigh, NC: North Carolina State University. pp. 1–18.

Soule, J. D. and Piper, J. K. (1992). *Farming in Nature's Image*. Washington DC: Island Press.

Staver, C., Guharay, F., Monterroso, D. and Muschler, R.G. (2002). Designing pest-suppressive multistrata perennial crop systems: shade grown coffee in Central America. *Agroforestry Systems* (in press).

Stiling, P., *et al.* (2003). Associational resistance mediated by natural enemies. *Ecological Entomology*, **28**, 587–92.

Sunderland, K. and Samu, F. (2000). Effects of agricultural diversification on the abundance, distribution, and pest control potential of spiders: a review. *Entomologia Experimentalis et Applicata*, **95**, 1–13.

Thies, C. and Tscharntke, T. (1999). Landscape structure and biological control in agroecosystems. *Science*, **285**, 893–95.

Tilman, D., Wedin, D. and Knops, J. (1996). Productivity and sustainability influenced by biodiversity in grassland ecosystems. *Nature*, **379**, 718–20.

Van Driesche, R.G. and Bellows, T.S., Jr. (1996). *Biological Control*. New York: Chapman and Hall.

Vandermeer, J. (1995). The ecological basis of alternative agriculture. *Annual Review of Ecology and Systematics*, **26**, 210–24.

Vandermeer, J. and Perefecto, I. (1995). *Breakfast of Biodiversity*. Oakland, CA: Food First Books.

Vandermeer, J. (1997) Syndromes of production: an emergent property of simple agroecosystem dynamics. *Journal of Environmental Management*, **51**, 59–72.

van Emden, H.F. and Williams, G.F. (1974). Insect stability and diversity in agroecosystems. *Annual Review of Entomology*, **19**, 455–75.

van Emden, H.F. (1966). Studies on the relations of insect and host plant. III. A comparison of the reproduction of *Brevicoryne brassicae* and *Myzus persicae* (Hemiptera: Aphididae) on brussels sprout plants supplied with different rates of nitrogen and potassium. *Entomologia Experimentalis et Applicata*, **9**, 444–60.

Victor, T.J. and Reuben, R. (2000). Effects of organic and inorganic fertlisiers on mosquito populations in rice fields on southern India. *Medical and Veterinary Entomology*, **14**, 361–8.

Wolfe, M. (2000). Crop strength through diversity. *Nature*, **406**, 681–2.

Zhu, Y., Fen, H., Wang, Y. *et al.* (2000). Genetic diversity and disease control in rice. *Nature*, **406**, 718–72.

15

Applications of molecular ecology to IPM: what impact?

P. J. DE BARRO, O. R. EDWARDS AND P. SUNNUCKS

15.1 Introduction

The use of molecular markers in biology, phylogeny and ecology has a long and distinguished history. Variable heritable protein markers (mostly allozymes) formed the basis of numerous studies in population biology and genetics (see Loxdale and den Hollander, 1989; Symondson and Liddell, 1996 for reviews). They enabled boundaries between functionally independent populations to be identified and allowed estimation of gene flow (and inferred migration) between them. Their use has been largely replaced by the use of DNA-based (often referred to as "molecular") techniques and it is here where we focus our discussion. We use "molecular ecology" to mean the application of molecular biology to population ecology (see reviews in Avise, 1994; Schierwater *et al.*, 1994; Moritz and Lavery, 1996; Carvalho, 1998; Sunnucks, 2000; see Loxdale and Lushai, 1999; MacDonald and Loxdale, 2004 for reviews on entomological applications).

Reasons for the shift to DNA-based methods for studying pests include the following.

a. Technological factors, such as the fact that usable DNA can be obtained from very small specimens (e.g. tiny grape phylloxera *Daktulosphaira vitifoliae*, Downie, 2000; individual aphid eggs, Sloane *et al.*, 2001) and preserved specimens such as pinned museum collections or ethanol-preserved suction-trapped insects.
b. Issues of resolution: DNA techniques often yield sufficient variation where allozymes and morphology cannot because invading populations have lost variability (e.g. Davies *et al.*, 1999*a*; *b*), the taxa have low allozyme variation (e.g. aphids, De Barro *et al.*, 1995*a*; *b*; *c*;

Sunnucks *et al.*, 1996) or increasingly the awareness that relationships between population structure, arthropod behaviour, and biology may be explainable by the presence of morphologically indistinguishable cryptic species (e.g. *Bemisia tabaci*, De Barro *et al.*, 2005). In these cases, molecular techniques offer the researcher a perspective of the underlying influences on population and community ecology that may not be available through other approaches (Moritz and Lavery, 1996).

Molecular techniques have been widely applied to address questions of ecological significance. However, closer inspection shows that their application to ecological questions in integrated pest management (IPM) has been relatively small. The types of questions addressed have not changed significantly over the past 25 years or so. In the study of IPM these questions may be grouped as follows:

- Basic biology of pest species, such as reproductive mode
- Origin of pest incursions and refuges for pests and their natural enemies
- Dispersal and gene flow within and between habitat patches and on regional or continental scales
- Diagnosing and identifying biotypes and sibling species
- Genetic variation underlying phenotypic traits, primarily host utilization and pesticide resistance
- Spatial and temporal variation in population structure on various geographic scales.

However, in terms of the overall research output between 1985 and January 2005, the publication of fewer than 300 papers, representing <60 pests and 20 beneficial species suggests that the impact has been minor, and restricted to a narrow range of pest (aphids and fruitflies represent 50% of pest taxa), and beneficial taxa (70% parasitic Hymenoptera). To address why this may be so, we first look at the history of molecular ecology in the broad context of IPM and then pose reasons and present prospects.

15.2 The history of molecular ecology in IPM

15.2.1 Non-polymerase chain reaction approaches

The first molecular approaches involved restriction fragment length polymorphism (RFLP) within the mitochondrial DNA (mtDNA) genome. Their use in studies of population structure was pioneered by Avise *et al.* (1979), but their application to insects of agricultural significance only began in earnest a decade later. Examples include attempts to diagnose aphid biotypes

(Powers *et al.*, 1989), to identify the source of incursions of Mediterranean fruit fly, *Ceratitis capitata* (Sheppard *et al.*, 1992), to distinguish between Asian and non-Asian forms of gypsy moths, *Lymantria dispar* (Bogdanowicz *et al.*, 1993), to study clonal diversity in the pea aphid, *Acyrthosiphon pisum* (Barrette *et al.*, 1994) and to identify the corn and rice strains of fall armyworm, *Spodoptera frugiperda* (Lu and Adang, 1996). Owing to low levels of genetic variation and the difficulty of getting sufficient mtDNA from small animals, these approaches had limited impacts on questions of ecology. In the case of *Lymantria dispar*, the ability to separate the European and Asian forms contributed to the understanding of natural migration of the Asian form from east to west across Europe (Roy *et al.*, 1995).

During this period of development of mtDNA marker systems, Wyman and White (1980) discovered the presence of long sequences of highly variable repetitive DNA in the human genome. The significance of this discovery was not evident until 1985 when a paper in the journal *Nature* coined the term "DNA fingerprint" (Jeffreys *et al.*, 1985). These fingerprints reveal the presence in nuclear DNA of multiple regions or loci containing "hypervariable" DNA, revealed by probing DNA ladders with probes that detect certain "core sequences". Variability is so great that the only humans that share identical patterns are monozygotic twins. The basis of the high levels of variation in these systems is the number of tandemly repeated copies of core sequences, and these vary greatly among and within individuals. The result looks much like a supermarket bar code. Jeffreys' immediate use for this was paternity testing in humans, but ecologists and biologists were quick to see the technique's wider potential. DNA fingerprinting was born and it precipitated a revolution in the way we address ecological questions dealing with parentage and relatedness, population structure, movement and origin, and species and biotype identification. Conceptually and methodologically similar techniques can be used to reveal complex banding patterns from sequences other than Jeffreys' ones, such as variation in the tandem repeats in nuclear ribosomal DNA. We refer to this whole group of approaches as multilocus nuclear DNA fingerprinting.

The first applications of multilocus nuclear DNA fingerprinting to insects of agricultural importance investigated variation between individuals from different strains, populations, origins and hosts. In the target organisms, traditional methods to distinguish individuals were ineffective, e.g. green peach aphid, *Myzus persicae*, and grain aphid, *Sitobion avenae* (Carvalho *et al.*, 1991), Mediterranean fruit fly, *Ceratitis capitata* (Haymer *et al.*, 1992), greenbug, *Schizaphis graminum* (Shufran *et al.*, 1992; Shufran and Wilde, 1994), fall armyworm, *Spodoptera frugiperda* (Lu *et al.*, 1992; 1994), Asian rice gall midge, *Orseolia oryzae* (Ehtesham *et al.*, 1995), leafroller moths, *Planotortrix* spp. (Sin *et al.*, 1995) and African woolly pine aphid, *Pineus pini* (Blackman *et al.*, 1995). Blackman *et al.* (1995) also demonstrated that this variation could be used to identify the source of recent incursions of exotic pest species. These papers demonstrated that DNA variation was a powerful tool with which to demonstrate variation between individuals and/or populations, but as with mtDNA RFLP, few went on to apply the tools to questions of ecological significance.

One of the first research programs to apply DNA techniques to obtain detailed ecological information about a pest was that of Shufran *et al.* (1991) and Shufran and Wilde (1994). The hypervariable intergenic spacer region of the ribosomal RNA gene was used to study spatial and temporal variation in *Schizaphis graminum* on wheat and sorghum. These studies indicated that aphid populations were homogeneously distributed in space and showed no host-based genetic differentiation. De Barro *et al.* (1994) used information from a tandemly repeated oligonucleotide $(GATA)_4$ to analyze the process of colonization and spread of *Sitobion avenae* within a wheatfield over the course of a season. The results complemented those obtained through direct observation, but avoided the problem of the observer's influencing aphid behavior and at the same time provided a level of spatial and temporal structure that was not easy to obtain by conventional ecological techniques. De Barro *et al.* (1995*a*) then used multilocus nuclear DNA fingerprinting to demonstrate genetic variation at different geographic scales and indicated that *Sitobion avenae* lineages exhibit differential host relations. Prior to that study, conventional studies of aphids in the laboratory had demonstrated considerable genetic variability in performance across different hosts, but had never before been able to demonstrate that this had any influence on population structure in the field.

These early studies revealed some challenges to the emerging technologies. In sexually reproducing species, variation within a population is often as great as that found between populations, making the demonstration of population level variation virtually impossible. Technical aspects of DNA fingerprints are also not ideal. The approach requires relatively large quantities of DNA, and a time-consuming and costly process that required the use of radioisotopes. Many single insects are too small to provide the quantity of DNA needed. Further, the cost and time involved in obtaining the fingerprints meant that sample sizes were often very small. The invention of the polymerase chain reaction (PCR) (Saiki *et al.*, 1988) and the development of efficient DNA sequencing (Avise, 1994; Hillis *et al.*, 1996) largely overcame these problems.

15.2.2 *Polymerase chain reaction (PCR) approaches*

The development of PCR is a key cause for the rapid expansion and adoption of molecular techniques. PCR techniques are relatively simple and the equipment readily accessible. There are a number of different techniques in current use (see Loxdale and Lushai, 1998). These fall into two main groups. Multilocus techniques simultaneously amplify many DNA regions (most commonly randomly amplified polymorphic DNA PCR or RAPD-PCR, and amplified fragment length polymorphism or AFLP). These approaches reveal arbitrary and unknown segments of DNA scattered through the genome. In contrast, single-locus techniques (e.g. PCR of known DNA regions such as mitochondrial DNA (mtDNA), nuclear introns, and microsatellites also known as simple sequence repeats) amplify specific DNA regions that are then compared with respect to their lengths and/or DNA sequences. MtDNA is something of a special case, in part because it is technically very accessible through the development of PCR primers that work on most arthropods (Simon *et al.*, 1994). In addition, the fundamental biological properties of mtDNA lend it particular strengths. Because mtDNA is haploid, usually maternally inherited and non-recombining, it becomes diagnostic of taxonomic entities (populations, biotypes, species, etc) more quickly than do nuclear markers. In addition, its maternal inheritance can show patterns of gene flow attributable to differences between the sexes, sometimes with unexpected circumstances. For example, some *Sitobion avenae* lineages in southern England were shown to arise through previously undetected hybridization of *Sitobion avenae* with another pest aphid *Sitobion fragariae* (Sunnucks *et al.*, 1997*a*). Application of mtDNA showed that hybridization between the species was asymmetrical: female *Sitobion fragariae* reproduced successfully with male *Sitobion avenae* far more than male *Sitobion fragariae* with female *Sitobion avenae*. Given differences in host use and basic biology of the species, this information could be of substantial use in IPM. Most importantly, apparent intraspecific host specificity in these aphids needed to be reassessed. The molecular data showed that genotypes collected on cockfoot grass (*Dactylis glomerata*), but not on wheat, were a set of interspecific avenae/fragariae hybrids (interestingly adopting the wheat-avoiding tendency of *S. fragariae*) rather than a host race of *S. avenae*.

15.3 Contributions of molecular techniques to IPM

Given that the molecular techniques now being applied in IPM are based on PCR, we will use these to illustrate contributions of molecular techniques to IPM.

15.3.1 *Diagnosing and identifying species, subspecies, and biotypes*

In IPM and allied areas of biological control, DNA-based diagnostics have become an increasingly useful tool for aiding identification of individuals where conventional systematics is either cumbersome or not sufficiently developed. An example is the whitefly, *Bemisia tabaci*, a complex of morphologically indistinguishable populations that varies considerably in host range and ability to transmit viruses and damage crops. Here RAPD-PCR and in later years mtDNA and ribosomal DNA were used to distinguish the different biotypes (Perring *et al.*, 1993; Gawel and Bartlett, 1993; De Barro and Driver, 1997; Kirk *et al.*, 2000; Brown *et al.*, 2002; Abdullahi *et al.*, 2003; Zanic *et al.*, 2005). The capacity to distinguish two biotypes was essential in teasing apart the complex interactions found in mixed populations (De Barro and Hart, 2000; De Barro *et al.*, 2005). RAPD-PCR has also been used to identify the egg parasitoids of pod-sucking pentatomid bugs (Aljanabi *et al.*, 1998) and larvae of cerambycid beetles (Yulin *et al.*, 1998) and to distinguish between Asian and North American populations of gypsy moth, *Lymantria dispar* (Garner and Slavicek, 1996). The same technique together with a simple diagnostic test based on RFLP of a mtDNA PCR fragment was used to determine that aphids damaging subterranean clover pastures were a genetically and biologically distinct biotype of the spotted alfalfa aphid *Therioaphis trifolii* (Sunnucks *et al.*, 1997*b*).

Restriction fragment length variation in an amplified portion of the ribosomal RNA gene was used to separate Asian and European gypsy moths (Pfeifer *et al.*, 1995). Similarly, RFLP within amplified portions of mtDNA were used to separate species of spider mite *Tetranychus* (Lee and Lee, 1997), ermine moth, *Yponomeuta* spp. (Sperling *et al.*, 1995) and larval *Melolonthini* cane grubs (Miller *et al.*, 1999).

MtDNA has also been used to investigate reproductive compatibility between different populations of *Bemisia tabaci* (Calvert *et al.*, 2001; Maruthi *et al.*, 2004), species complex variability in *Diadegma*, a group of diamondback moth parasitoids (Wagener, 2004), variability within the genus of predatory bugs *Orius* (Muraji *et al.*, 2004), invasive leafminers *Liriomyza* spp. (Scheffer *et al.*, 2001) and the parasitoid genus *Eretmocerus* spp. (De Barro *et al.*, 2000). Within species variation has been further examined for greenbug, *Schizaphis graminum* (Aikhionbare and Mayo, 2000) and different populations of *Rhabdoscelus obscurus* that produce different aggregation pheromones (Giblin-Davis *et al.*, 2000). Various regions within the ribosomal RNA gene region have similarly been used to distinguish between a range of morphologically similar species of predatory bugs *Orius* spp. (Hinomoto *et al.*, 2004), parasitic hymenoptera *Eretmocerus* spp. (De Barro *et al.*, 2000), *Aphelinus hordei* and *Aphidius colemani* (Zhu *et al.*, 2000), *Encarsia* spp. (Schmidt *et al.*, 2001), *Trichogramma* spp. (Chang *et al.*, 2001; Li *et al.*, 2004; Borghuis *et al.*, 2004) and *Encarsia meritoria* species complex (Polaszek *et al.*, 2004), pests thrips (Toda and Komazaki, 2002), and invasive leafminers *Liriomyza* spp. (Scheffer and Lewis, 2001)

Molecular markers have also been contributed to quality control in insect mass-rearing facilities (Landry *et al.*, 1993), to distinguish between the species and strains of parasitoids imported to the USA as part of the *Bemisia tabaci* biological control program (Legaspi *et al.*, 1996), to improve the understanding of the relationship between Cecidophyopsis mites and their various currant hosts (Fenton *et al.*, 1995; 1996) and to distinguish between the larvae of Tephritid fruitflies (McKenzie *et al.*, 1999).

Single-stranded conformational polymorphism (SSCP, review of applications in molecular ecology in Sunnucks *et al.*, 2000) was used to investigate the possible relationship between host aphid choice and genetic variation in the parasitoid *Diaeretiella rapae* (Baer *et al.*, 2004). In this case, no such relationship was found, suggesting that host selection had not been responsible for formation of host races.

15.3.2 *Basic biology*

The basic biology of some pest insects is not well known for various reasons, including that they have only recently come to attention, that they are very small or have complex life cycles, or that their biology in newly invaded regions differs from that of their source areas. Molecular techniques can often supply ready solutions to these challenges and yield information that is crucial in IPM. At a fundamental level, molecular genetics have been used to show that diverse morphs of insects actually belong to the same species (e.g. for aphids, Stern *et al.*, 1997). Indeed, aphid biology has been a rich arena of unexpected findings concerning basic biology, owing to their complex life cycles and variable phenology. The typical life cycle of aphids is cyclic parthenogenesis (many parthenogenetic generations per year followed by a single sexual generation to produce overwintering eggs). Parts of the cycle are induced by environmental control, and this leads to very plastic reproductive modes, including during colonization of new regions. Highly resolving molecular techniques, especially single-locus ones such as microsatellites, have allowed resolution of the reproductive mode of aphid lineages, producing evidence for unexpected patterning of reproductive modes (Sunnucks *et al.*, 1996; 1997*a*; Simon *et al.*, 1999*a*; Wilson *et al.*, 1999).

Similar progress has recently been made for other important pests with substantial parthenogenetic phases, including grape phylloxera (Corrie *et al.*, 2002; 2003). Since sexual and asexual lines have different demographics, spatiotemporal patterning and likely evolutionary characteristics, understanding reproductive mode is of obvious potential applied significance.

Given the importance of clonal diversity and trait variation in cyclically parthenogenetic pests, it is perhaps surprising to find that the most basic genetics of many pest species is unknown. For example, only recently have the genetic processes and outcomes during sexual and asexual reproduction in pest aphids been verified by the application of highly resolving molecular markers (Sloane *et al.*, 2001; Hales *et al.*, 2002*a*; *b*). Similarly, the consequences of sexual and asexual reproduction on the genetic diversity of the aphids *Myzus persicae* and *Rhopalosiphum padi*, and the overall decline in diversity as the season progresses, give insights into the selection forces that may underpin such factors as the development of host plant resistance (Delmotte *et al.*, 2002; Fenton *et al.*, 2003; Guillemaud *et al.*, 2003). Further, it has only recently been shown that a pest mite of citrus, *Brevipalpus phoenicis*, is haploid, and is maintained in parthenogenetic state by infection with a bacterium other than *Wolbachia* (Weeks *et al.*, 2001). These findings could hold the clues to understanding and controlling this pest. Molecular markers have enabled a better understanding of the levels of remating in species controlled using the release of sterile males. Bonizzoni *et al.* (2003) have shown that for Mediterranean fruit fly *Ceratitis capitata*, up to 21% were able to re-mate, a level that is of critical importance in determining the numbers of sterile males released if effective suppression is to be achieved.

When combined with information on traits like intrinsic rates of increase under different conditions, reproductive modes, and virus transmission, the potential for predicting the trajectories of pest populations is considerable (Simon *et al.*, 1999*a*; Terradot *et al.*, 1999). In this context, identifying variation in important traits itself can benefit from application of molecular markers. Edwards and Hoy (1995*a*) found in mixed laboratory populations of resistant and wild-type walnut aphid parasitoids, *Trioxys pallidus*, that the resistant biotype was heavily favored, irrespective of pesticide treatment, suggesting a high degree of laboratory adaptation. RAPD markers were used to discriminate between parents and their hybrid offspring and suggested the selective advantage of the resistant biotype may have been due to partial mating incompatibility between the biotypes. From the data it could then be inferred that the incompatibility was due either to wild-type females mating with resistant males and producing only male offspring, suggesting cytoplasmic incompatibility or wild-type females being unable to find mates, suggesting a behavioral incompatibility. Associations between geographic origin and cold tolerance in the aphid parasitoid *Lysiphlebus testaceipes* indicated the existence of a close correlation between the two (Shufran *et al.*, 2004) and is useful in determining efficacy in the context of introducing biological control agents.

15.3.3 Geographic origin

Knowledge of the geographic origin of pest species may be extremely useful in pinpointing areas to search for potential biological control agents and host plant resistance. In addition, it may enable better predictions to be made in determining from where particular pest outbreaks and incursions may have originated. For example *Micronotus hyperodae*, a parasitoid of the Argentine stem weevil *Listronotus bonariensis*, may be more effective against hosts with which they have co-evolved locally rather than comparable ones merely from similar climates (Goldson *et al.*, 1997).

The most powerful indicators of geographic affinities will be provided by highly resolving suites of single-locus, co-dominant genetic markers including microsatellites, SNPs (single nucleotide polymorphisms), nuclear introns and anonymous markers and mtDNA variation (Sunnucks, 2000). For all these marker types (with appropriate quality control) allelic variation can be identified and compared unambiguously in studies across the globe and through the decades. Usually mtDNA will have resolution over broader geographic ranges, for example indicating a continental region from which a pest might have originated (for grape phylloxera, Downie *et al.*, 2001; Corrie *et al.*, 2003). Unambiguous conclusions about finer scales will usually require several or many highly resolving single-locus markers. For example, the application of a common suite of microsatellites to the cereal aphid *Sitobion avenae* has led to major advances in knowledge of the population biology and demographics of this pest (De Barro *et al.*, 1995*b*; *c*; Sunnucks *et al.*, 1997*a*; Simon *et al.*, 1999*a*; Haack *et al.*, 2000; Llewellyn *et al.*, 2003; Papura *et al.*, 2003; Llewellyn *et al.*, 2004; Figueroa *et al.*, 2005). Here, associations between host-based genetic variation, host selection, and fitness not only provided insights into short- and long-range dispersal. Together these studies provided a biological underpinning to the observed patterns in population structure, host utilization and movement at a number of geographic scales. The work even detected cryptic species and cryptic hybridization, necessitating a re-evaluation of earlier reports of intraspecific host specialization, much of which is actually most likely interspecific. The studies have shown that even in the face of high mobility, clinal variation in reproductive mode results from strong selective responses to winter temperature (Simon *et al.*, 1999*a*; Llewellyn *et al.*, 2003). Despite the advantages of sexual reproduction in producing winter eggs, parthenogens can be very persistent (e.g. around half the genotypes detected in England by Llewellyn *et al.* (2003), had been seen in earlier years in England or France). Indeed aphid asexual "superclones" can dominate on crops: for example, a common clone found by Haack *et al.* (2000) was also predominant in Chile at around the same time (Figueroa *et al.*, 2005). The phenomenon seems to be widespread: superclones have been found in several aphid species including the important pest *Myzus persicae* (Vorburger *et al.*, 2003) and in grape phylloxera (Corrie *et al.*, 2002) (review in Wilson *et al.*, 2003).

It is important to note that tracking in space and time of individual genotypes is possible only by the application of the same suite of highly resolving single locus markers (usually microsatellites) with good quality control and availability of genotype data bases. For some pest species, suites of microsatellites have been published, and for some taxa, a subset may be successfully applied to species other than those for which they were first developed (e.g. Wilson *et al.*, 2004). Single-locus markers also have a major strength of being added to in the future if more resolution is required.

Armed with the tools to follow the fortunes of lineages in the field, it becomes possible to investigate the genotypic and phenotypic basis of highly successful genotypes of pests. Vorburger (2005) demonstrated that genotypes of *Myzus persicae* apparently do not suffer from negative fitness correlations (some genotypes can be "masters of all trades"), perhaps owing to genetic variation in the ability to obtain resources and a cost of resource acquisition. Even with a relatively few loci, microsatellite heterozygosity predicted genotypes' fitnesses in the laboratory, which in turn were significantly associated with abundance on crops. Thus, in addition to their other uses, selectively neutral genetic markers (microsatellites) may offer a potential monitoring/predictive tool for pest demographics.

The origin of invading Mediterranean fruit fly *Ceratitis capitata* has attracted a great deal of research effort. Apart from the studies mentioned earlier using non-PCR approaches (Sheppard *et al.*, 1992; Haymer *et al.*, 1992), several (Haymer *et al.*, 1997; Villablanca *et al.*, 1998; Gomulski *et al.*, 1998; Davies *et al.*, 1999*a*; *b*; He and Haymer, 1999; Meixner *et al.*, 2002; Reyes and Ochando, 2004), have investigated the genetic relationships between possible origins and incursions using PCR-based techniques inclusive of single-locus, co-dominant marker suites. These latter approaches had the resolution required to determine sources of incursions, even though much of the genetic variation (the basis of the information content of the genetic marker systems) had been eroded through population bottlenecks caused by small numbers of founders. Most recently, microsatellite markers (Bonizzoni *et al.*, 2004) have been used to identify the origins of incursions into Australia and their subsequent spread, an insight that is basic to developing effective incursion response strategies and areawide freedom. These insights are equally important to several other fruitfly species *Anastrepha fraterculus* (Alberti *et al.*, 2003), *Bactrocera* spp. (Yu *et al.*, 2001; Muraji and Nakahara, 2002; Gilchrist *et al.*, 2004), *Ceratitis rosa* and *Ceratitis fasciventris* (Baliraine *et al.*, 2004) that have also been studied to determine geographic origins and structuring of populations.

Other studies that have investigated the origins of incursion pests include Frohlich *et al.* (1999) and De Barro *et al.* (2000) that used molecular markers to generate a phylogeny of the pest whitefly, *Bemisia tabaci*, collected from different countries. Since then a range of studies have investigated various aspects of population structure and disease transmission for this species (Legg *et al.*, 2002; Viscarret *et al.*, 2003; Berry *et al.*, 2004; Abdullahi *et al.*, 2004). Understanding the relationships between these various groups is the platform upon which conceptions of disease transmission, incursion potential, susceptibility to natural enemy attack, importance of host choice, capacity to disperse, etc. are based. Morris and Mound (2004) compared populations of *Scirtothrips aurantii* from South Africa with those from a recent incursion into Australia and concluded that there were no differences between the populations, indicating a possible origin of the invasion. The pattern of spread of the invasive small hive beetle *Aethina tumida* in the USA was able to determine inferences to be made about the likely pathway of introduction, but also the number of introductions (Evans *et al.*, 2003).

15.3.4 *Population structure, dispersal and gene flow and host utilization*

Molecular markers have perhaps had their greatest influence in estimating population structure. In the cases of Spackman and McKechnie (1995) and Chang *et al.* (1997), mtDNA from the cotton bollworm *Helicoverpa armigera* and diamondback moth *Plutella xylostella* was used to study genetic variation at various geographic scales. In both cases it was thought that DNA markers might provide useful insights into local patterns of movement in addition to the wider migratory flights. In both cases evidence for clear geographic variation was scant. Part of the reason may be due to widespread gene flow consistent with known patterns and levels of migration, but also small sample sizes and the use of inbred laboratory cultures may have contributed in one of the studies to the faint pattern.

Gene flow estimates are invaluable for modeling the development and spread of insecticide resistance (Loxdale *et al.*, 1998), to determine whether individuals are moving at low numbers between populations on different hosts (Via, 1999), the development of areawide control programs, and an overall understanding of a landscape-scale systems approach to pest management. Molecular markers have also revealed a better understanding of patterns of host utilization by phytophagous insects. McMichael and Prowell (1999), and more recently Prowell *et al.* (2004), demonstrated that populations of fall armyworm *Spodoptera frugiperda* from corn were shown to be genetically distinct from those found in grass pastures. In a similar study, Shufran *et al.* (2000) found biotypes of *Schizaphis graminum* feeding on wheat and sorghum were genetically distinct from those on other grasses. Coates *et al.* (2004) used mt COI and COII to identify restriction fragment length polymorphisms that enabled populations of the European corn borer *Ostrinia nubilalis* on the Atlantic coast to be differentiated from those in the midwestern USA and to further enable the separation of uni- and bivoltine populations. De Leon and Jones (2004) similarly showed geographically separated populations of glassy winged sharpshooter *Homalodisca coagulata* were genetically distinct. Local movement plant to plant dispersal in the western corn rootworm, Diabrotica *virgifera virgifera* and the acceptability of crop and non-crop hosts (Hibbard *et al.*, 2004; Clark and Hibbard, 2004) were studied as part of the development of resistance management strategies for transgenic corn. Dispersal in terms of migration into eradication zones was investigated for boll weevil *Anthonomus grandis* indicated that there was considerable genetic separation between eastern and western and populations separated by <300 km experienced high levels of gene flow between them (Kim and Sappington, 2004*b*). Corrie *et al.* (2002) and later Corrie and Hoffmann (2004), by investigating the clonal structure of the grape phylloxera in vineyards, demonstrated limited non-random distribution where distribution of clones followed rows. Hufbauer *et al.* (2004) compared the genetics of an exotic aphid parasitoid *Aphidus ervi* in its introduced range with that in its native range and determined the two populations were distinct. They further showed that the separation was due to a bottleneck event associated with the founders. Similarly, Baker *et al.* (2004) demonstrated a significant bottleneck associated with the introduction of the aphid parasitoid *Diaeretiella rapae*. However, in both cases the consequences of this in terms of efficacy and impact were unclear. There have been few studies relating to the variability of insects that infest glasshouses and other protected cropping systems. One study by Fuller *et al.* (1999) showed that the consequences of limited gene flow, founder events, and drift on clonal variability in glasshouses. These processes demonstrate one underlying reason resistance to insecticides is often more rapidly generated in protected cropping than in the field.

Genetic insights into spatiotemporal patterning of pest populations are particularly useful when seeking to understand the processes of crop colonization and spread of insecticide resistance. Host based genetic differentiation in *Thrips tabaci* (Brunner *et al.*, 2004) identified three major lineages; two associated with leek and the third with tobacco. In contrast, host-based differentiation did not explain the structure observed in pinyon pine beetle *Ips confusus* populations in North America; instead post-glaciation events were better correlated with the observed structure (Cognato *et al.*, 2003). Host based genetic divergence has also been shown for the parasitoid *Microctonus aethiopoides* associated with different *Sitona* species was demonstrated by Vink *et al.* (2003). Here genetic variation was better associated with host choice than morphological variation. Host-based genetic partitioning of populations has also been determined in the lettuce root aphid, *Pemphigus bursarius* (Miller *et al.*, 2003), grape phylloxera (in which there is also evidence for a root-specific suite of genotypes, Corrie *et al.*, 2003; Corrie and Hoffmann, 2004), *Bemisia tabaci* (Abdullahi *et al.*, 2003). The pea aphid, *Acyrthosiphum pisum*, also demonstrates host-based genetic differentiation (Simon *et al.*, 2003) with individuals on pea, clover, and alfalfa being genetically divergent and so constituting distinct host races. The study also demonstrated that there was a strong association between these races and their symbiotic microflora. Similarly, greenbug *Schizaphis graminum* showed clear genetic structuring with regard to a range of different grass hosts (Anstead *et al.*, 2002). Considerable clonal variability in the capacity to feed on resistant lucerne cultivars has been demonstrated for the pea aphid *Acyrthosiphon pisum* and highlights the importance of considering aphid genetic diversity in breeding programs for resistance in cultivated plants (Bournoville *et al.*, 2000).

The ability to detect and monitor insecticide resistance is a key part of pest management. Molecular markers have enabled resistance genes to be monitored both directly and indirectly. In the case of resistance in *Myzus persicae*, PCR led to the identification of the resistance genes involved and their mode of expression (Field *et al.*, 1996). In the same species Fuentes-Contreras *et al.* (2004) determined that insecticide resistance in Chile was linked to the introduction of one clonal line of a tobacco-feeding race, while Wilson *et al.* (2002) used microsatellites to investigate clonal structure in Australian populations and discovered that resistant clones often contained an autosomal 1,3 translocation (and which appears to be distinct from the A1,3 translocation associated with resistance in Europe, Sloane *et al.*, 2001). In other cases, the actual resistance mechanism and the genes involved are unknown. In these cases molecular markers linked to the resistance phenotype have been developed for pests such as diamondback moth, *Plutella xylostella* (Heckel *et al.*, 1995) enabling resistance to be tracked indirectly. A case in point is Edwards and Hoy (1995*a*; *b*) where the spread of resistant *Trioxys pallidus* was monitored after their release into walnut orchards. The studies were able to link observable increases in resistance with an increase in the frequency of the resistance gene within the parasitoid population. In a novel approach, McKenzie and Batterham (1998) used mutant sheep blowflies, *Lucilia cuprina*, to predict likely resistance mechanisms to novel insecticides before they evolved in the field. Being able to predict the resistance mechanisms will not only enable resistance to be monitored, but also may enable strategies to be developed that minimize the risk of resistance evolving.

The study of predator–prey dynamics has involved the use of monoclonal antibodies and certain marker immunoglobulins (IgGs). Recently, molecular markers have been developed to identify the presence of *Helicoverpa armigera* in the gut contents of predators (Agusti *et al.*, 1999). This approach has much potential to provide insights into the role of predation in pest management. Edwards and Hopper (1999) used molecular markers to study super-parasitism in *Macrocentrus cingulum*, a parasitoid of European corn borer, *Ostrinia nubialis*. In this case, markers were able to distinguish between clonal individuals derived from a single egg and those derived from different eggs. The results provided insights into the underlying phenology of parasitism in this pest. In a similar vein, Hoogendoorn and Heimpel (2001) identified the remains of *Ostrinia nubilalis* in the guts of both juvenile and adult *Coleomegilla maculata*, Agusti *et al.* (2003) *Collembola* in the guts of spiders and Cuthbertson *et al.* (2003) apple-grass aphid *Rhopalosiphum insertum* in the guts of the predatory mite *Anystis baccarum*.

15.3.5 *Why have molecular techniques had relatively limited impact upon IPM?*

Despite major advances in technology over the past 15 years, molecular techniques are only just being wholeheartedly applied to understanding the ecology of agricultural ecosystems, and there seems to be some way to go before they are generally implemented under a unifying conceptual approach to managing pests. Several factors may have contributed to this missed opportunity. In some cases, molecular markers have been difficult to develop, for example in agriculturally important species of Hymenoptera (Graur, 1985), Hemiptera (e.g. Loxdale *et al.*, 1985; Steiner *et al.*, 1985*a*; *b*) and Lepidoptera (Meglécz *et al.*, 2004; Zhang, 2004), contributing to low utility and awareness of the techniques for IPM practitioners. For certain species, suitable genetic markers have been developed but further population studies have not been published (e.g. Ehtesham *et al.*, 1995; Sin *et al.*, 1995; Aljanabi *et al.*, 1998). In some cases this may reflect that most practitioners of IPM are looking for immediate management results in isolation from understanding underlying processes, and the attitudes of funding agencies driven by short term demands to address growers' problems.

The limited availability of PCR primers for highly variable regions in some species still hinders the utility of single-locus PCR techniques such as PCR-RFLP and EPIC-PCR (Exon-Primed Intron-Crossing Polymerase Chain Reaction) (Simon *et al.*, 1993; Sunnucks, 2000). This problem can be marked for invertebrates in which the wide range of phylogenetic diversity makes it less likely that PCR primers can be transferred between taxa. This accounts for the earlier popularity of multilocus PCR techniques such as RAPDs, where prior knowledge of the genetic makeup of the target organism is not necessary. However, RAPDs are hampered by technical and fundamental shortcomings including high incidences of artefacts, non-target contaminating DNA, and limitations in analyses (Hurme and Savolainen, 1999; Isabel *et al.*, 1999; Rabouam *et al.*, 1999; Brodeur and Rosenheim, 2000; Sandstrom *et al.*, 2001). These can be overcome by converting RAPDs to single locus, codominant markers, a process that will both verify their origin and greatly increase the level of information yielded (e.g. Simon *et al.*, 1999*b*). Multilocus, dominant AFLP analysis is also amenable to such conversions and is technically more reliable than RAPD (Mueller and Wolfenbarger, 1999).

In the past five years the screening of variation at multiple microsatellites loci has become increasingly important. These studies tend to focus on population structure and species movement and so are highly relevant to IPM. In the past, the cost of developing and then utilizing microsatellites has been high, but the marked decline in the cost of consumables and the development of faster, mass throughput technology has seen the cost fall considerably. This has led to a sharp increase in the number of species for which microsatellites are available. Species of agricultural importance for which microsatellite markers have been developed include lettuce root aphid, *Pemphigus bursarius* (Miller *et al.*, 2000), Mediterranean fruit fly *Ceratitis capitata* (Casey and Burnell, 2001) and related species (Baliraine *et al.*, 2003), the bird cherry-oat aphid *Rhopalosiphum padi* (Simon *et al.*, 2001), the aphid parasitoids *Aphidius ervi* (Hufbauer *et al.*, 2001) and *Diaeretiella rapae*

(MacDonald *et al.*, 2003), gypsy moth *Lymantria dispar* (Koshio *et al.*, 2002), spruce bark beetle *Ips typographus* (Salle *et al.*, 2003), the ground beetles *Carabus problematicus* (Gaublomme *et al.*, 2003), *Abax parallelepipedus* (Keller and Largiadèr, 2003) and *Carabus insulicola* (Takami and Katada, 2001), Queensland fruit fly *Bactrocera tryoni* (Kinnear *et al.*, 1998; Zhou, 2003), *Bemisia tabaci* (De Barro *et al.*, 2003; Tsagkarakou and Roditakis, 2004), groundnut seed beetle *Caryedon serratus* (Sembene *et al.*, 2003), pine shoot beetle *Tomicus piniperda* (Kerdelhue *et al.*, 2003), western flower thrips *Frankliniella occidentalis* (Brunner and Frey, 2004), oriental fruit fly *Bactrocera dorsalis* (Dai *et al.*, 2004), stable fly *Stomoxys calcitrans* (Gilles *et al.*, 2004), African armyworm *Spodoptera exempta* (Ibrahim *et al.*, 2004), boll weevil *Anthonomus grandis* (Kim and Sappington, 2004*a*), the parasitoid complex, *Horismenus* (Aebi *et al.*, 2004), house fly *Musca domestica* (Chakrabarti *et al.*, 2004), Caribbean fruit fly *Anastrepha suspense* (Fritz and Schable, 2004), pea aphid *Acyrthosiphon pisum* (Caillaud *et al.*, 2004; Kurokawa *et al.*, 2004), *Helicoverpa armigera* (Scott *et al.*, 2004), coffee berry borer *Hypothenemus hampei* (Gauthier and Rasplus, 2004), black aphid, *Aphis fabae* (D'Acier *et al.*, 2004), glassy winged sharpshooter *Homalodisca coagulata* (de Leon and Jones, 2004) and European corn borer *Ostrinia nubilalis* (Dopman *et al.*, 2004).

This list above is not exhaustive, but is included to illustrate the range of important pest species for which microsatellites are now available. As single-locus, co-dominant marker developments grow in number, it becomes increasingly likely that researchers working on "new" species will be able to get a suite of useful markers "off the shelf". The extent to which individual microsatellite loci work well for species other than the one for which they were developed will vary among taxa, but can be quite successful. For aphids, Wilson *et al.* (2004) tested cross-species amplification of 48 aphid loci and found considerable cross-amplification including in taxa from different aphid tribes and subfamilies. When seeking primers, it is worthwhile exploring a number of sources of information. These include online databases (e.g. at the Molecular Ecology Notes website, and there are many taxon-specific ones – none is exhaustive), searches of literature (*Current Contents and Biological Abstracts*) and contacting email groups such as Evoldir.

The greatest resource of genetic information useful in distinguishing between individuals, populations etc. is the variation that occurs at nucleotides in DNA. Single-base differences known as single nucleotide polymorphisms (SNPs) can be powerful tools for distinguishing closely related individuals, and are likely to become popular for insect population studies (Morin *et al.*, 2004). Uptake of SNPs has been hindered by the lack of inexpensive, effective methods for finding them, and then screening them in large numbers of individuals. For example, SNPs can be found using AFLP-based approaches, then screened using single stranded conformation polymorphism (SSCP) (Zhang *et al.*, 1999; Sunnucks *et al.*, 2000) but when large numbers of individuals and loci are to be screened, newer emerging technologies may prove more cost- and time-efficient, including fluorescent primer extension techniques (Morin *et al.*, 2004), matrix-assisted laser desorption ionization time of flight (MALDI-TOF) mass spectrometry, and microarray technology.

15.3.6 *MALDI-TOF mass spectrometry*

MALDI-TOF mass spectrometry identifies single-base pair variations in DNA sequences by the difference in their mass (Haff and Smirnov, 1997). SNPs are detected using PCR of a short section of DNA just upstream from the target polymorphic locus (Figure 15.1). The mass of the resulting oligonucleotide is used to determine the genotype. This technology has been used to detect 12 SNPs simultaneously (Ross *et al.*, 1998). The matrix-assisted input allows the rapid processing of large numbers of DNA from individual organisms. Thus, this method can rapidly score both multiple loci and multiple samples, as would be required in population genetic studies.

15.3.7 *Microarray technology*

Microarray technology is only in its infancy, but already it is revolutionizing the field of molecular biology (Service, 1998). Its impact on science may rival the invention of microcomputers (Stipp, 1997). DNA microarrays, or "DNA chips", bear nanoliter quantities of up to 100,000 different DNAs (oligonucleotides, cDNAs) arranged in an array pattern, usually on glass slides, where they can act as targets for samples to be analyzed. Fluorescent markers and high-resolution fluorescent scanners are used to detect and quantify the level of hybridization of sample to target in just minutes.

One type of oligonucleotide array, the "SNP chip", can be used to detect SNPs in test samples (Wang *et al.*, 1998; Sapolsky *et al.*, 1999). These samples were hybridized to a mixture of perfect matching and mismatching oligonucleotides (Figure 15.2); the difference in hybridization among the "probes" on the chip identifies the SNP state of the sample (Wang *et al.*, 1998; Sapolsky *et al.*, 1999). Thousands of SNP loci can be characterized in a single experiment (Cho *et al.*, 1999; Sapolsky *et al.*, 1999), and single chips can be used to screen pooled population samples to determine the frequency of alleles at one or more SNP loci (Bang-Ce *et al.*, 2004).

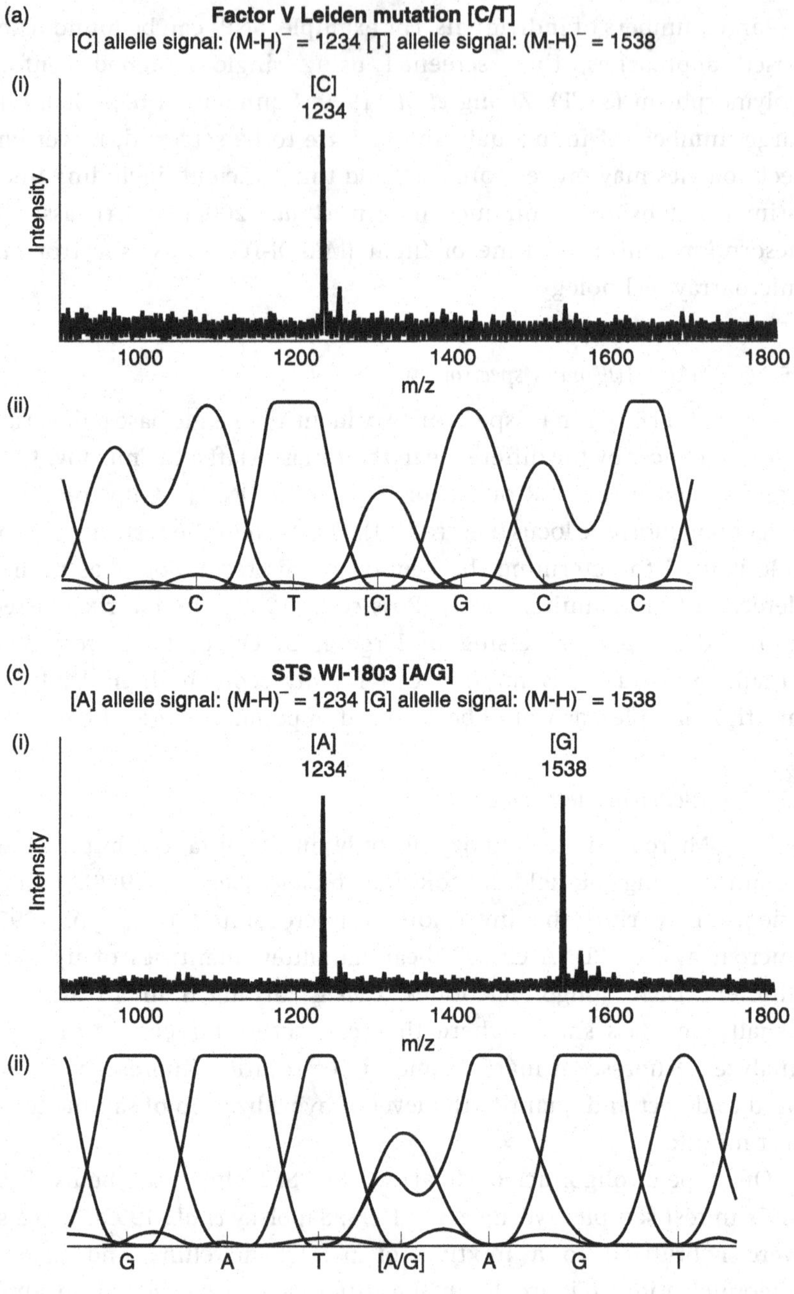

Figure 15.1. Single nucleotide polymorphisms (SNPs) at four loci (i) detected using MALDI-TOF mass spectrometry, and (ii) confirmed by sequence analysis. A and B represent homozygous individuals whereas C and D are heterozygotes (from Griffin *et al.*, 1999).

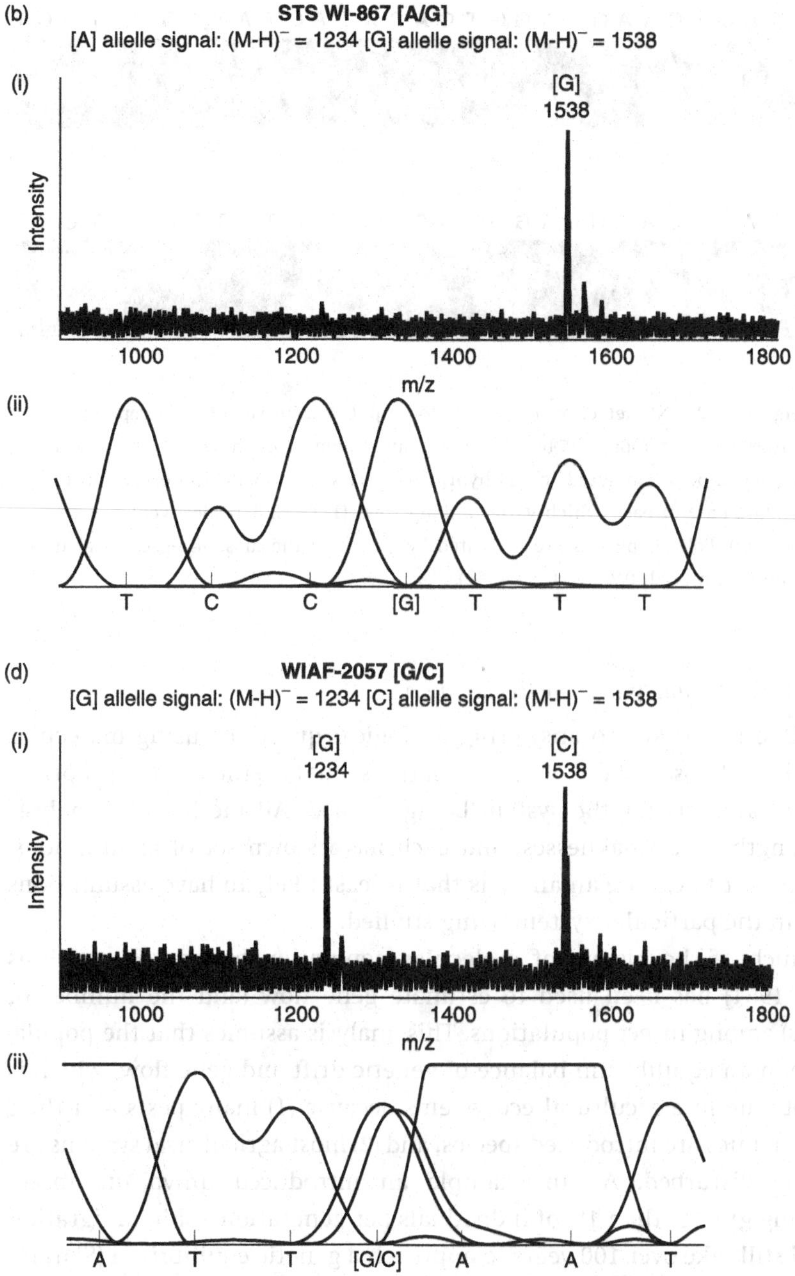
(b)
STS WI-867 [A/G]
[A] allele signal: (M-H)⁻ = 1234 [G] allele signal: (M-H)⁻ = 1538
(i)
[G]
1538
Intensity
1000
1200
1400
1600
1800
m/z
(ii)
T
C
C
[G]
T
T
T
(d)
WIAF-2057 [G/C]
[G] allele signal: (M-H)⁻ = 1234 [C] allele signal: (M-H)⁻ = 1538
(i)
[G]
1234
[C]
1538
Intensity
1000
1200
1400
1600
1800
m/z
(ii)
A
T
T
[G/C]
A
A
T

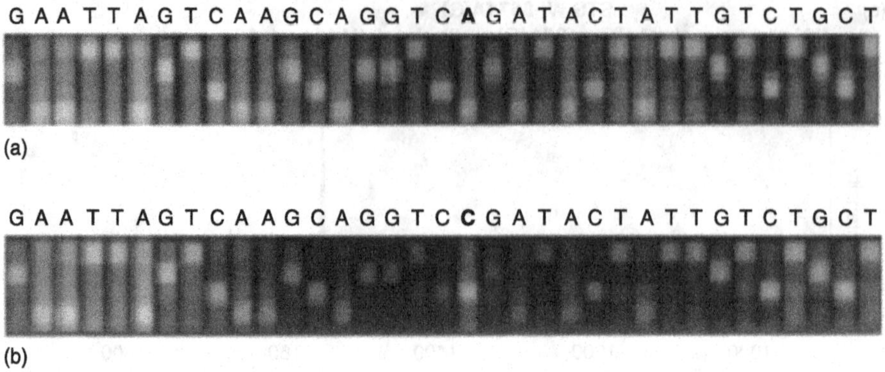

Figure 15.2. SNP detection using a DNA chip. Consecutive columns represent successive overlapping 25-base oligomer subsequences of the known consensus sequence being analyzed in the hybridized test sample, with the exception of the middle base of each which varies among rows (T, G, C, A respectively, top to bottom). Polymorphisms are indicated by a shift in the most intense signal in a column (in bold) (Wang *et al.*, 1998).

15.3.8 *Statistical analysis of molecular data*

One major key to answering ecological questions using molecular markers is to choose not only the best markers, but also the most appropriate method of analysis for the system being studied. All methods of analysis have strengths and weaknesses, and each has its own set of assumptions. It is important to choose an analysis that is least likely to have assumptions violated in the particular system being studied.

For much of the history of molecular population biology, F_{ST} analysis (Wright, 1951) has been used to estimate gene flow (and the number of migrants) among insect populations. This analysis assumes that the populations are in an equilibrium balance of genetic drift and gene flow, which is often not true in agricultural ecosystems because (1) many pests and their natural enemies are introduced species, and (2) most agricultural systems are frequently disturbed. As an example an introduced univoltine insect exchanging greater than 1% of individuals per generation, a high migration rate, will still take over 100 years to approach a genetic equilibrium (Slatkin, 1985). Perturbations like pesticide applications, harvesting of crops, and ploughing of fields result in population bottlenecks and possibly extinction/recolonization cycles that disturb the equilibrium between drift and migration. The time required to return to equilibrium will usually be greater than the time between disturbances, so the populations may never achieve genetic equilibrium. Moreover, because the relationship between migration/gene flow and F_{ST} is non-linear, above about 4–5 genetically effective migrants per generation) even orders of magnitude difference in gene flow cannot be detected (Templeton, 1998).

In recent years, there has been a vast improvement in the variety and quality of statistical methods readily applicable to analyze the types of molecular data that are now readily available and applied in IPM; typically a suite of microsatellites and single copy mtDNA or nuclear marker sequence data. As a result, a wider range of ecological questions can now be addressed. While the following is far from complete, we present outlines of some analytical approaches that are potentially of particular use in IPM. Because IPM generally deals with relatively short-term processes (on human timescales) and is often concerned with generation-by-generation mobility and recent gene flow, the most useful techniques are at the genotypic end of the hierarchy genotypic-genic-genealogical (sensu Sunnucks, 2000). That is, they capture the information held in individual genotypic arrays; in sexual organisms these arrays turn over at every generation (by genetic segregation and usually recombination), hence they hold signals of very recent and finescale population processes such as recent dispersal. Nonetheless, we also canvas a genealogical approach (Nested Clade Analysis) which is able to add a temporal and geographically explicit dimension.

15.3.9 *Assignment tests: extracting the information held in genotypes*

Most population genetics applied to IPM to date has focused on genic analyses, i.e. tracking spatial and temporal changes in frequencies of individual genes. Because genotypes (estimated from arrays of single loci) are reshuffled each generation in sexually reproducing species, the shortest- and finest-scale population processes (e.g. tracking the movements of individuals, relatedness/parentage estimation) can be estimated more effectively using genotypic analyses (Sunnucks, 2000). Assignment tests are a relatively new group of analyses that harness the information held in the associations among loci within individuals, i.e. individual genotypes.

Assignment tests were originally developed to address two forensic applications: (1) testing for human paternity (Chakraborty *et al.*, 1988); and (2) testing if fish were illegally caught (Waser and Strobeck, 1998). Assignment tests are used to determine the likelihood that an individual genotype is derived from various potential sources that must first be identified and characterized. The allele frequencies at each locus must be estimated for each potential source population, which is usually extrapolated using a random sample of the population and assuming Hardy–Weinberg equilibrium. Rannala and Mountain (1997) suggest a more sophisticated approach using Bayesian statistics (review in Shoemaker *et al.*, 1999) to generate posterior probability distributions for the allele frequencies at each locus, thereby removing the confounding influences of sampling error (Cornuet *et al.*, 1999).

Maximum likelihood methods are usually used to compare among sources, either as inclusion or exclusion tests (Cornuet *et al.*, 1999). The likelihood that a genotype is derived from each of the potential sources is calculated for each locus. The overall likelihood is the product of the likelihoods over all loci, thus introducing the assumption of linkage equilibrium. The comparison statistic is usually the difference between the log likelihoods of each population pair (Rannala and Mountain, 1997; Marshall *et al.*, 1998), but the results can also be interpreted graphically (Figure 15.3).

Assignment tests are more effective at identifying the correct source of individuals than methods based on genetic distances (Cornuet *et al.*, 1999), which have been used in the past to determine the source of insect populations (Mendel *et al.*, 1994; Williams *et al.*, 1994; Goldson *et al.*, 1997). Assignment tests have been shown to detect dispersal events when genetic differentiation is not detectable (Castric and Bernatchez, 2004). Assignment tests are appropriate when the individual to be assigned is the immigrant or a recent descendant of the immigrant (Marshall *et al.*, 1998); phylogenetic methods (see below) are more appropriate when identifying the source of

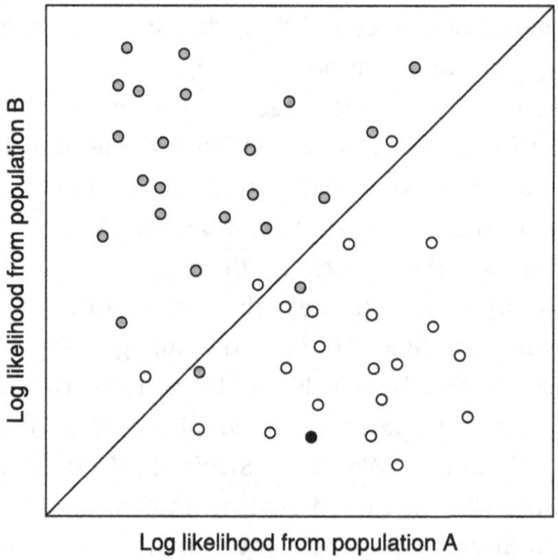

Figure 15.3. Graphical representation of a hypothetical assignment test. Individuals from population A (white) and population B (grey) are plotted based on the overall log likelihood that they were derived from each population. An individual of unknown origin (black) appears to be derived from population A.

historical immigration events (Templeton, 1998). The two methods can sometimes be complementary (Mun *et al.*, 2003). Assignment tests have been used in insect systems, for example to investigate the origin of Mediterranean fruit fly outbreaks in the United States (Davies *et al.*, 1999*b*), and to compare Asian and North American gypsy moth populations (Bogdanowicz *et al.*, 1997). In the latter study, some Asian moths were misassigned to North American populations. Some authors have taken advantage of misassigned individuals by using them as an indicator of the genetic similarity of populations (Paetkau *et al.*, 1995; Haig *et al.*, 1997; Kjaer *et al.*, 1996; Andersen *et al.*, 1997). It is important to note that some misassigned individuals may in fact be immigrants (Paetkau *et al.*, 1995), which was not a reasonable interpretation in the gypsy moth study. If the number of misassigned individuals is balanced in all directions, the resulting dispersal data can be quantitatively similar to mark-recapture studies (Berry *et al.*, 2004).

A major advantage of assignment tests over classical population genetic analyses is that they can be applied more readily to non-equilibrium populations. The level of differentiation among the potential sources (Davies *et al.*, 1999*a*) limits the discriminatory power of the tests, but this limitation can be overcome by examining more loci (Cornuet *et al.*, 1999). Multilocus assignment tests assume no linkage among loci, so if linkage is known multiple single locus assignment tests may be more appropriate (Silva *et al.*, 2003). Accurate estimates of dispersal among non-equilibrium populations should make assignment tests particularly effective for IPM systems. Using these methods, dispersal rates can be compared between sexes (Favre *et al.*, 1997; de Meeus *et al.*, 2002), between morphs, or between individuals possessing different traits.

15.3.10 Spatial autocorrelation

The spatial scales over which genetic similarity changes significantly can be analyzed using spatial autocorrelation analysis. This examines how the value of a variable fluctuates across geographic space by measuring the degree to which the correlation in values changes with increased distance. Statistical significance is assigned by comparing observed population structures with randomized ones generated from the same data set, i.e. Mantel tests (Mantel, 1967). Spatial autocorrelation analysis was originally developed to analyze geological data, but Sokal and Oden (1978) recognized its utility for biological systems. The analysis usually calculates Moran's I-statistic at various distances which results in a graph known as a correlogram (Cliff and Ord, 1981) (Figure 15.4).

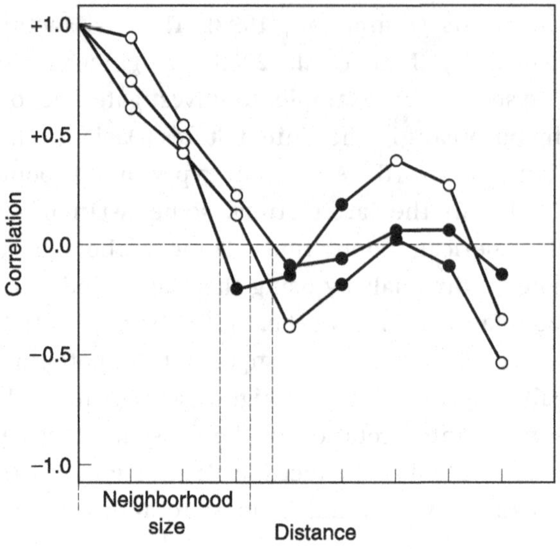

Figure 15.4. Hypothetical correlogram showing spatial autocorrelation analysis of three populations. Open circles represent distances within populations with statistically significant positive or negative autocorrelations. The neighborhood size is defined as the minimum distance at which significant positive autocorrelation disappears.

Spatial autocorrelation of population allele frequency data is more effective at showing differentiation at smaller spatial scales than is F_{ST} (Epperson and Li, 1996). Using allele frequencies of haploid (0, 1) or diploid (0, 0.5, 1) individuals (Heywood, 1991) rather than those of populations improves the discriminatory power of the analysis (Epperson, 1995*a*). Even more discrimination can be accomplished using a join-count statistic (reviewed in Epperson, 1993) instead of Moran's I-statistic. This alternative method compares numbers of each type of single locus genotype pair between samples to generate the correlogram (Epperson, 1995*b*). A user-friendly computer program GenAlEx is freely available for individual-based spatial autocorrelation (Peakall and Smouse, 2001).

Estimating gene flow from spatial autocorrelation, like from F_{ST}, assumes that the populations have approached a genetic equilibrium. Because the time necessary to achieve equilibrium decreases with population size and increased gene flow, equilibrium is approached faster at smaller spatial scales. Simulation studies indicate that equilibrium of spatial autocorrelation statistics occurs at a rate similar to F_{ST}, the rate increasing with increased gene flow and reduced population size (Hardy and Vekemans, 1999). The major advantage of spatial autocorrelation is that population structure is detectable when gene flow is sufficiently high that the populations cannot be distinguished by F_{ST} (Epperson and Li, 1996). Thus, spatial autocorrelation should be effective for estimating insect movement at the smaller spatial scales typical of IPM studies.

In most cases of spatial autocorrelation analysis on insect population genetic data, the insects examined have not been in agricultural ecosystems (Stone and Sunnucksm, 1993; Eber and Brandl, 1994; Chapuisat *et al.*, 2004). De Barro *et al.* (1995*a*) used spatial autocorrelation to compare *S. avenae* populations on wheat and roadside grasses but rather than comparing allele frequencies for each locus generated by their $(GATA)_4$ probe, they used mean population values for the ADBM (average distance of band migration). Robinson *et al.* (2002) used spatial autocorrelation to investigate geographic patterns of clonal diversity of a parthenogenetic earth mite. More recently, spatial autocorrelation analysis was used to demonstrate low levels of phylloxera dispersal, and a tendency to disperse along rows of vines rather than between them (Corrie and Hoffmann, 2004).

Spatial autocorrelation analysis should prove very effective for examining the spatial structure of insect populations (Smouse and Peakall, 1999). Mathematical models available for co-dominant markers can be easily adapted for dominant markers (Smouse and Peakall, 1999), making them useful for RAPDs and AFLPs. This approach to the analysis of spatial autocorrelation has recently been used to estimate gene flow in butterflies (Harper *et al.*, 2003) and mosquitoes (Foley *et al.*, 2004).

15.3.11 *Multivariate analyses*

Many classical analyses of population genetic structure lose much information by reducing data from many loci and individuals into composite measures of genetic difference between populations (e.g. Smouse, 1998). One solution is to use multivariate analyses to uncover patterns in data sets closer to the original individual genotypes. A common procedure is to use an analogue of ANOVA, analysis of molecular variance (AMOVA, Excoffier *et al.*, 1992) using the program ARLEQUIN (Schneider *et al.*, 1997). AMOVA partitions genotypic variation into within and among population components, testing by simulation the significance of differences in the extent and pattern of population differentiation. This kind of analysis is common in molecular ecology, and has been used in studies of pest species (e.g. Downie, 2000).

Analyses that explicitly force data into hierarchical patterns may not be appropriate for some situations. Ordination analyses, part of the family of multivariate analyses including principal component analysis, examine the positions of individuals within a multidimensional space (Lessa, 1990).

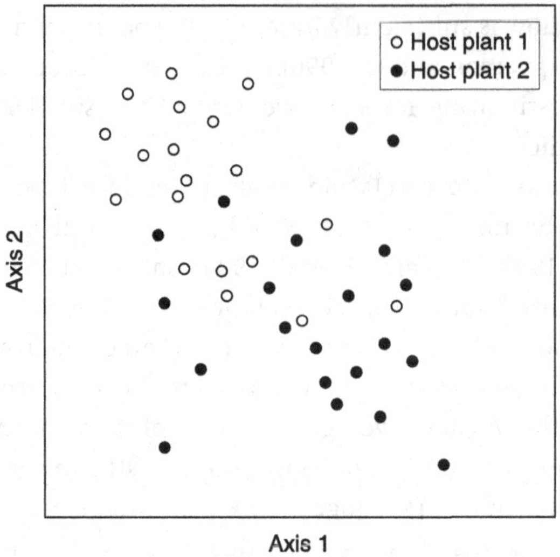

Figure 15.5. Hypothetical example of non-metric multidimensional analysis (NMDS) of various individuals in a population. Individuals are arranged in the two-dimensional space such that the distance between each is proportional to a measured genetic distance. The influence of other biological or ecological attributes (host plant in this example) can then be assessed.

For example, non-metric multidimensional scaling (NMDS, implemented in SYN-TAX, Podani, 1995) positions individuals in (usually) two dimensions to best fit a data matrix of genotypes based on multiple loci (Figure 15.5). NMDS does not assume normality of data nor any particular underlying pattern. This type of analysis is starting to be used more widely, and has been applied to some invading pests (e.g. O'Hanlon *et al.*, 1999, Wilson *et al.*, 1999).

Canonical correspondence analysis is another statistical approach more familiar to ecologists than population geneticists, and has been applied very rarely on insect genetic data (Simon *et al.*, 1999*a*; Brouat *et al.*, 2004). This approach has a number of characteristics that could be very powerful in an IPM context. It is a multivariate approach that allows the estimation of the relative importance of the different environmental factors affecting genetic–demographic patterning in populations (Angers *et al.*, 1999). It also provides a firm framework in which phenotypic and genotypic characteristics of individuals can be assessed simultaneously (Figure 15.6), yielding intuitive plots of character space, and allowing the identification of particular alleles that are predictive of suites of characters (Simon *et al.*, 1999*a*). This last attribute has obvious major implications for predictive modeling of pest dynamics.

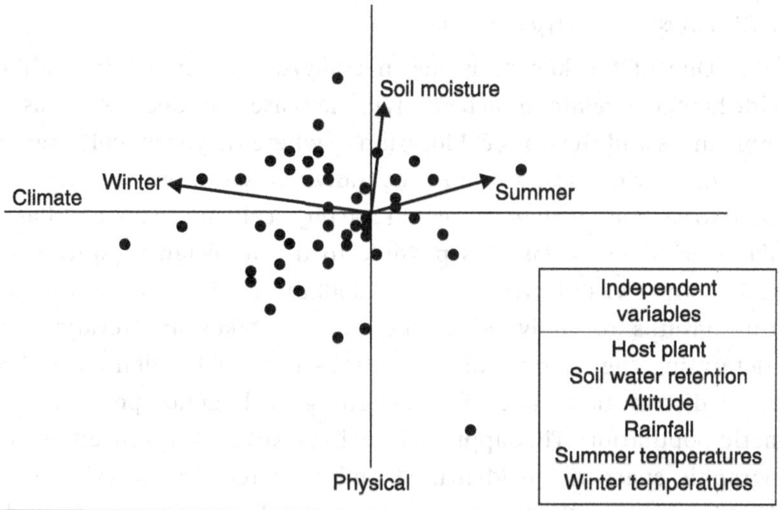

Figure 15.6. Hypothetical example of canonical correspondence analysis (CCA) of various individuals in a population. This example extends the analysis shown in Figure 15.5 by also considering the additional influence of various physical and climatic factors. Two of the factors tested (altitude, rainfall) were found not to influence the distribution of individuals. The arrows indicate the influence of increasing the value of each of the remaining three factors on the position of individual genotypes.

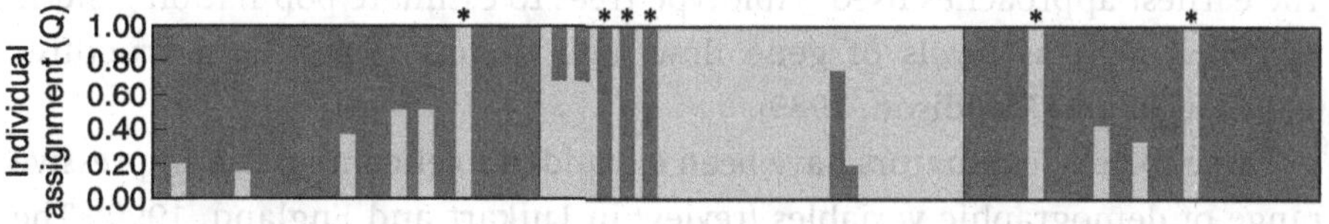

Figure 15.7. A hypothetical example of a Bayesian clustering analysis. The different shades identify those individuals (indicated as bars along the x axis) that group together into genetically discrete populations. Individuals marked with an * are geographically located among individuals of one genetic population, but derive their entire genetic origins from another genetic population and so are more likely to be immigrants. The shade of the bar indicates their most likely source. Individuals identified by bars composed of two shades have a mixed genetic origin and so gives an indication of the degree of mixing, i.e. gene flow between genetic populations. The influence of other biological or ecological attributes that could be driving the genetic structure can then be assessed.

15.3.12 *Bayesian genotypic approaches*

One of the key decisions in analyzing groups of individuals is to decide how they relate to each other. In may cases this decision is based on the sample units and the physical location of where they were collected, but this places an a priori assumption that individuals located together are more related to each other than those that are physically more distant. Rather than make a priori decisions it is possible to use Bayesian clustering methods (e.g. Structure, v2.0; Pritchard *et al.*, 2000) to help identify genetic structure within groups of individuals. The method takes an iterative approach to determine the number of populations present by identifying breaks in genetic distributions and then assigning each genotype to a particular genetic population. This approach enables patterns of movement within a geographic space to be identified and to determine whether movement is leading to an exchange of genetic material. The capacity to understand movement and the implications of dispersal and immigration is essential in understanding the contributions that various elements of a landscape make to pest and beneficial species dynamics.

15.3.13 *Phylogenetic methods*

For most of the history of population genetic analysis, estimates of genetic differentiation have been based on allele frequencies. Harnessing the additional information inherent in genealogies (primarily DNA sequences) has received a great deal of attention recently, because it adds, uniquely, a temporal/historical dimension to genetic analysis (Templeton, 1998). The earliest approaches used haplotype trees to estimate population genetic variables such as levels of gene flow, and to test population structures (e.g. Slatkin and Maddison, 1989).

These sorts of estimators have been expanded to encompass an impressive range of demographic variables (review in Luikart and England, 1999). The next sort of development was the incorporation of estimates of divergence among alleles/haplotypes into genetic distances among samples: e.g. the phi Φ statistics of Excoffier *et al.* (1992). These approaches increase the likelihood of detecting genetic differentiation and increase the precision of its extent. Finally, a number of procedures are being developed that use gene genealogies in a hypothesis-testing framework to infer the processes that have led to current population genetic structures, and even the spatial and temporal juxtaposition of these historical forces. The highest profile of these analyses is nested clade analysis, NCA (Figure 15.8) (implemented by GeoDis, Posada *et al.*, 2000), that is starting to be applied widely in molecular

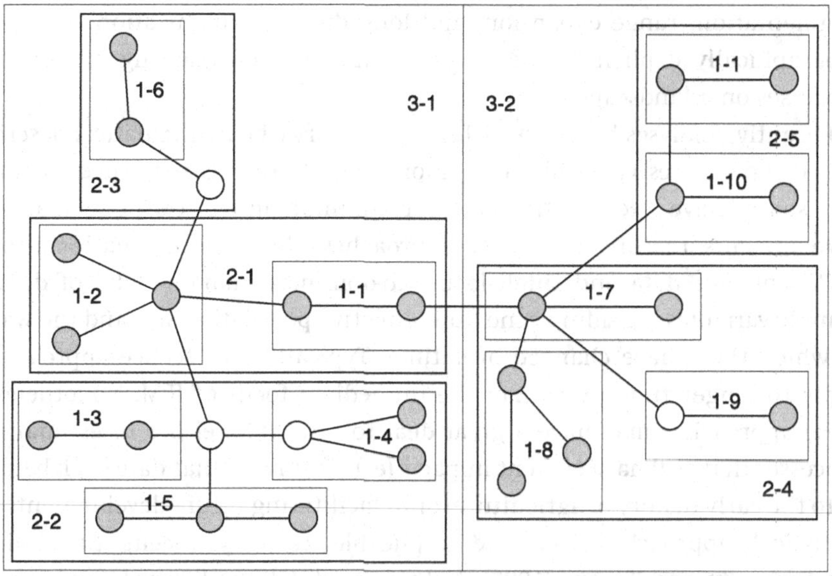

Figure 15.8. Hypothetical example of hierarchical organization of haplotypes in nested clade analysis (NCA). Haplotypes are linked by lines to those differing by only a single mutation, including both those collected (shaded) and uncollected transition haplotypes (open), then are grouped into a hierarchy. The relationships between geographic and genetic distance among haplotypes within and among the different levels of the hierarchy form the basis for hypothesis testing.

ecology (Templeton, 1998) including recent examples with insects (Althoff and Pellmyr, 2002; de Brito *et al.*, 2002; Fairley *et al.*, 2002; Johnson *et al.*, 2002; Schultheis *et al.*, 2002; Turgeon and McPeek, 2002; Contreras-Diaz *et al.*, 2003; Fordyce and Nice, 2003; Contreras-Moya *et al.*, 2004; Kauwe *et al.*, 2004; Monteiro *et al.*, 2004; Segrave and Pellmyr, 2004). This approach can provide valuable information on the history of pest invasions, especially to determine whether new genotypes have been introduced after the pest was already established (Mun *et al.*, 2003). NCA utilizes intuitive concepts, e.g. that under restricted gene flow new variants will tend to be concentrated with the geographical range of the lineage from which they evolved, that range expansion will lead to radiation from the point of origination, and that long distance colonization will result in variants far from their nearest relatives. At present NCA can be applied only to single genes, other than qualitative comparisons of inferences from multiple markers. The major strengths of NCA, particularly in the context of IPM, are that is can distinguish between past historic events such as population fragmentation, range expansion, and long distance colonization, and it is geographically explicit in the sense of allowing the marking of inferred processes on a landscape map.

Recently, analyses based on coalescence theory (the mathematical description of how ancestry within and among populations is patterned in time and space) have become tractable for common usage, owing in part to developments in software. These approaches allow the estimation (from DNA sequence data and single-locus, co-dominant marker data) of demographic variables including gene flow, effective population size and the ways in which these have changed over time. Typically coalescence approaches apply to longer timescales than the immediate focus of IPM. Nonetheless, these approaches may have a great deal to offer IPM, either in estimating processes that still have current impacts (e.g. host races that date back before or to the early history of agriculture) or in facilitating future developments of statistical approaches that are applicable to very recent timescales. See Donnelly and Tavare (1995), Kuhner *et al.* (1998), Edwards and Beerli (2000), Beerli and Felsenstein (2001), Beaumont (2004), Knowles (2004) for introductions to the field.

15.3.14 *Genomics and ecogenomics: molecular approaches to functional and community ecology*

Our focus here has been predominantly on the application of genetic markers and population genetic approaches to understanding population biology. Nonetheless, studies of functional genetic variation make profound contributions to the understanding of pest species. A classic example is the determination of the molecular and chromosomal bases of insecticide resistance discussed earlier.

The profound insights that can be achieved from understanding the role of even single functional genes in single species are well exemplified by the studies of "general protein 9" in the fire ant *Solenopsis invicta* (Kreiger and Ross, 2002). This protein appears to be a pheromone receptor that influences the social behavior of this highly invasive and economically important insect. Polygyne (multi-queen) colonies are rare in the native range of the species (Argentina) but many invasive populations form immensely destructive polygyne supercolonies with quite different individual and group ecology and behavior, owing to genetic changes at this single locus.

Many methods can be applied to producing genetic maps that will allow the exploration of the genetic basis of traits. Approaches include microsatellites (in *Myzus persicae*, Sloane *et al.*, 2001) and RAPDs (in this case screened with SSCP) in a wasp *Bracon hebetor* and a mosquito *Aedes aegypti*

(Antolin *et al.*, 1996), and AFLPs are now commonly applied for genetic mapping (Parsons and Shaw, 2002). As genetic marker availability improves, discovery of the genetic basis of ecologically important traits is becoming more tractable (e.g. pheromone-based sexual isolation in two different strains of European cornborer *Ostrinia nubilalis* (Dopman *et al.*, 2004). A number of pest species are, or will be, the focus of complete genome projects: the sequencing of the genome of the pea aphid is now underway. We can expect an explosion of information about functional and non-functional genetic variation in pest insects in the next few years and certainly in the next decade.

Ecogenomics is used to describe the application of genomics tools to ecology, and is based on "the belief that the interactions among species and their environments can be understood in the same terms as the complex interactions of genes and proteins at the cellular level" (Chapman, 2001). Some of the best examples of "functional" ecogenomics research are coming from agricultural ecosystems, where genomics approaches are being used to investigate how insect herbivore and host plant genes interact, not only to achieve resistance or susceptibility, but also to affect other insects in the community (see review by Ferry *et al.*, 2004). In the best studied of these systems herbivore-induced defences in tobacco have been shown to respond directly to herbivore salivary components (Roda *et al.*, 2004), the response differing between sucking and chewing insects (Voelckel and Baldwin, 2004*a*; Voelckel *et al.*, 2004) and between specialists and generalists (Voelckel and Baldwin, 2004*b*). In addition, the plant response to one herbivore affects how it responds to other herbivores (Voelckel and Baldwin, 2004*a*), the level of competition among herbivores (Kessler and Baldwin, 2004; Kessler *et al.*, 2004), and the interaction between herbivores and their natural enemies (Kessler and Baldwin, 2001; Baldwin *et al.*, 2002). Studies such as these help to explain some of the complexity observed in multitrophic interactions, and may in the future lead to the identification of novel strategies to achieve insect control.

Future ecogenomics research may not be restricted to single species, but may instead target insect guilds or communities. Microbial communities are already being examined using "comparative" ecogenomics methods, whereby microarrays or other genome screening methods quantify species diversity by detecting variability in particular gene sequences (Torsvik and Øvreås, 2002; Bodrossy *et al.*, 2003; Greene and Voordouw, 2003; Stine *et al.*, 2003). Often, genes encoding functional enzymes are used on these microarrays (functional gene arrays, FGAs) to target a particular microbial guild (Zhou, 2003). This technology should be transferable to the analysis of insect guilds or communities in the not too distant future.

"Functional" ecogenomics approaches can also be used to examine the global response of an ecological community to environmental change, and has already been used successfully for microbial communities (Dennis *et al.*, 2002). The success of this method at the moment depends greatly on the level of homology among species for the genes being targeted (Wu *et al.*, 2001), and as such technological advances may be necessary before applications to insect communities can be envisioned. However, the idea of screening for functional variation in order to select a target species for study, rather than vice versa, should be an appealing concept for insect functional ecologists.

15.4 Conclusions

There has been progress in addressing questions relevant to IPM using molecular techniques, but the overall research impact has been less than might have been anticipated. Availability of DNA markers has improved substantially for many taxa, and many IPM questions can be addressed efficiently because of improvements in methods of analysis. Individual-based assignment tests and Bayesian clustering analyses should come into common use for determining the source of recent pest incursions, while historic pest incursions can be examined in retrospect using phylogenetic/geographic methods. Assignment tests and other spatially explicit analyses will provide more meaningful descriptions of insect dispersal in agricultural ecosystems, and realms of influence of particular characters or genes. Spatial autocorrelation and multivariate analyses can provide valuable insight into the distribution of genetic variation in populations that cannot be studied using classical methods, and form a predictive framework that encompasses phenotype and genotype. The question of subspecies or biotype diagnostics will be improved by examining their evolutionary history using phylogenetic methods, and is greatly facilitated by advances in rapid screening of highly variable markers.

The next revolution in molecular ecology is now starting as a result of new technologies that are inexpensive with respect to consumable reagents and allow screenings of populations for single-base pair mutations. However, the low availability of the expensive equipment needed to analyze the samples is currently a limitation. The expected impact of these new technologies on the biological sciences through functional genomics and proteomics research raises the hope that access to equipment will improve. These innovations should dramatically increase the future applications of molecular ecology to IPM research.

References

Abdullahi, I., Winter, S., Atiri, G.I. and Thottappilly, G. (2003). Molecular characterization of whitefly, *Bemisia tabaci* (Hemiptera: Aleyrodidae) populations infesting cassava. *Bulletin of Entomological Research*, **93**, 97–106.

Abdullahi, I., Atiri, G.I., Thottappilly, G. and Winter, S. (2004). Discrimination of cassava-associated *Bemisia tabaci* in Africa from polyphagous populations, by PCR-RFLP of the internal transcribed spacer regions of ribosomal DNA. *Journal of Applied Entomology*, **128**, 81–7.

Aebi, A., Alvarez, N., Butcher, R.D.J. *et al.* (2004). Microsatellite markers in a complex of Horismenus sp. (Hymenoptera: Eulophidae), parasitoids of bruchid beetles. *Molecular Ecology Notes*, **4**, 707–9.

Agusti, N., De Vincente, C. and Gabarra, R. (1999). Development of sequence amplified characterized region (SCAR) markers of *Helicoverpa armigera*: a new polymerase chain reaction-based technique for predator gut analysis. *Molecular Ecology*, **8**, 1467–74.

Agusti, N., Shayler, S.P., Harwood, J.D. *et al.* (2003). Collembola as alternative prey sustaining spiders in arable ecosystems: prey detection within predators using molecular markers. *Molecular Ecology*, **12**, 3467–75.

Aikhionbare, F.O. and Mayo, Z.B. (2000). Mitochondrial DNA sequences of greenbug (Homoptera: Aphididae) biotypes. *Biomolecular Engineering*, **16**, 199–205.

Alberti, A.C., Rodriguero, M.S., Cendra, P.G., Saidman, B.O. and Vilardi, J.C. (2003). Evidence indicating that argentine populations of *Anastrepha fraterculus* (Diptera: Tephritidae) belong to a single biological species. *Annals of the Entomological Society of America*, **95**, 505–12.

Aljanabi, S.M., Loiacono, M.S., Lourenco, R.T., Borges, M. and Tigano, M.S. (1998). RAPD analysis revealing polymorphism in egg parasitoids of soybean stink bugs (Hemiptera; Pentatomidae). *Anais da Sociedade Entomologica do Brasil.*, **27**, 413–20.

Althoff, D.M. and Pellmyr, O. (2002). Examining genetic structure in a bogus yucca moth: a sequential approach to phylogeography. *Evolution*, **56**, 1632–43.

Andersen, L.W., Holm, L.-E., Siegismund, H.R. *et al.* (1997). A combined DNA-microsatellite and isozyme analysis of the population structure of the harbour porpoise in Danish waters and West Greenland. *Heredity*, **78**, 270–6.

Angers, B., Magnan, P., Plante, L. and Bernatchez, M. (1999). Canonical correspondence analysis for estimating spatial and environmental effects on microsatellite gene diversity in brook charr (*Salvelinus fontinalis*). *Molecular Ecology*, **8**, 1043–55.

Anstead, J.A., Burd, J.D. and Shufran, K.A. (2002). Mitochondrial DNA sequence divergence among *Schizaphis graminum* (Hemiptera: Aphididae) clones from cultivated and non-cultivated hosts: haplotype and host associations. *Bulletin of Entomological Research*, **92**, 17–24.

Antolin, M.F., Bosio, C.F., Cotton, J. *et al.* (1996). Intensive linkage mapping in a wasp (*Bracon hebetor*) and a mosquito (*Aedes aegypti*) with single-strand conformation polymorphism analysis of Random Amplified Polymorphic DNA markers. *Genetics*, **143**, 1727–38.

Avise, J.C., Lansman, R.A. and Shade, R.O. (1979). The use of restriction endonucleases to measure mitochondrial DNA sequence relatedness in natural populations. I. Population structure and evolution in the genus Peromyscus. *Genetics*, **92**, 279–95.

Avise, J.C. (1994). *Molecular Markers, Natural History and Evolution*. London: Chapman and Hall.

Baer, C.F., Tripp, D.W., Bjorksten, T.A. and Antolin, M.F. (2004). Phylogeography of a parasitoid wasp (*Diaeretiella rapae*): no evidence of host-associated lineages. *Molecular Ecology*, **13**, 1859–69.

Baker, D.A., Loxdale, H.D. and Edwards, O.R. (2004). Genetic variation and founder effects in the parasitoid wasp, *Diaeretiella rapae* (M'intosh) (Hymenoptera: Braconidae: Aphidiidae), affecting its potential as a biological control agent. *Molecular Ecology*, **12**, 3303–11.

Baldwin, I.T., Kessler, A. and Halitschke, R. (2002). Volatile signaling in plant–plant–herbivore interactions: what is real? *Current Opinion in Plant Biology*, **5**, 351–4.

Baliraine, F.N., Bonizzoni, M., Guglielmino, C.R. *et al.* (2004). Population genetics of the potentially invasive African fruit fly species, *Ceratitis rosa* and *Ceratitis fasciventris* (Diptera: Tephritidae). *Molecular Ecology*, **13**, 683–95.

Baliraine, F.N., Bonizzoni, M., Osir, E.O. *et al.* (2003). Comparative analysis of microsatellite loci in four fruit fly species of the genus *Ceratitis* (Diptera: Tephritidae). *Bulletin of Entomological Research*, **93**, 1–10.

Bang-Ce, Y., Peng, Z., Bincheng, Y. and Songyang, L. (2004). Estimation of relative allele frequencies of single-nucleotide polymorphisms in different populations by microarray hybridization of pooled DNA. *Analytical Biochemistry*, **333**, 72–8.

Barrette, R.J., Crease, T.J. and Hebert, P.D.N. (1994). Mitochondrial DNA diversity in the pea aphid *Acyrthosiphon pisum*. *Genome*, **37**, 858–65.

Beaumont, M.A. (2004). Recent developments in genetic data analysis: what can they tell us about human demographic history? *Heredity*, **92**, 365–79.

Beerli, P. and Felsenstein, T.J. (2001). Maximum likelihood estimation of a migration matrix and effective population sizes in n subpopulations by using a coalescent approach. *Proceedings of the National Academy of Sciences USA*, **98**, 4563–8.

Berry, S.D., Fondong, V.N., Rey, C., Rogan, D., Fauquet, C.M. and Brown, J.K. (2004*a*). Molecular evidence for five distinct *Bemisia tabaci* (Homoptera: Aleyrodidae) geographic haplotypes associated with cassava plants in sub-Saharan Africa. *Annals of the Entomological Society of America*, **97**, 852–9.

Berry, O., Tocher, M.D. and Sarre, S.D. (2004*b*). Can assignment tests measure dispersal? *Molecular Ecology*, **13**, 551–61.

Black, W.C. IV. (1993). PCR with arbitrary primers: approach with care. *Insect Molecular Biology*, **2**, 1–6.

Blackman, R.L., Watson, G.W. and Ready, P.D. (1995). The identity of the African pine woolly aphid: a multidisciplinary approach. *Bulletin OEPP*, **25**, 337–41.

Bodrossy, L., Stralis-Pavese, N., Murrell, J.C. *et al.* (2003). Development and validation of a diagnostic microbial microarray for methanotrophs. *Environmental Microbiology*, **5**, 566–82.

Bogdanowicz, S.M., Wallner, W.E., Bell, J., Odell, T.M. and Harrison, R.G. (1993). Asian gypsy moths (Lepidoptera: Lymantriidae) in North America: evidence from molecular data. *Annals of the Entomological Society of America*, **86**, 710–15.

Bogdanowicz, S.M., Mastro, V.C., Prasher, D.C. and Harrison, R.G. (1997). Microsatellite DNA variation among Asian and North American gypsy moths (Lepidoptera: Lymantriidae). *Annals of the Entomological Society of America*, **90**, 768–75.

Bonizzoni, M., Katsoyannos, B.I., Marguerie, R. *et al.* (2003). Microsatellite analysis reveals remating by wild Mediterranean fruit fly females, *Ceratitis capitata*. *Molecular Ecology*, **11**, 1915–21.

Bonizzoni, M., Guglielmino, C.R., Smallridge, C.J. *et al.* (2004). On the origins of medfly invasion and expansion in Australia. *Molecular Ecology*, **13**, 3845–55.

Borghuis, A., Pinto, J.D., Platner, G.R. and Stouthamer, R. (2004). Partial cytochrome oxidase II sequences distinguish the sibling species *Trichogramma minutum* Riley and *Trichogramma platneri* Nagarkatti. *Biological Control*, **30**, 90–4.

Bournoville, R., Simon, J.C., Badenhausser, I. *et al.* (2000). Clones of pea aphid, *Acyrthosiphon pisum* (Hemiptera: Aphididae) distinguished using genetic markers, differ in their damaging effect on a resistant alfalfa cultivar. *Bulletin of Entomological Research*, **90**, 33–9.

Brodeur, J. and Rosenheim, J.A. (2000). Intraguild interactions in aphid parasitoids. *Entomologia Experimentalis et Applicata*, **97**, 93–108.

Brouat, C., Chevallier, H., Meusnier, S., Noblecourt, T. and Rasplus, J.-Y. (2004). Specialization and habitat: spatial and environmental effects on abundance and genetic diversity of forest generalist and specialist *Carabus* species. *Molecular Ecology*, **13**, 1815–26.

Brown, S., McLaughlin, W., Jerez, I.T. and Brown, J.K. (2002). Identification and distribution of *Bemisia tabaci* (Gennadius) (Homoptera: Aleyrodidae) haplotypes in Jamaica. *Tropical Agriculture*, **79**, 140–9.

Brunner, P.C. and Frey, J.E. (2004). Isolation and characterization of six polymorphic microsatellite loci in the western flower thrips *Frankliniella occidentalis* (Insecta: Thysanoptera). *Molecular Ecology Notes*, **4**, 599–601.

Brunner, P.C., Chatzivassiliou, E.K., Katis, N.I. and Frey, J.E. (2004). Host-associated genetic differentiation in *Thrips tabaci* (Insecta: Thysanoptera), as determined from mtDNA sequence data. *Heredity*, **93**, 364–70.

Caillaud, M.C., Mondor-Genson, G., Levine-Wilkinson, S. *et al.* (2004). Microsatellite DNA markers for the pea aphid *Acyrthosiphon pisum*. *Molecular Ecology Notes*, **4**, 446–8.

Calvert, L.A., Cuervo, M., Arroyave, J.A. *et al.* (2001). Morphological and mitochondrial DNA marker analyses of whiteflies (Homoptera: Aleyrodidae) colonizing cassava and beans in Colombia. *Annals of the Entomological Society of America*, **94**, 512–19.

Carvalho, G.R., Maclean, N., Wratten, S.D., Carter, R.E. and Thurston, J.P. (1991). Differentiation of aphid clones using DNA fingerprints from individual aphids. *Proceedings of the Royal Society London B series*, **243**, 109–14.

Carvalho, G.R. (ed.) (1998). *Advances in Molecular Ecology*. IOS Press, Amsterdam.

Casey, D.G. and Burnell, A.M. (2001). The isolation of microsatellite loci in the Mediterranean fruitfly *Ceratitis capitata* (Diptera: Tephritidae) using a biotin/streptavidin enrichment technique. *Molecular Ecology Notes*, **1**, 120–2.

Castric, V. and Bernatchez, L. (2004). Individual assignment test reveals differential restriction to dispersal between two salmonids despite no increase of genetic differences with distance. *Molecular Ecology*, **13**, 1299–312.

Chakrabarti, S., Kambhampati, S., Grace, T. and Zurek, L. (2004). Characterization of microsatellite loci in the house fly, *Musca domestica* L. (Diptera: Muscidae). *Molecular Ecology Notes*, **4**, 728–30.

Chakraborty, R., Meagher, T.R. and Smouse, P.E. (1988). Parentage analysis with genetic markers in natural populations. I. The expected proportion of offspring with unambiguous paternity. *Genetics*, **52**, 922–7.

Chang, S.C., Hu, N.T., Hsin, C.Y. and Sun, C.N. (2001). Characterization of differences between two Trichogramma wasps by molecular markers. *Biological Control*, **21**, 75–8.

Chang, W.X.Z., Tabashnik, B.E., Artelt, B. *et al.* (1997). Mitochondrial DNA sequence variation among geographic strains of diamondback moth (Lepidoptera: Plutellidae). *Annals of the Entomological Society of America*, **90**, 590–5.

Chapman, R.W. (2001). EcoGenomics – a consilience for comparative immunology? *Developmental and Comparative Immunology*, **25**, 549–51.

Chapuisat, M., Bocherens, S. and Rosset, H. (2004). Variable queen number in ant colonies: no impact on queen turnover, inbreeding, and population genetic differentiation in the ant *Formica selysi*. *Evolution*, **58**, 1064–72.

Cho, R.J., Mindrinos, M., Richards, D.R. *et al.* (1999). Genome-wide mapping with biallelic markers in *Arabidopsis thaliana*. *Nature Genetics*, **23**, 203–7.

Clark, T.L. and Hibbard, B.E. (2004). Comparison of nonmaize hosts to support western corn rootworm (Coleoptera: Chrysomelidae) larval biology. *Environmental Entomology*, **33**, 681–9.

Cliff, A.D. and Ord, J.K. (1981). *Spatial Processes. Models and Applications*. London: Pion Ltd.

Coates, B.S., Sumerford, D.V. and Hellmich, R.L. (2004). Geographic and voltinism differentiation among North American *Ostrinia nubilalis* (European corn borer) mitochondrial cytochrome c oxidase haplotypes, *Journal of Insect Science*, **4**, 35.

Cognato, A.I., Harlin, A.D. and Fisher, M.L. (2003). Genetic structure among pinyon pine beetle populations (Scolytinae: *Ips confusus*). *Environmental Entomology*, **32**, 1262–70.

Contreras-Diaz, H.G., Moya, O., Oromi, P. and Juan, C. (2003). Phylogeography of the endangered darkling beetle species of Pimelia endemic to Gran Canaria (Canary Islands). *Molecular Ecology*, **12**, 2131–43.

Cornuet, J.M., Piry, S., Luikart, G., Estoup, A. and Solignac, M. (1999). New methods employing multilocus genotypes to select or exclude populations as origins of individuals. *Genetics*, **153**, 1989–2000.

Corrie, A.M., Crozier, R.H., Van Heeswijck, R. and Hoffmann, A.A. (2002). Clonal reproduction and population genetic structure of grape phylloxera, *Daktulosphaira vitifoliae*, in Australia. *Heredity*, **88**, 203–11.

Corrie, A.M., van Heeswijck, R. and Hoffmann, A.A. (2003). Evidence for host-associated clones of grape phylloxera *Daktulosphaira vitifoliae* (Hemiptera: Phylloxeridae) in Australia. *Bulletin of Entomological Research*, **93**, 193–201.

Corrie, A.M. and Hoffmann, A.A. (2004). Fine-scale genetic structure of grape phylloxera from the roots and leaves of Vitis. *Heredity*, **92**, 118–27.

Cuthbertson, A.G.S., Fleming, C.C. and Murchie, A.K. (2003). Detection of *Rhopalosiphum insertum* (apple-grass aphid) predation by the predatory mite *Anystis baccarum* using molecular gut analysis. *Agricultural and Forest Entomology*, **5**, 219–25.

D'Acier, A.C., Sembene, M., Audiot, P. and Rasplus, J.Y. (2004). Polymorphic microsatellites loci in the black Aphid, *Aphis fabae* Scopoli, 1763 (Hemiptera: Aphididae). *Molecular Ecology Notes*, **4**, 306–8.

Dai, S.M., Lin, C.C. and Chang, C. (2004). Polymorphic microsatellite DNA markers from the oriental fruit fly *Bactrocera dorsalis* (Hendel). *Molecular Ecology Notes*, **4**, 629–31.

Davies, N., Villablanca, F.X. and Roderick, G.K. (1999*a*). Determining the source of individuals: multilocus genotyping in nonequilibrium population genetics. *Trends in Ecology and Evolution*, **14**, 17–21.

Davies, N., Villablanca, F.X. and Roderick, G.K. (1999*b*). Bioinvasions of the medfly *Ceratitis capitata:* Source estimation using DNA sequences at multiple intron loci. *Genetics*, **153**, 351–60.

De Barro, P.J. and Driver, F. (1997). Use of RAPD PCR to distinguish the B biotype from other biotypes of *Bemisia tabaci* (Gennadius) (Hemiptera: Aleyrodidae). *Australian Journal of Entomology*, **36**, 149–52.

De Barro, P.J., Driver, F., Naumann, I.D. *et al.* (2000*a*). Descriptions of three species of *Eretmocerus* Haldeman (Hymenoptera: Aphelinidae) parasitising *Bemisia tabaci* (Gennadius) (Hemiptera: Aleyrodidae) and *Trialeurodes vaporariorum* (Westwood) (Hemiptera: Aleyrodidae) in Australia based on morphological and molecular data. *Australian Journal of Entomology*, **39**, 259–69.

De Barro, P.J., Driver, F., Trueman, J.W.H. and Curran, J. (2000*b*). Phylogenetic relationship of world populations of *Bemisia tabaci* (Gennadius) using ribosomal ITS1. *Molecular Phylogeny and Evolution*, **16**, 29–36.

De Barro, P.J. and Hart, P.J. (2000). Mating interactions between two biotypes of the whitefly, *Bemisia tabaci* (Hemiptera: Aleyrodidae) in Australia. *Bulletin of Entomological Research*, **90**, 103–12.

De Barro, P.J., Scott, K.D., Graham, G.C., Lange, C.L. and Schutze, M.K. (2003). Isolation and characterization of microsatellite loci in *Bemisia tabaci*. *Molecular Ecology Notes*, **3**, 40–3.

De Barro, P.J., Sherratt, T.N., Brookes, C.P., Markovic, O. and Maclean, N. (1995*c*). Spatial and temporal genetic variation in the population structure of the grain aphid, *Sitobion avenae* (F.) (Hemiptera: Aphididae) studied using RAPD-PCR. *Proceedings of the Royal Society, Series B, London*, **262**, 321–7.

De Barro, P.J., Sherratt, T.N., Carvalho, G.R. *et al.* (1994). An analysis of secondary spread by putative clones of *Sitobion avenae* within a Hampshire wheat field using the multilocus $(GATA)_4$ probe. *Insect Molecular Biology*, **3**, 253–60.

De Barro, P.J., Sherratt, T.N., Carvalho, G.R. *et al.* (1995*a*). Geographic and microgeographic genetic differentiation in two aphid species over southern England using the multilocus $(GATA)_4$ probe. *Molecular Ecology*, **4**, 375–82.

De Barro, P.J., Sherratt, T.N., Markovic, O. and Maclean, N. (1995*b*). An investigation of the differential performance of clones of the cereal aphid, *Sitobion avenae* on two host species. *Oecologia*, **104**, 375–85.

De Barro, P.J., Trueman, J.W.H. and Frohlich, D.R. (2005). *Bemisia argentifolii* is a race of *B. tabaci*: the molecular genetic differentiation of *B. tabaci* populations around the world. *Bulletin of Entomological Research*, **14**, 3695–718.

de Brito, R.A., Manfrin, M.H. and Sene, F.M. (2002). Nested cladistic analysis of Brazilian populations of *Drosophila serido*. *Molecular Phylogenetics and Evolution*, **22**, 131–43.

de Leon, J.H. and Jones, N.A. (2004). Detection of DNA Polymorphisms in *Homalodisca coagulata* (Homoptera: Cicadellidae) by polymerase chain reaction-based DNA fingerprinting methods. *Annals of the Entomological Society of America*, **97**, 574–85.

de Meeus, T., Beati, L., Delaye, C., Aeschlimann, A. and Renaud, F. (2002). Sex-biased genetic structure in the vector of Lyme disease, *Ixodes ricinus*. *Evolution*, **56**, 1802–7.

Delmotte, F., Leterme, N., Gauthier, J.P., Rispe, C. and Simon, J.C. (2002). Genetic architecture of sexual and asexual populations of the aphid *Rhopalosiphum padi* based on allozyme and microsatellite markers. *Molecular Ecology*, **11**, 711–23.

Dennis, P., Edwards, E.A., Liss, S.N. and Fulthorpe, R. (2002). Monitoring gene expression in mixed microbial communities by using DNA microarrays. *Applied and Environmental Microbiology*, **69**, 769–78.

Donnelly, P. and Tavare, S. (1995). Coalescents and genealogical structure under neutrality. *Annual Review of Genetics*, **29**, 401–21.

Dopman, E.B., Bogdanowicz, S.M. and Harrison, R.G. (2004). Genetic mapping of sexual isolation between E and Z pheromone strains of the European corn borer (*Oshinia nubilalis*). *Genetics*, **167**, 301–9.

Downie, D.A. (2000). Patterns of genetic variation in the native grape phylloxera on two sympatric host species. *Molecular Ecology*, **9**, 505–14.

Downie, D.A., Fisher, J.R. and Granett, J. (2001). Grapes, galls, and geography: the distribution of nuclear and mitochondrial DNA variation across host-plant species and regions in a specialist herbivore. *Evolution*, **55**, 1345–62.

Eber, S. and Brandl, R. (1994). Ecological and genetic spatial patterns of *Urophora cardui* (Diptera: Tephritidae) as evidence for population structure and biogeographical processes. *Journal of Animal Ecology*, **63**, 187–99.

Edwards, O.R. and Hopper, K.R. (1999). Using superparasitism by a stem borer parasitoid to infer a host refuge. *Ecological Entomology*, **24**, 7–12.

Edwards, O.R. and Hoy, M.A. (1995*a*). Monitoring laboratory and field biotypes of the walnut aphid parasite, *Trioxys pallidus*, in population cages using RAPD-PCR. *Biocontrol Science and Technology*, **5**, 313–27.

Edwards, O.R. and Hoy, M.A. (1995*b*). Random amplified polymorphic DNA markers to monitor laboratory-selected, pesticide-resistant *Trioxys pallidus* (Hymenoptera: Aphidiidae) after release into three California walnut orchards. *Environmental Entomology*, **24**, 487–96.

Edwards, S.V. and Beerli, P. (2000). Perspective: gene divergence, population divergence, and the variance in coalescence time in phylogeographic studies. *Evolution*, **54**, 1839–54.

Ehtesham, N.Z., Bentur, J.S. and Bennett, J. (1995). Highly repetitive DNA sequence elements from *Orseolia oryzae* (Wood–Mason) discriminate between the Indian isolates of the Asian rice gall midge and the paspalum midge. *Electrophoresis*, **16**, 1762–5.

Epperson, B.K. (1993). Spatial and space–time correlations in systems of subpopulations with genetic drift and migration. *Genetics*, **133**, 711–27.

Epperson, B.K. (1995*a*). Spatial structure of two-locus genotypes under isolation by distance. *Genetics*, **140**, 365–75.

Epperson, B.K. (1995*b*). Fine-scale spatial structure – correlations for individual genotypes differ from those for local gene frequencies. *Evolution*, **49**, 1022–6.

Epperson, B.K. and Li, T. (1996). Measurement of genetic structure within populations using Moran's spatial autocorrelation statistics. *Proceeding of the National Academy of Sciences USA*, **93**, 10528–32.

Excoffier, L., Smouse, P.E. and Quattro, J. (1992). Analysis of molecular variance from metric distance among DNA haplotypes: application to human mitochondrial DNA restriction data. *Genetics*, **131**, 479–91.

Evans, J.D., Pettis, J.S., Hood, W.M. and Shimanuki, H. (2003). Tracking an invasive honey bee pest: mitochondrial DNA variation in North American small hive beetles. *Apidologie*, **34**, 103–9.

Fairley, T.L., Povoa, M.M. and Conn, J.E. (2002). Evaluation of the Amazon River delta as a barrier to gene flow for the regional malaria vector, *Anopheles aquasalis* (Diptera: Culicidae) in northeastern Brazil. *Journal of Medical Entomology*, **39**, 861–9.

Favre, L., Balloux, F., Goudet, J. and Perrin, N. (1997). Female-biased dispersal in themonogamous mammal *Crocidura russula*: evidence from field data and microsatellite patterns. *Proceedings of the Royal Society of London Series B*, **264**, 127–32.

Fenton, B., Malloch, G., Jones, A.T. *et al.* (1995). Species identification of Cecidophyopsis mites (Acari: Eriophyidae) from different *Ribes* species and countries using molecular genetics. *Molecular Ecology*, **4**, 383–7.

Fenton, B., Jones, A.T., Malloch, G. and Thomas, W.P. (1996). Molecular Ecology of some Cecidophyopsis mites (Acari: Eriophyidae) on *Ribes* species and evidence for their natural cross colonization of blackcurrant (*R. nigrum*). *Annals of Applied Biology*, **128**, 405–14.

Fenton, B., Malloch, G., Navajas, M., Hillier, J. and Birch, A.N.E. (2003). Clonal composition of the peach-potato aphid *Myzus persicae* (Homoptera: Aphididae) in France and Scotland: comparative analysis with IGS fingerprinting and microsatellite markers. *Annals of Applied Biology*, **142**, 255–67.

Ferry, N., Edwards, M.G., Gatehouse, J.A. and Gatehouse, A.M.R. (2004). Plant–insect interactions: molecular approaches to insect resistance. *Current Opinions in Biotechnology*, **15**, 155–61.

Field, L.M., Crick, S.E. and Devonshire, A.L. (1996). Polymerase chain reaction-based identification of insecticide resistance genes and DNA methylation in the aphid *Myzus persicae* (Sulzer). *Insect Molecular Biology*, **5**, 197–202.

Figueroa, C.C., Simon, J.-C., Le Gallic, J.-F. *et al.* (2005). Genetic structure and clonal diversity of an introduced pest in Chile, the cereal aphid Sitobion avenae. *Heredity*, **95**, 24–33.

Foley, D.H., Russell, R.C. and Bryan, J.H. (2004). Population structure of the peridomestic mosquito *Ochlerotatus notoscriptus* in Australia. *Medical and Veterinary Entomology*, **18**, 180–90.

Fordyce, J.A. and Nice, C.C. (2003). Contemporary patterns in a historical context: phylogeographic history of the pipevine swallowtail, *Battus philenor* (Papilionidae). *Evolution*, **57**, 1089–99.

Fritz, A.H. and Schable, N. (2004). Microsatellite loci from the Caribbean fruit fly, *Anastrepha suspensa* (Diptera: Tephritidae). *Molecular Ecology Notes*, **4**, 443–5.

Frohlich, D.R., Torres-Jerez, I., Bedford, I.D., Markham, P.G. and Brown, J.K. (1999). A phylogeographical analysis of the *Bemisia tabaci* species complex based on mitochondrial DNA markers. *Molecular Ecology*, **8**, 1867–77.

Fuentes-Contreras, E., Figueroa, C.C., Reyes, M., Briones, L.M. and Niemeyer, H.M. (2004). Genetic diversity and insecticide resistance of *Myzus persicae* (Hemiptera: Aphididae) populations from tobacco in Chile: evidence for the existence of a single predominant clone. *Bulletin of Entomological Research*, **94**, 11–18.

Fuller, S.J., Chavigny, P., Lapchin, L. and Vanlerberghe-Masutti, F. (1999). Variation in clonal diversity in glasshouse infestations of the aphid, *Aphis gossypii* Glover in southern France. *Molecular Ecology*, **8**, 1867–77.

Garner, K.J. and Slavicek, J.M. (1996). Identification and characterization of a RAPD-PCR marker for distinguishing Asian and North American gypsy moths. *Insect Molecular Biology*, **5**, 81–91.

Gaublomme, E., Dhuyvetter, H., Verdyck, P. *et al.* (2003). Isolation and characterization of microsatellite loci in the ground beetle *Carabus problematicus* (Coleoptera: Carabidae). *Molecular Ecology Notes*, **3**, 341–3.

Gauthier, N. and Rasplus, J.Y. (2004). Polymorphic microsatellite loci in the coffee berry borer, *Hypothenemus hampei* (Coleoptera, scolytidae). *Molecular Ecology Notes*, **4**, 294–6.

Gawel, N.J. and Bartlett, A.C. (1993). Characterization of differences between whiteflies using RAPD-PCR. *Insect Molecular Biology*, **2**, 33–8.

Giblin-Davis, R.M., Gries, R., Crespi, B. *et al.* (2000). Aggregation pheromones of two geographical isolates of the New Guinea sugarcane weevil, *Rhabdoscelus obscurus*. *Journal of Chemical Ecology*, **26**, 2763–80.

Gilchrist, A.S., Sved, J.A. and Meats, A. (2004). Genetic relations between outbreaks of the Queensland fruit fly, *Bactrocera tryoni* (Froggatt) (Diptera: Tephritidae), in Adelaide in 2000 and 2002. *Australian Journal of Entomology*, **43**, 157–63.

Gilles, J., Litrico, I., Sourrouille, P. and Duvallet, G. (2004). Microsatellite DNA markers for the stable fly, *Stomoxys calcitrans* (Diptera: Muscidae). *Molecular Ecology Notes*, **4**, 635–7.

Goldson, S.L., Phillips, C.B., McNeill, M.R. and Barlow, N.D. (1997). The potential of parasitoid strains in biological control: observations to date on *Microctonus* spp. intraspecific variation in New Zealand. *Agriculture, Ecosystems Environment*, **64**, 115–24.

Gomulski, L.M., Bourtzis, K., Brogna, S. *et al.* (1998). Intron size polymorphism of the Adh1 gene parallels the worldwide colonization history of the Mediterranean fruit fly, *Ceratitis capitata*. *Molecular Ecology*, **7**, 1729–41.

Graur, D. (1985). Gene diversity in Hymenoptera. *Evolution*, **39**, 190–9.

Greene, E.A. and Voordouw, G. (2003). Analysis of environmental microbial communities by reverse sample genome probing. *Journal of Microbiological Methods*, **53**, 211–19.

Griffin, T.J., Hall, J.G., Prudent, J.R. and Smith, L.M. (1999). Direct genetic analysis by matrix-assisted laser desorption/ionization mass spectrometry. *Proceedings of the National Academy of Sciences USA*, **96**, 6301–6.

Guillemaud, T., Mieuzet, L. and Simon, J.C. (2003). Spatial and temporal genetic variability in French populations of the peach-potato aphid, *Myzus persicae*. *Heredity*, **91**, 143–52.

Haack, L., Gauthierm, J.-P., Plantegenest, M. and Dedryver, C.-A. (2000). Predominance of generalist clones in a cyclically parthenogenetic organism evidenced by combined demographic and genetic analyses. *Molecular Ecology*, **9**, 2055–66.

Haff, L.A. and Smirnov, I.P. (1997). Single-nucleotide polymorphism identification assays using a thermostable DNA polymerase and delayed extraction MALDI-TOF mass spectrometry. *Genome Research*, **7**, 378–88.

Haig, S.M., Gratto-Trevor, C.L., Mullins, T.D. and Colwell, M.A. (1997). Population identification of western hemisphere shorebirds throughout the annual cycle. *Molecular Ecology*, **6**, 412–27.

Hales, D.F., Sloane, M.A., Wilson, A.C.C. and Sunnucks, P. (2002*a*). Segregation of autosomes during spermatogenesis in the peach-potato aphid (*Myzus persicae*) (Sulzer), (Hemiptera: Aphididae). *Genetical Research*, **79**, 119–27.

Hales, D.F., Sloane, M.A., Wilson, A.C.C. *et al.* (2002*b*). A lack of detectable genetic recombination on the X chromosome during the parthenogenetic production of female and male aphids. *Genetical Research*, **79**, 203–9.

Hardy, O.J. and Vekemans, X. (1999). Isolation by distance in a continuous population: reconciliation between spatial autocorrelation analysis and population genetics models. *Heredity*, **83**, 145–54.

Harper, G.L., Maclean, N. and Goulson, D. (2003). Microsatellite markers to assess the influence of population size, isolation and demographic change on the genetic structure of the UK butterfly *Polyommatus bellargus*. *Molecular Ecology*, **12**, 3349–57.

Haymer, D.S., McInnis, D.O. and Arcangeli, L. (1992). Genetic variation between strains of the Mediterranean fruit fly, *Ceratitis capitata*, detected by DNA fingerprinting. *Genome*, **35**, 528–33.

Haymer, D.S., He, M. and McInnis, D.O. (1997). Genetic marker analysis of spatial and temporal relationships among existing populations and new infestations of the Mediterranean fruit fly (*Ceratitis capitata*). *Heredity*, **97**, 302–9.

He, M. and Haymer, D.S. (1999). Genetic relationships of populations and the origins of new infestations of the Mediterranean fruit fly. *Molecular Ecology*, **8**, 1247–57.

Heckel, D.G., Gahan, L.J., Tabashnik, B.E. and Johnson, M.W. (1995). Randomly amplified polymorphic DNA differences between strains of diamondback moth (Lepidoptera: Plutellidae) susceptible or resistant to *Bacillus thuringiensis*. *Annals of the Entomological Society of America*, **88**, 531–7.

Heywood. (1991). Spatial analysis of genetic variation in plant populations. *Annual Review of Ecology and Systematics*, **22**, 335–55.

Hibbard, B.E., Higdon, M.L., Duran, D.P., Chweikert, Y.M. and Ellersieck, M.R. (2004). Role of egg density on establishment and plant-to-plant movement by western corn rootworm larvae (Coleoptera: Chrysomelidae). *Journal of Economic Entomology*, **97**, 871–82.

Hillis, D.M., Moritz, C. and Mable, B.K. (1996). *Molecular Systematics*. Sinauer Associates, Sunderland MA.

Hinomoto, N., Muraji, M., Noda, T., Shimizu, T. and Kawasaki, K. (2004). Identification of five Orius species in Japan by multiplex polymerase chain reaction. *Biological Control*, **31**, 276–9.

Hoogendoorn, M. and Heimpel, G.E. (2001). PCR-based gut content analysis of insect predators: using ribosomal ITS-1 fragments from prey to estimate predation frequency. *Molecular Ecology*, **10**, 2059–67.

Hufbauer, R.A., Bogdanowicz, S.M., Perez, L. and Harrison, R.G. (2001). Isolation and characterization of microsatellites in *Aphidius ervi* (Hymenoptera: Braconidae) and their applicability to related species. *Molecular Ecology Notes*, **1**, 197–9.

Hufbauer, R.A., Bogdanowicz, S.M. and Harrison, R.G. (2004). The population genetics of a biological control introduction: mitochondrial DNA and microsatellie variation in native and introduced populations of *Aphidus ervi*, a parisitoid wasp. *Molecular Ecology*, **13**, 337–48.

Hurme, P. and Savolainen, O. (1999). Comparison of homology and linkage of random amplified polymorphic DNA (RAPD) markers between individual trees of Scots pine (*Pinus sylvestris* L.). *Molecular Ecology*, **8**, 15–22.

Ibrahim, K.M., Yassin, Y. and Elguzouli, A. (2004). Polymerase chain reaction primers for polymorphic microsatellite loci in the African armyworm, *Spodoptera exempta* (Lepidoptera: Noctuidae). *Molecular Ecology Notes*, **4**, 653–5.

Isabel, N., Beaulieu, J., Theriault, P. and Bousquet, J. (1999). Direct evidence for biased gene diversity estimates from dominant random amplified polymorphic DNA (RAPD) fingerprints. *Molecular Ecology*, **8**, 477–83.

Jeffreys, A.J., Wilson, V. and Thein, S.L. (1985). Hypervariable 'minisatellite' regions in human DNA. *Nature*, **314**, 67–73.

Johnson, K.P., Williams, B.L., Drown, D.M., Adams, R.J. and Clayton, D.H. (2002). The population genetics of host specificity: genetic differentiation in dove lice (Insecta: Phthiraptera). *Molecular Ecology*, **11**, 25–38.

Kauwe, J. S. K., Shizawa, D. K. and Evans, R. P. (2004). Phylogeographic and nested clade analysis of the stonefly *Pteronarcys californica* (Plecoptera: Pteronarcyidae) in the western USA. *Journal of the North American Benthological Society*, **23**, 824–38.

Keller, I. and Largiaderm, C. R. (2003). Five microsatellite DNA markers for the ground beetle *Abax parallelepipedus* (Coleoptera: Carabidae). *Molecular Ecology Notes*, **3**, 113–14.

Kerdelhue, C., Mondor-Genson, G., Rasplus, J. Y., Robert, A. and Lieutier, F. (2003). Characterization of five microsatellite loci in the pine shoot beetle *Tomicus piniperda* (Coleoptera: Scolytidae). *Molecular Ecology Notes*, **3**, 100–1.

Kessler, A. and Baldwin, I. T. (2001). Defensive function of herbivore-induced plant volatile emissions in nature. *Science*, **291**, 2141–4.

Kessler, A. and Baldwin, I. T. (2004). Herbivore-induced plant vaccination. Part I. The orchestration of plant defenses in nature and their fitness consequences in the wild tobacco Nicotiana attenuate. *The Plant Journal*, **38**, 639–49.

Kessler, A., Halitschke, R. and Baldwin, I. T. (2004). Silencing the jasmonate cascade: induced plant defenses and insect populations. *Science*, **305**, 665–8.

Kim, K. S. and Sappington, T. W. (2004*a*). Isolation and characterization of polymorphic microsatellite loci in the boll weevil/*Anthonomus grandis* Boheman (Coleoptera: Curculionidae). *Molecular Ecology Notes*, **4**, 701–3.

Kim, K. S. and Sappington, T. W. (2004*b*). Boll weevil (*Anthonomus grandis* Boheman) (Coleoptera: Curculionidae) dispersal in the southern United States: Evidence from mitochondrial DNA variation. *Environmental Entomology*, **33**, 457–70.

Kinnear, M. W., Bariana, H. S., Sved, J. A. and Frommer, M. (1998). Polymorphic microsatellite markers for population analysis of a tephritid pest species, *Bactrocera tryoni*. *Molecular Ecology*, **7**, 1489–95.

Kirk, A. A., Lacey, L. A., Brown, J. K. *et al.* (2000). Variation in the *Bemisia tabaci* s. l. species complex (Hemiptera: Aleyrodidae) and its natural enemies leading to successful biological control of Bemisia biotype B in the USA. *Bulletin of Entomological Research*, **90**, 317–27.

Kjaer, E. D., Siegismund, H. R. and Suangtho, V. (1996). A multivariate study on genetic variation in teak (*Tectona grandis* L.). *Silvae Genetica*, **45**, 361–8.

Knowles, L. L. (2004). The burgeoning field of statistical phylogeography. *Journal of Evolutionary Biology*, **17**, 1–10.

Koshio, C., Tomishima, M., Shimizu, K., Kim, H. S. and Takenaka, O. (2002). Microsatellites in the gypsy moth, *Lymantria dispar* L. (Lepidoptera: Lymantriidae). *Applied Entomology and Zoology*, **37**, 309–12.

Kreiger, M. J. B. and Ross, K. (2002). Identification of a major gene regulating complex social behavior. *Science*, **295**, 328–32.

Kuhner, M. K., Yamato, J. and Felsenstein, J. (1998). Maximum likelihood estimation of population growth rates based on the coalescent. *Genetics*, **149**, 429–34.

Kurokawa, T., Yao, I., Akimoto, S. I. and Hasegawa, E. (2004). Isolation of six microsatellite markers from the pea aphid, *Acyrthosiphon pisum* (Homoptera: Aphididae). *Molecular Ecology Notes*, **4**, 523–4.

Laffin, R.D., Langor, D.W. and Sperling, F.A.H. (2004). Population structure and gene flow in the white pine weevil *Pissodes strobe* (Coleoptera: Curculionidae). *Annals of the Entomological Society of America*, **97**, 949–56.

Landry, B.S., Dextraze, L. and Boivin, G. (1993). Random amplified polymorphic DNA markers for DNA fingerprinting and genetic variability assessment of minute parasitic wasp species (Hymenoptera: Mymaridae and Trichogrammatidae) used in biological control programs of phytophagous insects. *Genome*, **36**, 580–7.

Lee, M.L. and Lee, M.H. (1997). Amplified mitochondrial DNA identify four species of *Tetranychus* mites (Acarina: Tetranychidae) in Korea. *Korean Journal of Applied Entomology*, **36**, 30–6.

Legaspi, J.C., Legaspi, B.C., Carruthers, R.I. *et al.* (1996). Foreign exploration for natural enemies of *Bemisia tabaci* from Southeast Asia. *Subtropical Plant Science*, **48**, 43–8.

Legg, J.P., French, R., Rogan, D., Okao-Okuja, G. and Brown, J.K. (2002). A distinct *Bemisia tabaci* (Gennadius) (Hemiptera: Sternorrhyncha: Aleyrodidae) genotype cluster is associated with the epidemic of severe cassava mosaic virus disease in Uganda. *Molecular Ecology*, **11**, 1219–29.

Lessa, E.P. (1990). Multidimensional analysis of geographic genetic structure. *Systematic Zoology*, **39**, 242–52.

Li, Z.X., Zheng, L. and Shen, Z.R. (2004). Using internally transcribed spacer. 2 sequences to re-examine the taxonomic status of several cryptic species of *Trichogramma* (Hymenoptera: Trichogrammatidae). *European Journal of Entomology*, **101**, 347–58.

Llewellyn, K.S., Loxdale, H.D., Harrington, R. *et al.* (2003). Migration and genetic structure of the grain aphid (*Sitobion avenae*) in Britain related to climate and clonal fluctuation as revealed using microsatellites. *Molecular Ecology*, **12**, 21–34.

Llewellyn, K.S., Loxdale, H.D., Harrington, R., Clark, S.J. and Sunnucks, P. (2004). Evidence for gene flow and local clonal selection in field populations of the grain aphid (*Sitobion avenae*) in Britain revealed using microsatellites. *Heredity*, **93**, 143–53.

Loxdale, H.D., Brookes, C.P., Wynne, I.R. and Clark, S.J. (1998). Genetic variability within and between English populations of the damson-hop aphid, *Phorodon humuli* (Hemiptera: Aphididae), with special reference to esterases associated with insecticide resistance. *Bulletin of Entomological Research*, **88**, 513–26.

Loxdale, H.D. and den Hollander, J. (eds.) (1989). *Electrophoretic Studies on Agricultural Pests*. Systematics Association Special Volume No. 39. Oxford: Clarendon Press.

Loxdale, H.D. and Lushai, G. (1998). Molecular markers in entomology. *Bulletin of Entomological Research*, **88**, 577–600.

Loxdale, H.D. and Lushai, G. (1999). Slaves of the environment: the movement of herbivorous insects in relation to their ecology and genotype. *Philosophical Transactions of the Royal Society London, Series B*, **354**, 1479–95.

Loxdale, H.D., Tarr, I.J., Weber, C.P. *et al.* (1985). Electrophoretic study of enzymes from cereal aphid populations. III. Spatial and temporal genetic variation of populations of *Sitobion avenae* (F.) (Hemiptera: Aphididae). *Bulletin of Entomological Research*, **75**, 121–41.

Lu, Y. and Adang, M. J. (1996). Distinguishing fall armyworm (Lepidoptera: Noctuidae) strains using a diagnostic mitochondrial DNA marker. *Florida Entomologist*, **79**, 48–55.

Lu, Y. J., Adang, M. J., Isenhour, D. J. and Kochert, G. D. (1992). RFLP analysis of genetic variation in North American populations of the fall armyworm moth *Spodoptera frugiperda* (Lepidoptera: Noctuidae). *Molecular Ecology*, **1**, 199–208.

Lu, Y., Kochert, G. D., Isenhour, D. and Adang, M. J. (1994). Molecular characterization of a strain-specific repeated DNA sequence in the fall armyworm *Spodoptera frugiperda* (Lepidoptera: Noctuidae). *Insect Molecular Biology*, **3**, 123–30.

Luikart, G. and England, P. E. (1999). Statistical analysis of microsatellite data. *Trends in Ecology and Evolution*, **14**, 253–6.

MacDonald, C., Brookes, P., Edwards, K. J. *et al.* (2003). Microsatellite isolation and characterization in the beneficial parasitoid wasp *Diaeretiella rapae* (M'Intosh) (Hymenoptera: Braconidae: Aphidiinae). *Molecular Ecology Notes*, **4**, 601–3.

Macdonald, C. and Loxdale, H. D. (2004). Molecular markers to study population structure and dynamics in beneficial insects (predators and parasitoids). *International Journal of Pest Management*, **50**, 215–24.

Mantel, N. (1967). The detection of disease clustering and a generalised regression approach. *Cancer Research*, **27**, 209–20.

Marshall, T. C., Slate, J., Kruuk, L. E. B. and Pemberton, J. M. (1998). Statistical confidence for likelihood-based paternity inference in natural populations. *Molecular Ecology*, **7**, 639–55.

Maruthi, M. N., Colvin, J., Thwaites, R. M. *et al.* (2004). Reproductive incompatibility and cytochrome oxidase I gene sequence variability amongst host-adapted and geographically separate *Bemisia tabaci* populations (Hemiptera: Aleyrodidae). *Systematic Entomology*, **29**, 560–8.

McKenzie, J. A. and Batterham, P. (1998). Predicting insecticide resistance: mutagenesis, selection and response. *Philosophical Transactions of the Royal Society London, Series B*, **353**, 1729–34.

McKenzie, L., Dransfield, L., Driver, F. and Curran, J. (1999). *Molecular Protocols for Species Identification of Tephritid Fruit Flies*. Report to the Australian Quarantine Inspection Service Canberra, CSIRO Entomology.

McMichael, M. and Prowell, D. P. (1999). Differences in amplified fragment-length polymorphisms in fall armyworm (Lepidoptera: Noctuidae) host strains. *Annals of the Entomological Society of America*, **92**, 175–81.

Meglécz, E., Petenian, F., Danchin, E. *et al.* (2004). High similarity between flanking regions of different microsatellites detected within each of two species of Lepidoptera: *Parnassius apollo* and *Euphydryas aurinia*. *Molecular Ecology*, **13**, 1693–700.

Meixner, M. D., McPheron, B. A., Silva, J. G., Gasparich, G. E. and Sheppard, W. S. (2002). The Mediterranean fruit fly in California: evidence for multiple introductions and persistent populations based on microsatellite and mitochondrial DNA variability. *Molecular Ecology*, **11**, 891–9.

Mendel, Z., Nestle, D. and Gafny, R. (1994). Examination of the origin of the Israeli population of *Matsucoccus josephi* (Homoptera: Matsucoccidae) using random amplified polymorphic DNA-polymerase chain reaction method. *Annals of the Entomological Society of America*, **87**, 165–9.

Miller, L.J., Allsopp, P.G., Graham, G.C. and Yeates, D.K. (1999). Identification of morphologically similar canegrubs (Coleoptera: Scarabaeidae: Melolonthini) using a molecular diagnostic technique. *Australian Journal of Entomology*, **38**, 189–96.

Miller, N.J., Birley, A.J., Overall, A.D.J. and Tatchell, G.M. (2003). Population genetic structure of the lettuce root aphid, *Pemphigus bursarius* (L.), in relation to geographic distance, gene flow and host plant usage. *Heredity*, **91**, 217–23.

Miller, N.J., Birley, A.J. and Tatchell, G.M. (2000). Polymorphic microsatellite loci from the lettuce root aphid, *Pemphigus bursarius*. *Molecular Ecology*, **9**, 1951–2.

Monteiro, F.A., Donnelly, M.J., Beard, C.B. and Costa, J. (2004). Nested clade and phylogeographic analyses of the Chagas disease vector *Triatoma brasiliensis* in Northeast Brazil. *Molecular Phylogenetics and Evolution*, **32**, 46–56.

Morin, P.A., Luikart, G. and Wayne, R.K. (2004). SNPs in ecology, evolution, and conservation. *Trends in Ecology and Evolution*, **19**, 209–16.

Moritz, C. and Lavery, S. (1996). Molecular ecology: contributions from molecular genetics to population ecology. In R.B. Floyd, A.W. Sheppard and P.J. De Barro (eds.), *Frontiers of Population Ecology*. Melbourne: CSIRO Publishing. pp. 433–50.

Morris, D.C. and Mound, L.A. (2004). Molecular relationships between populations of South African citrus thrips (*Scirtothrips aurantii* Faure) in South Africa and Queensland, Australia. *Australian Journal of Entomology*, **43**, 353–8.

Moya, O., Contreras-Diaz, H.G., Oromi, P. and Juan, C. (2004). Genetic structure, phylogeography and demography of two ground-beetle species endemic to the Tenerife laurel forest (Canary Islands). *Molecular Ecology*, **13**, 3153–67.

Mueller, U.G. and Wolfenbarger, L.L. (1999). AFLP genotyping and fingerprinting. *Trends in Ecology and Evolution*, **14**, 389–94.

Mun, J., Bohonak, A.J. and Roderick, G.K. (2003). Population structure of the pumpkin fruit fly *Bactrocera depressa* (Tephritidae) in Korea and Japan: Pliocene allopatry or recent invastion? *Molecular Ecology*, **12**, 2941–51.

Muraji, M. and Nakahara, S. (2002). Discrimination among pest species of Bactrocera (Diptera: Tephritidae) based on PCR-RFLP of the mitochondrial DNA. *Applied Entomology and Zoology*, **37**, 437–46.

Muraji, M., Kawasaki, K., Shimizu, T. and Noda, T. (2004). Discrimination among Japanese species of the Orius flower bugs (Heteroptera: Anthocoridae) based on PCR-RFLP of the nuclear and mitochondrial DNAs. *Japan Agricultural Research Quarterly*, **38**, 91–5.

O'Hanlon, P.C., Peakall, R. and Briese, D.T. (1999). AFLP reveals introgression in weedy *Onopordum* thistles: hybridization and invasion. *Molecular Ecology*, **8**, 1239–46.

Paetkau, D., Calvert, W., Stirling, I. and Strobeck, C. (1995). Microsatellite analysis of population structure in Canadian polar bears. *Molecular Ecology*, **4**, 347–54.

Papura, D., Simon, J.C., Halkett, F. *et al.* (2003). Predominance of sexual reproduction in Romanian populations of the aphid *Sitobion avenae* inferred from phenotypic and genetic structure. *Heredity*, **90**, 397–404.

Parsons, Y.M. and Shaw, K.L. (2002). Mapping unexplored genomes: a genetic linkage map of the Hawaiian cricket Laupala. *Genetics*, **162**, 1275–82.

Pashley, D.P., McMichael, M. and Silvain, J.-F. (2004). Multilocus genetic analysis of host use, introgression, and speciation in host strains of fall armyworm (Lepidoptera: Noctuidae). *Annals of the Entomological Society of America*, **97**, 1034–44.

Peakall, R. and Smouse, P.E. (2001). *GenAlEx V5: Genetic Analysis in Excel. Population Genetic Software for Teaching and Research*. Canberra, Australia: Australian National University.

Pérez, T., Albornoz, J. and Domínguez, A. (1998). An evaluation of RAPD reproducibility and nature. *Molecular Ecology*, **7**, 1347–58.

Perring, T.M., Cooper, A.D., Rodriguez, R.J., Farrar, C.A. and Bellows, T.S. (1993). Identification of a whitefly species by genomic and behavioral studies. *Science*, **259**, 74–7.

Pfeifer, T.A., Humble, L.M., Ring, M. and Grigliatti, T.A. (1995). Characterization of gypsy moth populations and related species using a nuclear DNA marker. *Canadian Entomologist*, **127**, 49–58.

Podani, J. (1995). *SYN-TAX 5.02.Mac: Computer Programs for Multivariate Data Analysis on the Macintosh System*. Budapest: Scientia Publishing.

Polaszek, A., Manzari, S. and Quicke, D.L.J. (2004). Morphological and molecular taxonomic analysis of the *Encarsia meritoria* species-complex (Hymenoptera: Aphelinidae), parasitoids of whiteflies (Hemiptera: Aleyrodidae) of economic importance. *Zoologica Scripta*, **33**, 403–21.

Posada, D., Crandall, K.A. and Templeton, A.R. (2000). GeoDis: a program for the cladistic nested analysis of the geographical distribution of genetic haplotypes. *Molecular Ecology*, **9**, 487–8.

Powers, T.O., Jensen, S.G., Kindler, S.D., Stryker, C.J. and Sandall, L.J. (1989). Mitochondrial DNA divergence among Greenbug (Homoptera: Aphididae) biotypes. *Annals of the Entomological Society of America*, **82**, 298–302.

Pritchard, J.K., Stephens, M. and Donnelly, P. (2000). Inference of population structure using multilocus genotype data. *Genetics*, **155**, 945–59.

Prowell, D.P., McMichael, M. and Silvain, J.F. (2004). Multilocus genetic analysis of host use, introgression, and speciation in host strains of all armyworm (Lepidoptera: Noctuidae). *Annals of the Entomological Society of America*, **97**, 1034–44.

Rabouam, C., Comes, A.M., Bretagnolle, V. *et al.* (1999). Features of DNA fragments obtained by random amplified polymorphic DNA (RAPD) assays. *Molecular Ecology*, **8**, 493–504.

Rannala, B. and Mountain, J.L. (1997). Detecting immigration by using multilocus genotypes. *Proceedings of the National Academy of Sciences USA*, **94**, 9197–201.

Reyes, A. and Ochando, M.D. (2004). Mitochondrial DNA variation in Spanish populations of *Ceratitis capitata* (Wiedemann) (Tephritidae) and the colonization process. *Journal of Applied Entomology*, **128**, 358–64.

Robinson, M.T., Weeks, A.R. and Hoffmann, A.A. (2002). Geographic patterns of clonal diversity in the earth mite species Penthaleus major with particular emphasis on species margins. *Evolution*, **56**, 1160–7.

Roda, A., Halitschke, R., Anke, S. and Baldwin, I.T. (2004). Individual variability in herbivore-specific elicitors from the plant's perspective. *Molecular Ecology*, **13**, 2421–33.

Ross, P., Hall, L., Smirnov, I. and Haff, L. (1998). High level multiplex genotyping by MALDI-TOF mass spectrometry. *Nature Biotechnology*, **16**, 1347–551.

Roy, A.S., McNamara, D.G. and Smith, I.M. (1995). Situation of *Lymantria dispar* in Europe. *Bulletin OEPP*, **25**, 611–16.

Saiki, R.K., Gelfand, D.H., Stoffel, S. *et al.* (1988). *Science*, **219**, 487.

Salle, A., Kerdelhue, C., Breton, M. and Lieutier, F. (2003). Characterization of microsatellite loci in the spruce bark beetle *Ips typographus* (Coleoptera: Scolytinae). *Molecular Ecology Notes*, **3**, 336–7.

Sandstrom, J.P., Russell, J.A., White, J.P. and Moran, N.A. (2001). Independent origins and horizontal transfer of bacterial symbionts of aphids. *Molecular Ecology*, **10**, 217–28.

Sapolsky, R.J., Hsie, L., Berno, A. *et al.* (1999). High-throughput polymorphism screening and genotyping with high-density oligonucleotide arrays. *Genetic Analysis: Biomolecular Engineering*, **14**, 187–92.

Scheffer, S.J. and Lewis, M.L. (2001). Two nuclear genes confirm mitochondrial evidence of cryptic species within *Liriomyza huidobrensis* (Diptera: Agromyzidae). *Annals of the Entomological Society of America*, **94**, 648–53.

Scheffer, S.J., Wijesekara, A., Visser, D. and Hallett, R.H. (2001). Polymerase chain reaction-restriction fragment-length polymorphism method to distinguish *Liriomyza huidobrensis* from L-langei (Diptera: Agromyzidae) applied to three recent leafminer invasions. *Journal of Economic Entomology*, **94**, 1177–82.

Schierwater, B., Streit, B., Wagner, G.P. and DeSalle, R. (eds.) (1994). *Molecular Ecology and Evolution: Approaches and Applications*. Birkhäuser Verlag, Basel.

Schmidt, S., Naumann, I.D. and De Barro, P.J. (2001). *Encarsia* species (Hymenoptera: Aphelinidae) of Australia and the Pacific Islands attacking *Bemisia tabaci* and *Trialeurodes vaporariorum* (Hemiptera: Aleyrodidae) – a pictorial key and descriptions of four new species. *Bulletin of Entomological Research*, **91**, 369–87.

Schneider, C.C., Kueffer, J.M., Roessli, D. and Excoffier, L. (1997). ARLEQUIN, version 1.1: *A Software for Population Genetic Data Analysis*. Genera: Genetics and Biometry Laboratory, University of Geneva.

Schultheis, A.S., Weigt, L.A. and Hendricks, A.C. (2002). Gene flow, dispersal, and nested clade analysis among populations of the stonefly *Peltoperla tarteri* in the southern Appalachians. *Molecular Ecology*, **11**, 317–27.

Scott, K.D., Lange, C.L., Scott, L.J. and Graham, G.C. (2004). Isolation and characterization of microsatellite loci from *Helicoverpa armigera* Hubner (Lepidoptera: Noctuidae). *Molecular Ecology Notes*, **4**, 204–5.

Segraves, K.A. and Pellmyr, O. (2004). Testing the out-of-Florida hypothesis on the origin of cheating in the yucca-yucca moth mutualism. *Evolution*, **58**, 2266–79.

Sembene, M., Vautrin, D., Silvain, J. F., Rasplus, J. Y. and Delobel, A. (2003). Isolation and characterization of polymorphic microsatellites in the groundnut seed beetle, *Caryedon serratus* (Coleoptera: Bruchidae) *Molecular Ecology Notes*, **3**, 299–301.

Service, R. F. (1998). Microchip arrays put DNA on the spot. *Science*, **282**, 396–9.

Sheppard, W. S., Steck, G. J. and McPheron, B. A. (1992). Geographic populations of the medfly may be differentiated by mitochondrial DNA variation. *Experientia*, **48**, 1010–13.

Shoemaker, J. S., Painter, I. S. and Weir, B. S. (1999). Bayesian statistics in genetics: a guide for the uninitiated. *Trends in Genetics*, **15**, 354–8.

Shufran, K. A., Black, W. C. and Margolies, D. C. (1991). DNA fingerprinting to study spatial and temporal distributions of an aphid, *Schizaphis graminum* (Homoptera: Aphididae). *Bulletin of Entomological Research*, **81**, 303–13.

Shufran, K. A., Margolies, D. C. and Black, IV. W. C. (1992). Variation between biotype E clones of *Schizaphis graminum* (Homoptera: Aphididae). *Bulletin of Entomological Research*, **82**, 407–16.

Shufran, K. A. and Wilde, G. E. (1994). Clonal diversity in overwintering populations of *Schizaphis graminum* (Homoptera: Aphididae). *Bulletin of Entomological Research*, **84**, 105–14.

Shufran, K. A., Burd, J. D., Anstead, J. A. and Lushai, G. (2000). Mitochondrial DNA sequence divergence among greenbug (Homoptera: Aphididae) biotypes: evidence for host-adapted races. *Insect Molecular Biology*, **9**, 179–84.

Shufran, K. A., Weathersbee, A. A., Jones, D. B. and Elliott, N. C. (2004). Genetic similarities among geographic isolates of *Lysiphlebus testaceipes* (Hymenoptera: Aphidiidae) differing in cold temperature tolerances. *Environmental Entomology*, **33**, 776–8.

Silva, J. G., Meixner, M. D., McPheron, B. A., Steck, G. J. and Sheppard, W. S. (2003). Recent Mediterranean fruit fly (Diptera: Tephritidae) infestations in Florida – A genetic perspective. *Journal of Economic Entomology*, **96**, 1711–18.

Simon, J. C., Baumann, S., Sunnucks, P. *et al.* (1999*a*). Reproductive mode and population genetic structure of the cereal aphid *Sitobion avenae* studied using phenotypic and microsatellite markers. *Molecular Ecology*, **8**, 531–45.

Simon, J. C., Carre, S., Boutin, M. *et al.* (2003). Host-based divergence in populations of the pea aphid: insights from nuclear markers and the prevalence of facultative symbionts. *Proceedings of the Royal Society of London, Series B – Biological Science*, **270**, 1703–12.

Simon, C., Frati, F., Beckenbach, A. *et al.* (1994). Evolution, weighting, and phylogenetic utility of mitochondrial gene sequences and a compilation of conserved polymerase chain reaction primers. *Annals of the Entomological Society of America*, **87**, 651–701.

Simon, J. C., Leterme, N., Delmotte, F., Martin, O. and Estoup, A. (2001). Isolation and characterization of microsatellite loci in the aphid species, *Rhopalosiphum padi*. *Molecular Ecology Notes*, **1**, 4–5.

Simon, J. C., Leterme, N. and Latorre, A. (1999*b*). Molecular markers linked to breeding system differences in segregating and natural populations of the cereal aphid *Rhopalosiphum padi* L. *Molecular Ecology*, **8**, 965–73.

Simon, C., McIntosh, C. and Deniega, J. (1993). Standard restriction fragment length analysis of the mitochondrial genome is not sensitive enough for phylogenetic analysis or identification of 17-year periodical cicada broods (Hemiptera: Cicadidae): The potential for a new technique. *Annals of the Entomological Society of America*, **86**, 228–38.

Sin, F.Y.T., Suckling, D.M. and Marshall, J.W. (1995). Differentiation of the endemic New Zealand greenheaded and brownheaded leafroller moths by restriction fragment length variation in the ribosomal gene complex. *Molecular Ecology*, **4**, 253–6.

Slatkin, M. (1985). Gene flow in natural populations. *Annual Review of Ecology and Systematics*, **16**, 393–430.

Slatkin, M. and Maddison, W.P. (1989). A cladistic measure of gene flow inferred from the phylogenies of alleles. *Genetics*, **123**, 603–13.

Sloane, M.A., Sunnucks, P., Wilson, A.C.C., Hales, D.F. and Sunnucks, P. (2001). Microsatellite isolation, linkage group identification and determination of recombination frequency in the peach-potato aphid, *Myzus persicae* (Sulzer) (Hemiptera: Aphididae). *Genetical Research*, **77**, 251–60.

Smouse, P.E. (1998). To tree or not to tree. *Molecular Ecology*, **7**, 399–412.

Smouse, P.E. and Peakall, R. (1999). Spatial autocorrelation analysis of individual multiallele and multilocus genetic structure. *Heredity*, **82**, 561–73.

Sokal, R.R. and Oden, N.L. (1978). Spatial autocorrelation in biology. 1. Methodology. *Biological Journal of the Linnaean Society*, **10**, 199–228.

Spackman, M.E. and McKechnie, S.W. (1995). Assessing the value of mitochondrial DNA variation for detecting population subdivision in the cotton bollworm, *Helicoverpa armigera* (Lepidoptera: Noctuidae) in Australia. *Proceedings Beltwide Cotton Conference*, **2**, 811–13.

Sperling, F.A.H., Landry, J.F. and Hickey, D.A. (1995). DNA-based identification of introduced ermine moth species in North America (Lepidoptera: Yponomeutidae). *Annals of the Entomological Society of America*, **88**, 155–62.

Steiner, W.W.M., Voegtlin, D.J. and Irwin, M.E. (1985*a*). Genetic differentiation and its bearing on migration in North American populations of the corn leaf aphid, *Rhopalosiphum maidis* (Fitch) (Homoptera: Aphididae). *Annals of the Entomological Society of America*, **78**, 518–25.

Steiner, W.W.M., Voegtlin, D.J., Irwin, M.E. and Kampmeier, G. (1985*b*). Electrophoretic comparison of aphid species: detecting differences based on taxonomic status and host plant. *Journal of Comparative Biochemistry and Physiology B*, **81**, 295–9.

Stern, D.L., Aoki, S. and Kurosu, U. (1997). Determining aphid taxonomic affinities and life cycles with molecular data: a case study of the tribe *Cerataphidini* (Hormaphididae: Aphidoidea: Hemiptera). *Systematic Entomology*, **22**, 81–96.

Stine, O.C., Carnahan, A., Singh, R. *et al.* (2003). Characterization of microbial communities from coastal waters using microarrays. *Environmental Monitoring and Assessment*, **81**, 327–36.

Stipp, D. (1997). Gene chip breakthrough. *Fortune*, **135**, 56–66.

Stone, G.N. and Sunnucks, P. (1993). Genetic consequences of an invasion through a patchy environment – the cynipid gallwasp *Andricus quercuscalicis* (Hymenoptera: Cynipidae). *Molecular Ecology*, **2**, 251–68.

Sunnucks, P., England, P.R., Taylor, A.C. and Hales, D.F. (1996). Microsatellite and chromosome evolution of parthenogenetic *Sitobion aphids* in Australia. *Genetics*, **144**, 747–56.

Sunnucks, P., De Barro, P.J., Lushai, G., Maclean, N. and Hales, D. (1997*a*). Genetic structure of an aphid studies using microsatellites: cyclic parthenogenesis, differentiated lineages, and host specialisation. *Molecular Ecology*, **6**, 1059–73.

Sunnucks, P., Driver, F., Brown, W.V. *et al.* (1997*b*). Biological and genetic characterization of morphologically similar *Therioaphis trifolii* (Hemiptera: Aphididae) with different host utilization. *Bulletin of Entomological Research*, **87**, 152–62.

Sunnucks, P. (2000). Efficient genetic markers for population biology. *Trends in Ecology and Evolution*, **15**, 199–203.

Sunnucks, P., Wilson, A.C.C., Beheregaray, L.B. *et al.* (2000). SSCP is not so difficult: the application and utility of single-stranded conformation polymorphism in evolutionary biology and molecular ecology. *Molecular Ecology*, **9**, 1699–710.

Symondson, W.O.C. and Liddell, J.E. (eds.) (1996). *The Ecology of Agricultural Pests: Biochemical Approaches*. London: Chapman and Hall.

Takami, Y. and Katada, S. (2001). Microsatellite DNA markers for the ground beetle *Carabus insulicola*. *Molecular Ecology Notes*, **1**, 128–30.

Templeton, A.R. (1998). Nested clade analyses of phylogeographic data: testing hypotheses about gene flow and population history. *Molecular Ecology*, **7**, 381–97.

Terradot, L., Simon, J.C., Leterme, N. *et al.* (1999). Molecular characterization of clones of the *Myzus persicae* complex (Hemiptera: Aphididae) differing in their ability to transmit the potato leafroll luteovirus (PLRV). *Oecologia*, **108**, 121–9.

Toda, S. and Komazaki, S. (2002). Identification of thrips species (Thysanoptera: Thripidae) on Japanese fruit trees by polymerase chain reaction and restriction fragment length polymorphism of the ribosomal ITS2 region. *Bulletin of Entomological Research*, **92**, 359–63.

Torsvik, V. and Øvreås, L. (2002). Microbial diversity and function in soil: from genes to ecosystems. *Current Opinion in Microbiology*, **5**, 240–5.

Tsagkarakou, A. and Roditakis, N. (2003). Isolation and characterization of microsatellite loci in *Bemisia tabaci* (Hemiptera: Aleyrodidae). *Molecular Ecology Notes*, **3**, 196–8.

Turgeon, J. and McPeek, M.A. (2002). Phylogeographic analysis of a recent radiation of Enallagma damselflies (Odonata: Coenagrionidae). *Molecular Ecology*, **11**, 1989–2001.

Via, S. (1999). Reproductive isolation between sympatric races of pea aphids. I. Gene flow restriction and habitat choice. *Evolution*, **53**, 1446–57.

Villablanca, F.X., Roderick, G.K. and Palumbi, S.R. (1998). Invasion genetics of the Mediterranean fruit fly: variation in multiple nuclear introns. *Molecular Ecology*, **7**, 547–60.

Vink, C.J., Phillips, C.B., Mitchell, A.D., Winder, L.M. and Cane, R.P. (2003). Genetic variation in *Microctonus aethiopoides* (Hymenoptera: Braconidae). *Biological Control*, **28**, 251–64.

Viscarret, M.M., Torres-Jerez, I., De Manero, E.A. *et al.* (2003). Mitochondrial DNA evidence for a distinct new world group of *Bemisia tabaci* (Gennadius) (Hemiptera: Aleyrodidae) indigenous to Argentina and Bolivia, and presence of the Old World B biotype in Argentina. *Annals of the Entomological Society of America*, **96**, 65–72.

Voelckel, C. and Baldwin, I.T. (2004*a*). Herbivore-induced plant vaccination. Part II. Array-studies reveal the transiene of herbivore-specific transcriptional imprints and a distinct imprint from stress combinations. *The Plant Journal*, **38**, 650–63.

Voelckel, C. and Baldwin, I.T. (2004*b*). Generalist and specialist lepidopteran larvae elicit different transcriptional responses in *Nicotiana attenuata*, which correlate with larval FAC profiles. *Ecology Letters*, **7**, 770–5.

Voelckel, C., Weisser, W.W. and Baldwin, I.T. (2004). An analysis of plant–aphid interactions by different microarray hybridization strategies. *Molecular Ecology*, **13**, 3187–95.

Vorburger, C., Lancaster, M. and Sunnucks, P. (2003). Environmentally related patterns of reproductive modes in the aphid *Myzus persicae* and the predominance of two 'superclones' in Victoria, Australia. *Molecular Ecology*, **12**, 3493–504.

Vorburger, C. (2005). Positive genetic correlations among major life-history traits relative to ecological success in the aphid *Myzus persicae*. *Evolution*, **59**, 3493–3504.

Wagener, B., Reineke, A., Lohr, B. and Zebitz, C.P.W. (2004). A PCR-based approach to distinguish important *Diadegma* species (Hymenoptera: Ichneumonidae) associated with diamondback moth, *Plutella xylostella* (Lepidoptera: Plutellidae). *Bulletin of Entomological Research*, **94**, 465–71.

Wang, D., Fan, J., Siao, C. *et al.* (1998). Large-scale identification, mapping, and genotyping of single-nucleotide polymorphisms in the human genome. *Science*, **280**, 1077–82.

Waser, P.M. and Strobeck, C. (1998). Genetic signatures of interpopulation dispersal. *Trends in Ecology and Evolution*, **13**, 43–4.

Weeks, A.R., Marec, F. and Breeuwer, A.J. (2001). A mite species that consists entirely of haploid females. *Science*, **292**, 2479–82.

Wells, M.M. (1994). Small genetic distances among populations of green lacewings of the genus Chrysoperla (Neuroptera: Chrysopidae). *Annals of the Entomological Society of America*, **87**, 737–44.

Williams, C.L., Goldson, S.L., Baird, D.B. and Bullock, D.W. (1994). Geographical origin of an introduced insect pest, *Listronotus bonariensis* (Kuschel), determined by RAPD analysis. *Heredity*, **72**, 412–19.

Wilson, A.C.C., Sunnucks, P. and Hales, D.F. (1999). Microevolution, low clonal diversity and genetic affinities of parthenogenetic *Sitobion* aphids in New Zealand. *Molecular Ecology*, **8**, 1655–66.

Wilson, A.C.C., Sunnucks, P., Blackman, R.L. and Hales, D.F. (2002). Microsatellite variation in cyclically parthenogenetic populations of *Myzus persicae* in south-eastern Australia. *Heredity*, **88**, 258–66.

Wilson, A.C.C., Sunnucks, P. and Hales, D.F. (2003). Heritable genetic variation and potential for adaptive evolution in asexual aphids (Aphidoidea). *Biological Journal of the Linnean Society*, **79**, 115–35.

Wilson, A.C.C., Massonnet, B., Simon, J.C. *et al.* (2004). Cross-species amplification of microsatellite loci in aphids: assessment and application. *Molecular Ecology Notes*, **4**, 104–9.

Wright, S. (1951). The genetical structure of populations. *Annals of Eugenics*, **15**, 323–54.

Wu, L., Thompson, D.K., Li, G. *et al.* (2001). Development and evaluation of functional gene arrays for detection of selected genes in the environment. *Applied and Environmental Microbiology*, **67**, 5780–90.

Wyman, A.R. and White, R. (1980). A highly polymorphic locus in human DNA. *Proceedings of the National Academy of Science*, **77**, 6754–8.

Yu, H., Frommer, M., Robson, M.K. *et al.* (2001). Microsatellite analysis of the Queensland fruit fly *Bactrocera tryoni* (Diptera: Tephritidae) indicates spatial structuring: implications for population control. *Bulletin of Entomological Research*, **91**, 139–47.

Yulin, A., Caihua, D., Hongbing, Z. and Guoyao, J. (1998). RAPD assessment of three sibling species of *Monochamus guer* (Coleoptera: Cerambycidae). *Journal of Nanjing Forestry University*, **22**, 35–8.

Zanic, K., Cenis, J.L., Kacic, S. and Katalinic, M. (2005). Current status of *Bemisia tabaci* in coastal Croatia. *Phytoparasitica*, **33**, 60–4.

Zhao, J.T., Frommer, M., Sved, J.A. and Gillies, C.B. (2003). Genetic and molecular markers of the Queensland fruit fly, *Bactrocera tryoni*. *Journal of Heredity*, **94**, 416–20.

Zhang, A., Dunn, J.B. and Clark, J.M. (1999). An efficient strategy for validation of a point mutation associated with acetylcholinesterase sensitivity to azinphos-methyl in Colorado potato beetle. *Pesticide Biochemistry and Physiology*, **65**, 25–35.

Zhang, D.-X. (2004). Lepidopteran microsatellite DNA: redundant but promising. *Trends in Ecology and Evolution*, **19**, 507–9.

Zhou, J. (2003). Microarrays for bacterial detection and microbial community analysis. *Current Opinion in Microbiology*, **6**, 288–94.

Zhu, Y.C., Burd, J.D., Elliott, N.C. and Greenstone, M.H. (2000). Specific ribosomal DNA marker for early polymerase chain reaction detection of *Aphelinus hordei* (Hymenoptera: Aphelinidae) and *Aphidius colemani* (Hymenoptera: Aphidiidae) from *Diuraphis noxia* (Homoptera: Aphididae). *Annals of the Entomological Society of America*, **93**, 486–91.

16

Ecotoxicology: the ecology of interactions between pesticides and non-target organisms

P. C. JEPSON

16.1 Introduction

Ecotoxicology is a hybrid discipline that derives its principles and approaches from toxicology, chemistry and ecology. It has spawned numerous text books and manuals (e.g. Levin *et al.*, 1988; Calow, 1994*a*; *b*; Moriarty, 1999; Walker *et al.*, 1995) that attempt to draw together these constituent disciplines into a coherent enough whole to enable this applied science to evolve. The principles of ecotoxicology provide the underlying rationale for understanding, regulating and managing the impacts that toxic chemicals have on the environment. It is through its role in providing the technical and scientific foundation for regulatory toxicology that ecotoxicology has its greatest impact on IPM; and this impact is immense. All pesticides that are in use in the western world at least, are subjected to regulatory procedures that approve, restrict or deny use of these chemicals based upon the environmental, as well as the human health risks that they pose.

Ecological risk assessment (Suter, 1993) has developed as an elaborate set of procedures that address the environmental component of the risks posed by xenobiotic chemicals. These procedures are designed to systematically evaluate the probability that adverse effects may occur as a result of exposure to stressors, in this case pesticides. The objective of ecological risk assessment is to inform management decisions, which may include the decision to approve the use of a particular pesticide. Alternatively, as this chapter explores, it may also inform decisions that help managers determine which course of action to take in order to limit ecological risks and adverse effects when the pesticide has been approved and it is in commercial use.

The ecological risk assessment procedures that are used as a component of pesticide regulation tend to be hidden from the view of most IPM researchers and practitioners because they are applied before pesticides are marketed. They are nonetheless an intrinsic part of IPM because of the influence that they have on the pesticides that are available for use in agriculture. When used as part of the regulatory process that determines if pesticides are acceptable for commercialization, ecological risk assessment procedures adhere strictly to published guidelines (e.g. EPA, 1998) that govern the scientific integrity of the process. Although the details of regulatory risk assessment procedures may vary, the basic process tends to fall into three phases: a problem formulation phase, which develops hypotheses concerning how and why pesticides may cause particular ecological effects, an analysis phase where exposure to the pesticide and the possible effects of this exposure are characterized, and a risk characterization phase, where risks are calculated, through integration of exposure and effects data, and the likelihood of an adverse ecological impact is determined. The data that feed into this process may include actual effects data from the laboratory or the field, derived from standardized tests, or bioassays on a small number of test species that are selected to represent the groups of organisms or ecological processes that the measures aim to protect. The pesticide regulatory process is elaborate and extremely expensive to undertake, and decisions are often based upon a limited dataset that represents the minimum that is required to detect pesticide compounds or uses that might pose unacceptable risks.

Ecological risk assessment and pesticide regulatory processes have developed their own vocabularies and along with the apparent rigidity of their procedures they may not seem to be of particular value to agricultural scientists developing IPM systems. Aspects of ecological risk assessment procedures and principles may however feed directly into the development of IPM systems once pesticides are marketed because they may improve understanding and prediction of pesticide side-effects and therefore our ability to manage them.

Why might ecological risk assessment be needed following regulatory approval of a pesticide? It could logically be argued that regulatory approval of a pesticide provides a guarantee that adverse effects will not take place. It is important for IPM practitioners to understand that when a pesticide is approved by a regulatory authority, an adjudication has been made that the benefits of the pesticide outweigh any environmental risks that these chemicals pose. This does not represent a guarantee that the pesticide will pose no ecological risks, only that unacceptable, adverse impacts that have been defined in the problem formulation phase of risk assessment are unlikely to occur. If a pesticide is to be adopted within an IPM regime, the challenge for IPM practitioners is to determine that the benefits of the pesticide will outweigh the risks that it might pose locally, and that any risks that arise are either considered to be acceptable, or they can be mitigated to reduce the likelihood of adverse effects. Although potential environmental risks and specific risk avoidance or mitigation practices are normally outlined on pesticide labels, it is still possible for pesticides to be hazardous in an IPM context and for their legal use to result in environmental contamination, exposure to non-target organisms and consequent impacts.

Evidence for the broader environmental impacts of pesticides are not difficult to find. For example, in a 1992–2001 survey of pesticides in streams and ground water in the USA, more than half of the streams sampled had pesticide concentrations that exceeded benchmark concentrations for aquatic life (Gilliom *et al.*, 2006) (Table 16.1). These data are indicative of widespread environmental contamination by pesticides and the potential for ecological impacts, but the actual benchmarks that were being exceeded in the survey varied between sites, and it was not possible to translate the percentages of streams that exceeded the benchmark concentrations into a simple index of toxicity or impact. For example, insecticides tended to exceed benchmarks based upon acute or chronic effects on aquatic invertebrates, whereas herbicides tended to exceed benchmarks for acute or chronic effects on vascular and non-vascular plants. Ecotoxicologists are still developing methods to understand, interpret and predict the impacts of

Table 16.1. *Pesticides detected in US streams, 1992–2001*[a]

	Agricultural areas	Urban areas	Undeveloped areas	Mixed land uses
Occurrence	97	97	65	94
Concentrations exceeding human health benchmark[b]	9.6	6.7	0	1.5
Concentrations exceeding aquatic life benchmaks[c]	57	83	13	42

[a]From Gilliom *et al.*, 2006. All values are percentages of the number of stream samples taken in a survey of 51 major hydrologic systems.

[b]Percentage of samples in which concentrations exceeded values from standards and guidelines for drinking water, developed by the USEPA.

[c]Percentage of samples in which concentrations exceeded values from USEPA guidelines, toxicity values from USEPA risk assessments, or selected guidelines from other sources.

complex pesticide mixtures on communities of organisms, including those in fresh water systems (e.g. Posthuma *et al.*, 2002).

Ecological risk assessment procedures advance only slowly despite the evidence of widespread off-crop contamination, and may in some cases be simplified or limited in their scope when it is judged by regulatory agencies that the time and effort required for review of ecological risk data are not justified. An example of this was the decision in 1992 by the US Environmental Protection Agency to limit the use of field-derived data in pesticide regulatory review (Taub, 1997; EPA, 1998). A balance must be struck between the requirement for detailed data concerning pesticide impacts, and the costs associated with obtaining these data. The test for whether this tradeoff is being correctly struck is the degree to which pesticide regulations actually provide protection for the ecosystems that they address.

Studies of pesticide impacts in individual watersheds in the national survey of US streams (Gilliam *et al.*, 2006) provided some limited evidence of ecological impacts on aquatic organisms. For example, streams and drains in the Yakima River Basin in Washington State with the highest and most potentially toxic pesticide concentrations tended to have the highest numbers of pollution-tolerant benthic aquatic invertebrates, indicative of lower water quality (Fuhrer *et al.*, 2004). When viewed in the context of many other sources of environmental impact on ecosystems, the role that pesticides play is, however, very difficult to disentangle. It is evident, for example, that both direct toxic effects (e.g. Mineau, 2002; Mineau and Whiteside, 2006) and indirect effects operating through changes in food supply (e.g. Chamberlain *et al.*, 2000) may be contributing to the reported declines in farmland bird species (Krebs *et al.*, 1999). Similarly complex mechanisms underlie changes in the abundance and diversity of other flora and fauna in agroecosystems, and procedures that provide feedback from field data to regulatory systems to test their effectiveness have not emerged.

Conflicting evidence about pesticide impacts, uncertainty about the degree of environmental protection that is provided by pesticide regulation, lack of exposure to the concepts and procedures of ecological risk assessment and the inaccessibility of the jargon of ecotoxicology and regulatory toxicology leave IPM practitioners in the unenviable position of having to use their own professional judgment and experience to balance pesticide benefits and risks, with no clear scientific support. Although both regulatory toxicology, through its ecotoxicological foundation, and IPM contribute to rational pesticide management, they are rarely used in concert. This chapter takes an ecotoxicological perspective to analyze the impacts of pesticides on the natural enemies of crop pests. These are among the most serious side-effects of pesticides because they can disrupt pest limitation by predators and parasites, which may lead to pest resurgence and secondary pest outbreaks (Ripper, 1956; Croft, 1990; Hardin *et al.*, 1995; Trumper and Holt, 1998; Johnson and Tabashnik, 1999). A core principle of IPM is that pest management tactics complement each other, and the substitution of chemically induced mortality for mortality that results from predator–prey and parasite–host interactions not only violates this principle, but may increase the sensitivity of the crop to further pest damage. All pesticide applications are not however equal in the risks that they pose. Pesticides vary in their toxicity to natural enemies, and natural enemies vary in their susceptibility to pesticides (Croft and Brown, 1975; Croft, 1990). Populations of natural enemies may also recover rapidly following applications of short-persistence pesticides (Jepson and Thacker, 1990). There is no substitute for obtaining a detailed understanding of pesticide–natural enemy interactions in individual cropping systems, if the benefits of pesticides are to be maximized, and impacts on natural enemies restricted so that pest limitation by predation and parasitism is also maintained.

To achieve these dual goals, IPM practitioners have a number of options if a given IPM system requires pesticides to be used. They may, for example, select pesticides that are less toxic to natural enemies (termed physiological selectivity) or methods or times of application that avoid exposing natural enemies (termed ecological selectivity) to achieve a balance between chemical impacts on pests and their predators and parasites.

IPM practitioners may also provide critical feedback to the pesticide regulatory process by determining the degree to which IPM-based crop protection regimes enable pesticides to be used in a sustainable way. Where the properties of a pesticide prevent this from occurring, there needs to be a more effective process whereby this information is fed back to the regulatory process to enable improvements in pesticide risk assessment and regulatory procedures. Most IPM practitioners are aware that certain pesticides are incompatible with, and disruptive to, IPM procedures, and it is questionable whether these compounds should be approved for use in commodities where they cannot be employed in a sustainable way.

16.2 Ecotoxicology of pesticide–natural enemy interactions

Ecotoxicologists view the world in terms of the interactions between chemical substances and the abiotic and biotic environments. The outcome of this interaction may be an adverse ecological effect, but only if the chemical is biologically available to organisms, and a susceptible organism is exposed to it. An adverse impact is not simply therefore determined by how toxic a chemical happens to be, or how susceptible a natural enemy is to it; these two entities must occur together in time and space, the interaction must accumulate sufficient toxin at a target site for a toxicological effect to occur, and population and community processes must ensue to determine whether or not this effect leads to an ecologically adverse outcome. Adverse impacts may therefore be direct or indirect in nature. Direct natural enemy mortality may, for example, reduce pest limitation to such a degree that it leads to the indirect ecological effect of pest resurgence where pest population densities exposed to pesticides exceed the densities of those that are not exposed to pesticides (Hardin *et al.*, 1998; Trumper and Holt, 1998).

In order to determine whether exposure to a toxic chemical might result in a direct or indirect ecological effect on a particular organism, it is necessary to determine the proportion of the population that was exposed to the chemical, whether lethal or sublethal effects of exposure were significant enough to reduce population densities, and whether this reduction might persist or be of sufficient intensity for ecological processes to be disrupted. To help disentangle this complex hierarchy of questions, the ecotoxicological interactions between chemicals and organisms are often viewed as a progression through a series of stages that evolve from initial exposure of the organism and which ultimately encompass the overall impact, including long-term population impacts.

The processes that govern the exposure, uptake, toxic effects and population recovery of natural enemies exposed to pesticides, can be shown in this way to evolve over a sequence of temporal and spatial scales that encompass all the stages of ecotoxicological impact and the processes that drive them (Figure 16.1). The effects of the pesticide can be divided into an initial stage, which is dominated by the intimate interactions between chemical and organism, and a later stage which is dominated by population processes. As these stages evolve from one to the next, shifts take place not only in temporal and spatial context, but also in the level of biological organization at which the underlying mechanisms and processes are taking place. For example, it is only within the "micro"-scale (sensu Jepson, 1989; Figure 16.1), over hours following treatment and on the scale of the diurnal activity range of the organism, that direct toxicity and the factors controlling direct and indirect exposure of the natural enemy to the pesticide can be determined. Similarly, it is only on the "meso"-scale, over weeks to months and on the scale of the treated field or the "macro"-scale, over months to years and on the farm-scale and beyond, that the full ecological impact of this initial effect on the population in question may be determined

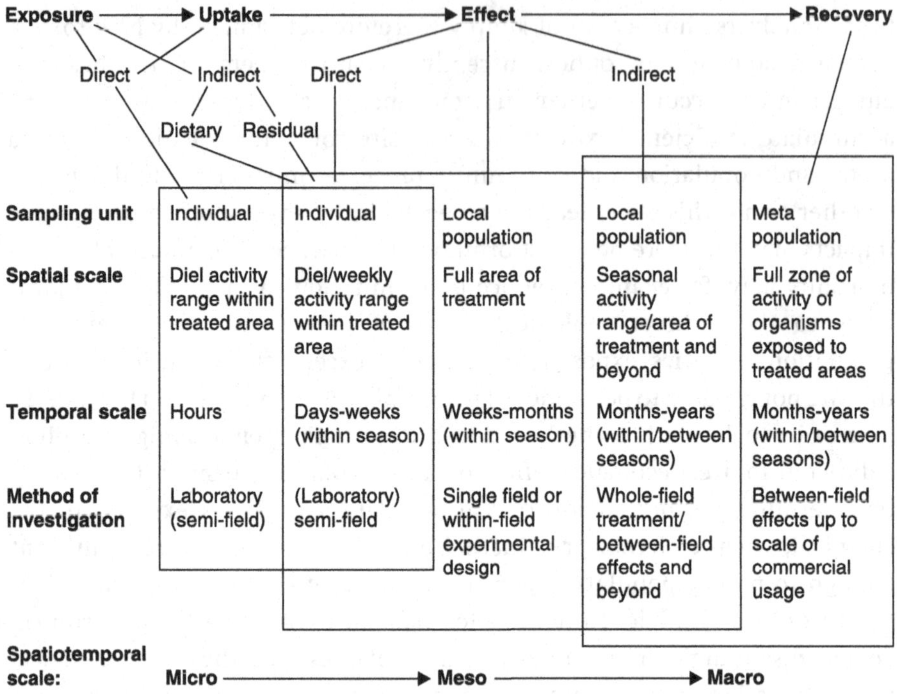

Figure 16.1. Temporal and spatial scales for the processes of contamination and biological effect of pesticides on biological control agents.

(Jepson, 1993*a*). Ultimately, population and community recovery may involve events on a scale well beyond the treated field, and the halo of influence of a chemical application may similarly be extensive.

The significance of this for the IPM practitioner is (1) that pesticide impacts can not simply be predicted from chemical toxicity data alone, and (2) that pesticide side-effects are context-specific, i.e. the ecological consequences of a given level of toxicity will depend upon the properties of system that is being exposed. Some systems may be resilient and relatively tolerant of certain impacts because there are abundant local reservoirs of natural enemies that re-colonize treated areas, whereas others may be far less ecologically competent because these reservoirs do not exist, or because other stressors are already causing significant disruption.

16.2.1 *Understanding and predicting short-term, micro-scale effects*

Processes that evolve within each of the scales defined within Figure 16.1 make significant contributions to the overall population impact of the pesticide exposure, and there is wide variability in values of the key parameters that govern them.

The key parameters that govern effects at the micro-scale, which defines the period immediately following pesticide application and the crop zone that is exposed, include for example (Jepson, 1993*a*):

1. The probability that the natural enemy is directly exposed to the spray cloud (a function of phenology, diurnal activity pattern, and the distribution of the organism within the crop canopy (Cilgi and Jepson, 1992; Everts *et al.*, 1991*a*; Kjaer and Jepson, 1995))
2. The contact rate of the natural enemy with residual deposits of the pesticide (a function of walking rate, the degree of body contact with the substrate and environmental conditions affecting the bioavailability of the pesticide (Salt and Ford, 1984; Jagers op Akkerhuis and Hamers, 1992; Jepson *et al.*, 1990; Wiles and Jepson, 1993a; Longley and Jepson, 1997))
3. The exposure of the natural enemy to the pesticide through its diet (a function of voracity, diet range and factors such as pesticide induced repellency (Wiles and Jepson, 1993*a*; Mullie and Everts, 1992))
4. The susceptibility of the natural enemy to the pesticide (a function of intrinsic physiological factors and physical conditions, including temperature and humidity that affect toxicological process and the survival rate of organisms that are suffering sublethal effects (Croft, 1990; Everts *et al.*, 1991*b*))

Even on the micro-scale there are very high levels of variation in each of these parameters. For example, 40-fold variation was detected in the direct exposure to pesticide spray drops of predatory beetles inhabiting winter wheat ears at the top of the crop canopy, compared with the same species underneath leaves where they were relatively well protected from exposure to the spray cloud (Cilgi and Jepson, 1992). There was also 17.7-fold variation in the ground deposition rate of pesticides throughout the crop growing season (Cilgi and Jepson, 1992) caused by variation in the level of ground cover that occurred as the crop developed (Gyldenkaerne *et al.*, 1999). These differences in insect distribution within the crop and changing characteristics of the crop as a target for pesticide sprays result in wide variation in the rates of direct pesticide exposure of natural enemies to pesticides, with significant consequences for the level of mortality that actually occurs (Jepson, 1993*a*; Wiles and Jepson, 1995). A pesticide that causes high levels of toxicity when the crop is in early growth stages and unable to provide refugia for natural enemies may have limited impact when applications take place in growth stages, where natural enemies are protected from direct exposure to the spray cloud.

There is also variation between natural enemy species in the degree of exposure they experience with a given pesticide deposit. For example, analysis of the relative exposure rates of walking predatory beetles to pesticide residues on soil demonstrated a 2.7-fold difference in the rate of exposure between *Tachyporus hypnorum* (Coleoptera: Staphylinidae) which has a high level of contact with the substrate including its abdomen and tarsi, and *Coccinella septempunctata* (Coleoptera: Coccinellidae) which has only limited contact via the ends of the tarsi (Wiles and Jepson, 1993*b*).

Even wider variation in exposure seems to be associated with differences in substrate type. For example, the toxicity of pyrethroid deposits to predatory beetles on soil may be 40–50 times lower than the toxicity of these compounds on cereal crop foliage at an equivalent dose rate (Wiles and Jepson, 1994). For synthetic pyrethroids, which have a half-life of toxicity of one day for predatory Coleoptera on soil, compared with 4 days on cereal foliage (Unal and Jepson, 1991), these differences could result in a 200-fold difference in pesticide exposure for soil inhabiting predators, compared with foliage-dwelling species that are exposed to synthetic pyrethroid spray deposits.

Predicting the relative impacts of a single spray on the range of natural enemy species inhabiting a crop canopy is therefore far from simple. Concealed natural enemies in dense crop canopies may avoid direct spray exposure and if deposits are repellent, and/or rendered unavailable by sorption processes, even highly toxic pesticides may have lower effects than would be predicted from toxicity data alone (Jepson *et al.*, 1995; Gyldenkjaerne *et al.*, 2000). It is not possible to extrapolate effects from one crop to another or between different natural enemies without significant risks of missing a critical difference in one of the key parameters underlying even the most intimate interactions between the pesticide and the natural enemy at the time of, or immediately following, spray application.

16.2.2 Understanding and predicting longer-term, "meso"- and "macro"-scale effects

If we were to "stop the clock" within a few hours or even a day following pesticide treatment, we would expect the most affected taxa from the natural enemy community to be those that were most susceptible to the pesticide and/or those that were most exposed to it. This ranking of relative impact is unlikely however to persist for long. Most pesticide residues decay rapidly and they may become less bioavailable to the natural enemies that re-colonize the crop canopy. Soon after treatment, some more tolerant taxa will re-invade the treated area, and these will be followed by more susceptible taxa as residues decay further. A few days after treatment, the most susceptible taxa may still be absent, but other natural enemies will have returned because they are highly mobile and re-colonize rapidly. Once most toxic residues have declined, the ranking of relative impact of the pesticide between species, may have as much to do with dispersal capacity and colonization rate as it does with susceptibility and exposure to the pesticide. This outcome is still the result of an ecotoxicological interaction, but we have to recognize that the phenomena that determine the impact of the pesticide include a suite of biological and ecological characteristics that have little to do with the pesticide and its toxicity or mode of action.

The key parameters that govern effects at the meso- and to a more limited extent the macro-scales in Figure 16.1, which encompass these biological and ecological drivers of the longer-term ecotoxicological response, include for example (Jepson 1993*a*):

1. The life history characteristics of the natural enemy (a function of reproductive strategy, fecundity, generation time, and the number of life stages exposed (Jepson, 1988; 1989))
2. The habitat requirements of the natural enemy (the degree of selectivity in habitat conditions, the distribution within crop and non-crop areas and the degree of fragmentation of the population (Jepson, 1988; 1989))
3. The dispersal rate of the natural enemy (the duration and extent of dispersal (Burn, 1992; Jepson, 1989; Sherratt and Jepson, 1993))
4. The diet range of the natural enemy (the degree to which the diet or host range of the natural enemy is restricted to species that are affected by the pesticide (Jepson, 1989))

Organisms that would be expected to be more susceptible to long-term effects which are particularly damaging to IPM systems include those with sensitive life histories (e.g. species with low fecundity and/or species with more than one life stage exposed to toxic pesticide treatments), limited habitat requirements (e.g. species with limited interactions with field boundaries, which provide a refuge from spray residues, and/or species with populations which are fragmented by landscape structure and isolated from reservoirs that might permit rapid recolonization), low dispersal rates (e.g. wingless species and/or species with limited periods of dispersal) and species with sensitive diet ranges (e.g. species that consume prey which are affected by the pesticide treatment).

The population responses of the multiple natural enemy species that inhabit crops are again highly variable, by virtue of the variation in life histories between species and the diversity and structure of crop and non-crop habitats that surround treated fields. Experimental data from the meso- and macro-scales portrayed in Figure 16.1 are rare, but lend support to the factors listed above as being of importance in determining the relative risks of longer-term effects (Jepson, 1993*a*).

Within-field investigations have for example, measured the relatively slow reinvasion rates of cursorial species, particularly Carabidae (Thacker and Jepson, 1991; Duffield and Aebischer, 1994) and long-term, multiseason, multifield experiments, have identified Carabidae as among the most susceptible taxa to long-term, population depletions in sprayed systems (Burn, 1992; Holland and Luff, 2000). Carabidae are not particularly susceptible to pesticides compared with many of the other natural enemies that they share the crop environment with, but they are disproportionately impacted because they experience slow rates of population recovery following pesticide treatments. In contrast, many natural enemy species, particularly those with high dispersal and reproductive rates, exhibit only short-term, transitory depletions following exposure to pesticides, followed by rapid population recovery. The distribution of effects among species is difficult to predict based upon experimental data that are often confined to a small spatial scale relative to the scale of population processes for the natural enemy species exposed to pesticide treatments. IPM practitioners are therefore required to consider long-term population monitoring of key natural enemy species within their systems to gain insights into their status at any given time, and trends in population densities that might indicate a transition into a pesticide use regime that limits the ability of key species to recover following pesticide exposure. Such adverse changes may occur, for example, if the toxicity of pesticides to these species increases, the frequency of spraying changes to coincide with critical stages in natural enemy life history or of the area sprayed increases, thus reducing the proportion of the landscape that provides refugia.

16.2.3 *Encompassing the scales of pesticide–natural enemy interactions and effects*

Progression through the various stages of the ecotoxicological process, and the transition between the scales portrayed in Figure 16.1, may be logical on paper, but very few research projects have encompassed this breadth. There are consequences for this, the most important of which is an excessive focus on phenomena that take place at the micro-scale, often to the exclusion of phenomena that evolve on the meso- and macro-scales. Thus, it was only in the 1990s, with the advent of multifield experimental regimes to investigate pesticide side-effects, that the types of natural enemies that are most susceptible to long-term population depletion in sprayed agroecosystems were actually first identified (e.g. Burn, 1992).

Experiments on inappropriately small scales are subject to rapid reinvasion of treated areas by natural enemies from untreated control plots (Jepson and Thacker, 1990; Thomas *et al.*, 1990) and they misrepresent or even overlook the phenomena that evolve from spray treatments to whole fields. The outcome of this has been an emphasis upon the susceptibility of taxa that are sensitive to effects at the micro-scale (largely foliage-borne, pest specialists with high reproductive rates and high dispersal rates), rather than the taxa with populations which are under the most severe threat of population depletion (largely soil and foliage borne polyphagous predators with low reproductive rates and limited dispersal powers). Clearly, IPM programs need to minimize pesticide impacts to both of these groups, but it could be argued from an ecological standpoint, that long-term depletion of a natural enemy population that may span more than one crop growing season, is a more severe impact than more transitory impacts, where populations recover within the crop growing season, to reach the densities they achieved prior to pesticide exposure. It may also be argued from an economic standpoint that the full spectrum of natural enemy types is required, including both of the groups referred to above, if long-term pest suppression is to be achieved.

The unintended focus of research into pesticide–natural enemy interactions on species that experience severe short term impacts, rather than those that suffer longer-term ecological impacts, has consequences for the way in which we regulate pesticides, and the management decisions we make in IPM to protect and preserve natural enemies. The next section examines the extent to which the measures that we take to protect organisms affected at the micro-scale are protective of the organisms most affected at the meso- and macro-scales. If by protecting the taxa that are immediately susceptible to pesticide spraying through effective regulation and effective IPM practices, we protect those taxa that exhibit impacts only in the longer-term, then questions concerning differentiation between long-term and short-term effects becomes an academic issue rather than a practical agricultural conundrum to solve.

16.3 How protective are pesticide regulatory regimes for natural enemies?

Internationally, we have entered a new world for pesticide regulation, where some regions (e.g. the European Union) require toxicological testing of pesticides against natural enemy species prior to registration or during re-registration processes for older pesticides (Barrett *et al.*, 1994; Campbell *et al.*, 2000; Candolfi *et al.*, 2001; Jepson, 1993*b*). These testing procedures have evolved over approximately 30 years, and reflect a rigorous attempt to establish effective protocols for test species that are sufficiently representative of natural enemy taxa for the resultant regulatory decisions to be broadly protective (Jepson, 1993*b*). The emphasis of the majority of these regulatory tests is to measure acute toxicological impact, either at field dose rate or at multiple dose-rates within dose-response bioassays that are undertaken in the laboratory.

These approaches have two major benefits. Firstly, they identify highly toxic, broad spectrum pesticides that may be unsuitable for IPM purposes and which are subject to label restrictions that limit their use, or even rejection by the regulatory authority. Secondly, they identify compounds of very limited or no toxicity to natural enemies, which may be highly compatible with IPM programs. If these procedures limit the number of broad spectrum, non-selective pesticides that are registered, or restrict their use, then they are likely to protect against widespread population impacts among natural enemies of all types, even though the procedures that identified them are based wholly upon measurements that take place at the micro-scale, and emphasize the physiological, rather than the ecological susceptibility of the test taxa. Similarly, broad protection for all natural enemy types may arise through the use of these testing regimes to identify non-toxic materials. It is highly unlikely, even impossible, for longer-term direct ecological effects to emerge at the meso- or macro-scales with non-toxic compounds, at least among taxa that are related to the test species that are used in the regulatory testing.

Problems with regulatory procedures that are driven solely by acute impacts against natural enemy taxa immediately following exposure to the pesticide may arise, however, where compounds of low to intermediate toxicity pass through the regulatory process without triggering concern (Stark *et al.*, 1995a). This is because species with the ecological attributes that underlie significant long-term population level effects may be harmed by relatively low pesticide-induced mortalities. It may therefore be argued that concern should not simply focus upon screening out the most toxic materials (Jepson and Sherratt, 1996), but also focus on less toxic pesticides that might cause severe impacts through frequent or widespread use (Sherratt and Jepson, 1993). Advances in regulatory testing that take the ecological susceptibility of natural enemies into account in determining acceptable levels of mortality during risk assessment, although scientifically justified, are probably not going to take place for a number of years. Until this occurs, it cannot be stated with any degree of confidence that current regulatory testing procedures are broadly protective of all types of natural enemy. It must also be borne in mind that, currently, this form of testing takes place only in Europe, and most regulatory systems internationally do not explicitly address the toxicity of pesticides to natural enemies, despite the logic of doing so.

Differences in basic life history and demographic attributes between organisms also underlie potentially large and counter-intuitive differences in the rates of population recovery following exposure to pesticides. This may also have implications for the degree to which the results of any laboratory-based toxicological testing can be extrapolated to determine the potential for impacts in the field (Stark *et al.*, 2004). There is still therefore, a general need for field-based evaluations of pesticide impacts following commercial release of pesticides, that are on a scale that is large enough to detect those situations where long-term, population effects may arise that significantly inhibit pest limitation by natural enemies. We still know far too little about those situations that trigger pest resurgence. Effects that persist for a whole season, or which continue into a following season, may be of far greater significance for pest limitation and IPM system development than effects that are transitory in nature. Of particular concern are those crops where repeated applications take place within a season, and those commodities where applications take place over large contiguous areas.

The results of these investigations should feed back to the design of regulatory test procedures to add further tiers of protection to natural enemies, if they are needed. They may also form the basis for cases to introduce regulatory testing of pesticides against natural enemies in countries where these procedures have not yet become established. In this regard, IPM researchers have an opportunity to feed information back to pesticide regulators that develop and undertake ecological risk assessments. If certain circumstances (i.e. combinations of pesticide, cropping system, pesticide use regimes and natural enemy community) repeatedly result in unacceptable limitation of natural enemy function, then it may be more appropriate to adjust the ecological risk assessment procedure to prevent that pesticide being used in these circumstances than to tolerate the disruption that is caused by continued use of this compound.

16.4 Alternative, ecologically based indexes of pesticide toxicity to natural enemies

This chapter has outlined the mechanisms and processes that underlie the ecotoxicological effects of pesticides on natural enemies, and it has explained the distinction between species that may suffer acute, short-term impacts but recover rapidly, and species that suffer disproportionately long population reductions even though they may not be particularly susceptible to the pesticide. Pesticide screening methods tend to rank organisms in terms of their intrinsic susceptibility to toxicological effects rather than ecological effects. These methods may be broadly protective of all natural enemy types where they detect the two extremes of either highly toxic, broad spectrum pesticides, where there is a clear potential for both short and long-term effects, and low or zero toxicity compounds, where in theory, neither form of impact should occur. For the majority of pesticides that fall between these extremes, it may be worth exploring approaches that permit pesticides to be evaluated in terms of both their toxicological and their ecological impacts.

Population recovery may represent a more attractive endpoint for the definition and quantification of unacceptable pesticide impacts than end-points which simply measure toxicity in the laboratory or the magnitude of population depletion soon after spraying in the field (Jepson and Thacker, 1990). This approach has also been advocated in a number of recent initiatives concerning the future of pesticide regulation for freshwater aquatic systems (e.g. Campbell *et al.*, 1999; Giddings *et al.*, 2002). Defining population recovery is however complex, and a number of definitions have been employed in risk assessment of pollutants, including pesticides (Maltby *et al.*, 2001). The procedures to measure recovery rates require whole fields to be available, or experimental plots that are equivalent in scale to commercial fields, so that recovery rates can be realistically estimated (Jepson and Thacker, 1990). They also entail extensive natural enemy sampling within treated areas, to track re-invasion by dispersal, as well as recovery that takes place through reproduction. When these procedures are followed however, the data that are collected in studies of this type can identify the mechanisms that underlie pest resurgence, and also permit comparisons of the recovery rates of different taxonomic groups of natural enemies in particular cropping systems (e.g. Duffield *et al.*, 1996).

It is unlikely that recovery rates can be measured directly in the field for large numbers of cropping systems and the suites of pesticides that they are exposed to. Can differences between pesticides in the rates at which they might accommodate the recovery of natural enemy populations following use, be estimated from chemical parameters and toxicological data, the same information-base that is currently employed in regulatory testing of natural enemies against pesticides? Van Straalen and Van Rijn (1998) distinguished between "ecotoxicological recovery time" (defined as the time when the pesticide dissipates to a level of bioavailability when it no longer has any adverse effect on the invertebrate community) and "ecological recovery time" (which is defined as the time that it takes for restoration of the original species composition and associated ecological functions after disturbance). The potential benefit of using ecotoxicological recovery time as a predictive tool for determining the likelihood of adverse impacts at the meso-scale and above (Figure 16.1), is that it can be estimated from laboratory-derived "community sensitivity statistics" (calculated from the susceptibilities of a number of test organisms to the pesticide in question, under conditions of controlled and uniform exposure) and the degradation rate of the pesticide (which is required data as a part of the pesticide regulatory process).

Van Straalen and Van Rijn (1998) compare predicted ecotoxicological recovery times for soil invertebrates, estimated from their laboratory data, with observed recovery times from published field studies for similar taxa. They found relatively close concordance for certain taxa between the predicted and observed recovery times, which, in the terminology developed in Figure 16.1, would imply similarity between effects measured at the micro-scale, and effects measured at the "meso" and macro-scales. The taxa that provided the closest correlations were however largely non-dispersive species and included Acari, Collembola and Lumbricidae. The differences between effects detected at the micro- and meso-scales might be expected to be small for species that recover through reproduction and localized dispersal from within the soil profile. This is because there will be limited delays in the recovery process for these species once pesticide residues have decayed to non-toxic concentrations because of their proximity to the treated area.

For more dispersive natural enemy taxa, including many natural enemies, population recovery may be delayed after the estimated "ecotoxicological recovery time" has been reached, because it is likely to be mediated by dispersal and colonization processes from beyond the treated area. If reservoirs of colonists are not locally available, then populations in the treated area may remain depleted after residues have decayed. If the treated area is large, there may also be delays caused by the time it takes for natural enemies to progress from reservoir habitats to the center of the treated area.

The process, of gradual "edge to center" recovery, once pesticide residues have dissipated is evident from field-derived data for pesticide impacts. For example, Unal and Jepson (1991) undertook leaf and soil in situ bioassays with carabid ground beetles after application of the organophosphate pesticide dimethoate to a cereal crop canopy. Dimethoate toxicity to Carabidae decayed completely after nine days on both leaf and soil substrates. In theory therefore, population recovery by Carabidae could occur once pesticide toxicity had declined. The period of toxicity for dimethoate did in fact correlate well with measurements of the minimum time for population recovery by Carabidae at the edges of dimethoate-sprayed cereal plots, which was also approximately nine days (Jepson and Thacker, 1990). However, Jepson and Thacker (1990) also demonstrated that time to recovery for Carabidae increased as a function of distance into the treated area, such that populations took over twenty days to recover at distances greater than 100 m into the sprayed area.

Population recovery rate is therefore a function of the ecological attributes that drive risk at the meso- and macro-scales, including dispersal characteristics of affected taxa and the scale of treatment. Population recovery for dispersive natural enemies can not simply be derived from the parameters that determine ecotoxicological recovery time, although these do enable the minimum time for recovery to be estimated. Thus ecological and ecotoxicological recovery time are unlikely to be correlated for more dispersive taxa and care must be taken to consider the proximity and ecological integrity of population reservoirs of beneficial taxa that lie beyond treated crops, and the scale of treatment when considering the recovery process and the ability of a cropping system to sustainably tolerate pesticide impacts.

The use of ecotoxicological recovery time as an indicator of potential risk to populations of natural enemies may still however have considerable value, despite the criticism that it may underestimate the true rate of ecological recovery for dispersive taxa. In situations where sprays are applied frequently for example, the concept could be used to estimate the level of population recovery by natural enemies that might take place between sprays. For natural enemies that are highly dispersive, artificially introduced, or for which productive refugia exist close to the treated area, ecotoxicological and ecological recovery time may be very similar. The use of ecotoxicological recovery times that are calculated from in situ bioassays with appropriately sensitive endpoints (Jepson, 1993*b*) could also permit efficient screening of numerous compounds for compatibility with IPM and biological control. Finally, van Straalen and van Rijn (1998) make an excellent case for the use of ecotoxicological recovery time for comparisons between pesticides, and demonstrate that for a group of 10 insecticides, fungicides and herbicides, this time may vary between 0 years (for non-toxic materials) to 3.8 years for toxic and persistent chemicals, and radically alter the relative rankings of pesticides compared with simple listings of toxicity. Importantly, in ecotoxicogical recovery time, they have also defined a statistic that provides a link between the micro- and meso-scales of the ecotoxicological interaction between pesticides and natural enemies, and a statistic that might be effectively incorporated within models of pesticide ecological impact.

16.5 Pesticide impacts at the agroecosystem scale: the macro-scale and beyond

The ecotoxicological approach to considering pesticide natural enemy interactions explicitly considered scale from the outset (Figure 16.1). In doing so it teased out a number of mechanisms that were exclusively associated with the three scale ranges that were identified. The scales where direct toxic effects arise and are played out were separated into a micro-scale, where direct and indirect exposure rates to organisms of differing susceptibilities within the crop canopy determine the relative toxicological impacts upon them, and a meso-scale, where residues and pesticide bioavailability are decaying, organisms are reinvading from beyond the treated area and reproductive and sublethal effects are delivering their impacts. Once toxicity has waned, effects may still be unfolding, but on the macro-scale where the ecological attributes of the organisms, and their interactions with habitat layout and habitat quality determine the long-lasting impacts that the initial toxic effects may have. This differentiation of mechanisms and processes between temporal and spatial scales is termed "scale dependency", and it is of particular significance for the investigation and management of pesticide impacts on natural enemies, because different types of natural enemies are sensitive to the types of effects that ensue from the underlying processes that are associated with different scales. Linyphiid spiders are, for example, far more susceptible to pesticides than carabid ground beetles, indicating higher sensitivity at the micro-scale, but their high rates of reproduction and dispersal provide for much more rapid population recovery than Carabidae, and it is the ground beetles that exhibit high sensitivity compared with Linyphiidae at the meso-scale and beyond (Jepson and Thacker, 1990; Sherratt and Jepson, 1993).

Does a similar discontinuity in the species that are most sensitive to impacts lie between the meso- and macro-scales, i.e. the transition between the field scale that is experimentally tractable, and the real world, where spray application takes place over extensive areas and throughout the life cycles of natural enemies? If such a discontinuity does exist, might there be yet another set of driving parameters that are unforeseen and that are driving currently unmeasured impacts on important natural enemy taxa? If such effects exist, and if they are driven by phenomena that evolve at this larger spatial scale, how important are they relative to micro- and meso-scale phenomena that are more readily measured and managed? If we are to effectively manage natural enemies in sprayed systems, or insert islands of organic production (that rely upon natural enemies) within seas of conventional pesticide usage, we need to understand the degree to which we affect regional dynamics of dispersive natural enemies by using pesticides.

Jepson and Thacker (1990) estimated time to recovery for populations of Carabidae exposed in a number of studies to the pesticide dimethoate in winter wheat crops. They demonstrated a positive correlation between time for population recovery to take place and experimental surface area among numerous investigations. In long-term experiments that have moved beyond the conventional, within-field experimental scale to whole-fields, there is evidence that population recovery can be so delayed that organisms are absent for several seasons. This constitutes local extinction (Burn, 1992), i.e. the persistent absence of a particular taxon from a treated field. This outcome of pesticide impacts against natural enemies is qualitatively and quantitatively distinct from the delayed recovery discussed above. We do not know how frequently local extinction events occur, but they form a distinct class of effect that is associated with whole-field or farm treatments, or applications on a larger scale that are compatible with the scale of population processes for affected taxa. This was termed the macro-scale in Figure 16.1 and the existence of an ecotoxicological endpoint, local extinction, that is unique to it confirms that scale dependency in the impacts of pesticides on natural enemies continues as scales move well beyond the treated field and farm to the agroecosystem.

Why should local extinction occur on these larger scales of exposure? The existence of boundaries, including roads, ditches and hedgerows between fields may enhance the severity of population impacts of pesticide spraying by delaying recovery processes. One mechanism underlying this apparent scale dependency in pesticide impacts may therefore be the additional habitat complexity that arises as the scale of exposure expands. This phenomenon was explored by Sherratt and Jepson (1993) in a mathematical model of the theoretical metapopulation dynamics of Carabidae exposed to pesticides on a whole-farm scale. They demonstrated that the chance of populations persisting within a sprayed field improved if neighboring fields were left unsprayed. Less intuitively, there appeared to be an optimal dispersal rate of carabids which maximized the chance of persistence in sprayed fields. They also demonstrated tradeoffs between the chance of local persistence in a field and pesticide toxicity, dispersal rate, reproductive rate and habitat structure, particularly the rate at which beetles crossed field boundaries.

At the farm scale therefore, a complex suite of factors may intensify or mitigate pesticide impacts, and the relative importance of these is dependent upon spatial scale. It is not possible to extrapolate simple population recovery models, derived from field experiments or estimates of ecotoxicological recovery time, to larger spatial scales where significant changes in habitat complexity take place. Habitat structure and its interaction with natural enemy population biology is likely therefore to provide additional parameters that mediate pesticide ecological risks on larger spatial scales (i.e. particularly the macro-scale in Figure 16.1).

Scaling up significantly, we are however faced with a lack of monitoring data for natural enemy taxa in agroecosystems that we can use to seek evidence for further tiers of scale dependency in the population impacts of pesticide use. Table 16.2 represents an attempt to examine the monitoring data that exist for Carabidae exposed to pesticides in temperate, European cropping systems.

Effects are greatest in intensity at the 10 square km scale (equivalent to a single farm in the UK), where intensive pesticide use within a narrow crop rotation of contiguous fields was sufficient to cause local extinction (Burn, 1992) (Table 16.2). Evidence for superabundance of some Carabidae, in contrast to the local extirpation of others in the farm scale experimental regime used by Burn, illustrates the importance of differentiating individual species life history attributes in determining population impacts (Jepson, 1989). Some Carabidae disperse by flight and have higher reproductive rates, enabling them to exploit fields from which other taxa, with lower dispersal and reproductive rates, have been lost. This finding also illustrates the importance of understanding predator–prey dynamics within the sprayed system, and the pattern of exploitation of newly available prey that have been temporarily released from predator limitation (Duffield *et al.*, 1996). For populations of polyphagous predators to become superabundant, there must be prey to exploit. Duffield *et al.* (1996) demonstrated, for example, the local resurgence of aphids and Collembola in the centers of sprayed areas that had suffered longer-term depletions and delayed recovery by the predatory fauna following pesticide use.

Table 16.2. *Impacts of agricultural practices, particularly pesticide application, on Carabidae, over a range of spatial and temporal scales*

Spatial/temporal scale of investigation	10^2 yr	10–20 yr	< 10 yr	< 1 yr	≪1 yr
10^2–10^3 km² *Dutch Carabidae in the twentieth century*	No detectable impact of modern agriculture[a]				
10–10^2 km² *30 km² of UK farmland over 20 years*		No decline detected[b]			
10 km² *Multiple fields over 7 years*			Local extinction, local superabundance[c]		
< 10 ha *Within-field, within season experiments*				Local recovery by diffusion[d]	
< 1 ha *Small-scale plot trials*					Evidence of toxicity over 7–10d[e]

[a]Hengeveld, 1985
[b]Aebischer, 1991
[c]Burn, 1992
[d]Jepson and Thacker, 1990
[e]Fischer and Chambon, 1987

At spatiotemporal scales below the farm scale, population recovery is artificially rapid and a function of the specific experimental design in use (i.e. small plot sizes, with contiguous untreated areas that provide reservoirs of natural enemies that rapidly re-invade treated areas). At spatiotemporal scales above this, but below 10 ha, where individual fields, or groups of fields are treated and gradual recovery takes place, it is likely that unsprayed and less cultivated habitats in the farmland matrix provide an adequate refuge from the impacts of pesticide use for populations of Carabidae to persist. This pattern will, however, be highly dependent upon the life history of the taxa in question, the land use and habitat distribution within the agroecosystem under investigation, and the use pattern and toxicity of pesticides.

Although no decline in Carabidae was detected by Aebischer (1991), in a large scale invertebrate monitoring survey, he did find significant declines in a number of taxa other than Carabidae across a farming system, including Araneae, Cryptophagidae, Lathridiidae, Lepidoptera, Lonchopteridae, Opiliones, Parasitica, Staphylinidae, Symphyta and Aphididae. As least some of these declines are attributed to pesticide use, although the mechanisms that underlie these effects are not known.

Finally, the largest spatiotemporal scale reported in Table 16.2, trapping data throughout the Netherlands for over a century failed to find a clear signal of impact from modern agriculture. Instead, on this scale, cyclical climatic variation was the dominant signal in the periodic changes in carabid communities. So, although on a local scale, extirpation may have been taking place as a result of pesticide exposure, this form of impact derived exclusively from agriculture was not evident at the scale of a much larger and more complex crop and habitat matrix subjected uniformly to the effects of climatic cycles.

Investigations of the spatial dynamics of the spider family Linyphiidae exposed to a range of agricultural practices, including pesticides (Stark *et al.*, 1995; Thomas and Jepson, 1997) have also led to models that explore population processes on large spatial scales, equivalent to agroecosystems (Halley *et al.*, 1996; Thomas *et al.*, 2003). These models demonstrated the importance of landscape heterogeneity for spider survival and abundance, and again suggest that there is an optimum dispersal strategy that maximizes the persistence of spiders in sprayed landscapes, as may also be the case for Carabidae. Weyman *et al.* (1995) also demonstrated that Linyphiidae, which are abundant in agriculture, disperse at rates which permit them to persist within the landscapes that Halley *et al.* (1996) modeled. It is obvious that agricultural landscapes will only contain the taxa that are adapted to survive within them, but this finding is significant because Linyphiidae are among the most susceptible natural enemies to pesticides, and the only reason they can persist in sprayed systems may be that they are adapted to colonize and exploit temporary habitat islands by having high rates of reproduction and dispersal. Many other susceptible natural enemies are less fortunate, and this form of experimentation, analysis and modeling could also be used to identify species that would return to agriculture if the toxicity and intensity of pesticide spraying was reduced, and appropriate natural enemy habitats were added back into the landscape.

In conclusion, there is evidence for a discontinuity between pesticide impacts on natural enemies that take place at the meso scale, associated with field experimentation and the macro-scale that is more consistent with

Table 16.3. *Applications of ecotoxicology to the management of pesticide–natural enemy interactions for IPM researchers and practitioners*[a]

ANALYSIS AND MANAGEMENT OF SHORT-TERM PESTICIDE IMPACTS

- Develop an inventory of natural enemies, including taxonomic composition, pest associations, phenology and patterns of distribution within the crop canopy and surrounding off-crop habitats
- Develop an inventory of pesticides used against key pests, including details of application rates, application method and timing
- Obtain literature data on susceptibility of the natural enemies that have been listed to the pesticides to which they may be exposed
- Develop and implement a program of bioassays to confirm toxicity values from the literature, fill knowledge gaps in natural enemy susceptibility and provide toxicological statistics for risk analysis
- Measure pesticide distribution through the crop canopy at different growth stages, using appropriate application techniques, and/or measure exposure of natural enemies to pesticides directly
- Undertake in situ bioassays to determine persistence of toxicity of the main pesticides to key natural enemies
- Calculate short-term risk from exposure and susceptibility data for key natural enemies exposed to the pesticides that are used when they are active in the crop canopy, and at the appropriate crop growth stages
- Exploit opportunities for physiological selectivity by ranking pesticides in terms of toxicity to key natural enemies, and include this analysis in procedures to select pesticides for use with the IPM program.

MITIGATION OF SHORT-TERM RISKS FOR PESTICIDES THAT ARE TOXIC TO NATURAL ENEMIES

- Exploit ecological selectivity by avoiding application of toxic pesticides at times when key natural enemies are active, including periods in the day when they are most exposed to pesticides, and periods in the season when they are most abundant
- Exploit ecological selectivity by improved targeting of sprays to reduce natural enemy exposure, or by strip treatments.

MANAGEMENT OF LONG-TERM RISKS FOR PESTICIDES THAT ARE TOXIC TO NATURAL ENEMIES

- Rank natural enemies in terms of the ecological attributes that underlie susceptibility to long-term ecological impacts
- Undertake field experiments that directly measure rates and patterns of recovery following chemical exposure
- Determine the mechanisms that enable recovery, including reproduction and colonization and identify sources or reservoirs of natural enemies within the agroecosystem
- If recovery for at-risk species takes place within the cropping season, and is considered sufficient to maintain natural enemy populations, initiate a program of monitoring that is sensitive enough to detect declines in the abundance of at-risk species between seasons

- If recovery does not take place within the cropping season, and is not considered sufficient to maintain the natural enemy population, consider expansion of natural enemy refugia in off-crop areas to enhance recovery rates
- If natural enemy recovery rates are inadequate and can not be enhanced within the cropping system when a given pesticide is in use, return to the start of the process and select alternative pesticides.

[a]A chronological sequence of steps that apply the principles and procedures outlined in this chapter for application to a specific crop setting. This outline assumes that exhaustive efforts have been undertaken to exploit alternatives to pesticides before these steps are undertaken.

actual agricultural practice. Impacts at this larger scale are affected by the pattern and intensity of spray application and interactions between natural enemies and farmland habitat types, mediated by their life histories and dispersal patterns. Without long-term monitoring data sets for natural enemies, we lack the information that is needed to determine how widespread phenomena such as extirpation are in affecting natural enemy species in sprayed farming systems. The conclusions for IPM that can be drawn from this analysis are clear. Choose your scale carefully. At the scales that are relevant to agriculture, individual farm managers and decision makers, natural enemy populations may experience long-term impacts of pesticide use. Pesticide management and risk mitigation practices are also most effectively implemented on this scale, and it could be argued than many of the severe impacts cited above can be avoided.

16.6 Conclusions for IPM researchers and practitioners

Pesticide impacts on natural enemies exhibit scale dependency that limits our ability to extrapolate from effects at one scale to effects at another. The differences in the spectrum of affected taxa between the micro- and meso-scales is sufficiently large and counter intuitive for us to expect similarly large differences between the meso- and macro-scales defined in Figure 16.1. Population monitoring, or even modeling, of natural enemy population dynamics in sprayed systems is, however, not widely practiced and we largely ignorant of the larger scale, longer-term impacts that pesticides have on natural enemy populations in agroecosystems. Enhanced knowledge of this would enable us to manage natural enemies more effectively, and to minimize pesticide impacts in sprayed agroecosystems.

An ecotoxicological perspective, that addresses patterns of exposure and susceptibility of natural enemies on a small experimental scale, can be used to explore more selective approaches to pesticide use, for example, use of less toxic materials, dose-reduction and pesticide spray targeting. It can also provide a clear mechanistic understanding of pesticide natural enemy interactions that can help to explain unexpected impacts in the field. This scale is also compatible with the lower tiers of pesticide screening for regulatory purposes and IPM researchers could profit by adopting some of the methods developed for ecological risk assessment in the development of more selective pesticide use practices. Table 16.3 provides a simple guideline for IPM researchers that outlines an ecotoxicological approach to pesticide management that minimizes the risks of both short-term and long-term pesticide impacts to natural enemies.

IPM practitioners and researchers are also in a position to feed their data on natural enemy impacts back to pesticide regulators in such a way that it can be used to tune the regulatory process and address questions of pesticide toxicity spectrum and persistence in the context of natural enemy impacts. If large scale population impacts are shown to be occurring in sprayed agroecosystems, it is perhaps time to consider regulatory restrictions on the pesticides, application patterns and practices that may underlie these effects.

16.7 An ecotoxicological reality check for IPM researchers and practitioners

Although a number of new pesticides have been marketed in the last few years, broad spectrum pesticides, including organophosphate, carbamate and synthetic pyrethroid compounds, are still widely applied in the USA and internationally. These compounds are among the most toxic to natural enemies, and they are associated with many of the ecotoxicological impacts referred to above, and most of the adverse consequences for IPM that result from damage to natural enemy communities. Crop pesticide use in the USA peaked at an estimated 263 million kg in 1997, and fell to 225 million kg in 2004, partly as a result of the pesticide reductions achieved following the introduction of genetically modified crops. The most recent data for the share of the insecticide market occupied by broad spectrum insecticides comes from the year 2000, when organophosphates, carbamates and pyrethroids accounted for 95% of insecticide applications (Osteen and Livingston, 2006). The problems referred to above are clearly not going to be swept away overnight by the introduction of a wave of selective insecticides. This puts the onus for action on the shoulders of IPM practitioners who can apply more analytical ecotoxicological approaches to understand and manage the impacts of broad spectrum insecticides on natural enemy populations in agricultural crops.

References

Aebischer, N.J. (1991). Twenty years of monitoring invertebrates and weeds in cereals in Sussex. In L. Firbank, N. Carter and G.R. Potts (eds.), *Ecology of Temperate Cereal Fields*. Oxford, UK: Blackwell Scientific Publishers. pp. 305–31.

Barrett, K.L., Grandy, N., Harrison, E.G., Hassan, S. and Oomen, P. (1994). *Guidance document on regulatory testing procedures for pesticides with non-target arthropods*. From the workshop on European Standard Characteristics of beneficials Regulatory Testing (ESCORT). Society of Environmental Toxicology and Chemistry (SETAC), Europe, Brussels, Belgium.

Burn, A.J. (1992). Interactions between cereal pests and their predators and parasites. In P. Greig-Smith, G.H. Frampton and A. Hardy (eds.), *Pesticides and the Environment: The Boxworth Study*. London, UK: HMSO. pp. 110–31.

Calow, P. (1994*a*; *b*). *Handbook of Ecotoxicology*. Volumes I and II. Oxford, UK: Blackwell Science Ltd.

Campbell, P.J., Arnold, D.J.S., Brock, T.C.M. *et al.* (1999). *Guidance Document on Higher-Tier Aquatic Risk Assessment for Pesticides (HARAP)*. Society of Environmental Toxicology and Chemistry (SETAC), Europe, Brussels, Belgium.

Campbell, P.J., Brown, K.C., Harrison, E.G. *et al.* (2000). A hazard quotient approach for assessing the risk to non-target arthropods from plant protection products under 91/414/EEC: hazard quotient trigger value proposal and validation. *Anzeiger fur Schadlingskunde (Journal of Pest Science)*, **73**, 117–24.

Candolfi, M.P., Barrett, K.L., Campbell, P.J. *et al.* (2001). *Guidance document on regulatory testing and risk assessment procedures for plant protection products with non-target arthropods*. From the ESCORT 2 workshop European standard characteristics of non-target arthropod regulatory testing. SETAC-Europe (ISBN 1 880611 52 X).

Chamberlain, D.E., Fuler, R.J., Bunce, R.G.H., Duckworth, J.C. and Shrubb, M. (2000). Changes in the abundance of farmland birds in relation to the timing of agricultural intensification in England and Wales. *Journal of Applied Ecology*, **37**, 771–88.

Cilgi, T. and Jepson, P.C. (1992). The direct exposure of beneficial invertebrates to pesticide sprays in cereal crops. *Annals of Applied Biology*, **121**, 239–47.

Croft, B.A. (1990). *Arthropod Biological Control Agents and Pesticides*. New York: Wiley Interscience.

Croft, B.A. and Brown, A.W.A. (1975). Responses of arthropod natural enemies to pesticides. *Annual Review of Entomology*, **20**, 285–335.

Duffield, S.J. and Aebischer, N.J. (1994). The effect of spatial scale of treatment with dimethoate on invertebrate population recovery in winter wheat. *Journal of Applied Ecology*, **31**, 263–81.

Duffield, S.J., Jepson, P.C., Wratten, S.D. and Sotherton, N.W. (1996). Spatial changes in invertebrate predation rate in winter wheat following treatment with dimethoate. *Entomologia Experimentalis et Applicata*, **78**, 9–17.

EPA. (1998). *Guidelines for Ecological Risk Assessment. U.S. Environmental Protection Agency*, EPA/630/R095/002F. Risk Assessment Forum, Washington, DC; EPA.

Everts, J.W., Aukema, B., Mullie, W.C. *et al.* (1991*a*). Exposure of the ground-dwelling spider, *Oedothorax apicatus* (Blackwall) (Erigonidae) to spray and residues of deltamethrin. *Archives of Environmental Contamination and Toxicology*, **20**, 13–19.

Everts, J.W., Willemsen, I., Stulp, M. *et al.* (1991*b*). The toxic effect of deltamethrin on linyphiid and erigonid spiders in connection with ambient temperature, humidity and predation. *Archives of Environmental Contamination and Toxicology*, **20**, 20–4.

Fischer, I. and Chambon, J.P. (1987). Faunistic inventory of cereal arthropods after flowering and incidence of insecticide treatments with deltamethrin, dimethoate and phosalone. *Med. Fak. Landbouw. Rijksuniv. Gent.*, **52**, 201–11.

Fuhrer, G.J., Morace, J.L., Johnson, H.M. *et al.* (2004). *Water Quality in the Yakima River Basin, Washington, 1999–2000*. US Geological Survey Circular 1237, Reston, Virginia, USA.

Giddings, J., Brock, T., Heger, W. *et al.* (2002). *Community Level Aquatic Systems Studies Interpretation Criteria (CLASSIC)*. Society of Environmental Toxicology and Chemistry (SETAC); Pensacola, Florida, USA.

Gilliom, R.J., Barbash, J.E., Crawford, C.G. *et al.* (2006). *The Quality of our Nation's Waters: Pesticides in the Nation's Streams and Ground Water, 1992–2001*. Circular 1291, Reston, VA: US Department of the Interior, US Geological Survey.

Gyldenkjaerne, S., Ravn, H.P. and Halling-Sorensen, B. (2000). The effect of dimethoate and cypermethrin on soil dwelling beetles under semi-field conditions. *Chemosphere*, **41**, 1045–57.

Gyldenkaerne, S., Secher, J.M. and Nordbo, E. (1999). Ground deposit of pesticides in relation to the cereal canopy density. *Pesticide Science*, **55**, 1210–6.

Halley, J.M., Thomas, C.F.G. and Jepson, P.C. (1996). A model of the spatial dynamics of linyphiid spiders in farmland. *Journal of Applied Ecology*, **33**, 471–92.

Hardin, M.R., Benrey, B., Coll, M. *et al.* (1995). Arthropod pest resurgence: an overview of potential mechanisms. *Crop Protection*, **14**, 3–18.

Hengeveld, R. (1985). Dynamics of Dutch beetle species during the 20th century. *Journal of Biogeography*, **12**, 389–411.

Holland, J.M. and Luff, M.L. (2000). The effects of agricultural practices on Carabidae in temperate agroecosystems. *Integrated Pest Management Reviews*, **5**, 109–29.

Jagers op Akkerhuis, G.A.J.M. and Hamers, T.H.M. (1992). Substrate-dependent bioavailability of deltamethrin for the epigeal spider *Oedothorax apicatus*. *Pesticide Science*, **36**, 59–68.

Jepson, P.C. (1988). Ecological characteristics and the susceptibility of non-target invertebrates to long-term pesticide side-effects. In *Field Methods for the Study of the Environmental Effects of Pesticides. BCPC Monographs*, **40**, 191–200.

Jepson, P.C. (ed.) (1989). *Pesticides and Non-Target Invertebrates*. Wimborne, UK: Intercept.

Jepson, P.C. (1993*a*). Ecological insights into risk analysis: the side-effects of pesticides as a case study. *Science of the Total Environment (Supplement, 1993) Part 2*. pp. 1547–66.

Jepson, P.C. (1993*b*). Insects, Spiders and Mites. In P. Calow (ed.), *Handbook of Ecotoxicology*. Oxford, UK: Blackwell Scientific Publications. pp. 299–325.

Jepson, P.C., Chaudry, A.G., Salt, D.W. *et al.* (1990). A reductionist approach towards short-term hazard analysis for terrestrial invertebrates exposed to pesticides. *Functional Ecology*, **4**, 339–49.

Jepson, P.C., Efe, E. and Wiles, J.A. (1995). The toxicity of dimethoate to predatory Coleoptera: developing an approach to risk analysis for broad-spectrum pesticides. *Archives of Environmental Contamination and Toxicology*, **28**, 500–7.

Jepson, P.C. and Sherratt, T.N. (1996). The dimensions of space and time in the assessment of ecotoxicological risks. In D.J. Baird, L. Maltby, P.W. Greig-Smith and P.E.T. Douben (eds.), *Ecotoxicology: Ecological Dimensions*. London, UK: Chapman Hall. pp. 43–54.

Jepson, P.C. and Thacker, J.R.M. (1990). Analysis of the spatial component of pesticide side-effects on non-target invertebrate populations and its relevance to hazard analysis. *Functional Ecology*, **4**, 349–58.

Johnson, M.W. and Tabashnik, B.E. (1999). Enhanced biological control through pesticide selectivity. In T. Bellows and T.W. Fisher (eds.), *Handbook of Biological Control*. San Diego, CA: Academic Press. pp. 297–318.

Kjaer, C. and Jepson, P.C. (1995). Estimation of direct pesticide exposure of a non-target weed dwelling chrysomelid beetle (*Gastrophysa polygoni*) in cereals. *Environmental Toxicity and Chemistry*, **14**, 993–9.

Krebs, J.R., Wilson, J.D., Bradbury, R.B. and Siriwadena, G.M. (1999). The second silent spring? *Nature*, **400**, 611–12.

Levin, S.A., Harwell, M.A., Kelly, J.R. and Kimball, K.D. (1988). *Ecotoxicology: Problems and Approaches*. New York: Spinger-Verlag.

Longley, M. and Jepson, P.C. (1997). Cereal aphid and parasitoid survival in a logarithmically diluted deltamethrin spray transect in winter wheat: field-based risk assessment. *Environmental Toxicology and Chemistry*, **16**, 1761–7.

Maltby, L., Kedwards, T.J., Forbes, V.E. *et al.* (2001). Linking individual responses and population-level consequences. In J. Baird and G. Allen-Burton (eds.), *Ecological Variability: Separating Natural from Anthropogenic Causes of Ecosystem Impairment*. SETAC Press, USA. pp. 27–82.

Mineau, P. (2002). Estimating the probability of bird mortality from pesticide sprays on the basis of the field study record. *Environmental Toxicology and Chemistry*, **21**, 1497–1506.

Mineau, P. and Whiteside, M. (2006). Lethal risk to birds from insecticide use in the United States – a spatial and temporal analysis. *Environmental Toxicology and Chemistry*, **25**, 1214–22.

Moriarty, F. (1999). *Ecotoxicology: The Study of Pollutants in Ecosystems*. San Diego, CA: Academic Press.

Mullie, W.C. and Everts, J.W. (1992). Uptake and elimination of 14C deltamethrin by Oedothorax apicatus (Arachnida: Erigonidae) with respect to bioavailability. *Pesticide Biochemistry and Physiology*, **39**, 27–34.

Osteen, C. and Livingston, M. (2006). Pest Management Practices. In K. Weibe and N. Gollehon (eds.), *Agricultural Resources and Environmental Indicators, 2006*. United States Department of Agriculture, Economic Information Bulletin No. (EIB-16), Washington, DC, USA. Chapter 4.3. pp. 107–15.

Posthuma, L., Suter, G.W. and Traas, T.P. (2002). *Species Sensitivity Distributions in Ecotoxicology*. Boca Raton, FL: Lewis Publishers.

Ripper, W. E. (1956). Effects of pesticides on the balance of arthropod pest populations. *Annual Review of Entomology*, **1**, 403–38.

Salt, D.W. and Ford, M.G. (1984). The kinetics of insecticide action part III. The use of stochastic modeling to investigate the pickup of insecticides from ULV-treated surfaces by larvae of *Spodoptera littoralis* Boisd. *Pesticide Science*, **15**, 382–410.

Sherratt, T.N. and Jepson, P.C. (1993). A metapopulation approach to modelling the long-term impact of pesticides on invertebrates. *Journal of Applied Ecology*, **30**, 696–705.

Stark, J.D., Jepson, P.C. and Mayer, D.F. (1995*a*). Limitations to use of topical toxicity data for predictions of pesticide side-effects in the field. *Journal of Economic Entomology*, **88**, 1081–8.

Stark, J., Jepson, P.C. and Thomas, C.F.G. (1995*b*). The effects of pesticides and spiders from laboratory to landscape. *Reviews in Pesticide Toxicology*, **3**, 83–110.

Stark, J.D., Banks, J.E. and Vargas, R. (2004). How risky is risk assessment: The role that life history strategies play in susceptibility of species to stress. *Proceedings of the National Academy of Sciences*, **101**, 732–6.

Suter, G.W., II, editor. (1993). Ecological Risk Assessment. Lewis Publishers, Boca Raton. Florida, USA.

Taub, F.B. (ed.) (1997). Are ecological studies relevant to pesticide registration decisions? *Ecological Applications*, **7**, 1086–132.

Thomas, C.F.G. and Jepson, P.C. (1997). Field-scale effects of farming practices on linyphiid populations in grass and cereals. *Entomologia Experimentalis et Applicata*, **84**, 59–69.

Thomas, C.F.G., Hol, E.H.A. and Everts, J.W. (1990). Modelling the diffusion component of dispersal during recovery of a population of linyphiid spiders from exposure to an insecticide. *Functional Ecology*, **4**, 357–68.

Thomas, C.F.G., Brain, P. and Jepson, P.C. (2003). Dispersal distances of ballooning spiders: field measurements of aerial activity and a simulation model. *Journal of Applied Ecology*, **40**, 912–27.

Trumper, E.V. and Holt, J. (1998). Modelling pest population resurgence due to recolonization of fields following an insecticide application. *Journal of Applied Ecology*, **35**, 273–85.

Unal, G. and Jepson, P.C. (1991). The toxicity of aphicide residues to beneficial invertebrates in cereal crops. *Annals of Applied Biology*, **118**, 493–502.

Van Straalen, N.M. and Van Rijn, J.P. (1998). Ecotoxicological risk assessment of soil fauna recovery from pesticide application. *Reviews of Environmental Contamination and Toxicology*, **154**, 83–141.

Walker, A., *et al.* eds. (1995). Pesticide Movement to Water: Proceedings of a Symposium organised by the British Crop Protection Council, and the Society of Chemical Industry. Held at the University of Warwick, Coventry, UK on 3-5 April 1995. British Crop Protection Council, Alton, Hampshire, UK. 414pp.

Weyman, G. S., Jepson, P. C. and Sunderland, K. D. (1995). Do seasonal changes in numbers of aerially dispersing spiders reflect population density on the ground or variation in ballooning motivation? *Oecologia*, **101**, 487–93.

Wiles, J. A. and Jepson, P. C. (1993*a*). The dietary effects of deltamethrin upon *Nebria brevicollis* (F.) (Coleoptera: Carabidae). *Pesticide Science*, **38**, 329–34.

Wiles, J. A. and Jepson, P. C. (1993*b*). An index of the intrinsic susceptibility of non-target invertebrates to residual deposits of pesticides. In H. Eijsackers, F. Heimbach and M. Donker (eds.), *Ecotoxicology of Soil Organisms*. Chelsea, USA: Lewis Publishers. pp. 287–302.

Wiles, J. A. and Jepson, P. C. (1994). Substrate mediated toxicity of deltamethrin residues to beneficial invertebrates: estimation of toxicity factors to aid risk assessment. *Archives of Environmental Contamination and Toxicology*, **27**, 384–91.

Wiles, J. A. and Jepson, P. C. (1995). Dosage reduction to improve the selectivity of deltamethrin between aphids and coccinellids in cereals. *Entomologia Experimentalis et Applicata*, **76**, 83–96.

Index

(Tables in *Italics*, Figures in **Bold**)

Printed in the United States
By Bookmasters